ENVIRONMENTAL RESEARCH ADVANCES

PLANT ECOLOGY AND EVOLUTION IN HARSH ENVIRONMENTS

Environmental Research Advances

Additional books in this series can be found on Nova's website
under the Series tab.

Additional e-books in this series can be found on Nova's website
under the e-book tab.

ENVIRONMENTAL RESEARCH ADVANCES

PLANT ECOLOGY AND EVOLUTION IN HARSH ENVIRONMENTS

NISHANTA RAJAKARUNA
ROBERT S. BOYD
AND
TANNER B. HARRIS
EDITORS

nova publishers
New York

For permission to use material from this book please contact us:
Telephone 631-231-7269; Fax 631-231-8175
Web Site: http://www.novapublishers.com

Additional color graphics may be available in the e-book version of this book.

Library of Congress Cataloging-in-Publication Data

Plant ecology and evolution in harsh environments / editors, Nishanta Rajakaruna (College of the Atlantic, ME, USA), Robert S. Boyd (Auburn University, AL, USA) and Tanner B. Harris (WRA, Inc., San Rafael, CA, USA).
pages cm -- (Environmental research advances)
Includes index.
ISBN 978-1-63321-955-7 (hardcover)
1. Plant ecology. 2. Plants--Evolution. 3. Plant-soil relationships. 4. Extreme environments. I. Rajakaruna, Nishanta, 1969-
QK901.P572 2014
581.7--dc23

2014034858

Published by Nova Science Publishers, Inc. † New York

CONTENTS

PREFACE

Harsh environments occur across the globe, often appearing as barren, desolate places incapable of supporting life. In reality, harsh environments are hotspots for biological diversity, supporting numerous rare and endemic organisms (Harrison & Rajakaruna, 2011; Myers et al., 2000; Nagy & Grabherr, 2009; Ward, 2008; Wharton, 2002). These habitats range from arctic and alpine environments to Mediterranean regions and inland deserts. Included are habitats with extreme soils such as serpentine, saline, sodic, and gypseous soils, as well as mine tailings and other metal-enriched soils.

Organisms that live in these environments must cope with extreme climatic variables such as solar radiation, temperature, and water availability. They must also cope with other abiotic stressors such as nutrient limitation, elemental toxicity, or extreme pH values. These environmental extremes necessitate novel adaptations for survival and act as both drivers of evolution and ecological filters, resulting in unique assemblages of organisms. The study of these habitats and the organisms that inhabit them have taught us much about evolution, ecology, and biology, and has provided insights into organismal response to climate change and other environmental perturbations, as well as insights into exobiology, biotechnology, restoration, and conservation. Studies of microbes and small invertebrates have received the most attention (*e.g.,* Horikoshi & Grant, 1998; Kushner, 1978; Liebezeit et al., 2000; Seckbach, 1999; Wharton, 2002), and a recent book focused mainly on animal adaptations (Lubzens et al. 2010); however, there is an extensive set of literature on plants, algae, and fungi found in harsh environments.

Many of the books on plant ecology and evolution in harsh environments focus on either a single extreme habitat (*e.g.,* Harrison & Rajakaruna, 2011; Nagy & Grabherr, 2009; Ward, 2008) or a single stressor (*e.g.,* Aroca, 2012; Lüttge et al., 2011; Turkan, 2011). In this book, we bring together a wide range of topics on ecology and evolution in harsh environments of plants (as well as a wide range of other non-animal organisms), with chapters written by experts from around the world. We begin with the creation of harsh environments, focusing on the roles of bedrock geochemistry and soil evolutionary processes in generating habitats with extreme abiotic conditions for photosynthetic and chemosynthetic processes (Chapter 1). Later chapters discuss the biology, ecology, and evolution of bryophytes (Chapter 12), vascular plants (Chapters 4-11), lichens (Chapter 3), herbivores and pathogens (Chapter 10), mycorrhizal fungi (Chapter 2), and other beneficial microbes (Chapters 1, 10) found in a range of harsh environments, including alpine and arctic settings (Chapter 7), fire-prone Mediterranean climates (Chapter 8), serpentine outcrops (Chapters 2, 3, 6, 10, 11, 13-15), gypsum soils (Chapters 2, 5, 14), metal-rich mine tailings (Chapters 2, 3, 10, 11, 14, 15), and saline soils (Chapters 2, 4, 11). We highlight new tools and emerging techniques in high-throughput phenotyping, genomics, and phylogenetics and the role of these tools in developing our understanding of the patterns and processes of evolution in harsh environments (Chapters 4, 5, 9, 11). These techniques, combined with classical ecological approaches, including reciprocal transplant studies (Chapter 11), allow us to carefully examine adaptation to and evolution in harsh environments, even providing genomic insights to stress tolerance in plants with direct implications for agriculture, biotechnology, restoration, and conservation (Chapters 7, 9, 11, 13-15; Peleg et al., 2011). Finally, several lines of needed research are emphasized, from phylogenomics and population genomics to developmental genetics and comparative biology of non-model plants (Chapters 4, 5, 9, 11, 16). This research will improve our understanding of plant and fungal life found in extreme habitats, not only on Earth, but also extraterrestrial life that likely exists elsewhere in the universe. Due to the limited extent and patchy distribution of harsh environments, plants found in these habitats are particularly vulnerable to the effects of climate change (Chapters 7, 13) and other anthropogenic impacts (Chapters 14, 15). Many of these habitats have been given a protected status based on their unique biological diversity and have been subject to restoration efforts (Chapter 14); however, much work is needed to improve the efficacy of these conservation and restoration efforts. The studies summarized in this book highlight improvements in our understanding of the ecology and evolution of plants and fungi found in harsh environments, adding both tools and knowledge to our efforts to restore and protect these unique habitats and the organisms that occur there. In the final chapter of this book, we summarize additional areas of research needed to improve restoration and conservation efforts (Chapter 16).

REFERENCES

Aroca, R. (Ed.) (2012) *Plant responses to drought stress: From morphological to molecular features*. New York, NY: Springer.

Harrison, S. & Rajakaruna, N. (Eds.) (2011) *Serpentine: The evolution and ecology of a model system*. Berkeley, CA: University of California Press.

Horikoshi, K. & Grant, W. D. (1998) *Extremophiles: Microbial life in extreme environments*. New York, NY: Wiley.

Kushner, D. J. (1978) *Microbial life in extreme environments*. New York, NY: Academic Press.

Liebezeit, G., Dittmann, S. & Kröncke, I. (Eds.) (2000) *Life at interfaces and under extreme conditions. Developments in hydrobiology* Vol. 151. New York, NY: Springer.

Lüttge, U., Beck, E., & Bartels, D. (Eds.) (2011) *Plant desiccation tolerance*. New York, NY: Springer.

Lubzens, E., Cerdà, J. & Clark, M. S. (Eds.) (2010) *Dormancy and resistance in harsh environments*. New York, NY: Springer.

Myers, N., Mittermeier, R. A., Mittermeier, C. G., da Fonseca, G. A. B. & Kent, J. (2000) Biodiversity hotspots for conservation priorities. *Nature* 403, 853-858.

Nagy, L. & Grabherr, G. (2009) *Biology of alpine environments*. New York, NY: Oxford University Press.

Peleg, Z., Apse, M. P. & Blumwald, E. (2011) Engineering salinity and water-stress tolerance in crop plants: Getting closer to the field. *Advances in Botanical Research* 57, 405–443.

Seckbach, J. (Ed.) (1999) *Enigmatic microorganisms and life in extreme environments*. Boston, MA: Kluwer Academic Publishers.

Turkan, I. (Ed.) (2011) Plant responses to drought and salinity stress: Developments in a post-genomic era. *Advances in Botanical Research* Vol. 57. London: Elsevier.

Ward, D. (2008) *The biology of deserts*. New York, NY: Oxford University Press.

Wharton, D. A. (2002) *Life at the limits: Organisms in extreme environments*. Cambridge, MA: Cambridge University Press.

ACKNOWLEDGMENTS

We thank the following individuals for providing critical peer review of one or more chapters in this book:

Rose Andrew
School of Environmental and Rural Science, University of New England
Armidale, Australia

H. John B. Birks
Department of Biology, University of Bergen, Bergen, Norway

Sally Brown
School of Forest Resources, University of Washington
Washington, USA

Millie Burrell
Institute for Plant Genomics and Biotechnology, Department of Horticultural Sciences, Texas A&M University, Texas, USA

Christine Cargill
Centre for Australian National Biodiversity Research and Australian National Herbarium
Canberra, Australia

Alessandro Chiarucci
Dipartimento Scienze della Vita, Università degli Studi di Siena
Siena, Italy

Stephen R. Clayden
Botany and Mycology Section, New Brunswick Museum
New Brunswick, Canada

Ricardo da Cruz de Carvalho
Centro de Biologia Ambiental, Faculdade de Ciências da Universidade de Lisboa
Lisboa, Portugal

Naveed Davoodian
The New York Botanical Garden
New York, USA

Guillaume Echevarria
Phytoremediation Team, Laboratoire Sols et Environnement, Université de Lorraine - INRA
Vandoeuvre-les-Nancy, France

Adrián Escudero
Área de Biodiversidad y Conservación, Universidad Rey Juan Carlos, Madrid, Spain

John L. Freeman
Intrinsyx Technologies Corporation, Space BioSciences Division, NASA-Ames Research Center
California, USA

Kirsten Fristad
Exobiology Branch, NASA-Ames Research Center, California, USA

Alan M. Fryday
Herbarium, Department of Plant Biology, Michigan State University
Michigan, USA

Susana C. Gonçalves
Centro de Ecologia Funcional, Departamento de Ciências da Vida, Universidade de Coimbra
Coimbra, Portugal

Cristina Gonnelli
Department of Biology, University of Florence, Florence, Italy

Inger Greve Alsos
Tromsø University Museum
Tromsø, Norway

Sarah R. Hall
College of the Atlantic
Maine, USA

Susan P. Harrison
Department of Environmental Science and Policy, University of California, Davis, California, USA

Robert A. Ingle
Department of Molecular & Cell Biology, University of Cape Town
South Africa

Erik S. Jules
Department of Biological Sciences, Humboldt State University
California, USA

Todd Keeler-Wolf
Vegetation Program, California Department of Fish and Wildlife
California, USA

Jon E. Keeley
Western Ecological Research Center, U.S. Geological Survey
California, USA

Julie Kierstead Nelson
Shasta-Trinity National Forest, United States Forest Service
California, USA

Ian D. Medeiros
College of the Atlantic
Maine, USA

Joe Miller
Division of Environmental Biology, National Science Foundation
Arlington, Virginia

Glen Mittelhauser
Maine Natural History Observatory
Maine, USA

Matt Ogburn
Institute for Biospheric Studies, Yale University, Connecticut, USA

Fred C. Olday
College of the Atlantic
Maine, USA

Mel Oliver
Plant Genetics Research Unit, United States Department of Agriculture
Agricultural Research Service, University of Missouri, Columbia
Missouri, USA

Gregory L. Owens
Botany Department, University of British Columbia
British Columbia, Canada

Sara Palacio Blasco
Instituto Pirenaico de Ecología, Consejo Superior de Investigaciones Científicas
Huesca, Spain

Elizabeth Pilon-Smits
Biology Department, Colorado State University, Fort Collins
Colorado, USA

A. Joseph Pollard
Biology Department, Furman University
South Carolina, USA

Colin Quinn
United States Agency for International Development, Mozambique Office
Maputo, Mozambique

John J. Schenk
Department of Ecology and Evolutionary Biology, Tulane University, Louisiana, USA

Dylan Schwilk
Department of Biological Sciences, Texas Tech University
Texas, USA

José Gabriel Segarra-Moragues
Centro de Investigaciones sobre Desertificación (CIDE), Consejo Superior de Investigaciones Científicas (CSIC), Universitat de València
València, Spain

Miranda Sinnott-Armstrong
Department of Ecology and Evolutionary Biology,
Yale University
Connecticut, USA

Darlene Southworth
Department of Biology, Southern Oregon University
Oregon, USA

Sharon Y. Strauss
Department of Evolution and Ecology, University of California, Davis
California, USA

Jessica W. Wright
Pacific Southwest Research Station, USDA-Forest Service
California, USA

Nishanta Rajakaruna, Robert, S. Boyd and Tanner Harris

July 28 2014

In: Plant Ecology and Evolution in Harsh Environments ISBN: 978-1-63321-955-7
Editors: N. Rajakaruna, R. S. Boyd and T. B. Harris

Chapter 1

BEDROCK AND GEOCHEMICAL CONTROLS ON EXTREMOPHILE HABITATS

Dawn Cardace[1]*, D'Arcy R. Meyer-Dombard[2], Amanda A. Olsen[3] and M. Niki Parenteau[4]

[1]University of Rhode Island, Department of Geosciences, Kingston, RI, USA
[2]Department of Earth and Environmental Sciences, University of Illinois at Chicago, Chicago, IL, USA
[3]School of Earth and Climate Sciences, University of Maine Bryand Global Sciences Center, Orono, ME, USA
[4]SETI Institute and Exobiology Branch, NASA Ames Research Center, Moffett Field, CA, USA

ABSTRACT

Bedrock, sediments, and soils are in constant interaction and reaction with through-going water volumes and resident microbiology. Every parcel of solid Earth and water thus holds a geochemical inventory that is a cumulative product of past reactions and ongoing alteration, a biogeochemical ecology that at times must contend with extreme environmental gradients. In this contribution, we discuss some aspects of the evolution and dynamics of extreme environments in bedrocks and soils, and ways in which biology acts on and reacts to environmental gradients, including through weathering processes.

INTRODUCTION

Extremophiles are microorganisms that survive by successfully harvesting energy and carbon from their environment though they inhabit realms that are far from the terrestrial norm, however it be defined (Cavicchioli et al., 2011; Pikuta et al., 2007; Rothschild & Mancinelli, 2001). Variables like temperature, pressure, pH, salinity, and radiation load

* Corresponding author: cardace@mail.uri.edu.

interact in every environment, and control habitability. When environmental extremes are encountered, they create a selection pressure to which communities must respond. Some extreme environments that generate strong selection pressures on microbial communities include: high temperature sites (deep mines in areas with high geothermal gradients, hot springs, volcanic vents), low temperature sites (ice masses, permafrost), high pH/alkaline sites (soda lakes or playa lakes in evaporite basins, serpentinizing springs), low pH/acidic sites (hot springs, acid mine drainage-impacted sites, sulfur caves, etc.), relatively high pressure environments (subsurface habitats such as bedrock), ambient geochemistry (conditions of little/no water availability, aqueous systems with elevated salinity, problematic metal loads, etc.), and irradiated environments (such as those at high elevation and those associated with nuclear reactors/radioactive waste). In this chapter, we describe terrestrial environmental diversity from a geological perspective, considering how bedrock alteration and weathering generate diverse soil and shallow subsurface environments. We review information concerning how microbial life interacts with extreme environments, via photosynthesis and chemosynthesis, and consider how critical zone processes (*i.e.,* the living skin of planet Earth, extending roughly from the tallest vegetation down through soil blankets to bedrock) reflect broader patterns in how life and the environment co-evolve.

OVERVIEW OF ENVIRONMENTAL VARIATION IN EXTREME HABITATS

Bedrock geochemistry and subsequent soil evolution construct habitats with defined geochemical gradients and physical properties that constrain microbial community success. The spectrum of habitability can be defined through consideration of variables including temperature, pressure, pH, aqueous geochemistry and water activity, radiation load, etc. (Figure 1). Earth's crust and veneer of soils also interact with the modern atmosphere; the oxygen-, nitrogen-, and CO_2-rich atmosphere is a critical component of near surface biogeochemical cycles.

Microbes grow at temperatures ranging from the very hot (hyperthermophiles grow above 80°C and may thrive in deep mines, hot springs, volcanic vents, etc.) to the very cold (psychrophiles grow from ~ -20 to 20°C, *e.g.*, ice masses, permafrost; Konhauser et al., 2005), perhaps aided by cell membrane resilience or the secretion of protective chemicals as a layer supporting the cell membrane itself (Raymond & Fritsen, 2001). The upper temperature limit to life was for a long time estimated as between 110 and 200°C (Brock, 1967; Stetter et al., 1990), but is now verified empirically as at least 121°C (Kashefi & Lovley, 2003), though the absolute upper temperature is difficult to resolve (Daniel & Cowan, 2000). The upper limit is almost certainly lower than 140°C, the temperature at which biomolecules become unstable and also the energy required to maintain biomolecules becomes unsustainable (Jaenicke, 2000; Jaenicke & Bohm, 1998; Jaenicke & Sterner, 2006; White, 1984). In terms of the lower temperature limit for life, evidence exists for successful metabolism in supraglacial settings (Anesio & Laybourn-Parry, 2012; Hodson et al., 2013), including cryoconites (*i.e.*, microbe-mineral aggregate; Edwards et al., 2013), subglacial settings (Mitchell et al., 2013), polar lakes/ponds (Jungblut et al., 2012), and permafrost and related environments at temperatures near -15°C and -20°C (Montross et al., 2014).

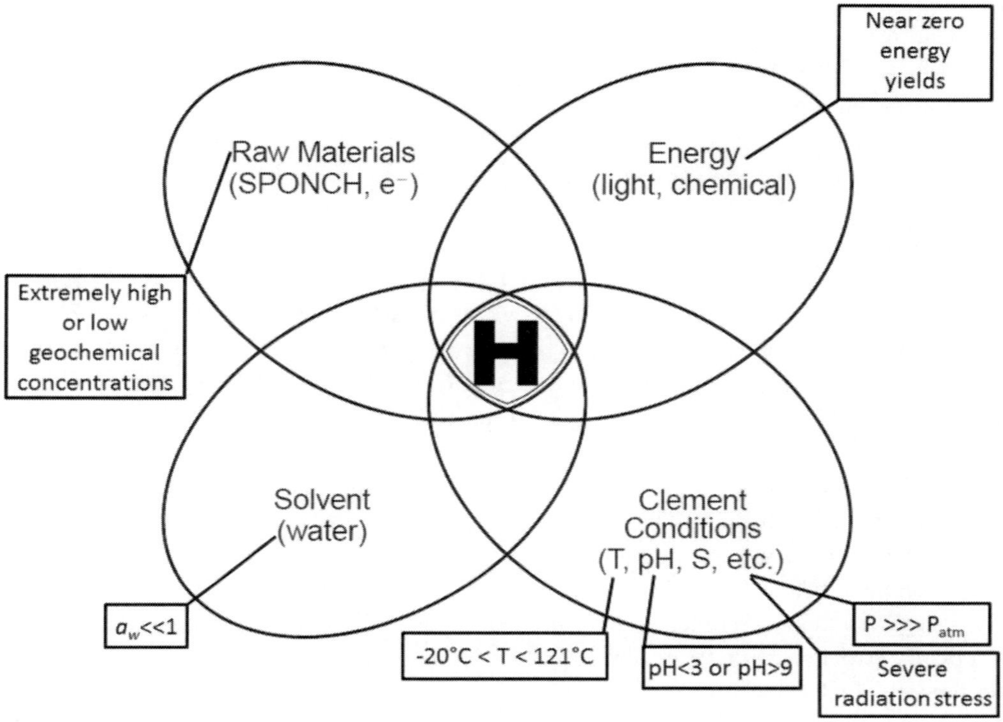

Figure 1. Schematic view of habitability, indicating that habitability (shown as the bold **H** in the intersection of all fields) can be defined as the region of accessible raw materials, energy, solvent, and clement conditions. Note that the environmental variables discussed in the text fall into the sets provided; boxed annotations indicate some extreme conditions that impact habitability in this space. Modified from Hoehler (2007).

Nonetheless, the true cold limit is not yet known—but may correspond with the lower bound for unicellular vitrification (not intracellular freezing, but a glass transition!), postulated at near -20°C (Clarke et al., 2013).

Pressure effects are surely important to microbial cell stability, yet do not appear to cause cell death across all domains of life (*i.e.*, Archaea, Bacteria, and Eucarya) at a defined point in pressure-temperature space. Cultivation of deep sea mud barophiles from ~2500 to ~6400 m depth has revealed optimal growth conditions of ~50MPa and 10°C, at which conditions microbes outperformed those grown at surface pressures (~0.1 MPa; Kato et al., 1995). Consider a 110 megapascal (MPa) pressure that is equivalent to an ocean depth of ~11,000 m; this pressure does not extinguish life, but spurs important membrane adaptations in microbial communities (Bartlett, 2002). Membrane lipid compositions shift to accommodate changing environmental conditions (not only in response to pressure, but also temperature and some geochemical gradients); the tuning of membrane lipid compositions to environmental parameters, in the interest of preserving membrane fluidity, is documented both in Bacteria (*e.g.*, Yano et al., 1998) and Archaea (*e.g.*, Oger & Cario, 2013). Additionally, there is evidence that within the membrane, it is not protein stability but protein-protein interaction that responds more readily to pressure (Meersman et al., 2008). Hazen et al. (2002) argue compellingly that modern barophiles serve as ready analogs for barophiles on the early Earth, showcasing pressure-adaptive strategies as critical to the success of early, ocean-cradled life; an ancient origin of barophily is also suggested by amino acid comparisons of model organisms (Yafremava et al., 2013).

The very presence of water and its geochemical inventory impact habitability in a fundamental sense—providing an ambient geochemistry that supports specific metabolic strategies. In particular, the effective concentration of water (*i.e.*, the thermodynamic activity of water; Lewis & Randall, 1923) impacts biological activity, and stresses due to low activities of water can be observed both in high salt and extremely dry situations—both can be described as environments of low water activity. Very low activities of water (a_w) in the environment stress cell membranes and thus constrain one of the limits to life in a fundamental sense. Pure water is taken to have $a_w = 1$, while a saturated salt solution has a_w ~0.75; very few xerophilic organisms can metabolize down to a_w ~0.61, akin to the a_w in honey (Grant, 2004). The true availability of water to biology can be reduced dramatically by freezing of liquid water (leaving little if any liquid water along grain boundaries, for example) or driving up salinities (leaving few water molecules partially free of aqueous complexes). In general, microbial life at high salt concentrations must cope with maintaining safe cytoplasm concentrations either through accumulation of KCl within the cell (the 'salt-in' strategy) or through exclusion of salt from the cytoplasm, perhaps by accumulation of organic compounds (the 'organic-solutes-in' strategy) that maintains osmotic balance though they may be energetically expensive for cells (Oren, 2011). How microbes survive even in hypersaline environments, such as deep hypersaline anoxic basins in Discovery Basin in the Eastern Mediterranean (van der Wielen et al., 2005), sub-zero brines near ice-sealed lakes of the Antarctic (Murray & Smith, 2009), or other NaCl-saturated and $MgCl_2$-rich settings (Bolhuis et al., 2004; Bolhuis et al., 2006) is not fully understood. Research on the square halophilic archaeon *Haloquadratum walsbyi* suggests that proteins involved in phosphate metabolism and protective substances are key to surviving what amounts to severe dessication stress (Bolhuis et al., 2006). Extremophiles may also adapt to very low water activity conditions by shifting enzyme expressions of 'extremozymes,' which can function even in settings with high salt, high pH, low temperatures, and non-water-based media (Karan et al., 2012).

Extreme pH values impact microbial life particularly in that cell membranes must safeguard near neutral intracellular pH (*i.e.*, pH homeostasis; Baker-Austin & Dopson, 2007) and may rely on adjusting cell membrane lipid components to maintain function (Boyd et al., 2013). Adaptations to extreme pH can be observed at very acidic sites, reviewed in Dopson & Johnson (2012) and Johnson (2012), including hot springs and active geothermal sites (*e.g.*, Plumb et al., 2007), acid mine drainage-impacted sites (Johnson & Hallberg, 2003), caves formed by interaction of H_2S with groundwater to produce H_2SO_4 (Barton & Luiszer, 2005; Hose et al., 2000), and very alkaline sites, including soda/playa lakes (Costa et al., 2008; Duckworth et al., 1996; Humayoun et al., 2003; Jones et al., 1998; Rees et al., 2004; Zhang et al., 2002; Zhilina & Zavarzin, 1994) and serpentinizing groundwaters (Cardace & Hoehler, 2011; Morrill et al., 2013; Szponar et al., 2013; Tiago & Verissimo, 2013). There is growing recognition of the importance of the Na^+/H^+ transporter in regulating intracellular pH under hyperalkaline conditions (Kitada et al., 2000; Krulwich, 1995; Krulwich et al., 1996; 2001).

Radiation stresses life to varying degrees, breaking bonds in DNA whether in instantaneous/acute or longer term/chronic radiation doses. Several mechanisms of successful defenses exist, including the synthesis of extremolytes (*e.g.*, scytonemin, mycosporine-like amino acids, shinorine, porphyra-334, palythine, bioperin and phlorotannin) that serve as radiation absorbers, thus shielding the organism's DNA (Gabani & Singh, 2013). Additionally there are DNA repair genes that are environmentally triggered, and speed DNA repair with excellent efficiency (Gabani & Singh, 2013). Remarkably, the model organism *Deinococcus radiodurans*, initially detected as a microbial contaminant in irradiated foods, survives 3000 times more ionizing radiation than a human cell and has a largely unknown biochemistry (Cox & Battista, 2005; Rew, 2003). Investigations into high-altitude Andean Lake microbial communities indicate that they may be employing triacylglycerol production to aid in UV-resistance in particular (Bequer Urbano et al., 2013). It is unclear what specific cell components are critical to survival following intense radiation stress. Pigments and protective biomolecules may assist in shielding cells from intense radiation (Edwards et al., 2006), but this is controversial (Ordonez et al., 2009). Of growing importance is the rapid and exceptionally error-free DNA break repair functions evident in laboratory experiments with *D. radiodurans*, which can splice together fragmented DNA in a matter of hours following intense irradiation (Cox & Battista, 2005; Cox et al., 2010), and also infrared radiation-resistant *E. coli* (Byrne et al., 2014).

THE ROLE OF WATER

All life as we know it requires liquid water. At the micro-scale, water-rich solutions deliver nutrients to and through cell membranes. At the mega-scale, our water-rich planet enjoys a diversity of geochemical environments tied to the planetary water cycle (*i.e.*, the cycling of water on, above, and below Earth's surface, which necessarily includes the subduction of multiple ocean volumes of Earth's water at convergent margins and the generation of volatile-rich magmas (Hirschmann, 2006). The chemistry of natural waters is derived from a cumulative history of water-rock interactions, with characteristics of local bedrock defining many (if not most) geochemical aspects of associated groundwaters. We discuss here, in broad strokes, the dominant geochemical impacts on shallow groundwaters and consider their importance to biology.

The chemical composition of waters that recharge subsurface environments is variable and impacts the aqueous geochemistry of the below-ground system. Total precipitation volumes and event frequency largely determine total influx of water. Then, reactions with soil and shallow bedrock cause shifts in aqueous composition through: (a) release of water-soluble components from soils/fractured rocks (*i.e.,* leaching); (b) biological activity both through evapotranspiration as water is cycled from soil solutions to the atmosphere and also through direct organic acid action on solids associated with roots; and (c) bulk soil mineral weathering (Langmuir, 1997). In soils and very shallow groundwater regimes, modern meteoric water (precipitation) often dominates water inputs. Regional geotherms (the rates of increasing temperature with increasing depth below Earth's surface) certainly impact solid/aqueous phase distributions: all else equal, hotter temperatures enhance the mobilization of most components as water passes through a rock unit. For example, the mobilization of salt from sedimentary rocks (through dissolution) or extremely saline brines held quasi-stably in porous horizons in subsurface rocks (through drilling-related disturbance or natural migration) can have immediate and strong impacts on groundwater chemistry. In addition to meteoric waters, infiltration of three other naturally occurring water types distinguishes some groundwaters:

magmatic, connate, and metamorphic waters. As described in White (1957), magmatic waters are those generated from magmatic activity, be it deep (*i.e.*, plutonic) or shallow (*i.e.*, volcanic), which transfers part of Earth's internal water reservoir (estimated as several volumes of the modern ocean volume at least; Hirshmann, 2006) back to the atmosphere and hydrosphere. Connate waters are those that were trapped in the rock during its formation, for example, in the pore spaces in sedimentary rocks or in vesicles in extrusive volcanic rocks. Metamorphic waters are those that result largely from the dehydration of hydrous minerals in a rock body during heat- or pressure-related alteration; this often occurs over great swaths of bedrock given tectonic and volcanic activity.

For soil-hosted waters, biogeochemical surface reactions with soil particles, both organic and inorganic, cause shifts in water chemistry. Specifically, as water infiltrates soil, it incrementally solubilizes some chemical components and stabilizes others, always in the presence of evapotranspirative transfer to the atmosphere. With increasing depth from the surface: (1) evapotranspiration likely concentrates Na^+, SO_4^{2-}, NO_3^-, and Cl^-; (2) $SiO_{2(aq)}$ increases in concentration as silicate minerals weather; (3) CO_2 pressure increases as biologically produced CO_2 settles downward, given its relatively high density; and (4) relatively modest changes in these components may be linked to stratified microbial activity, which could draw down SO_4^{2-}, NO_3^-, CO_2, and other chemical species (Langmuir, 1997).

When recharge waters interact with deeply sourced groundwaters, perhaps in/below soils, the geochemistry of the waters reflects also regional bedrock. The most easily dissolved minerals that are common in aquifer bedrocks include halite (NaCl), gypsum ($CaSO_4 \cdot 2H_2O$), and carbonates such as calcite ($CaCO_3$); if their presence is $\geq$1% in the host rock, these minerals exert strong control over groundwater compositions (Langmuir, 1997).

Earth's igneous rock types may bring distinctive chemistries to associated waters, although silicates and aluminosilicate minerals are not very soluble, thus play a lesser role in near surface environments. Igneous rocks are typically defined by SiO_2 content and trends in other major element oxides; broad groups of very low silica rocks are termed 'ultramafic' (such as mantle peridotites and pyroxenites), low silica rocks are termed 'mafic' or 'basaltic' or 'gabbroic,' intermediate silica contents locate a rock in the 'intermediate' or 'andesite/dacite' category, and high silica rocks are termed 'felsic' or 'rhyolitic' or 'granitic.' By convention, we use the fine-grained, volcanic rock name to identify the standard categories of igneous rocks: ultramafic (<45% SiO_2), basaltic (45 to 52% SiO_2), andesitic (52 to 63% SiO_2), dacitic (63 to 66% SiO_2), and rhyolitic (>66% SiO_2) (Hefferan & O'Brien, 2010). Deeply sourced groundwaters that have evolved over geologic time through reactions with bedrock (accessed for example in cratonic interiors or where faulting has generated complex conduits from great depth) may also yield geochemically distinctive waters, often with pH and dissolved gas contents distinct from other surface waters nearby. For reference, ultramafic-hosted groundwaters (such as those of serpentinites and peridotites) may have pH elevated to >12, with methane at saturation and significant quantities of dissolved hydrogen produced as water transforms Fe-silicates, as observed in California's Coast Range Ophiolite at the McLaughlin Reserve locality (Cardace et al., 2013) and the Cedars peridotite-hosted spring in Sonoma, CA (Morrill et al., 2013). Basalt-hosted groundwaters, as in the Columbia River Basalts (Stevens & McKinley, 2000), also have dissolved hydrogen levels higher than regional surface waters as Fe-silicates react with water. As water reacts with bedrock of rhyolitic composition, it evolves to a composition that may have elevated F^-, some heavy metals, and also an inventory of radioactive isotopes of Ra, Rn, Th and U that exceeds recommended limits (Banks et al., 1998).

Certainly the rate of water transport through a rock package also has an effect on the geochemistry of natural waters: rapid flow (high groundwater velocities) typically produces waters with low total dissolved solids (TDS) due to the shorter period of flowing water acting on rock surface area and also the greater net volume of water that interacts with a given volume of rock (*i.e.,* high water/rock ratio), while slow flow (slow groundwater velocities) yield water with relatively higher TDS (Langmuir, 1997).

When mixing of two solutions (or systems, perhaps) occurs, geochemical disequilibrium results, even when both solutions were at equilibrium prior to mixing. This means that in the 'mixing zone' associated with every aqueous interface, there is energy yielded to the surrounds, which may be harnessed by biology. In fact, most fundamentally, life takes advantage of energy available in the environment: community assemblages shift in response to geochemical gradients across many Earth environments (cf. Dinsdale et al., 2008). The most extreme environments host well-adapted communities that may have simple structures and very specialized functional components, all tied to the most efficient harvest of available energy. In the two sections that follow, we discuss photosynthesis and chemosynthesis in near surface settings to underscore the importance of microbial adaptations to environmental stresses.

PHOTOSYNTHESIS: THE HARNESSING OF SOLAR ENERGY BY LIFE IN SURFACE EARTH ENVIRONMENTS

Photosynthetic metabolism dominates modern surface environments. Hypotheses regarding the evolution of photosynthesis on the early Earth within the context of its geological and geochemical setting provide insights as to how it became so widespread. Despite its seeming ubiquity in modern aquatic and terrestrial habitats, environmental constraints do limit the distribution and diversity of photosynthetic organisms. This section examines stressors acting on the photosynthetic biosphere, which derive from the environment: such selection pressures include the availability of electron donors, temperature, pH, salinity, and concentration of inhibitory chemical species such as sulfide.

On the early Earth, primitive anoxygenic phototrophs likely used strong reductants emanating from volcanoes, such as hydrogen or hydrogen sulfide (H_2S) donating electrons to drive a single photosystem (Olson & Pierson, 1986). Due to the low redox potential of these compounds, electrons are easy to extract, but the energy yields of this type of photosynthesis are relatively low and require oxidation of large quantities of reductants to generate significant biomass. The volcanogenic hydrogen and H_2S were probably in limited supply on the early Earth and rapidly depleted. Consequently, the high demand for these compounds thus contributed to selection pressure for the development of a second photosystem that when linked to the first, would allow for the extraction of an electron from weaker but more abundant reductants such as water, producing oxygen as a waste product (Blankenship, 2010).

Due to the high redox potential of water, the energy yield of oxygenic photosynthesis is much greater than anoxygenic photosynthesis. This energy yield, coupled with the wide availability of water as an electron donor compared to the limited availability of the volcanogenic reductants, allowed oxygenic photosynthesis to spread easily and dominate aquatic and terrestrial habitats. Indeed, the evolution of oxygenic photosynthesis and the resulting oxygenation of the atmosphere and oceans was arguably one of the most important events on the early Earth. The so-called Great Oxidation Event occurred 2.45 to 2.2 billion years ago (Ga) (Bekker et al., 2004; Canfield, 2005), and set the stage for the evolution of higher life forms dependent on oxygen.

Oxygenic photosynthesis is the most productive type of photosynthesis, with Earth's nearly ubiquitous liquid water and CO_2 supplying ready sources of electrons and carbon, respectively. This metabolic scheme is currently responsible for fixing 100 to 115 petagrams of carbon per year (Field et al., 1998). In contrast, anoxygenic photosynthesis is usually limited by the flux of reductants from the environment; these types of communities might be 2 to 3 orders of magnitude less productive than modern oxygen-dependent ones (Des Marais, 2000). The energy yield from these two types of photosynthesis is likely important in determining the tolerance of these organisms to challenging environmental conditions because it is metabolically 'expensive' to synthesize or modify cellular components to confer stability to temperature, pH, and salinity. In general, oxygenic phototrophs such as unicellular algae and cyanobacteria slightly beat out the anoxygenic phototrophs in terms of tolerance of extreme environments, likely due to the higher energy yield of oxygenic photosynthesis that allows them to spend more metabolic energy on dealing with the extreme conditions.

The light reactions of oxygenic photosynthesis in plants, algae, and cyanobacteria use two linked photosystems: Photosystems I (PSI) and II (PSII). This linkage allows for the extraction of an electron from water, which has a very positive redox potential (O_2/H_2O pair; $E_0' = +0.87$ V). The electron is donated to the oxidized reaction center chlorophyll pigment in PSII. Following the absorption of a quantum of light, the chlorophyll molecule reduces the first carrier in the electron transport chain, and the electron flows down the carriers and leads to the reduction of $NADP^+$ to NADPH. A proton motive force is also generated, which allows for the synthesis of ATP (Ort & Yocum, 1996). These two compounds are in turn used in the dark reactions to fix CO_2 into cellular material via the Calvin cycle. Together, these biochemical transformations are the backbone of all modern oxygenic photosynthesis.

In contrast, anoxygenic photosynthesis displays a greater diversity of carbon fixation pathways, electron donors, photosystems, and pigments. There are several groups of Bacteria that perform anoxygenic photosynthesis: green sulfur bacteria, green non-sulfur bacteria (also known as filamentous anoxygenic phototrophs), purple sulfur bacteria, purple non-sulfur bacteria, heliobacteria and phototrophic Acidobacteria (Blankenship et al., 1995; Bryant et al., 2007). Some of these are capable of growing photoheterotrophically using reduced organic carbon sources, and thus do not require an external source of reductants. However, when growing photoautotrophically, these organisms require external electron donors to ultimately reduce $NADP^+$ to NADPH, or to reduce ferredoxin, which are then subsequently used to reduce CO_2 through various carbon fixation pathways (Calvin cycle, reverse citric acid cycle, or hydroxypropionate pathway). For metabolic work, anoxygenic phototrophs depend on redox couples that have a standard electrode potential more negative than that of water, such as hydrogen ($2H^+/H_2$ pair; $E_0' = -0.42$ V), sulfide (S_0/H_2S pair; $E_0' = -0.27$ V), or iron (Fe^{3+}/Fe^{2+}; $E_0' = +0.30$ V at circumneutral pH). They also use a single photosystem, which can be classified into two families, type I reaction centers (such as PSI in chloroplasts, cyanobacteria, green sulfur bacteria, heliobacteria, and phototrophic Acidobacteria) and type II reaction centers (such as PSII in chloroplasts, cyanobacteria, purple bacteria, and green non-sulfur bacteria) (Allen & Williams, 1998; Bryant et al., 2007). Rather than using chlorophyll pigments that absorb in the visible range of light, anoxygenic phototrophs utilize a diversity of bacteriochlorophyll pigments that absorb in the near-infrared (Blankenship et al., 1995). These traits enable organisms to exploit light niches that are unavailable to oxygenic phototrophs.

The distribution and diversity of photosynthetic organisms in surface oceans, lakes, marshes, hot springs, salty ponds, and terrestrial environments are dictated by environmental conditions such as temperature, pH, salinity, and the concentration of chemical species (*e.g.*, sulfide). In general, phototrophs flourish across a moderate range in these environmental variables; as environmental extremes are approached, diversity decreases. Specialized anoxygenic phototrophs, cyanobacteria, algae, and plants with adaptations to deal with extreme conditions then dominate. Other chapters in this book will examine the tolerance of plants (Chapters 4, 5, 6, 7, 8), lichens (Chapter 3) and mycorrhizal fungi (Chapter 2) to extreme environments. Here we focus on simple photosynthetic bacteria and unicellular algae, which surpass the aforementioned plants, lichens and fungi in terms of tolerance of extreme conditions. However, there are environmental limits to even these simple photosynthetic or phototrophic microbes, and beyond these limits chemosynthetic or chemotrophic microbes dominate. This is likely due to the dual metabolic expenses of synthesizing photosynthetic machinery, as well as synthesizing and modifying components such as lipid membranes and proteins to allow the cells to function in the extreme conditions. After the discussion of the environmental limits of phototrophs, brief examples of chemotrophs will be provided to contrast the difference in tolerance levels.

An excellent example of extreme environmental selection pressure is the very low microbial diversity observed in very high temperature terrestrial hot springs: hyperthermophiles are the only organisms that survive this thermal extreme. These hot springs result from rain and snow that percolate deep underground to be heated by hot or molten rock in volcanically/tectonically active areas such as subduction zones, extensional basins, or hot spots (Christiansen, 2001). The deep heated water is often very saline (a so-called 'parent brine') and as it ascends to lower pressure environments, begins to boil: driving off volatile magmatic gases such as H_2, H_2S and CO_2. The H_2S is oxidized to sulfate by oxygen-containing near-surface groundwater, which combines with water to form sulfuric acid (Fournier, 1989). This acidic water escapes at Earth's surface via fumaroles, mud pots, and acidic springs. The residual alkaline brine also makes its way to the surface and forms hot springs colonized by robust photosynthetic mats composed of cyanobacteria, anoxygenic phototrophs, and chemotrophs (Brock, 1967; Castenholz, 1969).

Interestingly, the most thermophilic phototrophs known include a cyanobacterium, *Synechococcus*, whose maximum growth temperature is 73°C (Peary & Castenholz, 1964); this temperature can be considered the upper limit for photosynthesis. The most thermophilic anoxygenic phototroph is the green non-sulfur bacterium *Chloroflexus*, whose upper temperature limit is 69°C when it is growing photoheterotrophically with *Synechococcus* (Pierson & Castenholz, 1995); when it is growing photoautotrophically on sulfide, the upper temperature limit is 66°C (Giovannoni et al., 1987). At temperatures hotter than the empirically observed upper limit for photosynthesis (*i.e.*, 73°C), only chemosynthetic microbes (chemotrophs) flourish. Photosynthetic machinery is able to operate at these higher temperatures due to the synthesis of more saturated fatty acids in the thylakoid membranes, which confers greater stability and allows the electron transport chain to remain functional (Nishiyama et al., 1993).

Extremely cold temperatures define another kind of extreme environment for phototrophs. Cyanobacteria that inhabit ice-based polar and alpine habitats are the dominant phototrophic primary producers (Jungblut et al., 2005). A diverse range of cyanobacteria can be found in permanently ice-covered lakes, meltwater ponds, cryoconite holes, streams, permafrost, and as endoliths in rocks (Zakhia, 2008). However, most of these phototrophs are only psychrotolerant (growth optima > 15°C) rather than psychrophilic (growth optima ≤ 15°C). A few true psychrophilic strains of *Oscillatoria* have been isolated that actively photosynthesize down to 2°C, but their growth rates are very slow (Nadeau & Castenholz, 2000). Most work has been focused on measuring photosynthetic rates *in situ*; little work has been done on characterizing specific cellular mechanisms related to cold tolerance. The purple non-sulfur bacterium *Rhodoferax antarcticus* is the first and only anoxygenic phototroph known to show cold tolerance; its growth optimum is 18°C, although it can grow down to 0°C in Antarctic Dry Valley lakes (Madigan et al., 2000). Below 0°C, chemotrophs have been shown to grow down to -20°C (*e.g.*, D'Amico et al., 2006).

Beyond temperature related stresses, acid tolerance enables phototrophic extremophiles to survive in many environments. The waters of the acid sulfate hot springs mentioned above are dominated by red unicellular algae belonging to the order Cyanidiales, which perform oxygenic photosynthesis. They are able to tolerate pH levels from 0.2 to 4.0 by pumping out protons that leak into the cell using a plasma membrane H^+-ATPase, and temperatures up to 56°C (Castenholz & McDermott, 2010). In contrast, cyanobacteria cannot withstand pH levels of less than 5.5 to 6.0 (Castenholz, 1988), and thus are not widespread community members of acidic environments. It is not known why cyanobacteria show a low tolerance of acid, but early studies revealed an increased maintenance energy requirement at a low external pH (Kallas & Castenholz, 1982). This large energy requirement inhibits growth, perhaps stumped by inefficient proton pumping mechanisms, as the cytoplasm is protected from acidification (Kallas & Castenholz, 1982). The most acid-tolerant anoxygenic phototrophs are the purple non-sulfur bacteria, which can withstand environmental pH levels of ~3, but whose growth optima are ~5 and above (Madigan et al., 2005; Pfennig, 1974). These purple non-sulfurs are typically found in stratified communities below a layer of Cyanidiales, and grow photoheterotrophically utilizing organic compounds generated by the overlying primary producers (Pfennig, 1974). Where phototrophs are stalled by very low environmental pH values, microbes such as the chemotroph *Picrophilus* yet thrive, as one of the most acidophilic microbes known, with a growth optimum of about pH 0.7 (Schleper et al., 1995).

A great diversity of cyanobacteria and anoxygenic phototrophs are found in alkaline pH settings such as soda lakes formed from the evaporation of Ca- and Mg-poor water from restricted basins (Boussiba et al., 2000; Madigan, 2003). These ecosystems generally show greater diversity than hydrothermal or acidic settings. As with the cold-tolerant phototrophs, most cyanobacteria are alkalitolerant, able to withstand alkaline pH values of 8.5 to 9, but they can also grow under neutral conditions. There are a few cyanobacteria that are true alkaliphiles, such as species of *Spirulina* that grow optimally at pH 9 to 10, and even exhibit significant photosynthetic activity to pH 11.5 (Vonshak, 1997). One of the challenges in growing in a carbonate-rich but CO_2-poor environment is the lack of CO_2 available for carbon fixation via the Calvin cycle. Cyanobacteria have evolved carbon concentrating mechanisms that sequester bicarbonate within the cell where it's converted to CO_2 near the Rubisco enzymes, allowing for growth in alkaline settings (*e.g.*, Kupriyanova et al., 2013). It is worth noting also that high pH soda lakes are often concurrently high in salinity, due to their evaporative history, and high salinity is another important environmental variable. Consider the purple sulfur bacterium *Ectothiorhodospira haloalkaliphila,* a phototroph with a growth optimum of pH 9 (and growth up to 10.5): it is dually adapted to high pH and a salinity range of 2.5 to 15% (Imhoff & Suling, 1996), and can be fairly called a 'haloalkaliphile,' with other similarly adapted microbes. For reference, seawater is ~ 3.5% salinity, and while water bodies with ample rain/fresh surface water inputs are ~ 0.5% (Huber et al., 2000).

Other hypersaline environments, including salterns and sabkhas formed from the evaporation of seawater, are also productive ecosystems with a variety of phototrophic representatives. Cyanobacteria such as *Aphanothece halophytica* can tolerate salt concentrations as high as 17.5%, and can even grow in gypsum crust composed of $CaSO_4 \cdot 2H_2O$, an evaporite mineral (Laloknam et al., 2006). Anoxygenic phototrophs such as the purple non-sulfur bacterium *Rhodovibrio sodomensis* can tolerate up to 12% NaCl, while the purple sulfur bacterium *Halochromatium salexigens* has a growth optimum of 8 to 11% NaCl (Madigan, 2003). In general, phototrophs are moderate halophiles at best; at levels of from 20 to 30% NaCl, extreme chemotrophic halophiles such as haloarchaea dominate. This is yet another case where chemotrophs survive even where environmental variables close out the habitable niche for phototrophs.

Aqueous geochemical species (that is, chemical constituents) in the environment constitute another critical selection pressure acting on photosynthetic organisms. Sulfide has been shown to poison PSII in cyanobacteria, effectively shutting off oxygenic photosynthesis (Oren et al., 1977). Some cyanobacteria from marine and hot spring environments are able to tolerate sulfide to varying degrees, and are even able to grow anoxygenically where sulfide donates an electron to PSI (Dewit & Vangemerden, 1987). However, if the sulfide levels are too high, such as occurs in some hot spring microbial mats, cyanobacteria are completely poisoned. In such as case, pure mats of *Chloroflexus* have been observed, growing photoautotrophically on sulfide (Giovannoni et al., 1987). Habitability may thus be controlled in a very strict sense for microbes that have stringent geochemical thresholds in the environment.

In sum, while oxygenic and anoxygenic phototrophs exhibit tolerance to a wide range of environmental characteristics, and are recognized as extremophiles in many settings, there are indeed hard limits to photosynthesis in terms of temperature, pH, salinity, and some aqueous geochemical variables. Beyond these environmental limits, chemotrophs replace phototrophs as key primary producers, and employ diverse metabolic strategies that are tuned to the extreme environment they experience.

Chemosynthesis: The Chemical Energy-Fueled Complement to Photosynthesis

While photosynthesis drives carbon fixation in most surface environments, exceptions exist in the subsurface and 'extreme' surface environments. Here, carbon fixation and carbon cycling is driven by chemosynthetic processes. Indeed in soil environments, while fixed carbon is provided via photosynthetic biomass, chemosynthetic organisms complete carbon cycling. Chemosynthetic organisms metabolize energy sources independent from sunlight - the deeper the ecosystem within the Earth's crust, the more divorced from sunlight the carbon cycle becomes. These organisms derive their energy by utilizing the oxidation or reduction of dissolved inorganic and organic compounds. Organisms metabolizing using chemical sources of energy may be referred to as 'chemolithotrophs.' Chemosynthetic carbon fixers, or 'chemolithoautotrophs,' utilize CO_2, or in some cases bicarbonate (depending on the pH of the ecosystem), in anabolic processes. Chemosynthetic heterotrophs cycle organic carbon that, in near surface ecosystems, likely had beginnings as photosynthetically fixed carbon.

In all surface and subsurface ecosystems, a virtual buffet of geochemical energy sources is available to chemotrophic microorganisms. This includes ecosystems with little or no free oxygen, such as saturated soils or the deeper subsurface. Common electron acceptors in anaerobic systems include sulfate, nitrate, carbon dioxide, and ferric iron. These may be reduced by another menu of electron donors, such as hydrogen gas, hydrogen sulfide gas, ferrous iron, ammonium/ammonia, methane, etc. In addition, the reactants in chemotrophic metabolism need not be dissolved, but also include solids such as elemental sulfur, or iron minerals such as hematite, goethite, magnetite, or amorphous iron oxides. In environments that are laden with 'toxins,' trace metals (*e.g.*, As, Hg, Se, U), and/or complex organic compounds (*e.g.,* jet fuel, polycyclic aromatic hydrocarbons), these may be used as metabolites as well.

The variety and diversity of geochemical environments possible are due largely in part to the above-discussed water-rock reactions, but also to water-gas, and even water-soil mineral reactions. The intersection of reduced fluids meeting oxygen-rich environments provides disequilibrium that can be harnessed for energy by chemotrophs. Importantly, energy can *only* be gained from driving reactions that are not in equilibrium. Further, the amount of energy available to chemotrophs for any given reaction in any environment can be quantified, allowing comparison of the energetic landscape with other environments or time points (Amend et al., 2003; Shock et al., 2010). This is accomplished using the following expression enabling the calculation of the overall Gibbs free energy of reaction:

$$\Delta G_r = \Delta G_r^\circ + RT \ln Q_r \qquad (1)$$

where R and T represent the gas constant and temperature (in Kelvin). In reaction 1, the standard state thermodynamic properties of the reactants and products are contributed by ΔG_r° given by the expression:

$$\Delta G_r^\circ = -2.303\, RT \log K_r \qquad (2)$$

where K_r is the equilibrium constant for the reaction. The second contribution of reaction 1 accounts for the activities of the reactants and products that are a reality in the environment of interest. This is accomplished by including the activity product, Q_r:

$$Q_r = \prod_i (a_i)^{vi,r} \qquad (3)$$

where a_i represents the activity of the ith compound in the reaction raised to its stoichiometric coefficient in the rth reaction, $v_{i,r}$. Comprehensive reviews of the calculation of ΔG_r can be found in Amend & Shock (2001). The convention is that reactions that are exergonic (yield energy) provide negative values of ΔG_r, while reactions that cannot provide energy have positive values of ΔG_r.

While the myriad of combinations of redox reactions in natural ecosystems is too expansive to review in total, a few key processes are worth mentioning with regard to near surface and surface extreme environments. Many surface environments and most subsurface environments face the reality of at least one 'extreme' environmental condition such as anaerobicity, extremes in pH, temperature, salinity, or elevated or depleted metal concentrations. Below, we highlight paired metabolic processes common in ecosystems with a redox boundary between aerobic and anaerobic conditions such as those that might be found in waterlogged soils or sediments.

Methanogenesis is a chemotrophic process that occurs in saturated soils in a range of environmental conditions, from temperate wetlands to high latitude permafrost. The production of CO_2 from the decomposition of organic carbon combined with anaerobic, reducing conditions provides conditions conducive to the microbial production of methane:

$$CO_{2(aq)} + 4H_{2(aq)} \rightarrow CH_{4(aq)} + 2H_2O_{(l)} \quad (4)$$

The extreme conditions in wetlands, permafrost, and peat bogs (*i.e.*, anaerobicity, acidic soils, or subzero temperatures) support active microbial communities that are key in biogeochemical cycling. Peat bogs sequester up to a third of global terrestrial organic carbon, which, when decomposed, produces CO_2, inducing increased CH_4 production and raising concerns about increased greenhouse gas emissions from high latitude soils. However, recent work suggests that communities of methane oxidizers may moderate the emission of methane from peat bogs (Kip et al., 2010). Functioning across the aerobic/anaerobic boundary, methanotrophic Bacteria consume CH_4 produced by methanogenesis by using O_2 produced by *Sphagnum* mosses: the mosses benefit by assimilating CO_2 released by the methanotrophs. This symbiotic relationship was demonstrated in peats worldwide by Kip et al. (2010). The pairing of methanogenesis and methanotrophy may be widespread in a variety of other 'extreme' environments, from marine sediments to hot spring environments.

Ferrous iron oxidation and ferric iron reduction is another pair of metabolic options possible across the aerobic-anaerobic boundary where opposing diffusion gradients of O_2 and Fe^{+2} exist in 'extreme' ecosystems such as groundwater, wetlands, cave walls, and hydrothermal vents or anaerobic zones of soils. In fact, iron oxidizing Bacteria appear to have the capacity to induce rapid microscale coupling of iron oxidation-reduction at the aerobic-anaerobic threshold (Roden et al., 2004). Further, Fe redox cycling is thought to be an ancient process, being rooted deeply in the phylogeny of microorganisms (Emerson et al., 2010), and occurring as early as the Archaean and early Proterozoic (Konhauser et al., 2005; Planavsky et al., 2009). Ferric oxyhydroxides (such as $Fe(OH)_3$ or FeOOH) are common and stable in aerobic and circumneutral pH environments. These insoluble and high-surface area minerals are subject to microbial reduction under anaerobic conditions, especially in non-sulfidogenic soils and sediments (Roden et al., 2004). Iron-reducing Bacteria are diverse, and couple the oxidation of organic carbon or H_2 to the reduction of both soluble and mineral phase Fe^{+3}

(Lovley et al., 2004). Perhaps most importantly, the balance between iron oxidation and reduction has implications for the mobility, abundance, and residence time of reactive iron in natural ecosystems, as well as the mobility of other metals and radionuclides with a high affinity for iron oxide surfaces (Roden et al., 2012).

Iron oxidation can also occur in anaerobic environments, via alternative electron acceptors such as nitrate. This has been demonstrated in numerous environments, including freshwater sediments (Coby et al., 2011) and wetlands (Weber et al., 2006). Iron cycling, nitrogen cycling, and sulfur cycling are often coupled in ecosystems in which aerobic-anaerobic thresholds are sharp and dynamic over time (Adams et al., 2007; Amend & Shock, 2001; Glazer & Rouxel, 2009; Straub et al., 2001).

We can use ΔG_r values to compare the metabolic potential of seemingly disparate ecosystems. To answer to the question, "Is ecosystem X a better habitat for methane oxidizing organisms than ecosystem Y?," calculated values of ΔG_r can provide a quantitative answer, particularly when the baseline metabolic needs of the microbial community are known. For example, in considering four reactions describing methane oxidation, ferrous iron oxidation, ferric iron reduction, and nitrate reduction, the amount of energy available in a hydrothermal vent vs. high pH seep are compared in Table 1. The energy available for each reaction varies as a function of the temperature and geochemical conditions of each ecosystem. As a result, there is more energy available to chemolithotrophs able to use methane oxidation, ferrous iron oxidation, and nitrate reduction in the shallow submarine hydrothermal vent ecosystem than the high pH seep ecosystem. However, metabolic processes based on ferric iron reduction are more favorable in the high pH seep system. Table 1 also drives home the point that the Gibbs free energy at standard state conditions, ΔG_r°, does not represent the true energy availability in these natural systems. Instead, actual measurements of environmental conditions are needed to evaluate the energetic landscape in a given environment.

In sum, photosynthetic processes may be inhibited by environmental conditions in many 'extreme' ecosystems, and chemosynthetic microorganisms may drive carbon fixation, carbon cycling, and the biogeochemical cycling of other key metabolites. Microbial metabolic processes directly affect the transformation and transport of metals and micronutrients. These considerations are key to understanding the function of any 'extreme' surface ecosystem. In fact, the impacts of sustained microbial activity on soil and bedrock strata, when integrated over long time scales, may be recorded as distinct zones of weathering.

BEDROCK WEATHERING: SUMMED INTERACTIONS BETWEEN LIFE AND ITS ENVIRONMENT

Chemical weathering is the destruction of rock-forming minerals and the creation of new, secondary minerals that are stable at Earth surface temperatures and pressures through interactions with natural waters; often biology plays an important role in mediating this weathering. Many major and trace nutrients needed by biota are derived from mineral weathering, as in the important case of phosphorus. Similarly, mineral weathering may release toxic heavy metals that may be dangerous to organismal health.

Table 1. Values of ΔG_r for four selected reactions used by chemotrophic microorganisms. We provide values for selected reactions in two different ecosystems: a pH 5, shallow marine hydrothermal vent system at 55°C ("The Grip" Hydrothermal "Vent") in Vulcano, Italy (Amend et al., 2003) and a pH 10.5, 25°C serpentinizing surface seep in the Zambales Ophiolite, Philippines ("Seep") (Cardace et al., 2013). Recall that $\Delta G_r < 0$ is the requirement for reactions to proceed spontaneously, thus across the Vent and Seep environments considered (ranging in pH and host rock geochemistry), the Gibbs Energy, and thus feasibility of the metabolisms, differs. Also shown in the two rightmost columns are values of ΔG_r° calculated with standard data for the given temperatures, following Amend and Shock (2001); the standard Gibbs Energy values diverge widely from those calculated with observed system geochemistry data, underscoring the importance of the environmental geochemistry in determining metabolic feasibility

Metabolism	Reaction	ΔG_r Vent[1]	ΔG_r Seep[2]	ΔG_r° STD 25°C[3]	ΔG_r° STD 55°C[4]
Methane oxidation	$CH_{4(aq)} + 2O_{2(aq)} \rightarrow CO_{2(aq)} + 2H_2O_{(l)}$	-96	-30	-859	-857
Ferrous iron oxidation	$3Fe^{+2} + 0.5O_{2(aq)} + 3H_2O_{(l)} \rightarrow Fe_3O_{4(s)} + 6H^+$	-71	< -1	-37	-43
Ferric iron reduction	$Fe_3O_{4(s)} + H_{2(aq)} + 6H^+ \rightarrow 3Fe^{+2} + 4H_2O_{(l)}$	-34	-80	-226	-218
Nitrate reduction	$NO_3^- + H_{2(aq)} \rightarrow NO_2^- + H_2O_{(l)}$	-78	-40	-176	-176

[1] pH 5, magmatic waters mixing with shallow fluids.
[2] pH 10.5, deeply sourced serpentinizing waters mixing with meteoric water.
[3] Calculated from standard data at 25°C.
[4] Calculated from standard data at 55°C.

Increasingly, research shows that this chemical weathering often, if not always, involves interactions between biota, water and minerals in the soil environment. Bacteria, fungi, lichens (Chapter 3), and plants have all been shown to alter chemical weathering in soil and uppermost (shallow) bedrock environments.

Bacteria enhance mineral weathering through a variety of mechanisms, facilitating nutrient mobilization from minerals in the process. Bacteria have been shown to mobilize nutrients from apatite (Lepleux et al., 2012), biotite (Shelobolina et al., 2012), olivine (Shirokova et al., 2012), hornblende (Liermann et al., 2000), pyrite (Mustin et al., 1992; Zhu & Reinfelder, 2012), smectite (Kostka et al., 1996; Zhu et al., 2011), carbonate minerals (Davis et al., 2007), and serpentine (Yao et al., 2013), as well as in whole rock studies of granite (Wongfun et al., 2014). Studies have shown that bacterial communities on mineral grain surfaces are distinct from communities in the bulk soil (Certini et al., 2004; Lepleux et al., 2012) as well as communities within mineral grains (McNamara et al., 2006). Relatively subtle changes in mineral chemistry have also been shown to change the bacterial communities present on the mineral surface (Gleeson et al., 2006).

In general, mineral weathering proceeds more quickly in the rhizosphere than in the bulk soil (Calvaruso et al., 2009). Complex bacterial communities exist around mycorrhizae; weathering by Bacteria in the rhizosphere has been documented in temperate forests (Lepleux et al., 2012; Lepleux et al., 2013) as well as desert (Goldstein et al., 1999; Wakelin et al., 2012), mangrove forest (Vazquez et al., 2000), and high latitude environments (Chapter 7; Frey et al., 2010; Summers et al., 2013; Wongfun et al., 2014). Interestingly, some studies have shown that most effective mineral-weathering Bacteria are present in the deepest soil horizons, suggesting they play a vital role in converting fresh bedrock to soil (Wang et al., 2014), although other studies show contradictory results (Huang et al., 2014).

Bacteria can contribute to mineral weathering through three main mechanisms: redox reactions, acidification, and chelation (Uroz et al., 2009a). In redox reactions, electron shuttling is the agent of weathering. For example, Bacteria can use iron as an electron acceptor; because iron is insoluble in solution under most environmental conditions, it is often easiest for Bacteria to access Fe at the surface of a Fe-rich mineral such as hematite and goethite (Newman, 2001; Uroz et al., 2009a) than from solution. Bacteria produce acids through multiple reactions, including respiration (carbonic acid) and nitrification (nitric and nitrous acids). These protons break metal-oxygen bonds on the mineral surface, releasing metals into soil solution (Uroz et al., 2009a). A last mechanism by which Bacteria participate in mineral dissolution is through the release of chelating compounds, such as siderophores (Uroz et al., 2009b), compounds which chelate Fe and can lead to enhanced dissolution rates (Haselwandter, 1995). Numerous Bacteria have been shown to release compounds that form complexes with Fe from mineral surfaces, allowing the Fe to be transported into the bacterial cells (Kalinowski et al., 2000; Liermann et al., 2000; Page & Huyer, 1984).

Laboratory studies have shown that Bacteria can mobilize nutrients from minerals, although the precise mechanism seems to vary for different minerals and Bacteria. Cyanogenic Bacteria have been shown to mobilize nutrients from minerals in granitic rocks when no other source of nutrients is available in experiments at low to neutral pH (Wongfun et al., 2014); similar results have been observed with other high-potassium rocks (Liu et al., 2012). Laboratory cultures of lithotrophic Fe-oxidizing Bacteria have been shown to grow in the presence of a biotite-rich medium (Shelobolina et al., 2012). Smectite clays, in the presence of *Bacillus mucilaginosus,* showed phase transformations resulting in the release of

Si and Al while increasing the amount of Fe and K in the mineral structure (Zhu et al., 2011). Similarly, dissolution experiments containing apatite with and without *Burkholderia glathei* and organic acids showed that Bacteria and organic acids increased the release of major and trace elements from the apatite by greater than an order of magnitude (Christophe et al., 2013). Together, these findings underscore the importance of microbiological interactions with the environment to large scale elemental cycling and fine scale surface alteration.

Though the influences of fungi and lichens on weathering have been extensively studied over the past decade, it is often difficult to separate the effect of fungi from that of other biota (Zhu & Reinfelder, 2012). Theoretical studies suggest that the quantitative effect of fungi on weathering is controlled by amount of biomass, the mineral surface area in contact with the fungal hyphae, and the ability of the fungi and associated roots to interact with the mineral or rock both physically and chemically (Taylor et al., 2009). The presence of vegetation on a landscape has been showed to accelerate weathering by four to ten times over non-vegetated landscapes (Bonneville et al., 2009; Martino & Perotto, 2010; Moulton et al., 2000). The roots of over 80% of plant species are colonized by mycorrhizal fungi (Chapter 2; Moulton et al., 2000) and these fungi are increasingly understood to enhance chemical weathering. Mycorrhizal fungi have been shown to grow in the direction of mineral nutrients that are essential for both their survival and that of their plant symbiont (Hoffland et al., 2003; Smits et al., 2005). Fungal hyphae exert both physical and chemical weathering stresses on soil minerals (Bonneville et al., 2009) and have been found to penetrate to depths of 4 mm in granite (Lamas et al., 1995), approximately 3 mm in sandstone (Wessels & Budel, 1995) and over 1 mm in quartzite (Cooks & Otto, 1990). Recent studies show that ectomycorrhizal fungi release nutrients from rocks including basalt, granite, and quartz at twice the rate of arbuscular mycorrhizal fungi (Quirk et al., 2014).

Extensive experimental work has examined the mobilization of phosphorus from apatite ($Ca_5(PO_4)_3(OH,F,Cl)$), a common source of mineral-derived phosphorus for plants and soil microbiota. *In situ* experimental and greenhouse studies have shown weathering tunnels most likely created by fungal hyphae in apatite mineral grains beneath forests dominated by both arbuscular and ectomycorrhizal fungi (Koele et al., 2014). Laboratory studies show that apatite weathering rates are three times faster in microcosms containing the fungi *Paxillus involutus* in symbiosis with *Pinus sylvestris* than in abiotic microcosms (Smits et al., 2012); *Paxillus involutus* in symbiosis with *Pinus sylvestris* also grew extensive hyphae and biolayers between 10 and 40 nm thick in microcosms with hornblende, biotite, and chlorite mineral grains (Gazze et al., 2012; Saccone et al., 2012). Similar field experiments showed a threefold increase in ectomycorrhizal fungi associated with Norway spruce over a five-year period following apatite additions to soils (Berner et al., 2012).

Fungi can affect chemical weathering in three different ways (Hoffland et al., 2004). First, fungi release low molecular weight organic acids (LMWOAs), which reduce local pH and mobilize nutrients from rocks (Gadd, 1999; Jarosz-Wilkolazka & Gadd, 2003; Landeweert et al., 2001). Fungi also produce protons during respiration that reduce the local pH of a microenvironment (Hoffland et al., 2004). Finally, fungi, like Bacteria, produce siderophores, which can lead to increased mineral dissolution by solubilizing Fe from minerals like goethite (Watteau & Berthelin, 1994).

LMWOAs are of particular importance in chemical weathering because they have high metal binding capacities (Hausrath et al., 2009; Hoffland et al., 2004; Neaman et al., 2005). Lichens are known to produce low molecular weight organic carboxylic acids such as oxalic, citric, gluconic and lactic acids (Adamo & Violante, 2000), which also act as complexing agents. They also produce slightly water-soluble polyphenolic compounds known collectively as 'lichen acids,' which are depsides and depsidones. Ectomycorrhizal fungi can secrete a variety of LMWOAs, including citrate and malate (Landeweert et al., 2001). Ericoid mycorrhizal fungi that are tolerant of heavy metals produce oxalate (Martino et al., 2003). Oxalate and citrate bind with di- and trivalent cations forming complexes. This lowers the activity of cations in solution, thereby possibly promoting dissolution by lowering the saturation state (Hoffland et al., 2004). Organic acids weather minerals more effectively than inorganic acids, a process that has been demonstrated repeatedly in laboratory studies across mineral groups (Drever & Stillings, 1997; Ganor et al., 2009; Hausrath et al., 2009; Olsen & Rimstidt, 2008; Schott et al., 2009). Plainly, at the mineral surface scale, there is intense work done by biology on solid Earth materials, and the successful mining of necessary nutrients and energy from those solids is what enables the survival of extremophiles.

Fungi also alter the metal availability in soils creating microenvironments that may be chemically different than that of the bulk soil (Chapter 2). In some cases, this may allow plants to grow in what would otherwise be inhospitable soils. Numerous studies have shown that mycorrhizal symbioses allow plants to thrive in soils with high metal toxicity, particularly Cu, Pb, and Zn (Mapelli et al., 2012; Martino & Perotto, 2010; Martino et al., 2003). Fungi have been showed to transform more available forms of Pb into chloropyromorphite, the most stable mineral form of Pb (Rhee et al., 2012). A few studies have specifically addressed the effect of fungi on metal availability in trace metal-rich serpentinites. Daghino et al. (2009) showed that the serpentine fungal species *V. leptobactrum* is more effective at removing structural ions from chrysotile than other fungi. The same group (Daghino et al., 2008) showed that serpentinicolous fungi release siderophores which they believe enhance Fe removal from the chrysotile structure (Daghino et al., 2006; Daghino et al., 2008).

Rocks covered by lichens have been shown to weather up to an order of magnitude faster than bare rocks (Stretch & Viles, 2002; Zambell et al., 2012), although some evidence suggests that lichens can also reduce weathering rates (Mottershead et al., 2003) by protecting the surface (Chapter 3; Arino et al., 1995). The secondary materials produced under lichens have been well-characterized, and include oxalates, iron oxides and hydroxides, siliceous relicts, amorphous aluminosilicates, and carbonates (Adamo & Violante, 2000). Weathering associated with lichen penetration is believed to be associated with exfoliation observed in the Antarctic dry valleys (Friedmann, 1982). Several researchers have suggested that physical processes rather than chemical processes, including freezing and thawing in lichen-colonized rocks play an important role in the alpine zone (Creveld, 1981), in temperate zones (Arino et al., 1997), and may play a role in exfoliation observed in Antarctica (Friedmann & Weed, 1987). Field research in recently deglaciated areas suggests that rocks colonized by lichens weather 200 to 300 times faster than rocks that have only been weathered abiotically (Matthews & Owen, 2008), although the effect is not always this strong (Guglielmin et al., 2012).

Interesting recent work has addressed the weathering ability of fungi isolated from sediments in a Damma glacial forefield in the Swiss Alps. Laboratory experiments using three zygomyceteous fungal species (*Mucor hiemalis, Umbelopsis isabellina* and *Mortierella alpina*) accelerated granite powder dissolution through the release of organic acids, including oxalate, citrate, and malate (Brunner et al., 2011). This work suggests that colonization by microorganisms could be key to the formation of early soils, as well as instrumental in their maturation, refining Earth materials in ways that may be subtle over the short term, but very influential over geologic time.

IMPLICATIONS

Billions of years of water-rock reactions on Earth have given rise to our modern, habitable planet, with extreme environmental niches that select for traits uncommon in the global surface/near surface biosphere. We have discussed the diversity of environmental extremes that underpin the habitability of all harsh environments, considering existing adaptation strategies at extremes of temperature, water activity, pressure, ambient geochemistry, and radiation. We have focused in turn on photosynthesis and chemosynthesis as strategies for successful metabolism, developing how each strategy places different energetic demands on microbes operating in extreme environments. Additionally, we have described how mineral-plant-microbe interactions in the critical zone, defined as the near-surface environment in which biota, rocks, water, and air interact, exert important controls on the evolution of the soil environment, which are propagated through entire ecosystems.

In compiling the material for this chapter, we identified several growing research areas that are pertinent to broad scale scientific inquiry into how life survives in harsh environments:

1) Limits to life at environmental extremes are periodically revised; as we continue to test and observe life's resiliency, the empirically observed limits to life expand—perhaps moving or blurring the edges of the habitable zone, as drawn in the intersection of circles in Figure 1.
2) Environmental gradients are implicitly driven by the reactions of Earth's bedrock, waters, and atmosphere, which have co-evolved over geologic time and will continue to shift in concert as the Earth system changes; as integration of environmental data and associated biological diversity progresses, we expect that key environmental stresses and prominent biological responses will be clarified.
3) Photosynthetic and chemosynthetic metabolisms respond simultaneously and sensitively to environmentally controlled energy availability, be it solar or chemical, resulting in metabolic strategies that are finely tuned to the environment; increasingly collaborative science, which applies complementary techniques in geochemistry, environmental microbiology, and genomics, will resolve how tightly coupled these metabolic systems may be.
4) Teasing apart the biologically driven and chemically driven aspects of weathering is experimentally challenging, and the state of current science leaves much unresolved; conventional wisdom indicates that these two strands may be interwoven at all scales.
5) Soil habitats host mineral, geochemical, and microbial diversity that together affect the diversity of surface plant communities; too few studies target more than one facet of the critical zone (such as surface features, shallow soils, soil solutions, fragmented shallow bedrock) and recognize the critical zone as a system of communicating parts.

Acknowledgments

This work was strengthened by support from the University of Rhode Island (Kingston, RI), the University of Illinois at Chicago (Chicago, IL), the University of Maine (Orono, ME), and the SETI Institute and NASA Ames Research Center. We are glad of the constructive comments of two anonymous reviewers also.

References

Adamo, P. & Violante, P. (2000) Weathering of rocks and neogenesis of minerals associated with lichen activity. *Applied Clay Science 16*, 229-256.

Adams, L. K., Harrison, J. M., Lloyd, J. R., Langley, S. & Fortin, D. (2007) Activity and diversity of Fe(III)-reducing bacteria in a 3000-year-old acid mine drainage site analogue. *Geomicrobiology Journal 24*, 295-305.

Allen, J. P. & Williams, J. C. (1998) Photosynthetic reaction centers. *FEBS Letters 438*, 5-9.

Amend, J. P. & Shock, E. L. (2001) Energetics of overall metabolic reactions of thermophilic and hyperthermophilic Archaea and Bacteria. *FEMS Microbiology Reviews 25*, 175-243.

Amend, J. P., Rogers, K. L., Shock, E. L., Inguaggiato, S. & Gurrieri, S. (2003) Energetics of chemolithoautotrophy in the hydrothermal system of Vulcano Island, southern Italy. *Geobiology 1*, 37-58.

Anesio, A. M. & Laybourn-Parry, J. (2012) Glaciers and ice sheets as a biome. *Trends in Ecology & Evolution 27*, 219-225.

Arino, X., Ortega-Calvo, J. J., Gomez-Bolea, A. & Saiz-Jimenez, C. (1995) Lichen colonization of the Roman pavement at Baelo Claudia (Cadiz, Spain): Biodeterioration vs. bioprotection. *The Science of the Total Environment 167*, 353-363.

Arino, X., Gomez-Bolea, A. & Saiz-Jimenez, C. (1997) Lichens on ancient mortars. *Biodeterioration and Biodegradation 40*, 217-224.

Baker-Austin, C. and Dopson, M. (2007) Life in acid: pH homeostasis in acidophiles. *Trends in Microbiology 15*, 165-171.

Banks, D., Reimann, C. & Skarphagen, H. (1998) The comparative hydrochemistry of two granitic island aquifers: The Isles of Scilly, UK and the Hvaler Islands, Norway. *Science of the Total Environment 209*, 169-183.

Bartlett, D. H. (2002) Pressure effects on in vivo microbial processes. *Biochimica et Biophysica Acta-Protein Structure and Molecular Enzymology 1595*, 367-381.

Barton, H. A. & Luiszer, F. (2005) Microbial metabolic structure in a sulfidic cave hot spring: Potential mechanisms of biospeleogenesis. *Journal of Cave and Karst Studies 67*, 28-38.

Bekker, A., Holland, H. D., Wang, P. L., Rumble, D., Stein, H. J., Hannah, J. L., Coetzee, L. L. & Beukes, N. J. (2004) Dating the rise of atmospheric oxygen. *Nature 427*, 117-120.

Bequer Urbano, S., Albarracín, V. H., Ordonez, O. F., Farías, M. E. & Alvarez, H. M. (2013) Lipid storage in high-altitude Andean Lakes extremophiles and its mobilization under stress conditions in *Rhodococcus* sp A5, a UV-resistant actinobacterium. *Extremophiles 17*, 217-227.

Berner, C., Johansson, T. & Wallander, H. (2012) Long-term effect of apatite on ectomycorrhizal growth and community structure. *Mycorrhiza 22*, 615-621.

Blankenship, R. E. (2010) Early evolution of photosynthesis. *Plant Physiology 154*, 434-438.

Blankenship, R. E., Madigan, M. T. & Bauer, C. E. (Eds.) (1995) *Anoxygenic photosynthetic Bacteria*. Dordrecht, The Netherlands: Kluwer.

Bolhuis, H., Palm, P., Wende, A., Falb, M., Rampp, M., Rodriguez-Valera, F., Pfeiffer, F. & Oesterhelt, D. (2006) The genome of the square archaeon *Haloquadratum walsbyi*: Life at the limits of water activity. *BMC Genomics 7*, 169-169.

Bolhuis, H., Poele, E. M. T. & Rodriguez-Valera, F. (2004) Isolation and cultivation of Walsby's square archaeon. *Environmental Microbiology 6*, 1287-1291.

Bonneville, S., Smits, M. M., Brown, A., Harrington, J., Leake, J. R., Brydson, R. & Benning, L. G. (2009) Plant-driven fungal weathering: Early stages of mineral alteration at the nanometer scale. *Geology 37*, 615-618.

Boussiba, S., Wu, X., Zarka, A. (2000) Alkaliphilic cyanobacteria. In: J. Seckbach (Ed.). *Journey to diverse microbial worlds: Adaptation to exotic environments* (pp. 221-224). Dordrecht: Kluwer.

Boyd, E. S., Hamilton, T. L., Wang, J., He, L. & Zhang, C. L. (2013) The role of tetraether lipid composition in the adaptation of thermophilic archaea to acidity. *Frontiers in Microbiology 4*, 62-62.

Brock, T. D. (1967) Life at high temperatures. Evolutionary, ecological, and biochemical significance of organisms living in hot springs is discussed. *Science 158*, 1012-1019.

Brock, T. D. (1978) *Thermophilic microorganisms and life at high temperatures*. New York, NY: Springer-Verlag.

Brunner, I., Plotze, M., Rieder, S., Zumsteg, A., Furrer, G. & Frey, B. (2011) Pioneering fungi from the Damma Glacier forefield in the Swiss Alps can promote granite weathering. *Geobiology 9*, 266-279.

Bryant, D. A., Costas, A. M. G., Maresca, J. A., Chew, A. G. M., Klatt, C. G., Bateson, M. M., Tallon, L. J., Hostetler, J., Nelson, W. C., Heidelberg, J. F. & Ward, D. M. (2007) *Candidatus Chloracidobacterium thermophilum*: An aerobic phototrophic acidobacterium. *Science 317*, 523-526.

Byrne, R. T., Klingele, A. J., Cabot, E. L., Schackwitz, W. S., Martin, J. A., Martin, J., Wang, Z., Wood, E. A., Pennacchio, C., Pennacchio, L. A., Perna, N. T., Battista, J. R. & Cox, M. M. (2014) Evolution of extreme resistance to ionizing radiation via genetic adaptation of DNA repair. *Elife 3*, 01322-01322.

Calvaruso, C., Mareschal, L., Turpault, M. P. & Leclerc, E. (2009) Rapid clay weathering in the rhizosphere of Norway Spruce and oak in an acid forest ecosystem. *Soil Science Society of America Journal 73*, 331-338.

Canfield, D. E. (2005) The early history of atmospheric oxygen: Homage to Robert A. Garrels. *Annual Review of Earth and Planetary Sciences 33*, 1-36.

Cardace, D. & Hoehler, T. M. (2011) Extremophiles in serpentinizing systems: Implications for life on the Early Earth and other lanets. In: S. Harrison & N. Rajakaruna (Eds.). *Serpentine: The evolution and ecology of a model system* (pp. 29-48). Berkeley, CA: University of California Press.

Cardace, D., Meyer-Dombard, D. R., Arcilla, C. A., Hoehler, T. M., McCollom, T. M. & Schrenk, M. O. (2013). Microbial metabolic landscapes derived from complementary mineralogical, aqueous geochemical, and gas data associated with high pH, actively serpentinizing springs in the Coast Range Ophiolite (CA, USA) and Zambales and

Palawan Ophiolites (Philippines). Fall Meeting of the American Geophysical Union, San Francisco, CA, 9-13 Dec.

Castenholz, R. (1969) Thermophilic blue-green algae and thermal environment. *Bacteriological Reviews 33*, 476-504.

Castenholz, R. W. (1988) Culturing methods for cyanobacteria. *Methods in Enzymology 167*, 68-93.

Castenholz, R. W. & McDermott, T. R. (2010) The Cyanidiales: Ecology, biodiversity, and biogeography. In: J. Seckbach & D. J. Chapman (Eds.). *Red algae in the genomic age* (pp. 357-371). Dordrecht: Springer.

Cavicchioli, R., Amils, R., Wagner, D. & McGenity, T. (2011) Life and applications of extremophiles. *Environmental Microbiology 13*, 1903-1907.

Certini, G., Campbell, C. D. & Edwards, A. C. (2004) Rock fragments in soil support a different microbial community from the fine earth. *Soil Biology & Biochemistry 36*, 1119-1128.

Christiansen, R. L. (2001). *The Quaternary and Pliocene Yellowstone Plateau volcanic field of Wyoming, Idaho, and Montana.* United States Geological Survey Professional Paper 729-G. Menlo Park, CA: US Geological Survey Western Region.

Christophe, C., Marie-Pierre, T., Pascale, F. K., Stephane, U., Marie-Claire, P., Zornitza, T. & Antoine, K. (2013) Increase of apatite dissolution rate by Scots pine roots associated or not with *Burkholderia glathei* PML1(12)Rp in open-system flow microcosms. *Geochimica et Cosmochimica Acta 106*, 287-306.

Clarke, A., Morris, G. J., Fonseca, F., Murray, B. J., Acton, E. & Price, H. C. (2013) A low temperature limit for life on Earth. *PloS One 8*: e66207-e66207.

Coby, A. J., Picardal, F., Shelobolina, E., Xu, H. & Roden, E. E. (2011) Repeated anaerobic microbial redox cycling of iron. *Applied and Environmental Microbiology 77*, 6036-6042.

Cooks, J. & Otto, E. (1990) The weathering effects of the lichen *Lecidea* aff. *sarcogynoides* (Koerb) on Magaliesberg quartzite. *Earth Surface Processes and Landforms 15*, 491-500.

Costa, K. C., Hallmark, J., Navarro, J. B., Hedlund, B. P., Moser, D. P., Labahn, S. & Soukup, D. (2008) Geomicrobiological changes in two ephemeral desert playa lakes in the Western United States. *Geomicrobiology Journal 25*, 250-259.

Cox, M. M. & Battista, J. R. (2005) *Deinococcus radiodurans* - The consummate survivor. *Nature Reviews Microbiology 3*, 882-892.

Cox, M. M., Keck, J. L. & Battista, J. R. (2010) Rising from the ashes: DNA repair in *Deinococcus radiodurans*. *PLoS Genetics 6*: e1000815-e1000815.

Creveld, M. (1981) Epilithic lichen communities in the alpine zone of Southern Norway. *Bibliotheca Lichenologica 17*, 1-287.

Daghino, S., Turci, F., Tomatis, M., Favier, A., Perotto, S., Douki, T., & Fubini, B. (2006) Soil fungi reduce the iron content and the DNA damaging effects of asbestos fibers. *Environmental Science and Technology 40*, 5793-5798.

Daghino, S., Martino, E., Vurro, E., Tomatis, M., Girlanda, M., Fubini, B. & Perotto, S. (2008) Bioweathering of chrysotile by fungi isolated in ophiolitic sites. *FEMS Microbiology Letters 285*, 242-249.

Daghino, S., Turci, F., Tomatis, M., Girlanda, M., Fubini, B. & Perotto, S. (2009) Weathering of chrysotile asbestos by the serpentine rock-inhabiting fungus *Verticillium leptobactrum*. *FEMS Microbiololgy Ecology 69*, 132-141.

D'Amico, S., Collins, T., Marx, J. C., Feller, G. & Gerday, C. (2006) Psychrophilic microorganisms: Challenges for life. *EMBO Reports 7*, 385-389.

Daniel, R. M. & Cowan, D. A. (2000) Biomolecular stability and life at high temperatures. *Cellular and Molecular Life Sciences 57*, 250-264.

Davis, K. J., Nealson, K. H. & Luttge, A. (2007) Calcite and dolomite dissolution rates in the context of microbe-mineral surface interactions. *Geobiology 5*, 191-205.

Des Marais, D. J. (2000) Evolution - When did photosynthesis emerge on earth? *Science 289*, 1703-1705.

Dewit, R. & Vangemerden, H. (1987) Oxidation of sulfide to thiosulfate by *Microcoleus-Chtonoplastes*. *FEMS Microbiology Ecology 45*, 7-13.

Dinsdale, E. A., Edwards, R.A., Hall, D., Angly, F., Breitbart, M., Brulc, J.M., Furlan, M., Desnues, C., Haynes, M., Linlin, L., McDaniel, L., Moran, M., Nelson, K. E., Nilsson, C., Olson, R., Paul, J., Brito, B. R., Yijun, R., Swan, B. K. & Stevens, R. (2008) Functional metagenomic profiling of nine biomes. *Nature 452(7187)*, 629-632.

Dopson, M. & Johnson, D. B. (2012) Biodiversity, metabolism and applications of acidophilic sulfur-metabolizing microorganisms. *Environmental Microbiology 14*, 2620-2631.

Drever, J. I. & Stillings, L. L. (1997) The role of organic acids in mineral weathering. *Colloids and Surfaces A-Physicochemical and Engineering Aspects 120*, 167-181.

Duckworth, A. W., Grant, W. D., Jones, B. E. & vanSteenbergen, R. (1996) Phylogenetic diversity of soda lake alkaliphiles. *FEMS Microbiology Ecology 19*, 181-191.

Edwards, H. G. M., Moeller, R., Jorge Villar, S. E., Horneck, G. & Stackebrandt, E. (2006) Raman spectroscopic study of the photoprotection of extremophilic microbes against ultraviolet radiation. *International Journal of Astrobiology 5*, 313-318.

Edwards, A., Pachebat, J. A., Swain, M., Hegarty, M., Hodson, A. J., Irvine-Fynn, T. D. L., Rassner, S. M. E. & Sattler, B. (2013) A metagenomic snapshot of taxonomic and functional diversity in an alpine glacier cryoconite ecosystem. *Environmental Research Letters 8*, 035003-035003.

Emerson, D., Fleming, E. J. & McBeth, J. M. (2010) Iron-oxidizing bacteria: an environmental and genomic perspective. *Annual Reviews of Microbiology 64*, 561-583.

Field, C. B., Behrenfeld, M. J., Randerson, J. T. & Falkowski, P. (1998) Primary production of the biosphere: Integrating terrestrial and oceanic components. *Science 281*, 237-240.

Fournier, R. O. (1989) Geochemistry and dynamics of the Yellowstone National Park hydrothermal system. *Annual Reviews of Earth and Planetary Science 17*, 13-53.

Frey, B., Rieder, S. R., Brunner, I., Plotze, M., Koetzsch, S., Lapanje, A., Brandl, H. & Furrer, G. (2010) Weathering-associated bacteria from the Damma Glacier Forefield: Physiological capabilities and impact on granite dissolution. *Applied and Environmental Microbiology 76*, 4788-4796.

Friedmann, E. I. (1982) Endolithic microorganisms in the Antarctic cold desert. *Science 215*, 1045-1053.

Friedmann, E. I. & Weed, R. (1987) Microbial trace fossil formation, biogenous, and abiotic weathering in the Antarctic cold desert. *Science 236*, 703-705.

Gabani, P. & Singh, O. V. (2013) Radiation-resistant extremophiles and their potential in biotechnology and therapeutics. *Applied Microbiology and Biotechnology 97*, 993-1004.

Gadd, G. M. (1999) Fungal production of citric and oxalic acid: Importance in metal speciation, physiology and biogeochemical processes. *Advances in Microbial Physiology 41*, 47-92.

Ganor, J., Reznik, I. J. & Rosenberg, Y. O. (2009) Organics in water-rock interactions. In: E. H. Oelkers & J. Schott (Eds.). *Thermodynamics and kinetics of water-rock interaction* (pp. 259-369). Washington, DC: Mineralogical Society of America, Geochemical Society.

Gazze, S. A., Saccone, L., Ragnarsdottir, K. V., Smits, M. M., Duran, A. L., Leake, J. R., Banwart, S. A. & McMaster, T. J. (2012) Nanoscale channels on ectomycorrhizal-colonized chlorite: Evidence for plant-driven fungal dissolution. *Journal of Geophysical Research-Biogeosciences 117*, G00N09, DOI: 10.1029/2012JG002016.

Giovannoni, S. J., Revsbech, N. P., Ward, D. M. & Castenholz, R. W. (1987) Obligately phototrophic *Chloroflexus* - Primary production in anaerobic hot-spring microbial mats. *Archives of Microbiology 147*, 80-87.

Glazer, B. T. & Rouxel, O. J. (2009) Redox speciation and distribution within diverse iron-dominated microbial habitats at Loihi Seamount. *Geomicrobiology Journal 26*, 606-622.

Gleeson, D. B., Kennedy, N. M., Clipson, N., Melville, K., Gadd, G. M. & McDermott, F. P. (2006) Characterization of bacterial community structure on a weathered pegmatitic granite. *Microbial Ecology 51*, 526-534.

Goldstein, A. H., Braverman, K. & Osorio, N. (1999) Evidence for mutualism between a plant growing in a phosphate-limited desert environment and a mineral phosphate solubilizing (MPS) rhizobacterium. *FEMS Microbiology Ecology 30*, 295-300.

Grant, W. D. (2004) Life at low water activity. *Philosophical Transactions of the Royal Society of London Series B-Biological Sciences 359*, 1249-1266.

Guglielmin, M., Worland, M. R., Convey, P. & Cannone, N. (2012) Schmidt Hammer studies in the maritime Antarctic: Application to dating Holocene deglaciation and estimating the effects of macrolichens on rock weathering. *Geomorphology 155*, 34-44.

Haselwandter, K. (1995) Mycorrhizal fungi: Siderophore production. *Critical Reviews in Biotechnology 15*, 287-291.

Hausrath, E. M., Neaman, A. & Brantley, S. L. (2009) Elemental release rates from dissolving basalt and granite with and without organic ligands. *American Journal of Science 309*, 633-660.

Hazen, R. M., Boctor, N., Brandes, J. A., Cody, G. D., Hemley, R. J., Sharma, A. & Yoder, H. S. (2002) High pressure and the origin of life. *Journal of Physics-Condensed Matter 14*, 11489-11494.

Hefferan, K. & O'Brien, J. (2010) *Earth materials*. Chichester: Wiley-Blackwell.

Hirschmann, M. M. (2006) Water, melting, and the Deep Earth H_2O Cycle. *Annual Review of Earth and Planetary Sciences 34*, 629–53.

Hodson, A., Paterson, H., Westwood, K., Cameron, K. & Laybourn-Parry, J. (2013) A blue-ice ecosystem on the margin's of the East Antarctic ice sheet. *Journal of Glaciology 59*, 255-268.

Hoffland, E., Giesler, R., Jongmans, A. G. & van Breemen, N. (2003) Feldspar tunneling by fungi along natural productivity gradients. *Ecosystems 6*, 739-746.

Hoffland, E., Kuyper, T. W., Wallander, H., Plassard, C., Gorbushina, A. A., Haselwandter, K., Holmstrom, S., Landeweert, R., Lundstrom, U. S., Rosling, A., Sen, R., Smits, M. M., van Hees, P. A. & van Breemen, N. (2004) The role of fungi in weathering. *Frontiers in Ecology and the Environment 2*, 258-264.

Hoehler, T. M. (2007) An energy balance concept for habitability. *Astrobiology 7*, 824-838.

Hose, L. D., Palmer, A. N., Palmer, M. V., Northup, D. E., Boston, P. J. & DuChene, H. R. (2000) Microbiology and geochemistry in a hydrogen-sulphide-rich karst environment. *Chemical Geology 169*, 399-423.

Huang, J., Sheng, X. F., Xi, J., He, L. Y., Huang, Z., Wang, Q. & Zhang, Z. D. (2014) Depth-related changes in community structure of culturable mineral weathering bacteria and in weathering patterns caused by them along two contrasting soil profiles. *Applied and Environmental Microbiology 80*, 29-42.

Huber, C., Klimant, I., Krause, C., Werner, T., Mayr, T. & Wolfbeis, O. S. (2000). Optical sensor for seawater salinity. *Fresenius' Journal of Analytical Chemistry 368*, 196-202.

Humayoun, S. B., Bano, N. & Hollibaugh, J. T. (2003) Depth distribution of microbial diversity in Mono Lake, a meromictic soda lake in California. *Applied and Environmental Microbiology 69*, 1030-1042.

Imhoff, J. F. & Suling, J. (1996) The phylogenetic relationship among Ectothiorhodospiraceae: A reevaluation of their taxonomy on the basis of 16S rDNA analyses. *Archives of Microbiology 165*, 106-113.

Jaenicke, R. (2000) Stability and stabilization of globular proteins in solution. *Journal of Biotechnology 79*, 193-203.

Jaenicke, R. & Bohm, G. (1998) The stability of proteins in extreme environments. *Current Opinion in Structural Biology 8*, 738-748.

Jaenicke, R. & Sterner, R. (2006) Life at high temperatures. In: M. Dworkin, S. Falkow, E. Rosenberg, H.-S. Schleifer & E. Stackebrandt (Eds.). *The prokaryotes. 2: Ecophysiology and biochemistry* (pp. 167-209). New York, NY: Springer.

Jarosz-Wilkolazka, A. & Gadd, G. M. (2003) Oxalate production by wood-rotting fungi growing in toxic metal-amended medium. *Chemosphere 52*, 541-547.

Johnson, D. B. (2012) Geomicrobiology of extremely acidic subsurface environments. *FEMS Microbiology Ecology 81*, 2-12.

Johnson, D. B. & Hallberg, K. B. (2003) The microbiology of acidic mine waters. *Research in Microbiology 154*, 466-473.

Jones, B. E., Grant, W. D., Duckworth, A. W. & Owenson, G. G. (1998) Microbial diversity of soda lakes. *Extremophiles 2*, 191-200.

Jungblut, A. D., Hawes, I., Mountfort, D., Hitzfeld, B., Dietrich, D. R., Burns, B. P. & Neilan, B. A. (2005) Diversity within cyanobacterial mat communities in variable salinity meltwater ponds of McMurdo Ice Shelf, Antarctica. *Environmental Microbiology 7*, 519-529.

Jungblut, A. D., Vincent, W. F. & Lovejoy, C. (2012) Eukaryotes in Arctic and Antarctic cyanobacterial mats. *FEMS Microbiology Ecology 82*, 416-428.

Kalinowski, B. E., Liermann, L. J., Brantley, S. L., Barnes, A. & Pantano, C. G. (2000) X-ray photoelectron evidence for bacteria-enhanced dissolution of hornblende. *Geochimica et Cosmochimica Acta 64*, 1331-1343.

Kallas, T. & Castenholz, R. W. (1982) Internal pH and ATP-ADP pools in the cyanobacterium *Synechococcus* sp. during exposure to growth-inhibiting low pH. *Journal of Bacteriology 149*, 229-236.

Karan, R., Capes, M. D. & DasSarma, S. (2012) Function and biotechnology of extremophilic enzymes in low water activity. *Aquatic Biosystems 8*, 1-15.

Kashefi, K. & Lovley, D. R. (2003) Extending the upper temperature limit for life. *Science 301*, 934-934.

Kato, C., Sato, T. & Horikoshi, K. (1995) Isolation and properties of barophilic and barotolerant bacteria from deep-sea mud samples. *Biodiversity and Conservation 4*, 1-9.

Kip, N., van Winden, J., Pan, Y., Bodrossy, L., Reichart, G.-J., Smolders, A., Jetten, M. S. M., Sinninghe Damste, J. A. & Op den Camp, H. J. M. (2010) Global prevalence of methane oxidation by symbiotic bacteria in peat-moss ecosystems. *Nature Geoscience 3*, 617-621.

Kitada, M., Kosono, S. & Kudo, T. (2000) The Na+/H+ antiporter of alkaliphilic *Bacillus* sp. *Extremophiles 4*, 253-258.

Koele, N., Dickie, I. A., Blum, J. D., Gleason, J. D. & de Graaf, L. (2014) Ecological significance of mineral weathering in ectomycorrhizal and arbuscular mycorrhizal ecosystems from a field-based comparison. *Soil Biology & Biochemistry 69*, 63-70.

Konhauser, K. O., Newman, D. K. & Kappler, A. (2005) The potential significance of microbial Fe(III) reduction during deposition of Precambrian banded iron formations. *Geobiology 3*, 167-177.

Kostka, J. E., Stucki, J. W., Nealson, K. H. & Wu, J. (1996) Reduction of structural Fe(III) in smectite by a pure culture of *Shewanella putrefaciens* strain MR-1. *Clays and Clay Minerals 44*, 522-529.

Krulwich, T. A. (1995) Alkaliphiles - Basic molecular problems of pH tolerance and bioenergetics. *Molecular Microbiology 15*, 403-410.

Krulwich, T. A., Ito, M., Gilmour, R., Sturr, M. G., Guffanti, A. A. & Hicks, D. B. (1996) Energetic problems of extremely alkaliphilic aerobes. *Biochimica et Biophysica Acta-Bioenergetics 1275*, 21-26.

Krulwich, T. A., Ito, M. & Guffanti, A. A. (2001) The Na+-dependence of alkaliphily in *Bacillus*. *Biochimica et Biophysica Acta-Bioenergetics 1505*, 158-168.

Kupriyanova, E. V., Sinetova, M. A., Cho, S. M., Park, Y. I., Markelova, A. G., Los, D. A. & Pronina, N. A. (2013) Specific features of the system of carbonic anhydrases of alkaliphilic cyanobacteria. *Russian Journal of Plant Physiology 60*, 465-471.

Laloknam, S., Tanaka, K., Buaboocha, T., Waditee, R., Incharoensakdi, A., Hibino, T., Tanaka, Y. & Takabe, T. (2006) Halotolerant cyanobacterium *Aphanothece halophytica* contains a betaine transporter active at alkaline pH and high salinity. *Applied and Environmental Microbiology 72*, 6018-6026.

Lamas, B. P., Brea, M. T. R. & Hermo, B. M. S. (1995) Colonization by lichens of granite churches in Galicia (Northwest Spain). *Science of the Total Environment 167*, 343-351.

Landeweert, R., Hoffland, E., Finlay, R. D., Kuyper, T. W. & van Breemen, N. (2001) Linking plants to rocks: Ectomycorrhizal fungi mobilize nutrients from minerals. *Trends in Ecology & Evolution 16*, 248-254.

Langmuir, D. (1997) *Aqueous environmental geochemistry*. Upper Saddle River, NJ: Prentice-Hall.

Lepleux, C., Turpault, M. P., Oger, P., Frey-Klett, P. & Uroz, S. (2012) Correlation of the abundance of betaproteobacteria on mineral surfaces with mineral weathering in forest soils. *Applied and Environmental Microbiology 78*, 7114-7119.

Lepleux, C., Uroz, S., Collignon, C., Churin, J. L., Turpault, M. P. & Frey-Klett, P. (2013) A short-term mineral amendment impacts the mineral weathering bacterial communities in an acidic forest soil. *Research in Microbiology 164*, 729-739.

Lewis, G. N. & Randall, M. (1923). *Thermodynamics and the free energy of chemical substances*. New York, NY: McGraw-Hill.

Liermann, L. J., Kalinowski, B. E., Brantley, S. L. & Ferry, J. G. (2000) Role of bacterial siderophores in dissolution of hornblende. *Geochimica et Cosmochimica Acta 64*, 587-602.

Liu, D. F., Lian, B. & Dong, H. L. (2012) Isolation of *Paenibacillus* sp. and assessment of its potential for enhancing mineral weathering. *Geomicrobiology Journal 29*, 413-421.

Lovley, D. R., Holmes, D. E. & Nevin, K. P. (2004) Dissimilatory Fe(III) and Mn(IV) reduction. *Advances in Microbial Physiology 49*, 219-286.

Madigan, M. T. (2003) Anoxygenic phototrophic bacteria from extreme environments. *Photosynthesis Research 76*, 157-171.

Madigan, M. T., Jung, D. O., Woese, C. R. & Achenbach, L. A. (2000) *Rhodoferax antarcticus* sp. nov., a moderately psychrophilic purple nonsulfur bacterium isolated from an Antarctic microbial mat. *Archives of Microbiology 173*, 269-277.

Madigan, M. T., Jung, D.O., Karr, E. A., Sattley, W. M., Achenbach, L. A. & van der Meer, M. T. J. (2005) Diversity of anoxygenic phototrophs in contrasting extreme environments. In: W. Inskeep & T. McDermott (Eds.). *Geothermal biology and geochemistry in Yellowstone National Park* (pp. 203-219). Bozeman, MT: Montana State University Publications.

Mapelli, F., Marasco, R., Balloi, A., Rolli, E., Cappitelli, F., Daffonchio, D. & Borin, S. (2012) Mineral-microbe interactions: Biotechnological potential of bioweathering. *Journal of Biotechnology 157*, 473-481.

Martino, E. & Perotto, S. (2010) Mineral transformations by mycorrhizal fungi. *Geomicrobiology Journal 27*, 609-623.

Martino, E., Perotto, S., Parsons, R. & Gadd, G. M. (2003) Solubilization of insoluble inorganic zinc compounds by ericoid mycorrhizal fungi derived from heavy metal polluted sites. *Soil Biology and Biochemistry 34*, 133-141.

Matthews, J. A. & Owen, G. (2008) Endolithic lichens, rapid biological weathering and Schmidt hammer R-values on recently exposed rock surfaces: Storbreen glacier foreland, Jotunheimen, Norway. *Geografiska Annaler Series A Physical Geography 90A*, 287-297.

McNamara, C. J., Perry, T. D., Bearce, K. A., Hernandez-Duque, G. & Mitchell, R. (2006) Epilithic and endolithic bacterial communities in limestone from a Maya archaeological site. *Microbial Ecology 51*, 51-64.

Meersman, F., Nies, E. & Heremans, K. (2008) On the thermal expansion of water and the phase behavior of macromolecules in aqueous solution. *Zeitschrift Fur Naturforschung Section B—A Journal of Chemical Sciences 63*, 785-790.

Mitchell, A. C., Lafreniere, M. J., Skidmore, M. L. & Boyd, E. S. (2013) Influence of bedrock mineral composition on microbial diversity in a subglacial environment. *Geology 41*, 855-858.

Montross, S., Skidmore, M., Christner, B., Samyn, D., Tison, J.-L., Lorrain, R., Doyle, S. & Fitzsimons, S. (2014) Debris-rich basal ice as a microbial habitat, Taylor Glacier, Antarctica. *Geomicrobiology Journal 31*, 76-81.

Morrill, P. L., Kuenen, J. G., Johnson, O. J., Suzuki, S., Rietze, A., Sessions, A. L., Fogel, M. L. & Nealson, K. H. (2013) Geochemistry and geobiology of a present-day serpentinization site in California: The Cedars. *Geochimica et Cosmochimica Acta 109*, 222-240.

Mottershead, D., Gorbushina, A., Lucas, G. & Wright, J. (2003) The influence of marine salt, aspect and microbes in the weathering of sandstone in two historic structures. *Building and Environment 38*, 1193-1204.

Moulton, K. L., West, A. J. & Berner, R. A. (2000) Solute flux and mineral mass balance approaches to the quantification of plant effects on silicate weathering. *American Journal of Science 300*, 539-570.

Murray, I. W. & Smith, F. A. (2009) Life in an extreme environment: The effects of heat on foraging and survival of woodrats (*Neotoma*) in Death Valley, California. *Integrative and Comparative Biology 49*, E279-E279.

Mustin, C., Berthelin, J., Marion, P. & Dedonato, P. (1992) Corrosion and electrochemical oxidation of a pyrite by *Thiobacilus ferrooxidans. Applied and Environmental Microbiology 58*, 1175-1182.

Nadeau, T. L. & Castenholz, R. W. (2000) Characterization of psychrophilic oscillatorians (cyanobacteria) from Antarctic meltwater ponds. *Journal of Phycology 36*, 914-923.

Neaman, A., Chorover, J. & Brantley, S. L. (2005) Element mobility patterns record organic ligands in soils on early Earth. *Geology 33*, 117-120.

Newman, D. K. (2001) Microbiology - How bacteria respire minerals. *Science 292*, 1312-1313.

Nishiyama, Y., Kovács, E., Lee, C. B., Hayashi, H., Watanabe, T. & Murata, N. (1993) Photosynthetic adaptation to high temperature associated with thylakoid membranes of *Synechococcus* PCC7002. *Plant and Cell Physiology 34*, 337-343.

Oger, P. M. & Cario, A. (2013) Adaptation of the membrane in Archaea. *Biophysical Chemistry 183*, 42-56.

Olsen, A. A. & Rimstidt, J. D. (2008) Oxalate-promoted forsterite dissolution at low pH. *Geochimica et Cosmochimica Acta 72*, 1758-1766.

Olson, J. M. & Pierson, B. K. (1986) Photosynthesis 3.5 thousand million years ago. *Photosynthesis Research 9*, 251-259.

Ordonez, O. F., Flores, M. R., Dib, J. R., Paz, A. & Farias, M. E. (2009) Extremophile culture collection from Andean lakes: Extreme pristine environments that host a wide diversity of microorganisms with tolerance to UV radiation. *Microbial Ecology 58*, 461-473.

Oren, A. (2011) Thermodynamic limits to microbial life at high salt concentrations. *Environmental Microbiology 13*, 1908-1923.

Oren, A., Padan, E. & Avron, M. (1977) Quantum yields for oxygenic and anoxygenic photosynthesis in cyanobacterium *Oscillatoria-Limnetica. Proceedings of the National Academy of Sciences USA 74*, 2152-2156.

Ort, D. R. & Yocum, C.F. (1996) *Oxygenic photosynthesis: The light reactions*. Dordrecht: Kluwer.

Page, W. J. & Huyer, M. (1984) Derepression of the *Azotobacter vinelandii* siderophore system, using iron-containing minerals to limit iron repletion *Journal of Bacteriology 158*, 496-502.

Peary, J. A. & Castenholz, R.W. (1964) Temperature strains of a thermophilic blue-green alga. *Nature 202*, 720-721.

Pfennig, N. (1974) *Rhodopseudomonas globiformis*, sp. n., a new species of the Rhodospirillaceae. *Archives of Microbiology 100*, 197-206.

Pierson, B. K. & Castenholz, R. W. (1995) Taxonomy and physiology of filamentous anoxygenic phototrophs. In: R. E. Blankenship, M. T. Madigan & C. E. Bauer (Eds.). *Anoxygenic photosynthetic bacteria* (pp. 31-47). Dordrecht: Kluwer Academic.

Pikuta, E. V., Hoover, R. B. & Tang, J. (2007) Microbial extremophiles at the limits of life. *Critical Reviews in Microbiology 33*, 183-209.

Planavsky, N., Rouxel, O. J., Bekker, A., Shapiro, R., Fralick, P. & Knudsen, A. (2009) Iron-oxidizing microbial ecosystems thrived in late Paleoproterozoic redox-stratified oceans. *Earth and Planetary Science Letters 286*, 230-242.

Plumb, J. J., Haddad, C. M., Gibson, J. A. E. & Franzmann, P. D. (2007) *Acidianus sulfidivorans* sp nov., an extremely acidophilic, thermophilic archaeon isolated from a solfatara on Lihir Island, Papua New Guinea, and emendation of the genus description. *International Journal of Systematic and Evolutionary Microbiology 57*, 1418-1423.

Quirk, J., Leake, J. R., Banwart, S. A., Taylor, L. L. & Beerling, D. J. (2014) Weathering by tree-root-associating fungi diminishes under simulated Cenozoic atmospheric CO2 decline. *Biogeosciences 11*, 321-331.

Raymond, J. A. & Fritsen, C. H. (2001) Semipurification and ice recrystallization inhibition activity of ice-active substances associated with Antarctic photosynthetic organisms. *Cryobiology 43*, 63-70.

Rees, H. C., Grant, W. D., Jones, B. E. & Heaphy, S. (2004) Diversity of Kenyan soda lake alkaliphiles assessed by molecular methods. *Extremophiles 8*, 63-71.

Rew, D. A. (2003) *Deinococcus radiodurans*. *European Journal of Surgical Oncology 29*, 557-558.

Rhee, Y. J., Hillier, S. & Gadd, G. M. (2012) Lead transformation to pyromorphite by fungi. *Current Biology 22*, 237-241.

Roden, E. E., Sobolev, D., Glazer, B. T. & Luther, G. W., III (2004) Potential for microscale bacterial Fe redox cycling at the aerobic-anaerobic Interface. *Geomicrobiology Journal 21*, 379-391.

Roden, E. E., McBeth, J. M., Blothe, M., Percak-Dennett, E. M., Fleming, E. J., Holyoke, R. R., Luther, G. W., III,, Emerson, D. & Schieber, J. (2012) The microbial ferrous wheel in a neutral pH groundwater seep. *Frontiers in Microbiology 3*, DOI: 10.3389/fmicb.2012.00172.

Rothschild, L. J. & Mancinelli, R. L. (2001) Life in extreme environments. *Nature 409*, 1092-1101.

Saccone, L., Gazze, S. A., Duran, A. L., Leake, J. R., Banwart, S. A., Ragnarsdottir, K. V., Smits, M. M. & McMaster, T. J. (2012) High resolution characterization of ectomycorrhizal fungal-mineral interactions in axenic microcosm experiments. *Biogeochemistry 111*, 411-425.

Schleper, C., Puehler, G., Holz, I., Gambacorta, A., Janekovic, D., Santarius, U., Klenk, H. P. & Zillig, W. (1995) *Picrophilus* gen-nov, fam-nov - A novel aerobic, heterotrophic, thermoacidophilic genus and family comprising Archaea capable of growth around pH-0. *Journal of Bacteriology 177*, 7050-7059.

Schott, J., Pokrovsky, O. S., & Oelkers, E. H. (2009) The link between mineral dissolution/precipitation kinetics and solution chemistry. *Reviews in Mineralogy and Geochemistry 70*, 207-258.

Shelobolina, E., Xu, H. F., Konishi, H., Kukkadapu, R., Wu, T., Blothe, M. & Roden, E. (2012) Microbial lithotrophic oxidation of structural Fe(II) in biotite. *Applied and Environmental Microbiology 78*, 5746-5752.

Shirokova, L. S., Benezeth, P., Pokrovsky, O. S., Gerard, E., Menez, B. & Alfredsson, H. (2012) Effect of the heterotrophic bacterium *Pseudomonas reactans* on olivine dissolution kinetics and implications for CO2 storage in basalts. *Geochimica et Cosmochimica Acta 80*, 30-50.

Shock, E. L., Holland, M. E., Meyer-Dombard, D. R., Amend, J. P., Osburn, G. R. & Fisher, T. (2010) Quantifying inorganic sources of geochemical energy in hydrothermal ecosystems, Yellowstone National Park, USA. *Geochimica et Cosmochimica Acta 74*, 4005-4043.

Smits, M. M., Bonneville, S., Benning, L. G., Banwart, S. A. & Leake, J. R. (2012) Plant-driven weathering of apatite - The role of an ectomycorrhizal fungus. *Geobiology 10*, 445-456.

Smits, M. M., Hoffland, E., Jongmans, A. G. & van Breemen, N. (2005) Contribution of mineral tunneling to total feldspar weathering. *Geoderma 125*, 59-69.

Stetter, K. O., Fiala, G., Huber, G., Huber, R. & Segerer, A. (1990) Hyperthermophilic microorganisms. *FEMS Microbiology Letters 75*, 117-124.

Stevens, T. O. & McKinley, J. P. (2000) Abiotic controls on H-2 production from basalt-water reactions and implications for aquifer biogeochemistry. *Environmental Science & Technology 34*, 826-831.

Straub, K. L., Benz, M. & Schink, B. (2001) Iron metabolism in anoxic environments at near neutral pH. *FEMS Microbiology Ecology 34*, 181-186.

Stretch, R. & Viles, H. (2002) The nature and rate of weathering by lichens on lava flows on Lanzarote. *Geomorphology 47*, 87-94.

Summers, S., Whiteley, A. S., Kelly, L. C. & Cockell, C. S. (2013) Land coverage influences the bacterial community composition in the critical zone of a sub-Arctic basaltic environment. *FEMS Microbiology Ecology 86*, 381-393.

Szponar, N., Brazelton, W. J., Schrenk, M. O., Bower, D. M., Steele, A.& Morrill, P. L. (2013) Geochemistry of a continental site of serpentinization, the Tablelands Ophiolite, Gros Morne National Park: A Mars analogue. *Icarus 224*, 286-296.

Taylor, L. L., Leake, J. R., Quirk, J., Hardy, K., Banwart, S. A. & Beerling, D. J. (2009) Biological weathering and the long-term carbon cycle: integrating mycorrhizal evolution and function into the current paradigm. *Geobiology 7*, 171-191.

Tiago, I. & Verissimo, A. (2013) Microbial and functional diversity of a subterrestrial high pH groundwater associated to serpentinization. *Environmental Microbiology 15*, 1687-1706.

Uroz, S., Calvaruso, C., Turpault, M. P. & Frey-Klett, P. (2009a) Mineral weathering by bacteria: Ecology, actors and mechanisms. *Trends in Microbiology 17*, 378-387.

Uroz, S., Calvaruso, C., Turpault, M. P., Sarniguet, A., de Boer, W., Leveau, J. H. J. & Frey-Klett, P. (2009b) Efficient mineral weathering is a distinctive functional trait of the bacterial genus *Collimonas*. *Soil Biology & Biochemistry 41*, 2178-2186.

van der Wielen, P., Bolhuis, H., Borin, S., Daffonchio, D., Corselli, C., Giuliano, L., D'Auria, G., de Lange, G. J., Huebner, A., Varnavas, S. P., Thomson, J., Tamburini, C., Marty, D., McGenity, T. J., Timmis, K. N. & BioDeep Scientific Party (2005) The enigma of prokaryotic life in deep hypersaline anoxic basins. *Science 307*, 121-123.

Vazquez, P., Holguin, G., Puente, M. E., Lopez-Cortes, A. & Bashan, Y. (2000) Phosphate-solubilizing microorganisms associated with the rhizosphere of mangroves in a semiarid coastal lagoon. *Biology and Fertility of Soils 30*, 460-468.

Vonshak, A. (1997) *Spirulina Platensis Arthrospira: Physiology, cell-biology and biotechnology*. Boca Raton, FL: CRC Press.

Wakelin, S. A., Anand, R. R., Reith, F., Gregg, A. L., Noble, R. R. P., Goldfarb, K. C., Andersen, G. L., DeSantis, T. Z., Piceno, Y. M. & Brodie, E. L. (2012) Bacterial communities associated with a mineral weathering profile at a sulphidic mine tailings dump in arid Western Australia. *FEMS Microbiology Ecology 79*, 298-311.

Wang, Q., Cheng, C., He, L. Y., Huang, Z. & Sheng, X. F. (2014) Characterization of depth-related changes in bacterial communities involved in mineral weathering along a mineral-rich soil profile. *Geomicrobiology Journal 31*, 431-444.

Watteau, F. & Berthelin, J. (1994) Mineral dissolution of iron and aluminium from soil minerals: efficiency and specificity of hydroxamate siderophores compared to aliphatic acids. *European Journal of Soil Biology 30*, 1-9.

Weber, K. A., Urrutia, M. M., Churchill, P. F., Kukkadapu, R. K. & Roden, E. E. (2006) Anaerobic redox cycling of iron by freshwater sediment microorganisms. *Environmental Microbiology 8*, 100-113.

Wessels, D. C. J. & Budel, B. (1995) Epilithic and cryptoendolithic cyanobacteria of Clarens Sandstone Cliffs in the Golden-Gate-Highlands-National-Park, South-Africa. *Botanica Acta 108*, 220-226.

White, D. E. (1957) Magmatic, connate, and metamorphic waters. *Geological Society of America Bulletin 68*, 1659-1682.

White, R. H. (1984) Hydrolytic stability of biomolecules at high-temperatures and its implication for life at 250-degrees-C. *Nature 310*, 430-432.

Wongfun, N., Plotze, M., Furrer, G. & Brandl, H. (2014) Weathering of granite from the Damma Glacier area: The contribution of cyanogenic bacteria. *Geomicrobiology Journal 31*, 93-100.

Yafremava, L. S., Di Giulio, M. & Caetano-Anolles, G. (2013) Comparative analysis of barophily-related amino acid content in protein domains of *Pyrococcus abyssi* and *Pyrococcus furiosus*. *Archaea—An International Microbiological Journal,* DOI: 10.1155/2013/680436.

Yano, Y., Nakayama, A., Ishihara, K. & Saito, H. (1998) Adaptive changes in membrane lipids of barophilic bacteria in response to changes in growth pressure. *Applied and Environmental Microbiology 64*, 479-485.

Yao, M. J., Lian, B., Teng, H. H., Tian, Y. C. & Yang, X. Q. (2013) Serpentine dissolution in the presence of bacteria *Bacillus mucilaginosus*. *Geomicrobiology Journal 30*, 72-80.

Zakhia, F., Jungblut, A-D., Taton, A., Vincent, W. F., & Wilmotte, A. (2008) Cyanobacteria in cold ecosystems. In: R. Margesin, F. Schinner, J-C. Marx & C. Gerday (Eds.). *Psychrophiles: from biodiversity to biotechnology* (pp. 121-132). Berlin: Springer-Verlag.

Zambell, C. B., Adams, J. M., Gorring, M. L. & Schwartzman, D. W. (2012) Effect of lichen colonization on chemical weathering of hornblende granite as estimated by aqueous elemental flux. *Chemical Geology 291*, 166-174.

Zhang, W. Z., Xue, Y. F., Ma, Y. H., Zhou, P. J., Ventosa, A. & Grant, W. D. (2002) *Salinicoccus alkaliphilus* sp nov., a novel alkaliphile and moderate halophile from Baer

Soda Lake in Inner Mongolia Autonomous Region, China. *International Journal of Systematic and Evolutionary Microbiology 52*, 789-793.

Zhilina, T. N. & Zavarzin, G. A. (1994) Alkaliphilic anaerobic community at pH-10. *Current Microbiology 29*, 109-112.

Zhu, W. Y. & Reinfelder, J. R. (2012) The microbial community of a black shale pyrite biofilm and its implications for pyrite weathering. *Geomicrobiology Journal 29*, 186-193.

Zhu, Y., Li, Y., Lu, A. H., Wang, H. R., Yang, X. X., Wang, C. Q., Cao, W. Z., Wang, Q. H., Zhang, X. L., Pan, D. M. & Pan, X. H. (2011) Study of the interaction between bentonite and a strain of *Bacillus mucilaginosus. Clays and Clay Minerals 59*, 538-545.

In: Plant Ecology and Evolution in Harsh Environments ISBN: 978-1-63321-955-7
Editors: N. Rajakaruna, R. S. Boyd and T. B. Harris

Chapter 2

THE ECOLOGY AND EVOLUTION OF MYCORRHIZAL FUNGI IN EXTREME SOILS

Shannon Schechter* and Sara Branco
Department of Plant and Microbial Biology, University of California, Berkeley, CA, USA

ABSTRACT

Mycorrhizal fungi are critical components of terrestrial ecosystems. Most plants interact with the soil environment through mycorrhizal fungi and depend on them to improve establishment and growth in stressful environments. In this chapter, we review the ecology and evolution of arbuscular mycorrhizal fungi and ectomycorrhizal fungi living in three types of extreme soils: serpentine, saline, and gypsum soils. These soils have a large but context-dependent effect on mycorrhizal fungi. The response of mycorrhizal fungal communities to extreme soils vary; however, there is evidence that tolerance to extreme soils may be widespread in mycorrhizal fungi and that stress-tolerant fungal species may confer stress tolerance to host plants. We identify broad ecological patterns identified through field studies of mycorrhizal fungal communities and describe areas in need of further investigation. We also describe experimental evidence for mycorrhizal fungal adaptation to extreme soils, as well as plant-fungal co-evolutionary patterns and processes, and discuss future research that may elucidate evolutionary patterns related to extreme soils.

INTRODUCTION

Soils are one of the major determinants of biodiversity in terrestrial ecosystems and are closely linked to the ecology and evolution of many organisms, mainly by promoting divergence and speciation (Kruckeberg, 2002; Schadt et al., 2003). Soil environmental factors such as chemistry, moisture, and temperature shape biological diversity by influencing the colonization and persistence of species. Inhospitable soil chemistry and its role in shaping

* Corresponding author: shannonp.schechter@gmail.com.

biological diversity is an active topic of investigation. Extreme soils can occur naturally, such as soils derived from rocks rich in heavy metals, or can result from anthropogenic activities such as mining. Like other inhospitable environments, extreme soils tend to be characterized by depauperate communities of organisms with specialized adaptations for coping with the particular abiotic stresses of the environment in which they occur (Rothschild & Mancielli, 2001).

The vast majority of species are symbiotic and rely on partners for survival (Douglas, 1994). Symbionts play a crucial role in the establishment and persistence of host species; however, they are very rarely considered in studies on adaptation or tolerance to extreme environments. Symbiotic species frequently colonize unsuitable habitats only in the presence of their partners (Gross, 2001; Marquez et al., 2007), making it particularly important to study symbiotic systems in these environments as a whole.

Mycorrhizal symbioses between plants and fungi are among the most ubiquitous mutualisms (Smith & Read, 2008). Fossil evidence indicates that the mycorrhizal symbiosis was important for land plant establishment (Redecker et al., 2000), with fungi giving plants nutrients and water in exchange for fixed carbon. Although simple in form and function, this symbiosis has major ecological and evolutionary importance (Likar, 2011).

There are two main mycorrhizal fungal groups: arbuscular mycorrhizal fungi (AMF) and ectomycorrhizal fungi (EMF). Arbuscular mycorrhizal fungi are the predominant group, consisting of fungi in the Glomeromycota and forming symbioses with the majority of plant families (92%) and species (80%) (Wang & Qiu, 2006). The Glomeromycota is an ancient fungal group that consists solely of symbiotic species. The dependency of host plants on the arbuscular mycorrhizal symbiosis ranges from facultative to highly dependent, but AMF are wholly dependent on their plant hosts for carbon and cannot be cultured without a host (Smith & Read, 2008). Plants can host a range of AMF species, and these can link neighboring plants to form common mycorrhizal networks (Walder et al., 2012).

Ectomycorrhizal fungi belong almost exclusively to the Basidiomycota and Ascomycota and associate with around 10% of plant families, including the Pinaceae, Fagaceae, Betulaceae, and Dipterocarpaceae (Smith & Read, 2008). These fungal phyla include diverse lineages that contain ectomycorrhizal as well as saprobic and parasitic fungi. Unlike AMF, EMF are polyphyletic, and one genus (*Endogone*) belongs to the Zygomycota, a basal fungal group. Also unlike AMF, some EMF can be cultured *in vitro* without a host, and the vast majority of species exhibit little host-specificity in nature.

Arbuscular Mycorrhizal Fungi

Arbuscular mycorrhizal fungal communities consist of colonized plant roots, soil hyphae, and soil-borne spores. Traditionally, soil-extracted spores identified by morphological characteristics have served as the primary method for characterizing AMF communities. The application of molecular techniques (*e.g.*, amplifying and sequencing genomic regions using AMF-specific primers) has allowed for the identification of root-associated AMF taxa (Helgason et al., 1998). These species are defined as operational taxonomic units (OTUs) based on sequence similarity and/or phylogenetic groupings. However, no universal primer set or defining genetic marker have been established for AMF. Genomic regions used to define AMF vary widely among studies and may represent different taxonomic scales (*e.g.*, genera to genotypes).

In "normal", or non-extreme soils, AMF community structure and composition are shaped by soil factors (Johnson et al., 1992; Landis et al., 2004; Lekberg et al., 2007), host-specificity (Helgason et al., 2002; Vandenkoornhuyse et al., 2003), and spatial autocorrelation (related to dispersal limitation) (Dumbrell et al., 2010; Lekberg et al., 2007). Arbuscular mycorrhizal fungi improve the establishment and growth of host plants in stressful environments by enhancing nutrient and water uptake (Smith & Read, 2008). The more nutrient-limiting the environment is, the more dependent plants are on AMF for nutrient acquisition and growth (Habte & Manjunath, 1987; Yost & Fox, 1979). Therefore, plant traits (*e.g.*, requirement for and response to AMF) and fungal traits (*e.g.*, tolerance of or adaptation to edaphic stress) that affect symbiotic functioning in extreme soils can contribute to the co-evolution of adapted hosts and symbionts under extreme edaphic conditions.

Ectomycorrhizal Fungi

Ectomycorrhizal fungi are hyperdiverse organisms that assemble in complex and dynamic communities. As such, documenting EMF communities is not easy. Molecular tools are the most efficient approach, with the internal transcribed spacer (ITS) region as the DNA marker of choice for detecting EMF species (Schoch et al., 2012). Sequence similarity thresholds (typically 95-97%) define OTUs that may be used as surrogates for species. Traditionally, the detection of EMF is accomplished by sampling root tips (*e.g.*, Branco & Ree, 2010; Peay et al., 2007), but other methods such as ingrowth bags or direct soil sequencing have also been used (Branco et al., 2013; Koide et al., 2005; Talbot et al., 2014; Wallander et al., 2013). Whatever the sampling method used, it is clear that EMF communities are characterized by high richness and rarity, making it challenging to achieve complete community descriptions (Branco, 2011). Little information exists on the environmental requirements and limitations of EMF species, and accurate distribution data is lacking for most species. However, it has been shown that environmental factors such as nitrogen deposition (Avis et al., 2008) and atmospheric CO_2 and O_3 (Andrew & Lilleskov, 2009) affect EMF community composition.

Mycorrhizal Fungi in Extreme Soils

Single-species experiments have shown that both plant hosts and mycorrhizal partners have physiological mechanisms for coping with harsh soils and that tolerance to toxic factors can vary between populations (Brady et al., 2005; Gadd, 1993; Meharg & Cairney, 1999). However, it is not known how variation in tolerance to extreme soil factors between the partners may affect associations, co-evolution, or above- and below-ground community ecology. Species may be intolerant or tolerant of, or may be specialized to, extreme soils. The evolution of edaphic tolerance or specialization depends upon the strength of the selective agent as well as the age of the environment. Specialization can range from local adaptation to complete speciation, with the formation of endemic species associated with harsh environments (Williams, 1966).

Several scenarios can be hypothesized concerning the adaptation, co-evolution, and community ecology of mycorrhizal partners in extreme soils. Meharg & Cairney (1999) suggested evolutionary strategies for mycorrhizal colonization in metal-contaminated soil with none, one, or both partners showing edaphic specialization. We elaborate and expand on this hypothesis to include any extreme soil chemistry.

In our first scenario, both plants and fungi might show edaphic tolerance (*i.e.*, have the ability to survive in extreme soils without any type of specialization). In this scenario challenging soil chemistry does not constitute a selective agent for either plants or fungi, and there would be no differentiation between the populations growing in extreme or benign soils. In this case, individuals are plastic and tolerate both hospitable and inhospitable conditions.

Alternatively, one of the two partners may be specialized to an extreme edaphic condition. When only one symbiont is specialized (*e.g.*, locally adapted or even endemic) the adapted partner may act as a buffer and enable the persistence of a non-tolerant plant or fungus, thereby expanding its range to include otherwise inhospitable soils. In these asymmetric cases it is not clear whether there is greater probability for one or the other partner to be adapted to extreme soil chemistry. The partner with a shorter generation time has greater probability of evolving specialized adaptations; however, virtually no data exist on the generation time of mycorrhizal fungi, making it difficult to make comparisons with the generation times of plants.

Lastly, both partners may be specialized to harsh soil chemistry. In this case, soil chemistry contributes to evolutionary change and co-evolution in both the plant and the fungus. In this scenario, both partners are adapted to extreme conditions presented as either local adaptation or endemism, often demonstrating higher fitness in the extreme environment relative to more benign environments.

Variation in tolerance to soil conditions between partners can affect beta diversity between extreme and non-extreme soils. The prevalence of specialized species determines how distinctive community assemblages found in extreme soils are relative to those found in non-extreme soils. Often, chemically harsh soil environments (*e.g.*, serpentine soils or mine spoils) host depauperate communities (Gadd, 1993; Hartley et al., 1997; Leyval et al., 1997). These impoverished communities may be composed of a small subset of tolerant species from the communities of adjacent, non-extreme soils, may be communities of distinct specialized species, or may be a combination of these.

In the first scenario, soil chemistry is not an effective barrier to colonization, implying that either selection is not sufficiently strong to lead to speciation or the habitat is too young to have allowed the formation of endemics. This does not preclude the existence of ecotypes, however. Alternatively, soils may be so toxic that only highly specialized species can tolerate them. In this case, only endemic species would be found, and no overlap with surrounding communities would occur. In an intermediate scenario, harsh soils may host both endemic and tolerant species and therefore share species from communities on adjacent, non-extreme soils.

For mycorrhizal fungi, these diversity scenarios may be further complicated due to the fact that many fungal species have low host-specificity. Fungal and plant host-diversity in extreme soils, therefore, may not be directly linked. Although there is extensive literature on the flora and vegetation associated with chemically harsh soils, the study of fungi in those environments is still in its infancy.

Polluted Soils

Anthropogenic activities such as mining, industry, and agriculture can lead to the accumulation of specific elements in the soil and cause chemical imbalances that affect various forms of life (Chapters 14, 15). When in excess, metals such as Cd, Cr, Cu, Hg, Ni, Pb, and Zn are toxic to many organisms, disrupting normal cell homeostasis primarily through protein denaturation (Gadd & Griffiths, 1978). Metal-uptake is regulated and coordinated with detoxification and storage to insure that cytoplasmatic concentrations do not exceed toxic or lethal thresholds (Colpaert et al., 2011). High concentrations of environmental metals challenge these regulatory mechanisms and, therefore, can select for tolerance via more robust processes for maintaining homeostasis. Metal-tolerance depends on extra- and intracellular mechanisms involved in reducing metal-uptake and neutralizing toxic effects of metals in the cell, either by binding metals to other compounds, by vacuolar storage, or through antioxidant detoxification systems. These strategies can be inherent or they can be induced by high levels of heavy metals in the cytoplasm (Colpaert et al., 2011).

Metal-contaminated sites are unique in that anthropogenic activities drive rapid evolution that can span just centuries or decades, rather than the thousands or millions of years of exposure to extreme geologic features that are typical in natural environments.

Arbuscular Mycorrizal Fungi from Polluted Soils

Disturbance has a large effect on AMF communities (Schnoor et al., 2011). Clearing vegetation, disturbing, covering, or compacting the soil, can reduce AMF community diversity (Entry et al., 2002). Polluted soils are disturbed areas that vary in size, level of disturbance, and distance from naturally vegetated areas. Spatial factors impact AMF community diversity and structure (Husband et al., 2002; Lekberg et al., 2007), but are generally not measured in studies of AMF communities on polluted soils. Distance from vegetation source, size of disturbance, and other spatially explicit or disturbance related factors should be included in such studies to distinguish between the impact of disturbance and spatial factors from the impact of pollution alone.

Mining activities, which lead to metal-contaminated soils, have been shown to have a negative impact on AMF spore density, richness, and diversity (Shetty et al., 1994). Metal contamination also decreases the diversity of native AMF (Hassan et al., 2011). However, plant hosts found in metal-contaminated soils have been shown to have higher AMF colonization rates than those of plant hosts found in non-contaminated soils (Khade & Adholeya, 2009; Leyval et al., 1997; Tarafdar & Rao, 1997). Moreover, AMF colonization has been shown to improve the growth and establishment of host plants in mine soils relative to the non-mycorrhizal condition (Leyval et al., 1997; Shetty et al., 1994). Using molecular methods, Hassan et al. (2011) found that AMF community structure was distinct between contaminated and uncontaminated sites. These AMF community patterns have led authors to suggest that heavy metal-exposure can promote the development of metal-tolerance in AMF species (Hassan et al., 2011; Khade & Adholeya, 2009; Leyval et al., 1997).

Plant hosts exhibit a variety of adaptive responses to metal-contaminated soils; however, most studies support a strong role for AMF in the growth and establishment of host plants in contaminated soils (Leyval et al., 1997; Meharg, 2003; Meharg & Cairney, 1999). The responses of mycorrhizal fungi to heavy metals vary from metal exclusion to increased metal uptake, indicating diversity in tolerance mechanisms (Meharg & Cairney, 1999). The diversity of physiological responses to metal contaminated soil suggests that AMF may be able to colonize a wide range of contaminated microhabitats (Meharg, 2003).

Pollutant resistance in AMF can also be conferred to host plants (Leyval et al., 1997; Meharg & Cairney, 1999). For example, AMF found on mine spoils were resistant to arsenate and were shown to confer resistance to both resistant and nonresistant host species via reduced arsenate uptake (Gonzalez-Chavez et al., 2002). Similarly, mycorrhiza attenuated the negative effects of Cd by sequestration of the metal within the fungal partner (Hutchinson et al., 2004; Rivera-Becerril et al., 2002), however, this effect differed by plant genotype. Arsenic-resistance in the plant host may not be directly conferred by AMF (Kelly et al., 2005; Knudson et al., 2003); colonization by metal-tolerant AMF may not confer added resistance to the host, but instead enhance the growth of host plants solely through improved nutrition, similar to mycorrhizal associations in non-contaminated sites (Meharg & Cairney, 1999).

The contribution of AMF to the metal-tolerance of host plants is likely context-dependent (Glassman & Casper, 2012). The range of AMF taxa, host species, metal contaminants, and other site-specific edaphic factors varies so widely between studies that clear evolutionary patterns have remained elusive. However, it is clear that AMF have the capacity to evolve tolerance or resistance to metal-contamination and to confer resistance to host plants through direct (*e.g.*, metal sequestration) and/or indirect (*e.g.*, growth enhancement) mechanisms.

Ectomycorrhizal Fungi from Polluted Soils

Although there is evidence for shifts in EMF communities associated with metal contaminated sites (Colpaert, 2008; Ruotsalainen et al., 2009; Staudenrausch et al., 2005), no metal-specific endemic taxa (at the species or at higher taxonomic levels) have been documented (Blaudez et al., 2000; Krpata et al., 2008). In fact, many EMF, including *Hebeloma* spp., *Pisolithus* spp., *Scleroderma* spp. and *Amanita muscaria*, have widespread constitutive metal tolerance and are able to colonize contaminated sites without any detectable differentiation (Gast et al., 1988; Jones & Hutchinson, 1986; Jourand et al., 2010). Despite this lack of specialization, adaptive metal tolerance has been suggested for a few EMF such as *Pisolithus tinctorius* and *Suillus* spp. (Colpaert et al., 2000; Colpaert el al., 2004; Egerton-Warburton & Griffin, 1995; Krznaric et al., 2009). Metal-tolerant ecotypes have been found to be adapted to high levels of Al, Zn, Cd, and Cu, and some species have been found to have adapted to different metals independently (Colpaert et al., 2011). This suggests that some species may be more prone to locally adapt to chemically inhospitable edaphic environments. For example, *Suillus luteus* and *S. bovinus* have known ecotypes adapted to Cd-, Cu-, and Zn-contaminated soils (Adriaensen et al., 2005; Colpaert et al., 2000; 2011; Ruytinx et al., 2013). These differentiated populations accumulate lower metal concentrations in their mycelia, and accumulate metals at a slower rate relative to sensitive ecotypes, through a mechanism that efficiently exports metals outside of the cell and prevents accumulation in fungal tissues.

Despite such examples of local adaptation, there is no evidence that high concentrations of heavy metals in soils are involved in promoting speciation or high taxon divergence in EMF. The lack of edaphic endemism in EMF indicates that heavy-metal contaminated soils either are a relatively weak selection agent or are too recent to allow formation of new species.

Ectomycorrhizal fungi enhance plant resource acquisition by expanding the absorption area of root systems (Smith & Read, 2008) and have been shown to play an important role in plant establishment in metal-contaminated soils (Panaccione et al., 2001) by reducing plant metal uptake (Hartley et al., 1997) and improving heavy metal-tolerance in plants (Wilkinson & Dickinson, 1995). Some fungi have been shown to be more efficient at improving plant fitness (Colpaert & Van Assche, 1992) and tolerant ecotypes have been suggested to be good metal filters, playing a particularly important ecological role in metalliferous soils (Colpaert et al., 2011). However, there is no evidence for specialized mechanisms of metal tolerance in EMF fungi. On the contrary, these fungi appear to have intrinsic mechanisms that enable widespread tolerance of high metal concentrations. It is possible, however, that a closer look at EMF fungal communities found in metal-contaminated soils coupled with experimental approaches aimed at identifying metal tolerance mechanisms will reveal more examples of local adaptation or speciation linked to chemically harsh soils. Approaches such as common garden experiments, coupled with genomic tools, will shed light on the physiological mechanisms involved in such tolerance.

Serpentine Soils

Serpentine soils are harsh environments known to host specialized plant communities. They have unique chemical and physical characteristics, including low levels of essential macronutrients; absence of some micronutrients; toxic concentrations of elements such as Mg, Ni, Co, or Cr; low Ca:Mg ratios, as well as a tendency for poor water retention. The combination of these features has been referred to as the 'serpentine syndrome' (Brady et al., 2005) and makes these soils challenging in a multifaceted way. Serpentine substrates cover about 1% of the planet and are patchily distributed worldwide (Roberts & Proctor, 1992). They host depauperate floras with low productivity, sparse plant cover, and high plant endemism (Alexander et al., 2007; Baker et al., 1992; Brady et al., 2005; Brooks, 1987).

Serpentine plant adaptation has been extensively studied, with numerous examples of adaptive divergence documented through reciprocal transplant studies (*e.g.*, Kruckeberg, 1950; also see Chapter 6). Serpentine plant ecotypes tend to be small and exhibit slow growth rates. They tend to show higher fitness on serpentine soils, not because they cannot withstand non-serpentine conditions, but because of a tolerance-competitive ability trade-off that prevents establishment when in competition with non-specialized plants (Brady et al., 2005; Kruckeberg, 1954). Serpentine tolerance and specialization have evolved in phylogenetically distant plant lineages, as well as independently within the same species (Mengoni et al., 2003; Nyberg Berglund et al., 2001; 2004; Patterson & Givnish, 2004; Turner et al., 2010).

Serpentine soils have a long geologic history, existing in landscapes for thousands to millions of years (Kruckeberg, 1984). Within this time period, serpentine-tolerant and serpentine-restricted plant species have evolved (Brady et al., 2005; Kruckeberg, 1984). Serpentine soils may have provided similar selective pressure on AMF and EMF resulting in stable ecotypes, and perhaps even endemic taxa; however, additional investigation is needed to determine whether this is true.

Arbuscular Mycorrhizal Fungi from Serpentine Soils

Serpentine environments can occur as large, isolated ecological islands or fine-scale mosaics of serpentine and non-serpentine habits (Kruckeberg, 1984). In mosaic habitats, no dispersal limitation has been found for AMF within distances of 50 m to 2 km between sites (Schechter & Bruns, 2008; 2012). However, spatial patterns may be more apparent in larger serpentine islands or at greater distances (Lekberg et al., 2007).

Arbuscular mycorrhizal colonization has been shown to be abundant in serpentine soils, even on plants which are typically non-mycorrhizal (Hopkins, 1987). Using molecular methods, Schechter & Bruns (2012) compared serpentine and non-serpentine root-associated AMF communities and found no significant difference in species diversity, evenness, or richness between serpentine and non-serpentine soil types. However, when focused on AMF communities associated with roots of serpentine and non-serpentine host ecotypes, serpentine ecotypes were associated with significantly higher diversity and evenness of AMF species relative to non-serpentine host ecotypes (Schechter & Bruns, 2008).

Serpentine and non-serpentine AMF communities have been shown to be distinct (Schechter & Bruns 2008; 2012). However, it is difficult to separate edaphic selection from host-specific or plant community influences on AMF community composition and structure (Schechter & Bruns, 2008; 2012). Several studies have found that plant-fungal feedbacks influence the structure of AMF communities in serpentine habitats (Casper & Castelli, 2007; Casper et al., 2008; Castelli & Casper, 2003; Gustafson & Casper, 2004). Evidence based on multivariate variance partitioning suggests that both serpentine edaphic factors and host plant community contribute to the distinction between serpentine and non-serpentine AMF communities, but that soil factors have a stronger effect (Schechter & Bruns, 2012). Two greenhouse studies manipulating serpentine and non-serpentine AMF and host plant communities support a dominant role for soil type in determining AMF community structure (Ji et al., 2010; Schechter & Bruns, 2013). This research suggests that strong edaphic selection in serpentine environments may generate serpentine-tolerant or serpentine-adapted, as well as serpentine-intolerant, AMF taxa.

Few studies have specifically looked for serpentine-adaptation in AMF. Improved host-growth in serpentine soil inoculated with serpentine AMF isolates relative to host-growth in serpentine soil inoculated with non-serpentine AMF suggests AMF serpentine adaptation; however, the mechanism of such adaptation is unknown (Doubkova et al., 2012; Schechter, 2010). Benefits to host plants resulting from colonization by serpentine-associated AMF have been attributed to improved host-nutrition rather than improved tolerance to serpentine edaphic factors such as Ni, Ca or Mg concentrations (Doherty et al., 2008; Doubkova et al., 2011; 2012). Increased specificity between host plant and fungal symbiont could mediate plant adaptation to stress (Thrall et al., 2008). In this case, co-adaptation of host and symbiont to a specific environment results in adapted plant genotypes doing best with adapted symbiont genotypes (Thrall et al., 2008). However, despite ecological evidence of host-symbiont specificity between serpentine and non-serpentine adapted plant ecotypes and AMF taxa, a greenhouse common garden experiment showed no evidence of host ecotype-AMF specificity (Schechter & Bruns, 2008; 2013). Ecological studies and growth experiments indicate that there are serpentine-tolerant or serpentine-adapted AMF that improve plant growth on serpentine soils, but more research is needed to fully establish AMF serpentine adaptation and co-evolutionary mechanisms.

Ectomycorrhizal Fungi from Serpentine Soils

Little is known about EMF ecology and evolution in serpentine soils. Diversity studies have reported rich and phylogenetically broad EMF communities on serpentine soils relative to their non-serpentine counterparts (Branco & Ree, 2010; Moser et al., 2005; 2009; Urban et al., 2008). Unlike plants, EMF have not been shown to assemble in specialized communities in serpentine soils, and no serpentine-endemic EMF species have been identified (Southworth et al., 2013). This suggests that serpentine soils do not constitute a major selective agent for EMF. The few existing experimental studies have revealed widespread tolerance to serpentine within EMF (Branco, 2010; Gonçalves et al., 2007), indicating either constitutive tolerance or specialization below the species level. Limited evidence suggests the existence of local adaptation to serpentine in fungi. In a reciprocal transplant experiment, *Laccaria* sp. preferred serpentine soil (Branco, 2010). This species was recovered from oak root tips grown in serpentine and non-serpentine soils, and the species was more frequently found in root tips from the serpentine treatment, suggesting a preference for serpentine soils. A few studies using population screens and *in vitro* tests suggested serpentine specialization in *Cenococcum geophilum* (Gonçalves et al., 2007; 2009; Panaccione et al., 2001). However, limitations due to small sample size and the number of parameters included in the studies do not allow drawing definitive conclusions on serpentine local adaptation in this species.

Although local adaptation to serpentine has been extensively documented in plants (Chapters 6, 11), serpentine ecotypes are generally short-lived herbaceous plants, and EMF fungi generally associate with perennial woody plants. There is, however, evidence indicating serpentine differentiation in trees such as *Quercus ilex* (Fagaceae; Branco, 2009), *Pinus ponderosa* (Pinaceae; Wright & Stanton, 2007), *Larix kaempferi* (Pinaceae; Kayama et al., 2009), and *Picea glehnii* (Pinaceae; Kayama et al., 2006). These studies confirm that serpentine environments also act as a strong selective agent for long-lived, woody plant species.

Research on EMF from serpentine soils suggests asymmetrical evolution of plant and fungal partners, with plants being relatively constrained by the serpentine syndrome and their EMF partners being broadly tolerant. Further documentation of EMF communities found in serpentine soils and experiments targeting the fitness of EMF host plants in serpentine and non-serpentine environments are needed to confirm this hypothesis.

Saline and Gypsum Soils

High salinity levels are a serious and widespread problem for both plant and fungal growth (Chapter 4; Porcel et al., 2012). The sources of salt in saline environments vary widely, and time scales for exposure and adaptation can be hard to determine. Saline soils have high osmotic potentials and often contain toxic ion concentrations (Munns et al., 1983). Fungi cope with these conditions by compartmentalization, vacuolization, and the production of large amounts of glycerol and mannitol which act as non-toxic osmoregulators (Clipson & Jennings, 1992).

Gypsum soils are often found in arid climates where low rainfall limits leaching. Gypsum soils are known to affect plant growth not only due to imbalances in Ca:Mg and K:Ca ion ratios, but also due to high sulphate concentrations, low porosity, presence of hard crusts, and high mechanical instability (Chapter 5; Palacio et al., 2012). Gypsum deposits vary in origin and length of exposure; for example, gypsum deposits in the United States range in age from Silurian to Quaternary time periods (Stone, 1920), but some surface exposures may be more recent due to mining.

Arbuscular Mycorrhizal Fungi from Saline and Gypsum Soils

Little is known about how spatial factors affect AMF community composition and structure in gypsum, dolomitic, or saline soils. In spatially explicit sampling along a salinity gradient 80 meters long, spore numbers and colonization decreased at the highest salinity levels but no spatial pattern in AMF spore composition was detected (Roda et al., 2008). Spatial patterns in AMF communities have been found within a geographic distance of 25 meters (Dumbrell et al., 2010), but not within a one-kilometer distance (Schechter & Bruns, 2008). Thus, it is not clear at what distance AMF dispersal is limited or how it affects AMF community composition and structure.

Arbuscular mycorrhizal fungal spore numbers and host plant colonization have been shown to generally decrease with increasing soil salinity (Aliasgharzadeh et al., 2001; Hildebrandt et al., 2001; Johnson-Green et al., 2001; Roda et al., 2008). In saline soils, AMF colonize most species, including those from plant families which are typically non-mycorrhizal (Asghari et al., 2005; Hildebrandt et al., 2001); however, halophytes are generally not colonized by AMF in sites with the highest soil salinity (Aliasgharzadeh et al., 2001; Johnson-Green et al., 2001), suggesting that AMF may have lower salinity tolerance than halophytes (Johnson-Green et al., 2001).

Arbuscular mycorrhizal fungal communities in gypsum soils have been poorly studied. Arbuscular mycorrhizal fungal colonization and spore numbers have been shown to vary across gypsum sites, but have not been shown to limit mycorrhization (Alguacil et al., 2009a; b). Molecular studies of root-associated AMF in gypsum soils showed potential generalist and specialist AMF taxa, but AMF community composition showed host preference (Alguacil et al., 2009a; b). However, these studies focused solely on gypsum soils and gypsophilous plant communities (also see Chapter 5), which does not allow for measures of beta diversity associated with these environments.

Arbuscular mycorrhizal fungi have been shown to improve host plant biomass and nutrition in saline soils relative to non-mycorrhizal controls (Cantrell & Linderman, 2001; Dixon et al., 1993). In *Zea mays* (Poaceae), salt-adapted AMF benefitted host plants more than non-adapted AMF and conferred higher salinity-tolerance to host plants (Estrada et al., 2013). In *Lotus glaber* (Fabaceae), AMF improved growth, nutrition, and chlorophyll concentrations under saline conditions, but these benefits were more pronounced in salinity-tolerant host genotypes relative to salinity-sensitive ones (Sannazzaro et al., 2006).

Several studies have indicated mechanisms by which AMF mitigate salt stress in host plants (Bothe, 2012; Evelin et al., 2012; Giri & Mukerji, 2004; Hammer et al., 2011; Sannazzaro et al., 2006). The positive effects of mycorrhization may be the result of improved host plant nutrition, higher tissue K^+/Na^+ ratios, and better osmotic adjustment (Porcel et al., 2012). Arbuscular mycorrhizal fungi also regulate the expression of plant genes associated with maintaining tissue water status (Porcel et al., 2012).

Arbuscular mycorrhizal fungi isolated from gypsum mine spoils improved host plant growth and nutrition in the same soils (Rao & Tak, 2002). Colonization was higher on host plants that were gypsum generalists relative to specialists; however, specialists maintained higher nutrient concentrations despite lower colonization (Palacio et al., 2012). Higher AMF colonization in generalist species may help these non-specialists cope with gypsum soils (Palacio et al., 2012).

While these patterns in both saline and gypsum soils suggest AMF adaptation to these extreme soils, there is still too little known to confidently make such determinations. It is clear that AMF play a role in host plant growth and establishment in saline and gypsum soils, but more research is needed to establish co-evolutionary mechanisms.

Ectomycorrhizal Fungi from Saline and Gypsum Soils

Less is known about EMF from saline and gypsum soils relative to what is known of EMF from other harsh soils. Alkaline-saline soils host particularly poorly diverse EMF communities (Ishida et al., 2009). Basidiomycota are relatively salt-intolerant (Tresner & Hayes, 1971); however, certain genera such as *Laccaria* and *Hebeloma* are known to tolerate salts (Kernaghan et al., 2002). *In vitro* assays testing the effect of pH and salt concentration on fungal growth and survival revealed the existence of both sensitive and tolerant species (Kernaghan et al., 2002), the latter often being accompanied by reduced growth (Tang et al., 2009). *Suillus luteus* has been shown to be tolerant of high salt concentrations (Kernaghan et al., 2002). Assessing whether tolerance or specialization of EMF occurs in gypsum soils will enable comparisons with what is already known from other chemically challenging environments and lead to a better comprehension of how EMF evolve under stress.

CONCLUSION

Extreme soils have a large but context-dependent effect on mycorrhizal symbioses. The strength of selective agents and the age of the habitat may dictate evolutionary and ecological outcomes for both plant and fungal partners. It seems clear that fungal endemism is not a common occurrence in these environments. Mycorrhizal fungi colonize chemically inhospitable soils along with their plant partners, playing an important role in ecosystem dynamics. Arbuscular mycorrhizal fungi and EMF behave differently in polluted, serpentine, saline, gypsum, and other harsh soils. These habitats alter AMF community structure, and in most cases decrease AMF diversity, whereas the effect of extreme soils on EMF communities varies, with shifts occurring in some harsh soils, but not others. It is difficult to distinguish edaphic effects from host-specific or plant community effects on mycorrhizal fungal community structure and diversity. In most cases, plant communities also differ between extreme and normal soils, making a multivariate analysis approach necessary to distinguish the source of effects.

Tolerance to extreme soils may be widespread in the AMF and EMF in metal-contaminated, serpentine, saline, and gypsum soils; however, mycorrhizal fungal specialization (*i.e.*, endemism) to any extreme soil has only been suggested. More work at the population genetic level in well-defined species and within well-defined environmental conditions is needed to identify extreme soil specialists in mycorrhizal fungi. The increasing availability of fungal reference genomes and decreasing cost of sequencing make mycorrhizal fungi good candidates for population genomic approaches. These can unveil patterns of population genetic structure between different soil types and reveal if and how natural selection acts in populations on harsh soils.

There is evidence that tolerant AMF and EMF species confer tolerance to their host plants; however, the specific mechanisms involved are poorly understood and may depend on factors such as host genotype and host dependence on the symbiosis. Identifying specific genes and proteins involved in the process and testing for differential expression under different environments in both plant and fungal partners will enable a better understanding of mycorrhizal symbioses in extreme soils.

Although there is evidence that extreme soils impact mycorrhizal fungal ecology and evolution, consistent patterns are hard to detect. The number of AMF and EMF species and host plant species involved, the strength of selective agents, the number of site-specific edaphic factors, and the spatial and temporal scales vary so widely between studies that clear ecological and evolutionary patterns have remained elusive. Comparative studies across different extreme soils may help to clarify these trends. Another issue that is particularly problematic in deciphering patterns in AMF composition on extreme soils is the wide range of methods used to define communities and species. Future research should take careful measure of mitigating factors so as to help identify ecological and evolutionary patterns of mycorrhizal fungi in extreme soils that will aid in conservation, reclamation, and restoration efforts.

Acknowledgments

S. Branco is funded by National Science Foundation grant DBI 1046115.

References

Adriaensen, K., Vrålstad, T., Noben, J. P., Vangronsveld, J. & Colpaert, J. V. (2005) Copper-adapted *Suillus luteus*, a symbiotic solution for pines colonizing Cu mine spoils. *Applied and Environmental Microbiolology 71*, 7279–7284.

Alexander, E., Coleman, R., Keeler-Wolfe, T. & Harrison, S. (Eds.) (2007) *Serpentine geoecology of western North America: Geology, soils, and vegetation.* New York, NY: Oxford University Press.

Alguacil, M. M., Roldan, A. & Torres, M. P. (2009a) Assessing the diversity of AM fungi in arid gypsophilous plant communities. *Environmental Microbiology 11*, 2649-2659.

Alguacil, M. M., Roldan, A. & Torres, M. P. (2009b) Complexity of semiarid gypsophilous shrub communities mediates the AMF biodiversity at the plant species level. *Microbial Ecology 57,* 718-727.

Aliasgharzadeh, N., Rastin, N. S., Towfighi, H. & Alizadeh, A. (2001) Occurrence of arbuscular mycorrhizal fungi in saline soils of the Tabriz Plain of Iran in relation to some physical and chemical properties of soil. *Mycorrhiza 11,* 119-122.

Andrew, C. & Lilleskov, E. A. (2009) Productivity and community structure of ectomycorrhizal fungal sporocarps under increased atmospheric CO_2 and O_3. *Ecology Letters 12,* 813-822.

Asghari, H. R., Marschner, P., Smith, S. E. & Smith, F. A. (2005) Growth response of *Atriplex nummularia* to inoculation with arbuscular mycorrhizal fungi at different salinity levels. *Plant and Soil 273,* 245-256.

Avis, P. G., Mueller, G. M. & Lussenhop, J. (2008) Ectomycorrhizal fungal communities in two North American oak forests rspond to nitrogen addition. *New Phytologist 179*, 472-483.

Baker, A. J. M., Proctor, J. & Reeves, R. D. (Eds.) (1992) *The vegetation of ultramafic (serpentine) soils: Proceedings of the first international conference on serpentine ecology.* Hampshire, UK: Intercept, Ltd.

Blaudez, D., Jacob, C., Turnau, K., Colpaert, J. V., Ahonen-Jonnath, U., Finlay, R., Botton, B. & Chalot, M. (2000) Differential responses of ectomycorrhizal fungi to heavy metals in vitro. *Mycological Research 104*, 1366-1371.

Bothe, H. (2012) Arbuscular mycorrhiza and salt tolerance of plants. *Symbiosis 58,* 7-16.

Brady, K. U., Kruckeberg, A. R. & Bradshaw, H. D., Jr. (2005) Evolutionary ecology of plant adaptation to serpentine soils. *Annual Review of Ecology Evolution and Systematics 36,* 243-266.

Branco, S. (2009) Are oaks locally adapted to serpentine soils? *Northeastern Naturalist 16,* 329-340.

Branco, S. (2010) Serpentine soils promote ectomycorrhizal fungal diversity. *Molecular Ecology 19,* 5566-5576.

Branco, S. (2011) Fungal diversity – an overview. In: O. Grillo & G. Venora (Eds.). *The dynamical processes of biodiversity – case studies of evolution and spatial distribution* (pp. 211-226). InTech, DOI: 10.5772/1830.

Branco, S. & Ree, R. (2010) Serpentine soils do not limit mycorrhizal fungal diversity. *PLoS One* DOI: 10.1371/journal.pone.0011757.

Branco, S., Bruns, T. D. & Singleton, I. (2013) Fungi at a small scale: spatial zonation of fungal assemblages around single trees. *PLoS One* DOI: 10.1371/journal.pone.0078295.

Brooks, R. (1987) *Serpentine and its vegetation.* Portland, OR: Dioscorides Press.

Cantrell, I. C. & Linderman, R. G. (2001) Preinoculation of lettuce and onion with VA mycorrhizal fungi reduces deleterious effects of soil salinity. *Plant and Soil 233*, 269-281.

Casper, B. B. & Castelli, J. P. (2007) Evaluating plant-soil feedback together with competition in a serpentine grassland. *Ecology Letters 10,* 394-400.

Casper, B. B., Bentivenga, S. P., Ji, B. M., Doherty, J. H., Edenborn, H. M. & Gustafson, D. J. (2008) Plant-soil feedback: Testing the generality with the same grasses in serpentine and prairie soils. *Ecology 89,* 2154-2164.

Castelli, J. P. & Casper, B. B. (2003) Intraspecific AM fungal variation contributes to plant-fungal feedback in a serpentine grassland. *Ecology 84*, 323-336.

Clipson, N. J. W. & Jennings, D. H. (1992) *Dendryphiella salina* and *Debaryomyces hansenii*: models for ecophysiological adaptation to salinity by fungi that grow in the sea. *Canadian Journal of Botany 70,* 2097-2105.

Colpaert, J. V. (2008) Heavy metal pollution and genetic adaptations in ectomycorrhizal fungi. In: S. V. Avery, M. Stratford & P. Van West (Eds.). *Stress in yeasts and filamentous fungi* (pp. 157–174). Amsterdam: Academic Press.

Colpaert, J. V. & Van Assche, J. A. (1992) Zinc toxicity in ectomycorrhizal *Pinus sylvestris*. *Plant Soil 143,* 201–211.

Colpaert, J. V., Vandenkoornhuyse, P., Adriaensen, K. & Vangronsveld, J. (2000) Genetic variation and heavy metal tolerance in the ectomycorrhizal basidiomycete *Suillus luteus*. *New Phytologist 147,* 367–379.

Colpaert, J. V., Muller, L. A. H., Lambaerts, M., Adriaensen, K. & Vangronsveld, J. (2004) Evolutionary adaptation to zinc toxicity in populations of Suilloid fungi. *New Phytologist 162,* 549–559.

Colpaert, J. V., Wevers, J. H. L., Krnaric, E. & Adriaensen, K. (2011) How metal-tolerant ecotypes of ectomycorrhizal fungi protect plants from heavy metal pollution. *Annals of Forest Science 68*, 17-24.

Dixon, R. K., Garg, V. K. & Rao, M. V. (1993) Inoculation of leucaena and prosopis seedlings with glomus and rhizobium species in saline soil - rhizosphere relations and seedling growth. *Arid Soil Research and Rehabilitation 7*, 133-144.

Doherty, J. H., Ji, B. M. & Casper, B. B. (2008) Testing nickel tolerance of *Sorghastrum nutans* and its associated soil microbial community from serpentine and prairie soils. *Environmental Pollution 151,* 593-598.

Doubkova, P., Suda, J. & Sudova, R. (2011) Arbuscular mycorrhizal symbiosis on serpentine soils: the effect of native fungal communities on different *Knautia arvensis* ecotypes. *Plant and Soil 345,* 325-338.

Doubkova, P., Suda, J. & Sudova, R. (2012) The symbiosis with arbuscular mycorrhizal fungi contributes to plant tolerance to serpentine edaphic stress. *Soil Biology & Biochemistry 44,* 56-64.

Douglas, A. (1994) *Symbiotic interactions*. New York, NY: Oxford University Press.

Dumbrell, A. J., Nelson, M., Helgason, T., Dytham, C. & Fitter, A. H. (2010) Relative roles of niche and neutral processes in structuring a soil microbial community. *The ISME Journal 4*, 337-345.

Egerton-Warburton, L. & Griffin, B. (1995) Differential responses of *Pisolithus tinctorius* isolates to aluminium in vitro. *Canadian Journal Botany 73,* 1229–1233.

Entry, J. A., Rygiewicz, P. T., Watrud, L. S. & Donnelly, P. K. (2002) Influence of adverse soil conditions on the formation and function of arbuscular mycorrhizas. *Advances in Environmental Research 7*, 123-138.

Estrada, B., Barea, J. M., Aroca, R. & Ruiz-Lozano, J. M. (2013) A native *Glomus intraradices* strain from a Mediterranean saline area exhibits salt tolerance and enhanced symbiotic efficiency with maize plants under salt stress conditions. *Plant and Soil 366,* 333-349.

Evelin, H., Giri, B. & Kapoor, R. (2012) Contribution of *Glomus intraradices* inoculation to nutrient acquisition and mitigation of ionic imbalance in NaCl-stressed *Trigonella foenum-graecum*. *Mycorrhiza 22,* 203-217.

Gadd, G. M. (1993) Interaction of fungi with toxic metals. *New Phytologist 124,* 25-60.

Gadd, G. M. & Griffiths, A. J. (1978) Microorganisms and heavy metal toxicity. *Microbial Ecology 4,* 303-317.

Gast, C. H., Jansen, E., Bierling, J. & Haanstra, L. (1988) Heavy metals in mushrooms and their relationship with soil characteristics. *Chemosphere 17*, 789–799.

Giri, B. & Mukerji, K. (2004) Mycorrhizal inoculant alleviates salt stress in *Sesbania aegyptiaca* and *Sesbania grandiflora* under field conditions: evidence for reduced sodium and improved magnesium uptake. *Mycorrhiza 14,* 307-312.

Glassman, S. I. & Casper, B. B. (2012) Biotic contexts alter metal sequestration and AMF effects on plant growth in soils polluted with heavy metals. *Ecology 93*, 1550-1559.

Gonçalves, S., Portugal, A., Gonçalves, M. T., Vieira, R., Martins-Loucao, M. A. & Freitas, H. (2007) Genetic diversity and differential in vitro responses to Ni in *Cenococcum geophilum* isolates from serpentine soils in Portugal. *Mycorrhiza 17*, 677-686.

Gonçalves, S., Martins-Loucao, M. A. & Freitas, H. (2009) Evidence of adaptive tolerance to nickel in isolates of *Cenococcum geophilum* from serpentine soils. *Mycorrhiza 19,* 221–230.

Gonzalez-Chavez, C., Harris, P. J., Dodd, J. & Meharg, A. A. (2002) Arbuscular mycorrhizal fungi confer enhanced arsenate resistance on *Holcus lanatus*. *New Phytologist 155*, 163-171.

Gross, M. (2001) *Life on the edge: Amazing creatures thriving in extreme environments*. New York, NY: Perseus Publishing.

Gustafson, D. J. & Casper, B. B. (2004) Nutrient addition affects AM fungal performance and expression of plant/fungal feedback in three serpentine grasses. *Plant and Soil 259*, 9-17.

Habte, M. & Manjunath, A. (1987) Soil solution phosphorus status and mycorrhizal dependency in *Leucaena leucocephala*. *Applied and Environmental Microbiology 53*, 797-801.

Hammer, E. C., Nasr, H., Pallon, J., Olsson, P. A. & Wallander, H. (2011) Elemental composition of arbuscular mycorrhizal fungi at high salinity. *Mycorrhiza 21*, 117-129.

Hartley, J., Cairney, J. W. G. & Meharg, A. A. (1997) Do ectomycorrhizal fungi exhibit adaptive tolerance to potentially toxic metals in the environment? *Plant & Soil 189*, 303-319.

Hassan, S. E. D., Boon, E., St-Arnaud, M. & Hijri, M. (2011) Molecular biodiversity of arbuscular mycorrhizal fungi in trace metal-polluted soils. *Molecular Ecology 20*, 3469-3483.

Helgason, T., Daniell, T. J., Husband, R. A., Fitter, H. & Young, J. P. W. (1998) Ploughing up the wood-wide web? *Nature 394*, 431-431.

Helgason, T., Merryweather, J. W., Denison, J., Wilson, P., Young, J. P. W. & Fitter, A. H. (2002) Selectivity and functional diversity in arbuscular mycorrhizas of co-occurring fungi and plants from a temperate deciduous woodland. *Journal of Ecology 90,* 371-384.

Hildebrandt, U., Janetta, K., Ouziad, F., Renne, B., Nawrath, K. & Bothe, H. (2001) Arbuscular mycorrhizal colonization of halophytes in Central European salt marshes. *Mycorrhiza 10,* 175-183.

Hopkins, N. A. (1987) Mycorrhizae in a California serpentine grassland community. *Canadian Journal of Botany-Revue Canadienne De Botanique 65*, 484-487.

Husband, R., Herre, E. A., Turner, S. L., Gallery, R. & Young, J. P. (2002) Molecular diversity of arbuscular mycorrhizal fungi and patterns of host association over time and space in a tropical forest. *Molecular Ecology 11*, 2669-2678.

Hutchinson, J. J., Young, S. D., Black, C. R. & West, H. M. (2004) Determining uptake of radio-labile soil cadmium by arbuscular mycorrhizal hyphae using isotopic dilution in a compartmented-pot system. *New Phytologist 164,* 477-484.

Ishida, T. A., Nara, K., Ma, S., Takano, T. & Liu, S. (2009) Ectomycorrhizal fungal community in alkaline-saline soil in northeastern China. *Mycorrhiza 19,* 329-335.

Ji, B. M., Bentivenga, S. P. & Casper, B. B. (2010) Evidence for ecological matching of whole AM fungal communities to the local plant-soil environment. *Ecology 91*, 3037-3046.

Johnson, N. C., Tilman, D. & Wedin, D. (1992) Plant and soil controls on mycorrhizal fungal communities. *Ecology 73*, 2034-2042.

Johnson-Green, P., Kenkel, N. C. & Booth, T. (2001) Soil salinity and arbuscular mycorrhizal colonization of *Puccinellia nuttalliana*. *Mycological Research 105*, 1094-1100.

Jones, M. D. & Hutchinson, T. C. (1986) The effect of mycorrhizal infection on the response of *Betula papyrifera* to nickel and copper. *New Phytologist 102*, 429–442.

Jourand, P., Ducousso, M., Loulergue-Majorel, C., Hannibal, L., Santoni, S., Prin, Y. & Lebrun, M. (2010) Ultramafic soils from New Caledonia structure *Pisolithus albus* in ecotype. *FEMS Microbiology Ecology 72*, 238–249.

Kayama, M., Choi, D., Tobita, H., Utsugi, H., Kitao, M., Maruyama, Y., Nomura, M. & Koike, T. (2006) Comparison of growth characteristics and tolerance to serpentine soil of three ectomycorrhizal spruce seedlings in northern Japan. *Trees 20*, 430-440.

Kayama, M., Makoto, K., Nomura, M., Satoh, F. & Koike, T. (2009) Nutrient dynamics and carbon partitioning in larch seedlings (*Larix kaempferi*) regenerated on serpentine soil in northern Japan. *Landscape and Ecological Engineering 5*, 125-135.

Kelly, C. N., Morton, J. B. & Cumming, J. R. (2005) Variation in aluminum resistance among arbuscular mycorrhizal fungi. *Mycorrhiza 15*, 193-201.

Kernaghan, G., Hambling, B., Fung, M. & Khasa, D. (2002) In vitro selection of boreal ectomycorrhizal fungi for use in reclamation of saline-alkaline habitats. *Restoration Ecology 10*, 43-51.

Khade, S. W. & Adholeya, A. (2009) Arbuscular mycorrhizal association in plants growing on metal-contaminated and noncontaminated soils adjoining Kanpur Tanneries, Uttar Pradesh, India. *Water, Air, & Soil Pollution 202*, 45-56.

Knudson, J. A., Meikle, T. & DeLuca, T. H. (2003) Role of mycorrhizal fungi and phosphorus in the arsenic tolerance of basin wildrye. *Journal of Environmental Quality 32*, 2001-2006.

Koide, R. T., Xu, B. & Sharda, J. (2005) Contrasting below-ground views of an ectomycorrhizal fungal community. *New Phytologist 166,* 251-262.

Krpata, D., Peinter, U., Langer, I., Fitz, W. J. & Schweiger, P. (2008) Ectomycorrhizal communities associated with *Populus tremula* growing on a heavy metal contaminated site. *Mycological Research 112*, 1069-1079.

Kruckeberg, A. (1950) *An experimental inquiry into the nature of endemism on serpentine soils*. PhD Dissertation. Berkeley, CA: University of California, Berkeley.

Kruckeberg, A. (1954) The ecology of serpentine soils III. Plant species in relation to serpentine soils. *Ecology 35,* 267-274.

Kruckeberg, A. R. (1984) *California serpentines: Flora, vegetation, geology, soils, and management problems*. Berkeley, CA: University of California Press.

Kruckeberg, A. R. (2002) *Geology and plant life: The effects of landforms and rock types on plants*. Seattle, WA: University of Washington Press.

Krznaric, E., Verbruggen, N., Wevers, J. H. L., Carleer, R., Vangronsveld, J. & Colpaert, J. V. (2009) Cd-tolerant *Suillus luteus*: a fungal insurance for pines exposed to Cd. *Environmental Pollution 157*, 1581–1588.

Landis, F. C., Gargas, A. & Givnish, T. J. (2004) Relationships among arbuscular mycorrhizal fungi, vascular plants and environmental conditions in oak savannas. *New Phytologist 164,* 493-504.

Lekberg, Y., Koide, R. T., Rohr, J. R., Aldrich-Wolfe, L. & Morton, J. B. (2007) Role of niche restrictions and dispersal in the composition of arbuscular mycorrhizal fungal communities. *Journal of Ecology 95*, 95-105.

Leyval, C., Turnau, K. & Haselwandter, K. (1997) Effect of heavy metal pollution on mycorrhizal colonization and function: physiological, ecological and applied aspects. *Mycorrhiza 7,* 139-153.

Likar, M. (2011) Dark septate endophytes and mycorrhizal fungi of trees affected by pollution. In: A. M. Pirttilä & A. C. Frank (Eds.). *Endophytes of forest trees.* Forestry Sciences (Vol. 80). New York, NY: Springer.

Marquez, L. M., Redman, R. S., Rodriguez, R. J. & Roossinck, M. J. (2007) A virus in a fungus in a plant: three-way symbiosis required for thermal tolerance. *Science 315,* 513–515.

Meharg, A. A. (2003) The mechanistic basis of interactions between mycorrhizal associations and toxic metal cations. *Mycological Research 107*, 1253-1265.

Meharg, A. A. & Cairney, J. W. G. (1999) Co-evolution of mycorrhizal symbionts and their hosts to metal-contaminated environments. *Advances in Ecological Research 30*, 69-112.

Mengoni, A., Gonelli, C., Brocchini, E., Galardi, F. & Pucci, F. (2003) Chloroplast genetic diversity and biogeography in the serpentine endemic Ni-hyperacumulator *Alyssum bertolonii*. *New Phytologist 157*, 349-356.

Moser, A. M., Petersen, C. A., D'Allura, J. A. & Southworth, D. (2005) Comparison of ectomycorrhizas of *Quercus garryana* (Fagaceae) on serpentine and non-serpentine soils in southwest Oregon. *American Journal of Botany 92*, 224–230.

Moser, A. M., Frank, J. L., D'Allura, J. A. & Southwood, D. (2009) Ectomycorrhizal communities of *Quercus garryana* are similar on serpentine and nonserpentine soils. *Plant and Soil 315,* 185-194.

Munns, R., Greenway, H. & Kirst, G. O. (1983) Halotolerant eukaryotes. In: O. L. Lange, P. S. Nobel, C. B. Osmand & H. Ziegler (Eds.). *Physiological plant ecology III.* Berlin, Germany: Springer-Verlag.

Nyberg Berglund, A.-B., Saura, A. & Westerbergh, A. (2001) Genetic differentiation of a polyploid plant on ultramafic soils in Fennoscandia. *South African Journal of Science 97,* 533-535.

Nyberg Berglund, A.-B., Dahlgren, A. & Westerbergh, A. (2004) Evidence of parallel evolution and site-specific selection of serpentine tolerance in *Cerastium alpinum* during the colonization of Scandinavia. *New Phytologist 161,* 199-209.

Palacio, S., Johnson, D., Escudero, A. & Montserrat-Marti, G. (2012) Root colonisation by AM fungi differs between gypsum specialist and non-specialist plants: Links to the gypsophile behaviour. *Journal of Arid Environments 76,* 128-132.

Panaccione, D. G., Sheets, N. L., Miller, S. P. & Cumming, J. R. (2001) Diversity of *Cenococcum geophilum* isolates from serpentine and non-serpentine soils. *Mycologia 93,* 645-652.

Patterson, T. B. & Givnish, T. J. (2004) Geographic cohesion, chromosomal evolution, parallel adaptive radiations, and consequent floral adaptations in *Calochortus* (Calochortaceae): evidence from cpDNA phylogeny. *New Phytologist 161,* 253-264.

Peay, K. G., Bruns, T. D., Kennedy, P. G., Bergman, S. E. & Garbelotto, M. (2007) A strong species-area relationship for eukaryotic soil microbes: island size matters for ectomycorrhizal fungi. *Ecology Letters 10*, 470-480.

Porcel, R., Aroca, R. & Ruiz-Lozano, J. M. (2012) Salinity stress alleviation using arbuscular mycorrhizal fungi. A review. *Agronomy for Sustainable Development 32,* 181-200.

Rao, A. V. & Tak, R. (2002) Growth of different tree species and their nutrient uptake in limestone mine spoil as influenced by arbuscular mycorrhizal (AM)-fungi in Indian arid zone. *Journal of Arid Environments 51,* 113-119.

Redecker, D., Kodner, R. & Graham, L. E. (2000) Glomalean fungi from the Ordovician. *Science 289*, 1920-1921.

Rivera-Becerril, F., Calantzis, C., Turnau, K., Caussanel, J. P., Belimov, A. A., Gianinazzi, S., Strasser, R. J. & Gianinazzi-Pearson, V. (2002) Cadmium accumulation and buffering of cadmium-induced stress by arbuscular mycorrhiza in three *Pisum sativum* L. genotypes. *Journal of Experimental Botany 53,* 1177-1185.

Roberts, B. & Proctor, J. (1992) *The ecology of areas with serpentinized rocks - A world view*. Dordrecht, The Netherlands: Kluwer.

Roda, J. J., Diaz, G. & Torres, P. (2008) Spatial distribution of arbuscular mycorrhizal fungi in the rhizosphere of the salt marsh plant *Inula crithmoides* L. along a salinity gradient. *Arid Land Research and Management 22,* 310-319.

Rothschild, L. & Mancielli, R. (2001) Life in extreme environments. *Nature 409,* 1092-1101.

Ruotsalainen, A. L., Markkola, A. M. & Kozlov, M. V. (2009) Mycorrhizal colonisation of mountain birch (*Betula pubescens* ssp. *czerepanovii*) along three environmental gradients: Does life in harsh environments alter plant–fungal relationships? *Environmental Monitoring and Assessment 148,* 215–232.

Ruytinx, J., Nguyen, H., Van Hees, M., De Beeck, O., Vangronsveld, J., Carleer, R., Colpaert, J. V. & Adriaensen, K. (2013) Zinc export results in adaptive zinc tolerance in the ectomycorrhizal basidiomycete *Suillus bovinus*. *Metallomics 5,* 1225-1233.

Sannazzaro, A. I., Ruiz, O. A., Alberto, E. O. & Menendez, A. B. (2006) Alleviation of salt stress in *Lotus glaber* by *Glomus intraradices*. *Plant and Soil 285*, 279-287.

Schadt, C., Martin, A., Lipson, D. & Schmidt, S. (2003) Seasonal dynamics of previously unknown fungal lineages in tundra soils. *Science 301*, 1359–1361.

Schechter, S. P. (2010) *Testing the role of arbuscular mycorrhizal fungi in plant adaptation to serpentine soil.* PhD Dissertation. Berkeley, CA: University of California, Berkeley.

Schechter, S. P. & Bruns, T. D. (2008) Serpentine and non-serpentine ecotypes of *Collinsia sparsiflora* associate with distinct arbuscular mycorrhizal fungal assemblages. *Molecular Ecology 17,* 3198-3210.

Schechter, S. P. & Bruns, T. D. (2012) Edaphic sorting drives arbuscular mycorrhizal fungal community assembly in a serpentine/non-serpentine mosaic landscape. *Ecosphere* 3 DOI: 10.1890/ES12-00059.1

Schechter, S. P. & Bruns, T. D. (2013) A common garden test of host-symbiont specificity supports a dominant role for soil type in determining AMF assemblage structure in *Collinsia sparsiflora*. *PLoS One* DOI: 10.1371/journal.pone.0055507

Schoch, C. L., Seifert, K. A., Huhndorf, S., Robert, V., Spouge, J. L., Levesque, C. A., Chen, W. & Fungal Barcoding Consortium (2012) Nuclear ribosomal internal transcribed spacer (ITS) region as universal DNA barcode marker for fungi. *Proceedings of the National Academy of Sciences USA 109,* 6241-6246.

Schnoor, T. K., Lekberg, Y., Rosendahl, S. & Olsson, P. A. (2011) Mechanical soil disturbance as a determinant of arbuscular mycorrhizal fungal communities in semi-natural grassland. *Mycorrhiza 21*, 211-220.

Shetty, K. G., Hetrick, B. A. D., Figge, D. A. H. & Schwab, A. P. (1994) Effects of mycorrhizae and other soil microbes on revegetation of heavy-metal contaminated mine spoil. *Environmental Pollution 86,* 181-188.

Smith, S. E. & Read, D. J. (2008) *Mycorrhizal symbiosis.* Third edition. New York, NY: Academic Press.

Southworth, D., Tackleberry L. E. & Massicotte, H. B. (2013) Mycorrhizal ecology on serpentine soils. *Plant Ecology and Diversity 7*, 445-455

Staudenrausch, S., Kaldorf, M., Renker, C., Luis, P. & Buscot, F. (2005) Diversity of the ectomycorrhiza community at a uranium mining heap. *Biology and Fertility of Soils 41,* 439–446.

Stone, R. W. (1920) *Gypsum deposits of the United States*. Washington, DC: U.S. Government Printing Office.

Talbot, J. M., Bruns, T. D., Taylor, J. W., Smith, D. P., Branco, S., Glassman, S. I., Erlandson, S., Vilgalys, R., Liao, H-L. & Smith, M. (2014) Endemism and functional convergence across the North American soil mycobiome. *Proceedings of the National Academy of Sciences (USA) 111*, 6341-6346.

Tang, M., Sheng, M., Chen, H. & Zhang, F. F. (2009) In vitro salinity resistance of three ectomycorrhizal fungi. *Soil Biology & Biochemistry 41,* 948-953.

Tarafdar, J. C. & Rao, A. V. (1997) Mycorrhizal colonization and nutrient concentration of naturally grown plants on gypsum mine spoils in India. *Agriculture, Ecosystems & Environment 61,* 13-18.

Thrall, P. H., Bever, J. D. & Slattery, J. F. (2008) Rhizobial mediation of *Acacia* adaptation to soil salinity: Evidence of underlying trade-offs and tests of expected patterns. *Journal of Ecology 96,* 746-755.

Tresner, H. D. & Hayes, J. A. (1971) Sodium chloride tolerance of terrestrial fungi. *Applied Microbiology 22,* 210-213.

Turner, T., Bourne, E., Von Wettberg, E., Hu, T. & Nuzhdin, S. (2010) Population resequencing reveals local adaptation of *Arabidopsis lyrata* to serpentine soils. *Nature Genetics 42*, 260-263.

Urban, A., Puschenreiter, M., Strauss, J. & Gorfer, M. (2008) Diversity and structure of ectomycorrhizal and co-associated fungal communities in a serpentine soil. *Mycorrhiza 18,* 339–354.

Vandenkoornhuyse, P., Ridgway, K. P., Watson, I. J., Fitter, A. H. & Young, J. P. W. (2003) Co-existing grass species have distinctive arbuscular mycorrhizal communities. *Molecular Ecology 12,* 3085-3095.

Walder, F., Niemann, H., Natarajan, M., Lehmann, M. F., Boller, T. & Wiemken, A. (2012) Mycorrhizal networks: Common goods of plants shared under unequal terms of trade. *Plant Physiology 159,* 789-797.

Wallander, H., Eklab, A., Godbold, D. L., Johnson, D., Bahr, A., Baldrian, P., Björk, R. G., Kieliszewska-Rokicka, B., Kjoller, R., Kraigher, H., Plassard, C. & Rudawska, M. (2013) Evaluation of methods to estimate production, biomass and turnover of ectomycorrhizal mycelium in forest soils – a review. *Soil Biology and Chemistry 57,* 1034-1047.

Wang, B. & Qiu, Y. L. (2006) Phylogenetic distribution and evolution of mycorrhizas in land plants. *Mycorrhiza 16*, 299-363.

Wilkinson, D. & Dickinson, N. (1995) Metal resistance in trees: The role of mycorrhizae. *Oikos 72,* 298-300.

Williams, G. C. (1966) *Adaptation and natural selection*. Princeton, NJ: Princeton University Press.

Wright, J. W. & Stanton, M. L. (2007) *Collinsia sparsiflora* in serpentine and nonserpentine habitats: using F2 hybrids to detect the potential role of selection in ecotypic differentiation. *New Phytologist 173*, 354-366.

Yost, R. S. & Fox, R. L. (1979) Contribution of mycorrhizae to P nutrition of crops growing on an oxisol. *Agronomy Journal 71,* 903-908.

In: Plant Ecology and Evolution in Harsh Environments
Editors: N. Rajakaruna, R. S. Boyd and T. B. Harris
ISBN: 978-1-63321-955-7

Chapter 3

LICHENS ON METAL-RICH SUBSTRATES

Sergio Enrico Favero-Longo[1,2*]
[1]Dipartimento di Scienze della Vita e Biologia dei Sistemi,
Università di Torino, Torino, Italy
[2]Centro Interdipartimentale "G. Scansetti"
per lo Studio dell'Amianto e di altri Particolati Nocivi,
Università di Torino, Torino, Italy

ABSTRACT

Lichen-forming fungi have an outstanding capacity to tolerate environmental pressures, including extreme climates, but also the peculiar edaphic conditions of metal-rich substrates. This review summarizes current knowledge on the contribution of colonized rocks and soils to metal uptake by lichens, on the detrimental effects of heavy metals on the symbiotic partners, and on several tolerance mechanisms highlighted through laboratory analyses and/or field investigations in heavy-metal-rich sites. Ecological traits of lichen communities in mine, metal-contaminated, and ultramafic areas are also outlined, together with the significant role of lichens in surface processes on silicate and carbonate rocks.

INTRODUCTION

Lichens—the symbiotic phenotype of nutritionally specialized fungi (mycobionts) that are obligate biotrophs acquiring fixed carbon from a population of green-algal or cyanobacterial cells (photobionts)—have an outstanding capacity to survive under extreme environmental conditions (Honegger et al., 2013). Their tolerance of climate-related stress factors, including extreme ranges in temperature, water availability, and irradiation, allows them to occur across the planet, in almost all biomes (Grube, 2010). In addition to the interest in physiological adaptation to climatic extremes (Beckett et al., 2008) and climate change

* Corresponding author: sergio.favero@unito.it.

(Ellis & Yahr, 2011), another major focus in lichenology during the last half century has been on lichen responses to non-climate-related abiotic stress factors. The sensitivity of lichens to a wide range of atmospheric chemicals, including sulphur dioxide, ammonia, nitrogen oxides, fluorine, metals, radionuclides, chlorinated hydrocarbons, and other anthropogenic pollutants has spurred research on the use of these symbiotic organisms as indicators of air pollution (Nimis et al., 2002). Although lichens derive much of their nutrition from atmospheric deposition (Nash, 2008) and much research has focused on the impact of airborne metal pollution on epiphytic lichens (Bačkor & Loppi, 2009), lichen nutrition is also significantly supported by interactions with the substrate, which has long been known as a driving factor in the composition of lichen communities (Brodo, 1973).

In this context, the colonization of metal-rich substrates has been described as one of the various extremophilous behaviours of lichen-forming fungi (Purvis & Halls, 1996). More generally, many fungi have been found to tolerate, by a variety of active and incidental mechanisms, levels of metal ions that are usually toxic for other components of microbial populations, such as single-celled and filamentous bacteria. In addition, considerable interspecific and intraspecific variability in fungal resistance has been reported (Gadd et al., 2012). Advances in genomic and proteomic methods during the last decade have improved our ability to characterize microbial populations and have spurred research into the molecular basis of metal tolerance. The role of soil fungi in biogeochemical cycles and their ability to accumulate metals have attracted research interest due to the implications for agronomic and forest ecosystems and the potential for their use in bioremediation efforts (Gadd, 2006; Mishra & Malik, 2013). The mycorrhizal association has been a dominant topic in the investigation of ion mobilization and bioremediation in metal-polluted soils (Chapter 2; Martino & Perotto, 2010). Much research has focused on fungal exploitation of metal-rich surface environments, with a particular focus on implications for mineral weathering and early stages of pedogenesis (Adamo & Violante, 2000; Gadd et al., 2012). This has increased our knowledge of the colonization of rocks, mine spoils, and metal-polluted soils by saxicolous and terricolous lichen communities (Haas & Purvis, 2006; Purvis & Pawlik-Skowrońska, 2008). The integration of the different areas of research on lichens adapted to metal-rich substrates, both natural and anthropogenic, has helped to shift our attention from descriptive studies of lichen communities (Favero-Longo et al., 2004; Purvis & Halls, 1996) to studies aimed at revealing the physiological basis of adaptation.

Metal Acquisition

Contents of biologically significant metals in lichens—including Ca, Co, Cu, Fe, K, Mg, Mn, Mo, Na, Ni, V, and Zn—are notably variable and often associated with a remarkable accumulation of metals without a known metabolic function such as Al, Cd, Cr, Cs, Hg, Pb, Ti, and U (Bačkor & Fahselt, 2008; Nash, 2008; Tuominen & Jaakkola, 1973). Atmospheric deposition, mostly favoured by fog and dew, significantly contributes to the chemical contents of thalli as they lack a protective outer cuticle and thus absorb gases and water with dissolved metals and other compounds, incorporating exogenous particles over their whole surface. In the apoplastic space of their multilayered structure, lichens behave as metal accumulators, largely reflecting the metal availability of their surrounding atmosphere (Nash,

2008). Analysis of the metal contents of *in situ* and transplanted lichen thalli—mostly of foliose and fruticose epiphytic species—has been widely adopted as a tool for assessing atmospheric metal deposition both from natural sources such as volcanoes (Varrica et al., 2000) and from anthropogenic sources such as mines and smelters (Bačkor & Loppi, 2009; Brunialti & Frati, 2014). In contrast, studies on saxicolous and terricolous lichens have reported metal contents reflecting relative proportions of the rock or soil substrate (Armstrong, 1997; Bačkor & Fahselt, 2004a; Bačkor et al., 2009; Pawlik-Skowrońska et al., 2006; Purvis et al., 2011), although atmospheric deposition may contribute disproportionately to the metal content of their thalli (Garty et al., 1986) or may even be the dominant contributor (Chiarenzelli et al., 1997). Purvis et al. (2011) demonstrated differential element distribution related to thallus anatomical structure.

Entrapment of metal-rich mineral particles under lichen thalli following the disaggregation of the upper millimeter(s) of rock or soil is thought to be the primary contributor to metal uploading. Microscopic and spectroscopic analyses of crustose thalli have documented the occurrence of inorganic particles reflecting the mineral composition of the substrate (Edwards et al., 2002), including less recurring heavy-metal rich minerals and alloys (Favero-Longo et al., 2005a). A high capacity of hyphal structures of foliose thalli (rhizinae) to adhere to soil particles and to store large amounts of metals has been shown to contribute to lichen metal nutrition (Goyal & Seaward, 1981). A similar ability can be reasonably inferred for the Hyphal Penetration Component (HPC, *sensu* Favero-Longo et al., 2005a) of saxicolous crustose lichens, as suggested by myceliar explorative growth under toxic metal stress and nutrient poor conditions (Gadd, 2006).

The hyphal wall components—primarily chitin and chitosan—support the metal accumulation process in lichens by absorbing or binding positive cations onto their anionic functional groups, including hydroxylic, carboxylic, phosphate, amine, and sulphydryl groups (Brown & Brown, 1991; Purvis & Pawlik-Skowrońska, 2008). Dynamics of the ion exchange reactions have been experimentally explored for different lichen species and metals (mostly by sequential elution techniques; Branquinho et al., 1997a). These studies have demonstrated the dependence of these reactions on intrinsic factors such as thallus morphology and the selectivity of exchange sites on the cell wall as well as on extrinsic factors such as cation availability and pH (Branquinho et al., 2011; Haas & Purvis, 2006; Kłos et al., 2007; Puckett et al., 1973; Richardson, 1995). More-or-less soluble, excreted metabolites contain anionic functional groups on which metals in solution displace hydrogen ions or other weakly bound metals and are thus extracellularly immobilized (Purvis & Pawlik-Skowrońska, 2008). The combined chelating and/or acidic properties of the excreted low-molecular weight primary metabolites, such as oxalic and citric acids, and slightly water-soluble polyphenolic secondary metabolites, such as melanins and the so-called "Lichen Secondary Metabolites" (LSM; *e.g.*, depsides, depsidones, dibenzofurans, etc.), yield the formation of organic-metal complexes and organominerals (Adamo & Violante, 2000). Reports of lichen thalli with oxalic acid concentrations of 60% by dry weight (Syers & Iskandar, 1973 and refs. therein) highlight the contribution of these compounds to metal reservoirs in lichens and their role as an active mechanism employed by the mycobiont to avoid intracellular metal excesses. Hauck et al. (2007) highlighted that some of the secondary metabolites which were previously thought to be exclusively related to metal-exclusion may be involved in maintaining a more complex equilibrium of metal transport from the extra- to the intra-cellular compartment.

The nutrition of fungi under metal-deficient conditions is supplemented by the secretion of low molecular weight organic molecules that serve as chelators of biologically significant elements such as Fe (Winkelmann, 2007). By chelating Fe(III), siderophores (mostly hydroxamates) mediate its uptake in symbiotic and saprophytic fungi from insoluble hydroxides and from Fe adsorbed to solid surfaces (Gadd et al., 2012). Iron chelating molecules were shown to be secreted by endolithic lichen-forming fungi living on Fe-deficient substrates such as carbonate rocks (Favero-Longo et al., 2011). Extracellular Polysaccharide Substances (EPS) surrounding mycobiont hyphae are also thought to play a significant role in ion exchange (Purvis & Pawlik-Skowrońska, 2008), as has been widely demonstrated for other groups of microorganisms (Flemming & Wingender, 2010). Although EPS have been microscopically documented at the hyphae-rock interface (Barker & Banfield, 1996; de los Ríos et al., 2005), in the case of lichens these compounds have been poorly characterized, with little attention paid to their metal-binding properties.

Although spectroscopic and histochemical techniques have allowed us to identify intracellular ion localization (Rinino et al., 2005), the mechanisms by which metals reach an intracellular level within the mycobiont have not been fully clarified. Heavy metal accumulation was detected in mycobiont concentric bodies (Cd, U, Zn); plasmalemma (Cd); vacuoles (Cr); cytosol (Cd, Cr); polyphosphate granules (Mn); photobiont chloroplasts, close to the pyrenoid area (Cd); intracellular vesicles (Pb); and at the mycobiont-photobiont interface (Cd) (Álvarez et al., 2012; Haas et al., 1998; Paul et al., 2003; Sanità di Toppi et al., 2004, 2005; Spagnuolo et al., 2011). Sequential elution procedures highlighted that the intracellular fraction of trace elements in terricolous lichens found on metal polluted soils can reach elevated levels which vary by metal (65% Cd, 40% Pb, and 20% Zn) (Cuny et al., 2004a). In laboratory assays on epiphytic lichens, intracellular uptake of metals such as Cd, Cu, and Pb was suggestive of energy-requiring transmembrane carrier systems; the observed kinetics depended on the sensitivity to desiccation damage and metal toxicity, modifying membrane integrity and permeability and causing loss of ions (K and Mg) from the cell interior (Branquinho et al., 1997a, b).

METAL DETRIMENTAL EFFECTS

The toxicity of metals to lichens has long been recognized, but both the toxic effects and the physiological responses of the symbiotic partners appear variable, depending on the type of exposure (in both qualitative and quantitative terms), environmental conditions, and mycobiont and photobiont species (Álvarez et al., 2012; Branquinho et al., 1997a, 2011; Pawlik-Skowrońska et al., 2002, 2006). In general, metal-induced Reactive Oxygen Species (ROS) unbalance the cellular redox status and can affect cellular components and physiological processes by causing lipid peroxidation, membrane damage, enzyme inactivation, and DNA damage (Álvarez et al., 2012; Cuny et al. 2004a). Sensitivity of the photosynthetic apparatus to metal stress has been well documented, and the photosynthetic activity of the photobiont is considered a good indicator of metal stress in lichens (Bačkor & Fahselt, 2008). In the laboratory, chlorophyll fluorescence, which reflects the efficiency of the primary photochemical reactions of photosystem II, was shown to decrease under excessive intracellular uptake of Cd and Cu (Branquinho et al., 1997a; Sanità di Toppi et al., 2005).

Accordingly, Cu—but also Fe(II), Fe(III), Hg, Mn, Pb, Ti, and Zn—has been shown to negatively affect chlorophyll and carotenoid biosynthesis, yielding variations in the chlorophyll a:b ratio and in the phaeophytinization processes (Bačkor & Fahselt 2004b; Chettri et al., 1998; Nakajima et al., 2013). These metals have also been shown to affect lipid composition (Guschina & Harwood, 2006) and to promote peroxidation in the lipid bilayer of photosynthetic membranes, marked by increasing concentrations of malondialdehyde (MDA) (Carreras & Pignata, 2007). However, in some cases, no differences in chlorophyll a integrity were detected in lichens growing directly on the surface of copper mine tailings (Bačkor et al., 2009). Depending on species or strains, the effects of metal toxicity on photobionts may also include growth inhibition (Bačkor et al., 1998) and cytological damage such as the swelling of chloroplast thylakoids and mitochondrial cristae and the detachment of the plasmalemma from the cell wall (Sanità di Toppi et al., 2005; Tarhanen, 1998).

Although the photobionts have been often reported as the more metal-sensitive component of the lichen symbiosis (Bačkor et al., 2006; Bačkor & Fahselt, 2008), the metabolic activity of mycobionts has been shown to be even more affected by metal toxicity (Branquinho et al., 2011). In particular, the substantial loss of intracellular K—which is generally associated with lichen sensitivity to metal toxicity—may follow decreases in the ergosterol content of the mycobiont membrane, increasing its ion permeability (Tarhanen et al., 1999). Alternatively, K fluxes may be explained by an ion-dependent activation of cation channels, without requiring membrane damage (Cuny et al., 2004a). In aposymbiotic mycobionts, excessive Cu levels result in a strong inhibition of triphenyl formazan (TPF) production, which is directly linked to the mitochondrial respiratory chain, and thus, to the viability of the mycobiont (Bačkor et al., 2006). Exposure to metal solutions has also been shown to inhibit ascospore germination and to cause ascospore fragmentation, again demonstrating variable sensitivity by species (Pyatt, 1976). These responses, however, are not promoted by all heavy metals. For example, Pb mostly binds to extracellular sites, displacing exchangeable Ca and Mg; it generally does not alter membrane permeability and is found at relatively low concentrations at the intracellular level (Branquinho et al., 1997b).

MECHANISMS OF METAL TOLERANCE

Controlled exposures and/or incubations of thalli or aposymbiotically-cultured symbionts have increased our understanding of their physiological responses to metal toxicity. However, lichens growing on metal-rich substrates (mostly on metal-contaminated soils) have been regarded as a great opportunity to validate and expand upon laboratory studies regarding tolerance mechanisms.

The existence of metal-tolerant populations has been recognized within different taxa of *Trebouxia* photobionts. Beck (1999) found only two taxa of photobionts (two subspecies of *Trebouxia jamesii*) in ten saxicolous (9 crustose and 1 foliose) lichen species of a metallophytic community (Acarosporetum sinopicae). This result strongly differs from greater photobiont diversity detected in other lichen communities (5 photobiont species in 10 lichen species of *Physcietum adscendentis*). Notably, isolates of *T. jamesii* obtained from the metallophytic lichens were able to grow in Fe-rich (70 mM) liquid solutions, while isolates of the same photobiont species from a non-metallophytic lichen did not show the same tolerance. Cu-tolerant strains of *Trebouxia ericii* were obtained from the wild-type by gradually increasing the Cu concentration in culture medium (Bačkor & Váczi, 2002). A rather high number of photobionts, representing lineages with broad ecological amplitude and worldwide distribution, was instead characterized in foliose and fruticose lichens on mine-spoil heaps, suggesting that only the tight attachment of crustose thalli determines a high selective pressure on the photobiont partner (Bačkor et al., 2010).

Photobionts and mycobionts of lichens exposed to heavy metals (Cd, Pb, Zn) and As(V) in the laboratory were shown to produce phytochelatin (PC) and glutathione (GSH), respectively (Bačkor et al., 2007; Mrak et al., 2010; Pawlik-Skowrońska et al., 2002; Sanità di Toppi et al., 2008). Phytochelatins are glutathione-derived, cysteine-rich, metal-complexing peptides synthesized by plants to complex heavy metals and metalloids onto cysteine sulfhydryl groups, thereby preventing their free circulation in the cytosol. The role of GSH in lichen response to metal toxicity needs further study, as is the case with other low-molecular weight thiols; however, GSH is thought to play a major role in the mycobiont response to oxidative stress (Beckett et al., 2008). In aposymbiotically-grown mycobionts, GSH was reported as the main non-enzymatic antioxidant, likely acting as protection against reactive oxygen species or by directly chelating heavy metals via sulfhydryl coordination (Bačkor et al., 2006). Glutathione concentrations were higher in populations of the terricolous lichen *Diploschistes muscorum* growing on metal-contaminated soils than in specimens from non-polluted control sites (Cuny et al., 2004a).

The physiological response of algal cells to metal toxicity is related to the level of metal exposure and to the efficacy of other tolerance mechanisms. Terricolous lichens from Zn- and Pb-contaminated soils displayed higher levels of LSM, which are potentially involved in the extracellular binding of toxic metals, and showed significantly lower PC concentrations (Pawlik-Skowrońska & Bačkor, 2011). Glutathione and PC concentrations were analyzed in apothecia of *Lecanora polytropa* collected from an abandoned copper mine, where they displayed evidence of extracellular Cu complexation by carboxylic acids (oxalic, malic, citric) and secretion of LSM with chelating functions, and exposed to Cu in the laboratory. Glutathione was detected as an initial antioxidant under low Cu availability to algal cells. The presence of oxidized PC indicated its involvement in Cu detoxification, possibly by reduction of Cu(II) to Cu(I) (Pawlik-Skowrońska et al., 2006).

Amino acids (AA) are also potentially involved in metal tolerance at the cellular level, acting as ligands for heavy metals and free radicals (Bačkor & Fahselt, 2008). However, *Cladina mitis* from Cu-contaminated and control sites displayed only slight differences in AA concentrations, suggesting that other metabolic processes may have a larger role in metal tolerance (Bačkor et al., 2009). The role of stress proteins—including the protective antioxidants glutathione reductase (GR), superoxide dismutase (SOD), ascorbate peroxidase (APX), and catalase (CAT), as well as heat-shock-proteins (HSP)—has been investigated with regard to lichen metal tolerance (Bačkor & Fahselt, 2008). In the laboratory, GR, SOD, APX, and CAT were differently activated in thalli of *Dermatocarpon luridum* collected from a non-polluted site upon exposure to varying Cu concentrations (Monnet et al., 2006). A significant increase in GR, SOD, APX, CAT, and HSP70 was measured as a response to intracellular uptake of Pb in photobiont strains having less effective extracellular barriers (Álvarez et al., 2012). Accordingly, *D. muscorum* from Cd-, Pb-, and Zn-contaminated soils showed the highest SOD activity in thalli with the highest metal concentrations, but a significant correlation was only found for intracellular Cd, not for Pb or Zn (Cuny et al.,

2004a). A role for extracellular laccases in Fe-uptake and in suppressing Cu or Fe cytotoxicity has been suggested (Beckett et al., 2013) but remains poorly demonstrated.

In addition to contributing to our understanding of cellular detoxification processes, investigations of lichen communities occurring on natural and anthropogenic metal-rich substrates have revealed extracellular binding and biomineralization processes, as indicated by the presence of particular organometallic compounds, as the crucial avoidance strategy in lichens exposed to excessive levels of metals (Haas & Purvis, 2006; Purvis & Halls, 1996). Populations of *Peltigera didactyla* found on polluted soils demonstrated lower toxicity of Pb and Zn relative to other lichen species, likely because of 20-times higher contents of chitin in the mycobiont cell wall, favouring extracellular binding (Pawlik-Skowrońska et al., 2008). Notably, species within the Peltigerinae also display higher activity and greater diversity of cell wall redox enzymes, including laccases and tyrosinases, which are potentially involved in melanisation (Beckett et al., 2013). Production of melanin-like compounds and associated polysaccharides in apothecia of *Trapelia involuta* growing on U-bearing minerals was hypothesized as an adaptive response to protect ascospores due to the remarkable capacity of melanin to adsorb U, thereby mediating its toxicity (McLean et al., 1998; Purvis et al., 2004).

Fungi produce oxalic acid through: (a) oxidation of glucose to pyruvate, which is carboxylated to oxaloacetate (pyruvate carboxylase), which is hydrolyzed to oxalate and acetate (oxaloacetate hydrolase), or (b) through the isocitrate and glyoxalate cycle (Verrecchia et al., 2006). Ascorbate is another precursor of oxalate and may play an antioxidant role (Beckett et al., 2008). The reaction of the oxalic acid secreted by saxicolous lichen mycobionts with the underlying minerals yields the accumulation of oxalate deposits at the lichen-rock interface (Gadd et al., 2012). Oxalic acid secretion has been recognized and investigated as the dominant process involved in the chemical degradation of stone surfaces by lichens, being involved in both the deterioration of original mineral constituents of rocks and in biomineralization processes responsible for the formation of unpleasant patinas (Adamo & Violante, 2000; Gazzano et al., 2009). Extracellular metal complexation by oxalic acid is currently considered a response of lichens adapted to metal-rich substrates, allowing metal concentrations in thalli to accumulate to the range of plant hyperaccumulators [16% Cu by dry weight in *Acarospora rugulosa* (Chisholm et al., 1987); 32500 mg kg^{-1} Zn and 2900 mg kg^{-1} Pb by dry weight in *Diploschistes muscorum* (Sarret et al., 1998); and 109 μmol g^{-1} Cu by dry weight in apothecia of *Lecanora polytropa* (Pawlik-Skowrońska et al., 2006)]. The form of oxalate precipitate found at the lichen-substrate interface is generally related to the metal content of the substrate: calcium oxalates (monohydrate: whewellite; dehydrate: weddellite) characterize the interface between many calcicolous species (but not all) and their calcareous substrates (Adamo & Violante, 2000 and references therein). Magnesium (glushinskite; Wilson et al., 1981), Cu (mooloite; Chisholm, et al. 1987; Purvis et al., 2008a), Fe(III) (Ascaso et al., 1982), Mn (Wilson & Jones, 1984), Pb, and Zn (Sarret et al., 1998) oxalates were reported at the lichen-substrate interface of species growing on serpentinites, Cu-sulphide bearing rocks, Mn ore (cryptomelane + lithiophorite), Fe-rich dolomites, and Zn/Pb polluted soils, respectively. Solubility products of the different oxalates vary over a range of several orders of magnitude (*e.g.*, K_{ps} $4.83 \cdot 10^{-6}$ for glushinskite vs. K_{ps} $2.32 \cdot 10^{-9}$ for whewellite), accounting for a progressive substitution of more soluble salts with less soluble ones in environments with suitable cation supply (*e.g.*, Ca-rich runoff waters) (Favero-Longo et al., 2005a).

Secretion of citric, malic, and formic acids, which are known to play a role in metal-detoxification and metal-tolerance in both fungi and plants, has been reported for saxicolous lichens (Arocena et al., 2003; Pawlik-Skowrońska et al., 2006). However, how these low-weight organic ligands affect metal tolerance in lichens has yet to be clarified. This is likely because the production of these acids does not leave a biomineralization signature, unlike chelation by oxalic acid which produces oxalates which are easily detected, and therefore may have been overlooked in previous studies.

The fact that some of the polyphenolic LSM exclusively produced by lichens also display acidic and chelating functions has long suggested that they may play a significant role in metal scavenging/nutrition, mineral deterioration, and metal detoxification (Adamo & Violante, 2000; Huneck, 1999; Syers & Iskandar, 1973). Because of their low solubility in water, the mobilization of such compounds and their subsequent role in mineral deterioration has been questioned; however, spectroscopic analysis revealing these metabolites in the upper layers of sandstones underlying crustose species (Bjelland et al., 2002) and the remarkable chelation and dissolution properties demonstrated by compounds in laboratory incubations (Ascaso & Galvan, 1976) suggest that their biogeochemical activity at the lichen-substrate interface is worth further investigation. Because depsides, depsidones, dibenzofurans, and many other LSM often appear as precipitates encrusting the surface of medullary hyphae, their role in the extracellular binding of toxic metals has been more generally accepted (Pawlik-Skowrońska & Bačkor, 2011). Laboratory pharmaceutical studies have suggested that LSM may play a role as antioxidants through contrasting metal-induced lipid peroxidation (Brisdelli et al., 2013). However, this has been scarcely studied in the context of lichen interactions with metal-enriched substrates (Stepanenko et al., 2002). LSM-metal complexes were first reported from natural conditions in thalli growing on cupriferous substrates (Cu-norstictic acid complex: Purvis et al., 1985; Cu-psoromic acid complex: Purvis et al., 1990). The presence of Al, Cu, Fe, and Ni within crystals of usnic acid in the terricolous *Cladonia pleurota* growing on metal-rich historical ore-roasting beds also links complexing and detoxification (Bačkor & Fahselt, 2004b). However, the presence of one or more LSM in the cortex or the medulla of a particular lichen species does not imply that metal-acid complexes are detectable or that their role in detoxification can be established (Purvis et al., 1990; Purvis et al., 2004).

Reports of secondary metabolite concentrations in lichen species found on metal-enriched and control sites vary. Specimens of *Cladonia humilis* and *C. furcata* found on and off of metal-enriched substrates did not differ significantly in levels of atranorin and/or fumarprotocetraric acid; specimens of *C. pleurota* and *C. arbuscula* var. *mitis* did not differ significantly in concentrations of usnic acid. Specimens of *Stereocaulon japonicum* found on control sites had higher concentrations of atranorin and stictic and norstictic acids. These findings suggest that the protective action of these LSM against metal toxicity is a constitutive rather than inducible process (Bačkor & Fahselt, 2004b; Bačkor et al., 2011; Nakajima et al., 2013). This may be supported by the fact that these compounds are known to play other roles in lichens (*e.g.*, sun screening, antifeedant, control of medulla hydrophobicity) (Huneck, 1999). Other reports have found higher concentrations of LSM in lichens growing on metal-enriched sites relative to control sites. Specimens of *Hypocenomyce scalaris* and *Cladonia furcata* displayed higher concentrations of lecanoric acid and fumarprotocetraric acid, respectively, on Zn- and Pb-polluted soils relative to control sites. These compounds allow extracellular accumulation of toxic metals while limiting their availability to photobionts; this is supported by a lower rate of phytochelatin production (Pawlik-Skowrońska & Bačkor, 2011). Over the last decade, studies by Hauck and colleagues on metal homeostasis in epiphytic lichen communities exposed to metal air pollution and saxicolous lichens on metal-rich substrates have greatly contributed to our understanding of these apparently contrasting patterns. In particular, these results support the hypothesis that the type and abundance of LSM produced by lichen species influence their specificity for pH conditions and metal availabilities (Hauck et al. 2007, 2009a, 2010, 2013).

In the epiphytic *Hypogymnia physodes*, medullary depsidone physodalic acid was shown to reduce the intracellular uptake of Cu and Mn, but did not affect uptake of Fe(II) or Zn, indicating an ion-specific control of metal scavenging by lichen substances (Hauck, 2008). In the same species, however, other LSM (*e.g.*, atranorin, protocetraric acid, physodic acid) do not play a similar role in maintaining metal homeostasis (Hauck & Huneck, 2007). In thalli of *H. physodes* transplanted in the vicinity of a smelter, physodalic acid concentrations were significantly elevated relative to controls, while other metabolite concentrations were decreased (Białońska & Dayan, 2005). In an area not exposed to metal pollution, concentrations of physodalic acid were variable in thalli from different trees, but similar in thalli on the same tree, and were directly correlated with metal concentrations of the bark; other LSM showed an opposite trend (Hauck et al., 2013).

Although the mechanisms by which LSM contribute to metal homeostasis are still largely unknown (Hauck et al., 2013), their complexing activity may not be solely related to extracellular exclusion mechanisms but also to temporary chelation processes favouring metal nutrition (Hauck, 2008; Hauck et al., 2007). Because uncomplexed LSM are known to cross the plasmatic membrane, it can be reasonably hypothesized that metal complexes can also cross the plasmatic membrane, either through a similar direct transfer process or through a transporter-mediated process (Hauck et al., 2009a). Given the ion specificity of LSM, it may be advantageous to produce specific LSM on substrates with low availability of metals required for lichen nutrition, whereas it may be disadvantageous to produce the LSM on substrates with high levels of metals which could be toxic at high concentrations (*e.g.*, Hauck et al., 2007).

Experiments have shown that the uptake of Fe and other ions by saxicolous, epiphytic, and terricolous lichens is pH dependent (Paul et al., 2009a, b). Calcicolous/alkalinophytic species display more efficient Fe(III) and P uptake at pH 8 than do calcifuge/acidophytic species, but an excessive and toxic uptake of Fe(II) at pH 3 (Paul et al., 2009b). The formation of LSM-metal complexes has also been shown to be pH dependent (Hauck et al., 2009b, c), as pH controls the equilibrium of the water soluble fraction of LSM between the protonated and anion forms. On this basis, Hauck et al. (2010) determined that the ecological preferences of lichen species producing the depsidone norstictic acid and avoiding very acidic substrates were related to acidity tolerance patterns and metal nutrition requirements. Because norstictic acid has $pK_{a1}=4$, its occurrence at pH <4 would imply its dissociation and involvement in shuttling protons to the cytoplasm, causing toxic effects. Conversely, at pH >7.5, norstictic acid increases its affinity for Cu, Fe(III), Mn, and Zn—micro-nutrients which occur at low concentrations under natural conditions—whereas it does not affect the absorption of Mg or Ca, which are generally more available. Most species growing more or less exclusively on Fe-rich substrates are completely devoid of LSM or secrete LSM which do not have a significant effect on Fe(III) absorption or other LSM, such as rhizocarpic acid and norstictic acid, which significantly reduce the absorption of Fe(II) (Hauck et al., 2007).

Conversely, metabolites effective in Fe absorption are widespread in lichens that grow on substrates with low Fe content where the ability to absorb Fe may be useful (Hauck et al., 2007).

In general, there is a specific pH range within which LSM have a moderate to high affinity for specific metals. This pH range frequently corresponds with the pH range at which lichens that produce those specific LSM occur, suggesting a significant role for LSM in determining lichen substrate suitability (Hauck et al., 2009b). Lichens producing usnic acid (pK_{a1}=4.0) optimally occur at pH between 4.0 and 4.5, which favors the uptake of Mg and transition metals. These lichens are absent at lower pH levels because of excessive protonophoric action of the dissociated acid and are absent at higher pH levels because high usnic acid-metal affinity may yield stable complexes and prevent metal uptake (Hauck et al., 2009a, b). Conversely, parietin (pK_{a1}=3.9) and many pulvinic acid derivatives (pK_{a1} between 2.8 and 4), excluding rhizocarpic acid, increase metal uptake only at alkaline pH levels. As such, these LSM are generally characteristic of species found on calcareous substrates, but also of those found on acidic, nutrient-rich substrates (Hauck et al., 2009a).

The work highlighted here points toward a prominent role for LSM in maintaining metal homeostasis, determining pH preferences, and contributing to metal tolerance in lichens (Hauck et al., 2013). Nevertheless, major research focus on hypothesized LSM-mediated intracellular uptake mechanisms is still needed to definitely substantiate this eco-physiological overview.

Lichen Communities on Metal-Rich Substrates

Mine Spoils and Contaminated Sites

For more than half a century, studies have claimed that mining sites and metal-contaminated soils—where high metal concentrations limit colonization by vascular plants—support rich and specialized lichen communities (Purvis & Halls, 1996). Some "metallophyte" species—which are more or less restricted to the metal-rich substrates—are known to occur with many more widely-distributed lichens, in some cases displaying peculiar phenotypes such as metal-related coloring (Gilbert, 2000; Rajakaruna et al., 2011). In at least one case (Wales), the conservation of the old metal mine sites has been suggested to allow the survival of these habitat-restricted metallophyte lichens, which have been included in red data lists (Plantlife, 2012). In the unique relationship between saxicolous and terricolous lichens and their substrate and in the ability of such lichens to accumulate metals from their substrate, some authors (*e.g.*, Chettri et al., 1997) even recognized a potential for their use in geochemical explorations.

Rock type and/or ore chemistry exert a significant influence on lichen community composition given their primary role in determining substrate pH which leads to the selection of acidophilous/calcifuge vs. basiphilous/calcicole communities (Gilbert, 2000). Out of hundreds of species documented from mines (291 listed in 1999) (Cuny et al., 2004b), a small group of inconspicuous species with no LSM production (*e.g., Bacidia saxenii, B. viridescens, Coppinsia minutissima, Placynthiella hyporhoda, Sarcosagium campestre, Steinia geophana, Verrucaria bryoctona, Vezdaea acicularis, V. cobria, V. leprosa, V.*

retigera, and *V. rheocarpa*) particularly characterize heaps and veins in Pb and Zn mines, whose surface is periodically renewed, possibly restoring metal toxicity (Cuny et al., 2004b; Gilbert, 2000; Medeiros et al., 2014; Smith et al. 2009), and other peculiar Zn-rich substrates (Buck et al., 1999). Although these species have been suggested as potential indicators of certain metals, in most cases they are not restricted to metal-rich substrates (Smith et al., 2009). The few species exclusively reported for metal-rich habitats (*e.g.*, *C. minutissima*) may have been overlooked in less-explored, "banal" environments, and the paucity of knowledge on their metal tolerance suggests the necessity of further investigations to ascertain their metallophytic nature.

A terricolous association characterized by bryophytes and lichens tolerant of—but not restricted to—heavy-metal polluted soils (*Cladonietum rei*) has been reported in several metal-contaminated (Cu, Pb, Zn) slag dumps (Osyczka & Rola, 2013). *Cladonia rei* (a chemotype of *C. subulata*; Spier & Aptroot, 2007) displayed restrained accumulation patterns which were hypothesized to facilitate its higher colonization of polluted sites relative to other species such as some other *Cladonia* species or *Diploschistes muscorum* (Osyczka & Rola, 2013). At other Pb- and Zn-contaminated sites; however, only species displaying high levels of heavy metal accumulation, including *Diploschistes muscorum* and *Stereocaulon nanodes*, were associated with *Cladonia rei* (Cuny et al., 2004b).

Purvis & Halls (1996) described a community specific to Cu-rich environments (*Lecideion inopis*). The community is dominated by *Lecidea inops*—a species restricted to Cu-rich mine spoils and surfaces irrigated with Cu-rich groundwater—and is characterized by other species not found in other metal-rich communities and some yellow-green Cu-rich ecotypes (Huneck, 2006; Purvis & Halls, 1996). Species were reported with an anomalous colour due to the accumulation of Cu-rich compounds, including inorganic salts (*e.g.*, copper selenate/selenite in *Lecanora sierrae*; Purvis et al., 2011) and organic complexes (*e.g.*, oxalates in *Lecanora polytropa*; Pawlik-Skowrońska et al., 2006; Cu-norstictic acid complexes in *Acarospora smaragdula*; Purvis et al., 1987). These specimens were often initially recognized as new taxa (Purvis & Halls, 1996), but are now generally accepted as substrate-related "ecotypes" of other known species.

The development of typical communities on Fe-sulphide bearing mine spoils (described within the alliance *Acarosporion sinopicae*; Purvis & Halls, 1996 and references therein; Purvis 2014) was attributed to low pH, rather than atypical metal composition by Wirth (1972). Notably, these communities include species that do not secrete LSM or that produce LSM with a potential role in balancing Fe(II) and Fe(III) uptake and avoiding toxicity (Hauck et al. 2007). Some of the species in these communities have an obligate or facultative rusty appearance (*e.g.*, *Acarospora smaragdula*, *A. sinopica*, *Lecidea silacea*, *Rhizocarpon oederi*, *Tremolecia atrata*) due to Fe-rich extracellular deposits on the thallus surface. Analyses of the mineral composition of these deposits highlighted mixed sulphide and oxi-hydroxide phases with poor crystallinity, together with clay minerals; however, LSM-metal complexes were not detected (Purvis et al., 2008b).

Phylogenetic analyses of the *Acarospora smaragdula* complex (now within the genus *Silobia*) were performed on rusty populations from Fe-rich substrates, on greenish populations from Cu-rich substrates, and on non-metallophyte populations (Wedin et al., 2009). One clade (*A. smaragdula s. str.*), including both the copper-accumulating populations and non-metallophyte populations, produced norstictic acid, the secretion of which cannot be thus directly considered an adaptative response to Cu ions. Rusty populations, which did not secrete norstictic acid, represented three separate lineages, two of which were restricted to Fe-rich substrates. These findings support the hypothesis that LSM production is correlated to differential rates of mobilization and uptake of Fe(II) and Fe(III) by lichen thalli. In this way, LSM production may play a role in determining Fe-tolerance of different species, lineages, or individuals. Research on lichen species found in mines and other similar environments has thus not unveiled a wide set of metallophyte species *sensu stricto* (*i.e.*, edaphic endemics), but has contributed to our understanding of several physiological mechanisms allowing certain species to colonize metal-rich substrates.

Ultramafic Rocks and Soils

Ultramafic habitats have long been used as model systems for studying life adapted to high concentrations of heavy metals (*e.g.*, Co, Cr, Ni) or associated with other edaphic stress factors such as low Ca:Mg ratios or a paucity of essential nutrients as K, N, P or S (Chapter 6). Ultramafic habitats have also been valuable in the development of green technologies for the restoration of metal-contaminated areas (*e.g.*, Chapters 14, 15; Harrison & Rajakaruna, 2011). Although studies on cryptogams of ultramafic habitats are scarce relative to the large volume of literature on phanerogams, interest in lichen communities of ultramafic substrates has a long and persistent history. Little attention, however, has been paid to patterns of adaptation in these communities. In addition, most of the research has dealt with lichens found on serpentinized ultramafic rocks (*i.e.*, serpentinites) of ophiolite suites and their related soils, whereas lichens on non-serpentinized ultramafics (mainly peridotites *s.l.*) of layered-igneous complexes and subcontinental mantle intrusions have been poorly studied (Favero-Longo et al., 2004).

Whereas the phanerogam vegetation of ultramafic habitats is generally characterized by low species diversity, lichen diversity has repeatedly been reported to be higher on serpentinites than on different geologies of adjacent areas (Favero-Longo & Piervittori, 2009; Gilbert & James, 1987; Harris et al., 2007; Paukov, 2009; Rajakaruna et al., 2012; Sirois et al., 1987). Approximately 400 species have been reported from serpentine substrates worldwide. In the case of saxicolous communities, higher hardness and stability of serpentinite surfaces with respect to those of more friable, softer lithotypes may explain the frequently reported high species diversity and high cover values (often >90%), particularly in environments characterized by slow pedogenetic processes such as alpine habitats (Favero-Longo & Piervittori, 2009; Favero-Longo et al., 2004; Rajakaruna et al., 2012; Wirth, 1972). Higher weathering rates at the rock surface could account for low cover values reported on non-serpentinized peridotites (Favero-Longo, unpublished; Gilbert, 1983) but rarely on serpentinized ones (Sirois et al. 1987). Lower competition by plants is likely to favour lichen communities on serpentine soils (Favero-Longo & Piervittori, 2009). Low diversity and cover of lichen species on serpentine soils has only been reported for terricolous lichens found on the non-serpentinized ultramafics of the island of Rhum, Scotland (Gilbert, 1983).

The occurrence of endemics or the development of ecotypes in ultramafic habitats—which are characteristics of phanerogams—is not characteristic of lichen communities. Most of the species initially reported as restricted to ultramafic rocks have also been found on other basic-siliceous rocks (Favero-Longo et al., 2004 and references therein). *Porpidia nadvornikiana* is the only species for which records confirm a disjunct distribution restricted to ultramafic habitats (Fryday, 2005). Ecotypes demonstrating peculiar morphologies (*i.e.*, serpentinomorphoses such as stenophyllism or dwarfism) have only been reported in early works and have not been described recently (Favero-Longo et al., 2004). Rust-coloured forms can be explained by the accumulation of iron-rich minerals at the cortex layer and are not worthy of taxonomic recognition. Phylogenetic analyses on the ITS region of rDNA in *Candelariella vitellina* from ultramafic and non-ultramafic lithotypes in the Western Alps did not show a relationship between different lineages and substrates (Favero-Longo, 2006). This is contrary to findings for some mycorrhizal fungi (*e.g.*, Panaccione et al., 2001).

Nevertheless, a certain peculiarity of lichen communities on serpentinites relative to communities of other lithotypes has been often remarked (Rajakaruna et al., 2012). Several species occurring on adjacent mafic rocks are often absent; silicicolous and calcicolous species often co-occur—although the former generally dominate, as is often the case for phanerogams. Some species occur at their extreme distributional limits, and many rare species have been reported (Favero-Longo et al., 2004 and references therein; Medeiros et al., 2014). Investigations on the preference of certain species for the presence or absence of specific ions in specific concentrations in serpentinites vs. other lithotypes may likely clarify distributional trends observed in these habitats. A number of recent studies have examined lichen communities in association with rock chemistry (*e.g.*, Medeiros et al., 2014; Rajakaruna et al., 2012). The lack of comparable studies on lichen communities from non-ultramafic habitats and the lack of chemical analyses from lichen tissues have limited the contribution of these efforts in clarifying lichen colonization patterns. Although research on the chemical composition of serpentinicolous lichen thalli is still needed, the occurrence of Ni-rich minerals within the medulla, possibly exerting toxic effects on lichen physiology (Carreras & Pignata, 2007), has been documented through coupled microscopical and spectroscopical approaches (Favero-Longo et al., 2005a).

Because high concentrations of oxalates—including glushinskite (Mg-oxalate) and whewellite and weddellite (Ca-oxalates)—were reported in lichens found on serpentinites (Favero-Longo et al., 2005a; Wilson et al., 1981), the secretion of oxalic acid has been frequently reported as metal detoxification or tolerance mechanism. The absence of LSM production—which characterizes several species of the *Acarosporetum sinopicae* alliance (but also the "inconspicuous metallophyte" group) and possibly supports their tolerance of iron-rich acidic substrates (Hauck et al., 2007, 2009a)—characterizes 75 out of 245 species reported in the list of serpentinicolous species by Favero-Longo et al. (2004).[1] Differences in patterns of LSM production, experimentally evaluated by TLC in lichen communities of ultramafic (serpentinites and peridotites) and non-ultramafic (gabbros) rocks in the Western Alps, also suggest that the presence or absence of certain metabolites may explain species-specific serpentine tolerance (Favero-Longo, unpublished data).

Lichen colonization of ultramafics has been investigated over the last decade due to interest in lichen-driven deterioration of asbestos and asbestiform minerals associated with serpentinite rocks, where the fibres occur within different generations of metamorphic veins (Favero-Longo et al., 2013). Lichen colonization dynamics on asbestos-rich serpentinites were examined on the abandoned mine walls of the most important asbestos mine in Western Europe, a chrysotile mine in Balangero, NW Italy (Favero-Longo et al., 2006). The fibre

[1] This information is based on literature data (*e.g.*, Smith et al., 2009) on the LSM contents of each species and has not been experimentally evaluated directly on ultramafic populations.

rough texture was shown to improve surface bioreceptivity for lichens by increasing water retention and potentially promoting propagule deposition. Spectroscopic analyses of chrysotile fibres contacted by hyphae of different lichen species, including *Candelariella vitellina*, *Xanthoparmelia tinctina*, and *Lecanora rupicola*, highlighted a selective leaching of the octahedral Mg-rich layers of the minerals (Favero-Longo et al., 2005b, 2007). Secretion of oxalic acid by *X. tinctina* and *L. rupicola* and of the LSM pulvinic acid by *C. vitellina* accounted for the scavenging of Mg and of poorly coordinated Fe ions, replacing some Mg in the octahedral layer and being related to its surface reactivity (Favero-Longo et al., 2005b, 2013). As the induced chemical modification was associated with a partial decrease in the surface reactivity of chrysotile (and other fibrous minerals undergoing incongruent dissolution as balangeroite and fibrous tremolite), a bioattenuation role was suggested for lichens on asbestos-bearing rocks, although this effect cannot be generalized to all kinds of naturally occurring asbestos (Favero-Longo et al., 2013). The possibility of transplanting thalli of foliose lichens to promote recolonization of abandoned mine walls has been attempted (Favero-Longo & Piervittori, 2012); survival of thalli after ten years was encouraging, but low growth rates and losses from substrate instability and erosion from running water and debris suggested that lichens may be useful only for small-scale remediation work on relatively stable asbestos-rich surfaces.

Lichen-Metal Interactions on Silicate and Carbonate Rocks

The role of microorganisms as powerful agents of geomorphic change, especially in extreme environments, is increasingly recognized. However, information on their effects on geomorphological processes and on the temporal and spatial scale of their impact is still lacking (Viles, 2012). Saxicolous lichens have long been studied for their ability to deteriorate their rock substrate and to lead the first stages of soil formation. Mineral dissolution has been shown to be related to acidolysis and complexolysis processes driven by the primary and secondary metabolites secreted by the mycobiont (Jones, 1988). The neoformation of minerals at the lichen-rock interface depends on biomineralization processes, including the precipitation of organominerals, such as oxalates, but also the formation of alumino-silicates and Fe oxi-hydroxides (Adamo & Violante, 2000; Purvis et al., 2008b). Although a review of lichen deterioration of rocks is beyond the aim of this paper, it is worth noting here that lichen-metal interactions have a key role in diverse processes with pedological and/or geomorphological consequences, including the well documented metabolite-driven mineral dissolution (see Adamo & Violante, 2000), but also the lichen-related case-hardening of surfaces of silicate rocks and carbonate pitting by endolithic lichens.

In both arid and semi-arid warm landscapes and cold deserts, lichens are a dominant biotic component, living as endoliths within intrinsic discontinuities in the rock such as intergranular or intercrystalline voids and pores, where they are protected from excessive irradiation, wind, desiccation, and extreme temperatures (de los Ríos et al., 2005). Iron-rich diagenetic minerals, including Fe oxi-hydroxides (*e.g.*, goethite, hematite), are typically observed around the endolithic hyphae (Arocena et al., 2003; de los Ríos et al., 2005; Guglielmin et al., 2011; Souza-Egipsy et al., 2004) and are likely associated with the templating activity of hyphal EPS, involved in electrostatic interactions with multivalent cations (see Beech et al., 2005). These neoformation minerals were shown to be related to the hardening of the rock layers penetrated by the mycobiont hyphae (Smith, 2009) and to favour the development of peculiar geomorphologies (Guglielmin et al., 2011).

On carbonate rocks, a wide group of lichens is known to support the dissolution of the substrate through an active process that still has not been definitely clarified (Garvie et al., 2008). The secretion of chelating compounds potentially involved in the dissolution process has been recently reported (Favero-Longo et al., 2011). These metabolites were shown to be involved in Fe chelation; however, in Fe-poor substrates such as carbonates (where siderophore production by fungi is usually high) these metabolites may act as an unspecific scavenger of Ca, resulting in lichen-related pitting and etching processes. However, other non-mutually exclusive processes responsible for carbonate dissolution have been hypothesized, including the dissolution-promoting activity of carbonic anhydrase (Favero-Longo et al., 2011 and references therein). The relevance of lichen-related pitting to the development of mesoscale solution-basins in carbonate rocks, however, needs further investigation (McIlroy de la Rosa et al., 2012).

CONCLUSION

Lichens, together with non-lichenized fungi and free-living cyanobacteria and algae, are hypothesized to have developed extensive communities in terrestrial environments before the advent of plants (Honegger et al., 2013). Lichens still dominate terrestrial vegetation where phanerogam competition is absent or low, notably in high altitude and high latitude areas and in desert and steppe ecosystems. Hot-spots of high metal concentrations, such as mines and ultramafic sites, which represent harsh environments for plants, appear to promote high levels of lichen colonization and diversity. Lichens in these metal-rich environments, poor in nutrients and carbon, and often with significant heavy metal concentrations, may be thus considered living memories of ancient stages of terrestrial life and of related tolerance/adaptive strategies.

Numerous mechanisms have been demonstrated to support lichens in environments with high concentrations of metals. Several exclusion pathways have been characterized; however, other adaptive patterns, such as the recently suggested role of secondary metabolites in metal homeostasis, are still largely unexplored. Some of these strategies appear to be shared by species with wide distributional ranges, and ability to tolerate metal stress extends their distribution to metal-rich environments (tolerators). The frequent specialization of metallophytic plants to the unique edaphic conditions of mines and ultramafic habitats (*e.g.*, edaphic endemism) does not appear to occur in lichens. Although strain-specific strategies in maintaining metal homeostasis have been described for mycorrhizal fungi (Martino & Perotto, 2010), information on the intraspecific variability in the physiological behaviour of lichens in response to substrate-related stress is limited. As such, the potential existence (and taxonomic significance) of lineages not differing in their morphology, but rather in genotypic traits reflecting metabolic adaptation to metal-rich substrates, is likely to be an intriguing question in lichenology.

Acknowledgments

The author is grateful to Prof. Piervittori for her guidance in the study of lichen ecology, to Dr. Enrica Matteucci for helpful discussions and to the editors and anonymous reviewers for helpful comments. Aspects of this work were inspired by the participation in the project PROGEOPIEMONTE, co-founded by the University of Torino and Compagnia di San Paolo bank foundation.

References

Adamo, P. & Violante, P. (2000) Weathering of rocks and neogenesis of minerals associated with lichen activity. *Applied Clay Science 16*, 229-256.

Álvarez, R., del Hoyo, A., García-Brejo, F., Reig-Armiñana, J., del Campo, E. M., Guéra, A., Barreno, E. & Casano, L. M. (2012) Different strategies to achieve Pb-tolerance by the two *Trebouxia* algae coexisting in the lichen *Ramalina farinacea*. *Journal of Plant Physiology 169*, 1797-1806.

Armstrong, R. A. (1997) Are metal ions accumulated by saxicolous lichens growing in a rural environment? *Environmental and Experimental Botany 38*, 73-39.

Arocena, J. M., Zhu, L. P. & Hall, K. (2003) Mineral accumulations induced by biological activity on granitic rocks in Qinghai Plateau, China. *Earth Surface Processes and Landforms 28*, 1429-1437.

Ascaso, C. & Galvan, J. (1976) Studies on the pedogenetic action of lichen acids. *Pedobiologia 16*, 321-331.

Ascaso, C., Galvan, J. & Rodriguez-Pascal, C. (1982) The weathering of calcareous rocks by lichens. *Pedobiologia 24*, 219-229.

Bačkor, M. & Fahselt, D. (2004a) Using EDX-microanalysis and X-ray mapping to demonstrate metal uptake by lichens. *Biologia, Bratislava 59*, 39-45.

Bačkor, M. & Fahselt, D. (2004b) Physiological attributes of the lichen *Cladonia pleurota* in heavy metal-rich and control sites near Sudbury (Ont., Canada). *Environmental and Experimental Botany 52*, 149-159.

Bačkor, M. & Fahselt, D. (2008) Lichen photobionts and metal toxicity. *Symbiosis 46*, 1-10.

Bačkor, M. & Loppi, S. (2009) Interactions of lichens with heavy metals. *Biologia Plantarum 53*, 214-222.

Bačkor, M. & Váczi, P. (2002) Copper tolerance in the lichen photobiont *Trebouxia erici* (Chlorophyta). *Environmental and Experimental Botany 47*, 11-20.

Bačkor, M., Hudák, J. & Bačkorová, M. (1998) Comparison between growth responses of autotrophic and heterotrophic populations of lichen photobiont *Trebouxia irregularis* (Chlorophyta) on Cu, Hg and Cd chlorides treatment. *Phyton-Annales Rei Botanicae A, 38*, 239–250.

Bačkor, M., Pawlik-Skowrońska, B., Tomko, J., Budova, J., & Sanità di Toppi, L. (2006) Response to copper stress in aposymbiotically grown lichen mycobiont *Cladonia cristatella*: uptake, viability, ergosterol and production of non-protein thiols. *Mycological Research 110*, 994-999.

Bačkor, M., Pawlik-Skowrońska, B., Bud'ová, J., & Skowroński, T. (2007) Response to copper and cadmium stress in wild-type and copper tolerant strains of the lichen alga *Trebouxia erici*: metal accumulation, toxicity and non-protein thiols. *Plant Growth Regulation 52*, 17-27.

Bačkor, M., Klejdus, B., Vantová, I., & Kováčik, J. (2009). Physiological adaptations in the lichen *Peltigera rufescens* and *Cladina arbuscula* var. *mitis*, and the moss *Racomitrum lanuginosum* to copper-rich substrate. *Chemosphere*, *76*, 1340-1343.

Bačkor, M., Peksa, O., Škaloud, P., & Bačkorova, M. (2010) Photobiont diversity in lichens from metal-rich substrata based on ITS rDNA sequences. *Ecotoxicology and Environmental Safety 73*, 603-612.

Bačkor, M., Péli, E. R., & Vantová, I. (2011) Copper tolerance in the macrolichens *Cladonia furcata* and *Cladina arbuscula* subsp. *mitis* is constitutive rather than inducible. *Chemosphere 85*, 106-113.

Barker, W. W. & Banfield, J. E. (1996) Biologically versus inorganically mediated weathering reactions: relationships between minerals and extracellular microbial polymers in lithobiontic communities. *Chemical Geology 132*, 55-69.

Beck, A. (1999) Photobiont inventory of a lichen community growing on heavy-metal-rich rock. *The Lichenologist 31*, 501-510.

Beckett, R. P., Kranner, I. & Minibayeva, F. V. (2008) Stress physiology and the symbiosis. In: T. H. Nash III (Ed.) *Lichen biology*. Second edition (pp. 134-151). Cambridge: Cambridge University Press.

Beckett, R. P., Zavarzina, A. G. & Liers, C. (2013) Oxidoreductases and cellulases in lichens: possible roles in lichen biology and soil organic matter turnover. *Fungal Biology 117*, 431-438.

Beech, I. W., Sunner, J. A. & Hiraoka, K. (2005) Microbe-surface interactions in biofouling and biocorrosion processes. *International Microbiology 8*, 157-168.

Białońska, D. & Dayan, F. E. (2005) Chemistry of the lichen *Hypogymnia physodes* transplanted to an industrial region. *Journal of Chemical Ecology 31*, 2975-2991.

Bjelland, T., Sæbø, L. & Thorseth, I. H. (2002) The occurrence of biomineralization products in four lichen species growing on sandstone in western Norway. *The Lichenologist 34*, 429-440.

Branquinho, C., Brown, D. H. & Catarino, F. (1997a) The cellular location of Cu in lichens and its effects on membrane integrity and chlorophyll fluorescence. *Environmental and Experimental Botany 38*, 165-179.

Branquinho, C., Brown, D. H., Máguas, C. & Catarino, F. (1997b) Lead (Pb) uptake and its effects on membrane integrity and chlorophyll fluorescence in different lichen species. *Environmental and Experimental Botany 37*, 95-105.

Branquinho, C., Matos, P., Vieira, A. R. & Prestello Ramos, M. M. (2011) The relative impact of lichen symbiotic partners to repeated copper uptake. *Environmental and Experimental Botany 72*, 84-92.

Brisdelli, F., Perilli, M., Sellitri, D., Piovano, M., Garbarino, J. A., Nicoletti, M., Bozzi, A., Amicosante, G. & Celenza, G. (2013) Cytotoxic activity and antioxidant capacity of purified lichen metabolites: an *in vitro* study. *Phytotherapy Research 27*, 431-437.

Brodo, I. M. (1973) Substrate ecology. In: V. Ahmadjian & M. E. Hale (Eds.). *The lichens* (pp. 401-441). New York, NY: Academic Press.

Brown, D. H. & Brown, R. M. (1991) Mineral cycling and lichens: the physiological basis. *The Lichenologist 23*, 393-307.

Brunialti, G. & Frati, L. (2014) Bioaccumulation with lichens: the Italian experience. *International Journal of Environmental Studies 71*, 15-26.

Buck, W. R., Harris, R. C., Shaw, A. J., Piercey-Normore, M. D., Tabaee, A., Antonovics, J. & Crone, E. E. (1999) Unusual lichens under electricity pylons on zinc-enriched soils. *The Bryologist 102*, 130-132.

Carreras, H. A. & Pignata, M. L. (2007) Effects of the heavy metals Cu^{2+}, Ni^{2+}, Pb^{2+}, and Zn^{2+} on some physiological parameters of the lichen *Usnea amblyoclada. Ecotoxicology and Environmental Safety 67*, 59-66.

Chettri, M. K., Sawidis, T. & Karataglis, S. (1997) Lichens as a tool for biogeochemical prospecting. *Ecotoxicology and Environmental Safety 38*, 322-325.

Chettri, M. K., Cook, C. M., Vardaka, E., Sawidis, T. & Lanaras, T. (1998) The effect of Cu, Zn and Pb on the chlorophyll content of the lichens *Cladonia convoluta* and *Cladonia rangiformis*. *Environmental and Experimental Botany 39*, 1-10.

Chiarenzelli, J. R., Aspler, L. B., Ozarko, D. L. & Hall, G. E. M. (1997) Heavy metals in lichens, Southern District of Keewatin, Northwest Territories, Canada. *Chemosphere 35*, 1329-1341.

Chisholm, J. E., Jones, G. C. & Purvis, O. W. (1987) Hydrated copper oxalate, mooloite, in lichens. *Mineralogical Magazine 51*, 766-803.

Cuny, D., van Haluwyn, C., Shirali, P., Zerimech, F., Jérôme, L. & Haguenoer, M. (2004a) Cellular impact of metal trace elements in terricolous lichen *Diploschistes muscorum* (Scop.) R. Sant. – Identification of oxidative stress biomarkers. *Water, Air & Soil Pollution 152*, 55-69.

Cuny, D., Denayer, F.-O., de Focault, B., Schumacker, R., Colein, P. & van Haluwyn, C. (2004b) Patterns of metal soil contamination and changes in terrestrial cryptogamic communities. *Environmental Pollution 129*, 289-297.

de los Ríos, A., Wierzchos, J., Sancho, L. G., Green, T. G. A. & Ascaso C. (2005) Ecology of endolithic lichens colonizing granite in continental Antarctica. *The Lichenologist 37*, 383-395.

Edwards, H. G. M., Holder, J. M., Seaward, M. R. D. & Robinson, D. A. (2002) Raman spectroscopic study of lichen-assisted weathering of sandstone outcrops in the High Atlas Mountains, Morocco. *Journal of Raman Spectroscopy 33*, 449-454.

Ellis, C. J. & Yahr, R. (2011) An interdisciplinary review of climate change trends and uncertainties: lichen biodiversity, arctic-alpine ecosystems and habitat loss. In: T. R. Hodkinson, M. B. Jones, S. Waldren & J. A. N. Parnell (Eds.). *Climate change, ecology and systematics* (pp. 457-489). Cambridge: Cambridge University Press.

Favero-Longo, S. E. (2006) *Lichens for bioremediation: an ecological, microbiological and chemical analysis of vegetal and fungal interactions with serpentine substrata. The case of the Balangero and Corio disused asbestos mine (Torino, Italy).* PhD Dissertation: University of Torino, Italy.

Favero-Longo, S. E. & Piervittori, R. (2009) Measuring biodiversity of saxicolous lichens above timberline with reference to environmental factors: the case-study of a Natura 2000 site of western Alps. *Phytocoenologia 39*, 51-78.

Favero-Longo, S. E. & Piervittori, R. (2012) Cultivation of isidia and transplantation of adult thalli of *Xanthoparmelia tinctina* in an abandoned asbestos mine. *The Lichenologist 44*, 840-844.

Favero-Longo, S. E., Isocrono, D. & Piervittori, R. (2004) Lichens and ultramafic rocks: A review. *The Lichenologist 36*, 391-404.

Favero-Longo, S. E., Castelli, D., Salvadori, O., Belluso, E. & Piervittori, R. (2005a) Pedogenetic action of *Lecidea atrobrunnea, Rhizocarpon geographicum* gr. and *Sporastatia testudinea* on serpentinized ultramafic rocks in an alpine environment. *International Biodeterioration & Biodegradation 56*, 17-27.

Favero-Longo, S. E., Turci, F., Tomatis, M., Castelli, D., Bonfante, P., Hochella, M. F., Piervittori, R. & Fubini, B. (2005b) Chrysotile asbestos is progressively converted into a non-fibrous amorphous material by the chelating action of lichen metabolites. *Journal of Environmental Monitoring 7*, 764-766.

Favero-Longo, S. E., Siniscalco, C. & Piervittori, R. (2006) Plant and lichen colonization in an asbestos mine: Spontaneous bio-attenuation limiting fibre airborne dispersion. *Plant Biosystems 140,* 1-17.

Favero-Longo, S. E., Girlanda, M., Honegger, R., Fubini, B. & Piervittori, R. (2007) Interactions of sterile-cultured lichen-forming ascomycetes with asbestos fibres. *Mycological Research 111*, 473-481.

Favero-Longo, S. E., Gazzano, C., Girlanda, M., Castelli, D., Tretiach, M., Baiocchi, C. & Piervittori, R. (2011) Physical and chemical deterioration of silicate and carbonate rocks by meristematic microcolonial fungi and endolithic lichens (Chaetothyriomycetidae). *Geomicrobiology Journal 28*, 732-744.

Favero-Longo, S. E, Turci, F., Castelli, D., Fubini, B. & Piervittori, R. (2013) Lichen deterioration of asbestos and asbestiform minerals of serpentinite rocks in Western Alps. *International Biodeterioration & Biodegradation 84*, 342-350.

Flemming, H.-C. & Wingender, J. (2010) The biofilm matrix. *Nature Reviews - Microbiology 8*, 623-633.

Fryday, A. M. (2005) The genus *Porpidia* in northern and western Europe, with special emphasis on collections from the British Isles. *The Lichenologist 37*, 1-35.

Gadd, G. M. (2006) *Fungi in biogeochemical cycles.* Cambridge: Cambridge University Press.

Gadd, G. M., Rhee, J. R., Stephenson, K. & Wei, Z. (2012) Geomycology: Metals, actinides and biominerals. *Environmental Microbiology Reports 4*, 270-296.

Garty, J., Galun, M. & Hochberg, Y. (1986) The accumulation of metals in *Caloplaca aurantia* growing on concrete roof tiles. *The Lichenologist 18*, 257-263.

Garvie, L. A. J., Knauth, L. P., Bungartz, F., Klonowski, S. & Nash III, T. H. (2008) Life in extreme environments: Survival strategy of the endolithic desert lichen *Verrucaria rubrocincta*. *Naturwissenschaften 95*, 705-712.

Gazzano, C., Favero-Longo, S. E., Matteucci, E., Roccardi, A. & Piervittori, R. (2009) Index of Lichen Potential Biodeteriogenic Activity (LPBA): A tentative tool to evaluate the lichen impact on stonework. *International Biodeterioration & Biodegradation 63*, 836-843.

Gilbert, O. L. (1983) The lichens of Rhum. *Transactions of the Botanical Society of Edinburgh 44*, 139-149.

Gilbert, O. L. (2000) *Lichens*. London: Harper Collins Publishers.

Gilbert, O. L. & James, P. W. (1987) Field meeting on the Lizard Peninsula, Cornwall. *The Lichenologist 19*, 319-334.

Goyal, R. & Seaward, M. R. D. (1981) Metal uptake in terricolous lichens. I. Localization within the thallus. *New Phytologist 89*, 631-645.

Grube, M. (2010) Die hard: Lichens. In: J. Seckbach & M. Grube (Eds.). *Symbioses and stress: Joint ventures in biology* (pp. 509-523). Dordrecht: Springer.

Guglielmin, M., Favero-Longo, S. E., Cannone, N., Piervittori, R. & Strini, A. (2011) Role of lichens in granite weathering in cold and arid environments of continental Antarctica. In: I. P. Martini, H. M. French & A. Pérez Alberti (Eds.). *Ice-marginal and periglacial processes and sediments. Geological Society, London, Special Publications 354*, 195–203.

Guschina, I. A. & Harwood, J. L. (2006) Lead and copper effects on lipid metabolism in cultured lichen photobionts with different phosphorus status. *Phytochemistry 67*, 1731-1739.

Haas, J. R. & Purvis, O. W. (2006) Lichen biogeochemistry. In: G. M. Gadd (Ed.). *Fungi in biogeochemical cycles* (pp. 344-376). Cambridge: Cambridge University Press.

Haas, J. R., Bailey, E. H. & Purvis, O. W. (1998) Bioaccumulation of metals by lichens: uptake of aqueous uranium by *Peltigera membranacea* as a function of time and pH. *American Mineralogist 83*, 1494-1502.

Harris, T. B., Olday, F. C. & Rajakaruna, N. (2007) Lichens of Pine Hill, a peridotite outcrop in eastern North America. *Rhodora 109,* 430-447.

Harrison, S. & Rajakaruna, N. (Eds.) (2011) *Serpentine: The evolution and ecology of a model system*. Berkeley, CA: University of California Press.

Hauck, M. (2008) Metal homeostasis in *Hypogymnia physodes* is controlled by lichen substances. *Environmental Pollution 153*, 304-308.

Hauck, M. & Huneck, S. (2007) Lichen substances affect metal absorption in *Hypogymnia physodes*. *Journal of Chemical Ecology 33*, 219-223.

Hauck, M., Böning, J., Jacob, M., Dittrich, S., Feussner, I. & Leuschner, C. (2013) Lichen substance concentrations in the lichen *Hypogymnia physodes* are correlated with heavy metal concentrations in the substratum. *Environmental and Experimental Botany 85*, 58-63.

Hauck, M., Huneck, S., Elix, J. A. & Paul, A. (2007) Does secondary chemistry enable lichens to grow on iron-rich substrates? *Flora 202*, 471-478.

Hauck, M., Jürgens S.-R., Willenbruch, K., Huneck, S. & Leuschner, C. (2009a) Dissociation and metal-binding characteristics of yellow substances suggest a relationship with site preferences of lichens. *Annals of Botany 103*, 13-22.

Hauck, M., Willenbruch, K. & Leuschner, C. (2009b) Lichen substances prevent lichens from nutrient deficiency. *Journal of Chemical Ecology 35*, 71-73.

Hauck, M., Jürgens S.-R., Huneck, S. & Leuschner, C. (2009c) High acidity tolerance in lichens with fumarprotocetraric, perlatolic or thamnolic acids is correlated with low pK_{a1} values of these lichen substances. *Environmental Pollution 157*, 2776-2780.

Hauck, M., Jürgens S.-R. & Leuschner, C. (2010) Norstictic acid: correlations between its physico-chemical characteristics and ecological preferences of lichens producing this depsidone. *Environmental and Experimental Botany 68*, 309-313.

Honegger, R., Edwards, D. & Axe, L. (2013) The earliest records of internally stratified cyanobacterial and algal lichens from the Lower Devonian of the Welsh Borderland. *New Phytologist 197*, 264-275.

Huneck, S. (2006) Die Flechten der Kupferschieferhalden um Eisleben, Mansfeld und Sangerhausen. *Mitteilungen zur floristichen Kartierung Sachsen-Anhalt 4*, 1-66.

Huneck, S. (1999) The significance of lichens and their metabolites. *Naturwissenschaften 86*, 559-570.

Jones, D. (1988) Lichens and pedogenesis. In: M. Galun (Ed.). *Handbook of lichenology*. Volume III (pp. 109-124). Boca Raton, FL: CRC Press.

Kłos, A., Rajfur, M., Wacławek, M. & Wacławek, W. (2007) Heavy metal sorption in the lichen cationactive layer. *Bioelectrochemistry 71*, 60-65.

Martino, E. & Perotto, S. (2010) Mineral transformations by mycorrhizal fungi. *Geomicrobiology Journal 27*, 609-623.

McIlroy de la Rosa, J. P., Warke, P. A. & Smith, B. J. (2012) Microscale biopitting by the endolithic lichen *Verrucaria baldensis* and its proposed role in mesoscale solution basin development on limestone. *Earth Surface Processes and Landforms 37*, 374-384.

McLean, J., Purvis, O. W., Williamson, B. J. & Bailey, E. H. (1998) Role for lichen melanins in uranium remediation. *Nature 391*, 649-650.

Medeiros, I. D., Fryday, A. M. & Rajakaruna, N. (2014) Additional records and mineralogical data from metal-contaminated sites in Maine. *Rhodora 116,* 323-347.

Mishra, A. & Malik, A. (2013) Recent advances in microbial metal bioaccumulation. *Critical Reviews in Environmental Science and Technology 43*, 1162-1222.

Monnet, F., Bordas, F., Deluchat, V. & Baudu, M. (2006) Toxicity of copper excess on the lichen *Dermatocarpon luridum*: Antioxidant enzyme activities. *Chemosphere 65*, 1806-1813.

Mrak, T., Jeran, Z. & Batič, F. (2010) Arsenic accumulation and thiol status in lichens exposed to As(V) in controlled conditions. *Biometals 23*, 207-219.

Nakajima, H., Yamamoto, Y., Yoshitani, A. & Itoh, K. (2013) Effect of metal stress on photosynthetic pigments in the Cu-hyperaccumulating lichens *Cladonia humilis* and *Stereocaulon japonicum* growing in Cu-polluted sites in Japan. *Ecotoxicology and Environmental Safety 97*, 154-159.

Nash III, T. H. (2008). Nutrients, elemental accumulation, and mineral cycling. In: T. H. Nash III (Ed.). *Lichen biology*. Second edition (pp. 234-251). Cambridge: Cambridge University Press.

Nimis, P. L., Scheidegger, C. & Wolseley, P. A. (2002) *Monitoring with lichens – Monitoring lichens*. Dordrecht: Kluwer Academic Publishers.

Osyczka, P. & Rola, K. (2013) *Cladonia* lichens as the most effective and essential pioneers in strongly contaminated slag dumps. *Central European Journal of Biology 8*, 876-887.

Panaccione, D. G., Sheets, N. L., Miller, S. P. & Cumming, J. R. (2001) Diversity of *Cenococcum geophilum* isolates from serpentine and non-serpentine soils. *Mycologia 93*, 645-652.

Paukov, A. G. (2009) The lichen flora of serpentine outcrops in the Middle Urals of Russia. *Northeastern Naturalist 16 (Special Issue No. 5)*, 341-350.

Paul, A., Hauck, M. & Fritz, E. (2003) Effects of manganese on element distribution and structure in thalli of the epiphytic lichens *Hypogymnia physodes* and *Lecanora conizaeoides*. *Environmental and Experimental Botany 50,* 113-124.

Paul, A., Hauck, M. & Leuschner, C. (2009a) Iron and phosphate uptake in epiphytic and saxicolous lichens differing in their pH requirements. *Environmental and Experimental Botany 67*, 133-138.

Paul, A., Hauck, M. & Leuschner, C. (2009b) Iron and phosphate uptake explains the calcifuge-calcicole behavior of the terricolous lichens *Cladonia furcata* subsp. *furcata* and *C. rangiformis*. *Plant and Soil 319*, 49-56.

Pawlik-Skowrońska, B. & Bačkor, M. (2011) Zn/Pb-tolerant lichens with higher content of secondary metabolites produce less phytochelatins than specimens living in unpolluted habitats. *Environmental and Experimental Botany 72*, 64-70.

Pawlik-Skowrońska, B., Sanità di Toppi, L., Favali, M. A., Fossati, F., Pirszel, J. & Skowroński, T. (2002) Lichens respond to heavy metals by phytochelatin synthesis. *New Phytologist 156*, 95-102.

Pawlik-Skowrońska, B., Purvis, O. W., Pirszel, J. & Skowroński, T. (2006) Cellular mechanisms of Cu-tolerance in the epilithic lichen *Lecanora polytropa* growing at a copper mine. *The Lichenologist 38*, 267-275.

Pawlik-Skowrońska, B., Wójciak, H. & Skowroński, T. (2008) Heavy metal accumulation, resistance and physiological status of some epigeic and epiphytic lichens inhabiting Zn and Pb polluted areas. *Polish Journal of Ecology 56*, 195-207.

Plantlife (2012). *Heavy-metal lichens in Wales. A management guide*. Bangor: Plantlife Cymru.

Puckett, K. J., Nieboer, E., Gorzynski, M. J. & Richardson, D. H. S. (1973) The uptake of metal ions by lichens: a modified ion exchange process. *New Phytologist 72*, 329-342.

Purvis, O. W. (2014) Adaptation and interaction of saxicolous crustose lichens with metals. *Botanical Studies 55*, 23.

Purvis, O. W. & Halls, C. (1996) A review of lichens in metal-enriched environments. *The Lichenologist*, *28*, 571-601.

Purvis, O. W. & Pawlik-Skowrońska, B. (2008) Lichens and metals. In: S. V. Avery, M. Stratford & P. van West (Eds.). *Stress in yeasts and filamentous fungi* (pp. 175-200). Amsterdam: Elsevier.

Purvis, O. W., Gilbert, O. L. & James, P. W. (1985) The influence of copper mineralization on *Acarospora smaragdula*. *The Lichenologist 28*, 571-601.

Purvis, O. W., Elix, J. A., Broomhead, J. A. & Jones, G. C. (1987). The occurrence of copper-norstictic acid in lichens from cupriferous substrata. *The Lichenologist 19*, 193-203.

Purvis, O. W., Elix, J. A. & Gaul, K. L. (1990) The occurrence of copper-psoromic acid in lichens from cupriferous substrata. *The Lichenologist 22*, 345-354.

Purvis, O. W., Bailey, E. H., McLean, J., Kasama, T. & Williamson, B. J. (2004) Uranium biosorption by the lichen *Trapelia involuta* at a uranium mine. *Geomicrobiology Journal 21*, 159-167.

Purvis, O. W., Pawlik-Skowrońska, B., Cressey, G., Jones, G. C., Kearsley, A. & Spratt, J. (2008a) Mineral phases and element composition of the copper hyperaccumulator lichen *Lecanora polytropa*. *Mineralogical Magazine 72*, 607-616.

Purvis, O. W., Kearsley, A., Cressey, G., Crewe, A. T. & Wedin, M. (2008b) Mineralization in rust-coloured *Acarospora*. *Geomicrobiology Journal 25*, 142-148.

Purvis, O. W., Bennett, J. P. & Spratt, J. (2011) Copper localization, elemental content, and thallus colour in the copper hyperaccumulator lichen *Lecanora sierrae* from California. *The Lichenologist 43*, 165-173.

Pyatt, F. B. (1976) Lichen ecology of metal spoil tips: Effect of metal ions on ascospore viability. *The Bryologist 79*, 172–179.

Rajakaruna, N., Harris, T. B., Clayden, S. R., Dibble, A. C. & Olday, F. C. (2011) Lichens of the Callahan Mine, a copper- and zinc-enriched superfund site in Brooksville, Maine, U.S.A. *Rhodora 113*, 1-31.

Rajakaruna, N., Knudsen, K., Fryday, A. M., O'Dell, R. E., Pope, N., Olday, F. C. & Woolhouse, S. (2012) Investigation of the importance of rock chemistry for saxicolous lichen communities of the New Idria serpentinite mass, San Benito County, California, USA. *The Lichenologist 44*, 695-714.

Richardson, D. H. S. (1995) Metal uptake in lichens. *Symbiosis 18*, 119-127.

Rinino, S., Bombardi, V., Giordani, P., Tretiach, M., Crisafulli, P., Monaci, F. & Modenesi, P. (2005) New histochemical techniques for the localization of metal ions in the lichen thallus. *The Lichenologist 37*, 463-466.

Sanità di Toppi, L., Musetti, R., Marabottini, R., Corradi, M. G., Vattuone, Z., Favali, M. A. & Badiani, M. (2004) Responses of *Xanthoria parietina* thalli to environmentally relevant concentrations of hexavalent chromium. *Functional Plant Biology 31*, 329-338.

Sanità di Toppi, L., Musetti, R., Vattuone, Z., Pawlik-Skowrońska, B., Fossati, F., Bertoli, L., Badiani, M. & Favali, M. A. (2005) Cadmium distribution and effects on ultrastructure and chlorophyll status in photobionts and mycobionts of *Xanthoria parietina. Microscopy Research and Technique 66*, 229-238.

Sanità di Toppi, L., Pawlik-Skowrońska, B., Vurro, E., Vattuone, Z., Kalinowska, R., Restivo, F. M., Musetti, R. & Skowroński, T. (2008) First and second line mechanisms of cadmium detoxification in the lichen photobiont *Trebouxia impressa* (Chlorophyta). *Environmental Pollution 151*, 280-286.

Sarret, G., Manceau, A., Cuny, D., Van Haluwyn, C., Déruelle, S., Hazemann, J.-L., Soldo, Y., Eybert-Bérard, L. & Menthonnex, J.-J. (1998) Mechanisms of lichen resistance to metallic pollution. *Environmental Science & Technology 32*, 3325-3330.

Sirois, L., Lutzoni, F. & Grandtner, M. M. (1987) Les lichens sur serpentine et anphibolite du plateau du mont Albert, Gaspésie, Québec. *Canadian Journal of Botany 66*, 851-862.

Smith, B. J. (2009) Weathering processes and forms. In: A. J. Parsons & A. D. Abrahams (Eds.). *Geomorphology of desert environments.* Second edition (pp. 69-100). Berlin: Springer.

Smith, C. W., Aptroot, A., Coppins, B. J., Fletcher, A., Gilbert, O. L., James, P. W. & Wolseley, P. A. (2009) *The lichens of Great Britain and Ireland*. London: British Lichen Society.

Souza-Egipsy, V., Wierzchos, J., Sancho, C., Belmonte, A. & Ascaso, C. (2004) Role of biological soil crust cover in bioweathering and protection of sandstones in a semi-arid landscape (Torrollones de Gabarda, Huesca, Spain). *Earth Surface Processes and Landforms 29*, 1651-1661.

Spagnuolo, V., Zampella, M., Giordano, S. & Adamo, P. (2011) Cytological stress and elemental uptake in moss and lichen exposed in bags in urban area. *Ecology and Environmental Safety, 74* 1434-1443.

Spier, L. & Aptroot A. (2007) *Cladonia rei* is a chemotype and synonym of *Cladonia subulata*. *The Lichenologist 39*, 57-60.

Stepanenko, L. S., Krivoshchekova, O. E. & Skirina, I. F. (2002) Functions of phenolic secondary metabolites in lichens from Far East Russia. *Symbiosis 32*, 119-131.

Syers, J. K. & Iskandar, I. K. (1973) Pedogenetic significance of lichens. In: V. Ahmadjian & M. E. Hale (Eds.). *The lichens* (pp. 225-248). New York, NY: Academic Press.

Tarhanen, S. (1998) Ultrastructural responses of the lichen *Bryoria fuscescens* to simulated acid rain and heavy metal deposition. *Annals of Botany 82*, 735-746.

Tarhanen, S., Metsärinne, S., Holopainen, T. & Oksanen, J. (1999) Membrane permeability response of lichen *Bryoria fuscescens* to wet deposited heavy metals and acid rain. *Environmental Pollution 104*, 121-129.

Tuominen, Y. & Jaakkola, T. (1973) Absorption and accumulation of mineral elements and radioactive nuclides. In: V. Ahmadjian & M. E. Hale (Eds.). *The lichens* (pp. 185-223). New York, NY: Academic Press.

Varrica, D., Aiuppa, A. & Dongarrà, G. (2000) Volcanic and anthropogenic contribution to heavy metal content in lichens from Mt. Etna and Vulcano island (Sicily). *Environmental Pollution 108*, 153-162.

Verrecchia, E. P., Braissant, O. & Cailleau, G. (2006) The oxalate-carbonate pathway in soil carbon storage: the role of fungi and oxalotrophic bacteria. In: G. M. Gadd (Ed.). *Fungi in biogeochemical cycles* (pp. 289-310). Cambridge, UK: Cambridge University Press.

Viles, H. A. (2012) Microbial geomorphology: a neglected link between life and landscape. *Geomorphology 157-158*, 6-16.

Wedin, M., Westberg, M., Crewe, A. T., Tehler, A. & Purvis, O. W. (2009) Species delimitation and evolution of metal bioaccumulation in the lichenized *Acarospora smaragdula* (Ascomycota, Fungi) complex. *Cladistics 25*, 161-172.

Wilson, M. J. & Jones, D. (1984) The occurrence and significance of manganese oxalate in *Pertusaria corallina* (*Lichenes*). *Pedobiologia 26,* 373-379.

Wilson, M. J., Jones, D. & McHardy, W. J. (1981) The weathering of serpentinite by *Lecanora atra*. *The Lichenologist 13,* 167-181.

Winkelmann, G. (2007) Ecology of siderophores with special reference to the fungi. *Biometals 20,* 379-392.

Wirth, V. (1972) Die Silikatflechten Gemainschaften im ausseralpinen Zentraleuropa. *Dissertationes Botanicae 17,* 1-306.

In: Plant Ecology and Evolution in Harsh Environments ISBN: 978-1-63321-955-7
Editors: N. Rajakaruna, R. S. Boyd and T. B. Harris

Chapter 4

EVOLUTION OF SALT TOLERANCE IN ANGIOSPERMS: A PHYLOGENETIC APPROACH

C. Haris Saslis-Lagoudakis**, Camile Moray and Lindell Bromham

Centre for Macroevolution and Macroecology, Research School of Biology, Australian National University, Canberra, Australia

ABSTRACT

Salt tolerance in plants involves a number of physiological, anatomical, and morphological traits that mitigate the effects of osmotic and ionic stress of salinity. Despite the complexity of adaptations to environmental salinity, salt tolerance has arisen independently in many different lineages of flowering plants. In this chapter, we discuss phylogenetic perspectives on the study of salt tolerance. Although few angiosperm species are halophytes, salt tolerance appears to be evolutionarily labile, with relatively numerous instances of independent evolutionary origins. Salt tolerance evolves more often in some lineages than others, which may reflect the fact that it builds upon enabling pre-existing traits in those lineages that facilitate the evolution of tolerance to a wider range of environmental stresses. We describe phylogenetic comparative analyses that explore the evolutionary association between salt tolerance and other related ecophysiological strategies: C_4 photosynthesis, heavy metal tolerance, and alkali-tolerance. However, we discuss the limitations in identifying causality and direction of causality of these associations. We conclude that phylogenetic comparative studies can help us to understand the evolutionary dynamics of salt tolerance in angiosperms and potentially aid identification of shared mechanisms underlying tolerance to several environmental stresses. Despite outstanding challenges in conducting these studies, these comparative approaches have the potential to reveal evolutionary associations between different tolerances to environmental stress.

* Corresponding author: c.h.saslislagoudakis@gmail.com

INTRODUCTION

High levels of environmental salt can occur in most biogeographic regions both at land and sea, inland and at the coast, across a range of ecological settings, such as marshes, mangroves, deserts, salt lakes, and the ocean (Reimond & Queen, 1974). These areas of high salinity are often inhabited by plants that survive despite the twofold difficulties that salt poses to plant functions. Firstly, high salinity decreases plants' ability to take up water. Secondly, high concentrations of Na^+ and Cl^- within the plant impair metabolic processes and decrease photosynthetic efficiency, which in turn negatively affect plant growth (Flowers & Yeo, 1995; Mäser et al., 2002). Salt tolerance refers to various morphological, physiological, and biochemical adaptations that enable plants to survive and complete their life cycles in saline environments. Such plant species are often referred to as halophytes. Salt tolerance adaptations deal with both the osmotic stress, as well as the ionic stress caused by Na^+ (Munns & Tester, 2008; Sahi et al., 2006) and Cl^- (Teakle & Tyerman, 2010). The effects of osmotic stress are reduced by decreasing water loss while maximising water uptake and overall water use efficiency through reduced cellular expansion rates and tight packing of photosynthetic proteins (Munns & Tester, 2008; Reich et al., 1999; Wright et al., 2003). Ion toxicity is mitigated by maintaining a favourable osmotic gradient either through the compartmentalisation of toxic ions in specialised vacuoles and the production of compatible organic solutes (Blumwald, 2000; Glenn et al., 1999; Munns & Tester, 2008), or with the secretion of salt via specialised glands (Munns & Tester, 2008).

Although flowering plants are found in saline environments across the globe, halophytes are comparatively rare. In fact, only an estimated 0.25% of all flowering plant species are known to be halophytes (Flowers et al., 2010). However, halophytes are an extremely diverse group of species. They vary in their degree of salt tolerance, which they achieve through different traits (Flowers & Colmer, 2008), as well as in their life forms, which include annual and perennial herbs, shrubs, and trees (Rozema & Flowers, 2008). Furthermore, halophytes come from a wide range of angiosperm lineages, suggesting that the adaptations involved in salt tolerance have arisen repeatedly during angiosperm evolution (Flowers et al., 1977; Flowers et al., 2010).

Studies of the evolution of salt tolerance can address a series of evolutionary questions. Given that salt tolerance is a complex trait, how frequently has it evolved in angiosperms? How can we unveil the mechanisms underlying this evolutionary transition from glycophytes to halophytes? What conditions have facilitated this transition? This chapter explores what we can infer about evolutionary patterns of salt tolerance from phylogenetic analyses. We compare evolutionary patterns of salt tolerance with other adaptations to environmental stresses, such as modified photosynthetic routes, heavy metals and alkalinity, and discuss how these traits could be associated.

A PHYLOGENETIC VIEW OF SALT TOLERANCE IN ANGIOSPERMS

Analysing the phylogenetic distribution of halophytes is a key step in understanding the evolution of salt tolerance in angiosperms. This topic has been explored at the broad scale (Flowers et al., 1977), and we know that halophytes are found in at least 20 diverse angiosperm orders. Given the relationship between these orders, it is likely that salt tolerance has originated multiple times independently over the evolutionary history of angiosperms (Flowers et al., 2010). Within some groups like the chenopods (Kadereit et al., 2012) and sea grasses (Les et al., 1997), salt tolerance seems to have been gained early in the evolutionary history of those groups and lost rarely, if at all. However, a recent study inferred at least 70 independent origins of salt tolerance in the grass family (Bennett et al., 2013). Most reconstructed origins were near the tips of the phylogeny, and gave rise to few halophyte species (Bennett et al., 2013). This finding suggests that, although salt tolerance is a complex trait, it has evolved multiple times. The difference in the phylogenetic pattern of distribution of halophytes among different families shows that salt tolerance can have different evolutionary patterns in different lineages, which warrants further investigation.

Taxonomic Distribution of Halophytes

In this chapter, we explore the distribution of known halophytes across angiosperm families to investigate if halophytes are distributed randomly across angiosperms. We first recorded the angiosperm families recognised by the APG III and the number of species estimated in each family (http://www.mobot.org/MOBOT/ research/APweb/). We then found the number of known halophyte species in each angiosperm family recognised by the APG III, based on a published list of halophytes (Menzel & Lieth, 2003). This list provides approximately 2,600 names of plant species reported as halophytes in published studies based on ecological, physiological and anecdotal data (Menzel & Lieth, 2003). Although no published list of halophytes will be complete, due to poor knowledge of salt tolerance in certain families and geographical regions, we believe this is the most extensive published database of known halophytes.

We found the accepted name of each halophytic species in that list by searching The Plant List (2010) with the package "taxonstand" (Cayuela et al., 2012) in the program R (R Core Team, 2014). We then allocated each accepted halophyte species to its respective family using the taxonomic name resolution service [TNRS (Boyle et al., 2013)]. Using this method of estimation, we identified 1,653 halophytic species (Table 1). Based on this survey, we found that halophytes are distributed in 117 families and 34 orders. As expected based on previous studies (Flowers et al., 1977; Flowers et al., 2010), many of the families with the highest proportions of halophytes (Table 1) come from the orders Alismatales (including sea grasses) and Caryophyllales (including chenopods). However, there are several families with relatively high proportions of halophytes within the orders Malphigiales, Fagales, and Zygophyllales.

Phylogenetic Distribution of Halophytes

The distribution of halophytes among taxonomic groups shows that halophytes are found in at least a quarter of angiosperm families. However, we cannot assume that the 117 families with halophytes evolved salt tolerance independently. To estimate the number of origins of salt tolerance across angiosperm families, we carried out a phylogenetic investigation.

Table 1. Estimates for number and percentage of halophytes for 117 families recognised by APG III containing at least one known halophyte. Family names, orders, and estimated species numbers were taken from the APG website version 13 (http://www.mobot.org/MOBOT/ research/APweb/). Number of halophytes was derived from the set of accepted halophyte species included in Haloph v2 (Menzel & Lieth, 2003) based on The Plant List (2010), and their respective family affinities according to the Taxonomic Name Resolution Service (Boyle et al., 2013). We highlight families with more than 50 species in bold. Families are ranked alphabetically by the order to which they belong

Order	Family	Number of Species	Number of Halophytes	Percentage of Halophytes
Alismatales	**Alismataceae**	88	1	**1.14**
Alismatales	Cymodoceaceae	16	12	75.00
Alismatales	**Hydrocharitaceae**	116	13	**11.21**
Alismatales	Juncaginaceae	15	1	6.67
Alismatales	Posidoniaceae	9	3	33.33
Alismatales	**Potamogetonaceae**	102	7	**6.86**
Alismatales	Ruppiaceae	6	1	16.67
Alismatales	Zosteraceae	14	14	100.00
Apiales	**Apiaceae**	3780	9	**0.24**
Apiales	**Araliaceae**	1450	3	**0.21**
Arecales	**Arecaceae**	2361	29	**1.23**
Asparagales	**Amaryllidaceae**	1605	4	**0.25**
Asparagales	**Asparagaceae**	2480	9	**0.36**
Asparagales	**Iridaceae**	2025	4	**0.20**
Asparagales	**Xanthorrhoeaceae**	900	1	**0.11**
Asterales	**Asteraceae**	23600	117	**0.50**
Asterales	**Calyceraceae**	60	2	**3.33**
Asterales	**Campanulaceae**	2380	1	**0.04**
Asterales	**Goodeniaceae**	430	4	**0.93**
Brassicales	Bataceae	2	2	100.00
Brassicales	**Brassicaceae**	3710	21	**0.57**
Brassicales	**Capparaceae**	480	1	**0.21**
Brassicales	**Cleomaceae**	300	3	**1.00**
Brassicales	**Resedaceae**	75	1	**1.33**

Order	Family	Number of Species	Number of Halophytes	Percentage of Halophytes
Brassicales	Salvadoraceae	11	1	9.09
Caryophyllales	**Aizoaceae**	2035	36	**1.77**
Caryophyllales	**Amaranthaceae**	2275	393	**17.27**
Caryophyllales	Anacampserotaceae	32	1	3.13
Caryophyllales	Basellaceae	19	2	10.53
Caryophyllales	**Cactaceae**	1866	8	**0.43**
Caryophyllales	**Caryophyllaceae**	2200	17	**0.77**
Caryophyllales	Didiereaceae	16	2	12.50
Caryophyllales	**Frankeniaceae**	90	15	**16.67**
Caryophyllales	Halophytaceae	1	1	100.00
Caryophyllales	**Molluginaceae**	87	1	**1.15**
Caryophyllales	**Montiaceae**	226	3	**1.33**
Caryophyllales	**Nyctaginaceae**	395	7	**1.77**
Caryophyllales	**Plumbaginaceae**	836	28	**3.35**
Caryophyllales	**Polygonaceae**	1110	22	**1.98**
Caryophyllales	**Portulacaceae**	70	5	**7.14**
Caryophyllales	Sarcobataceae	2	1	50.00
Caryophyllales	Stegnospermataceae	3	1	33.33
Caryophyllales	Talinaceae	27	2	7.41
Caryophyllales	**Tamaricaceae**	90	28	**31.11**
Celastrales	**Celastraceae**	1400	8	**0.57**
Commelinales	Pontederiaceae	33	1	3.03
Cucurbitales	**Cucurbitaceae**	960	3	**0.31**
Dilleniales	**Dilleniaceae**	355	1	**0.28**
Ericales	**Ebenaceae**	548	2	**0.36**
Ericales	**Ericaceae**	3995	1	**0.03**
Ericales	**Lecythidaceae**	310	4	**1.29**
Ericales	**Primulaceae**	2590	10	**0.39**
Ericales	**Sapotaceae**	1100	2	**0.18**
Ericales	Tetrameristaceae	5	1	20.00
Fabales	**Fabaceae**	19500	113	**0.58**

Table 1. (Continued)

Order	Family	Number of Species	Number of Halophytes	Percentage of Halophytes
Fabales	Surianaceae	8	1	12.50
Fagales	**Casuarinaceae**	95	9	**9.47**
Gentianales	**Apocynaceae**	4555	20	**0.44**
Gentianales	**Gentianaceae**	1655	4	**0.24**
Gentianales	**Rubiaceae**	13150	4	**0.03**
Lamiales	**Acanthaceae**	4000	13	**0.33**
Lamiales	**Bignoniaceae**	800	8	**1.00**
Lamiales	**Lamiaceae**	7173	5	**0.07**
Lamiales	**Orobanchaceae**	2060	12	**0.58**
Lamiales	**Phrymaceae**	188	3	**1.60**
Lamiales	**Plantaginaceae**	1900	10	**0.53**
Lamiales	**Scrophulariaceae**	1800	10	**0.56**
Lamiales	**Verbenaceae**	918	12	**1.31**
Lamiales	**Boraginaceae**	2755	14	**0.51**
Laurales	**Lauraceae**	2500	2	**0.08**
Magnoliales	**Annonaceae**	2220	1	**0.05**
Malpighiales	**Calophyllaceae**	460	1	**0.22**
Malpighiales	**Chrysobalanaceae**	460	1	**0.22**
Malpighiales	**Clusiaceae**	595	1	**0.17**
Malpighiales	Elatinaceae	35	2	5.71
Malpighiales	**Euphorbiaceae**	5735	17	**0.30**
Malpighiales	**Hypericaceae**	560	1	**0.18**
Malpighiales	**Linaceae**	300	1	**0.33**
Malpighiales	**Phyllanthaceae**	1745	2	**0.11**
Malpighiales	**Putranjivaceae**	210	1	**0.48**
Malpighiales	**Rhizophoraceae**	149	19	**12.75**
Malpighiales	**Salicaceae**	1010	3	**0.30**
Malvales	**Malvaceae**	4225	27	**0.64**
Malvales	**Thymelaeaceae**	891	2	**0.22**
Myrtales	**Combretaceae**	500	10	**2.00**

Order	Family	Number of Species	Number of Halophytes	Percentage of Halophytes
Myrtales	**Lythraceae**	620	9	**1.45**
Myrtales	**Myrtaceae**	4620	22	**0.48**
Myrtales	**Onagraceae**	656	1	**0.15**
Nymphaeales	**Nymphaeaceae**	58	1	**1.72**
Pandanales	**Pandanaceae**	885	10	**1.13**
Picramniales	Picramniaceae	49	1	2.04
Piperales	**Piperaceae**	3615	1	**0.03**
Piperales	Saururaceae	6	1	16.67
Poales	**Bromeliaceae**	1770	2	**0.11**
Poales	**Cyperaceae**	5430	70	**1.29**
Poales	Flagellariaceae	4	1	25.00
Poales	**Juncaceae**	430	14	**3.26**
Poales	**Poaceae**	11160	212	**1.90**
Poales	**Restionaceae**	500	1	**0.20**
Poales	Typhaceae	25	6	24.00
Ranunculales	**Ranunculaceae**	2525	4	**0.16**
Rosales	Elaeagnaceae	45	2	4.44
Rosales	**Moraceae**	1125	5	**0.44**
Rosales	**Rhamnaceae**	925	5	**0.54**
Rosales	**Rosaceae**	2520	6	**0.24**
Rosales	Ulmaceae	35	1	2.86
Sapindales	**Anacardiaceae**	873	2	**0.23**
Sapindales	**Meliaceae**	615	3	**0.49**
Sapindales	Nitrariaceae	16	7	43.75
Sapindales	**Rutaceae**	2070	2	**0.10**
Sapindales	**Simaroubaceae**	110	1	**0.91**
Saxifragales	**Crassulaceae**	1370	1	**0.07**
Saxifragales	Cynomoriaceae	2	1	50.00
Solanales	**Convolvulaceae**	1625	14	**0.86**
Solanales	**Solanaceae**	2460	29	**1.18**
Vitales	**Vitaceae**	850	1	**0.12**
Zygophyllales	**Zygophyllaceae**	285	15	**5.26**

For this investigation we used the largest published tree of angiosperms, which contains over 56,000 angiosperm taxa and was constructed from publicly available sequences for six chloroplast and nuclear DNA markers (Smith et al., 2011). From this phylogenetic tree, we extracted a family-level phylogenetic tree, selecting one representative species for each family, randomly choosing between those species with the most sequence data in the alignment. We did not estimate branch lengths for this analysis, and used a phylogenetic tree with all branch lengths set to 1. We used the same list of halophytes described in the taxonomic analysis above, finding the accepted names of the species in a published list (Menzel & Lieth, 2003) according to The Plant List (2010), and using the TNRS (Boyle et al., 2013) to find family affinities. Using a parsimony ancestral state reconstruction method in Mesquite (Maddison & Maddison, 2006), we estimated that salt tolerance has evolved independently at least 59 times in the family-level phylogeny of angiosperms (Figure 1).

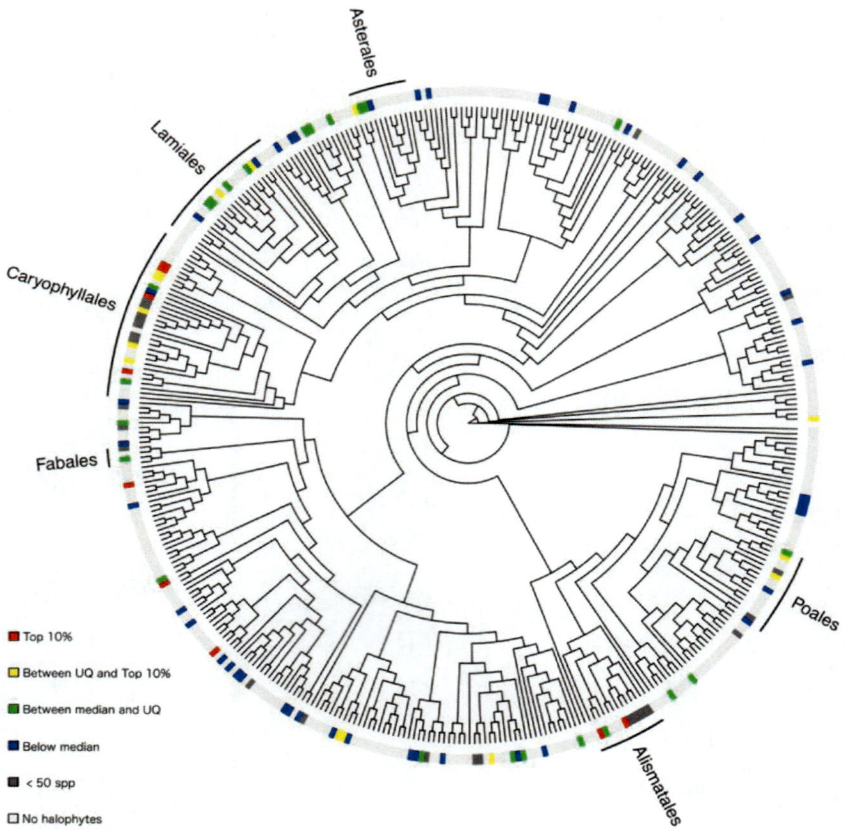

Figure 1. Phylogenetic tree of 401 APG III families extracted from a published angiosperm phylogeny (Smith et al., 2011). For each family, one representative taxon was selected based on maximum alignment length. Coloured tips represent families containing halophytes. Families were ranked by percentage of halophytes (see Table 1). Each tip is coloured based on the relative position of each family based on this ranking. We present families that were placed in top 10% of halophyte proportion (red), between the Upper Quartile (UQ) and the top 10% (yellow), between the median and the UQ (green), and below the median (blue). Families containing fewer than 50 species were not ranked and are shown in dark grey. Orders labelled on the phylogenetic tree contain at least 50 halophytes. The Figure was drawn with the R package "diversitree" (FitzJohn, 2012).

Although these origins are more prominent in some clades than others, they are dispersed on the phylogeny, with many close to the tips of the family-level tree, so are shared by only one or few families (Figure 1). Further, we explored the phylogenetic distribution of halophyte proportion within each family (Table 1). In Figure 1, we coloured the tips of the phylogeny according to halophyte proportion. We found that families with the highest proportion of halophytes do not appear to be clustered on the angiosperm family tree, but they are sometimes related to families with lower proportions of halophytes (Figure 1).

Of course, based on this result only, we cannot claim there have only been 59 origins of salt tolerance during the evolutionary history of angiosperms. Our analysis is at the family level and, although some families rarely lose salt tolerance [*e.g.*, chenopods (Kadereit et al., 2012), sea grasses (Les et al., 1997)], salt tolerance can be gained several times within a single family. For instance, in the Poaceae, which represent a single tip in our phylogenetic tree (Figure 1), we have identified over 70 origins of salt tolerance (Bennett et al., 2013). Therefore, if we expand our analysis to more shallow taxonomic levels, we expect that the number of estimated origins will only increase. However, it is not clear whether the labile evolutionary pattern of salt tolerance in the grasses is common across many families, or whether the factors driving salt tolerance evolution vary widely across lineages.

Future Phylogenetic Investigations of Salt Tolerance in Angiosperms

The findings presented in this chapter highlight areas of research that require further investigation with regard to phylogenetic patterns of salt tolerance in angiosperms. First, future investigations can combine measures of ancestral state reconstruction, phylogenetic clustering and estimations of speciation and extinction to characterise the evolutionary processes underlying the origin of salt tolerance in different lineages.

Second, future investigations can explore the idea that evolutionary innovation, such as ecological transition to harsh environments, can be assisted by pre-existing enabling traits that make adaptation to these environments easier by providing starting points to environmental tolerance (Donoghue, 2005). Therefore, macroevolutionary investigations of the factors that have enabled the adaptive evolution of lineages to tolerate environmental stress can reveal mechanisms facilitating these evolutionary transitions (Edwards & Donoghue, 2013). For instance, the transition from C_3 to C_4 photosynthesis in grasses was facilitated by particular pre-existing foliar anatomy (large bundle sheath cells) in lineages that evolved C_4 photosynthesis (Christin et al., 2013). Similarly, a particular pre-existing three-dimensional leaf venation pattern in certain lineages has facilitated the evolution of succulence in those lineages, through the decoupling of leaf water storage from hydraulic path length (Ogburn & Edwards, 2013). Salt tolerance involves various traits, and not all of these traits are present in all halophytes. Therefore, it would be interesting to identify which of these "salt tolerance traits" have evolved in different salt-tolerant lineages. Can the evolution of different traits in different lineages, such as the presence of salt glands or the production of different compatible solutes, explain the contrasting evolutionary patterns of salt tolerance observed among lineages? In turn, these approaches could shed more light on the processes that allow complex traits to evolve repeatedly, particularly those involved in ecophysiological strategies as adaptations to harsh environments.

Third, it is worth exploring whether ascertainment bias in recording salt tolerance has an effect on the evolutionary patterns that we have observed so far. Salt tolerance is often recorded as a binary trait in evolutionary studies, species scored as being either salt tolerant or salt sensitive. However, salt tolerance can demonstrate considerable intraspecific variation, depending on physiological (Chapter 11; Cuartero et al., 1992; Lowry et al., 2009) and genetic factors (Nguyen et al., 2004; Nguyen et al., 2009; Rajakaruna et al., 2003). Tolerance to salinity would be better expressed on a continuum: not all halophytes can tolerate the same level of salinity and a lot of glycophytes are not completely salt-intolerant. For example, the study of the evolution of succulence in plants has suffered from the same bias as it has been treated as a binary trait (Edwards & Donoghue, 2013), but a quantitative measure of succulence has recently enabled more explicit tests of its evolution (Ogburn & Edwards, 2012). Similarly, exploring salt tolerance as a continuous variable on phylogenies might provide a more complete picture of the evolution of salt tolerance in angiosperms.

Finally, one question that can be explored is whether a better understanding of the evolution of salt tolerance can provide information for breeding salt tolerant crops. Salinity affects up to a tenth of Earth's land surface, and approximately half of all irrigated land around the globe (Ruan et al., 2010), with serious economic costs for agriculture. The majority of crop species do not grow well in saline conditions. Over the past few decades, important advances have been made in breeding salt tolerance into crops (Flowers, 2004; Rozema & Flowers, 2008). However, important challenges remain, in order to develop crop species that can successfully tolerate soil salinity. Because salt tolerance has already evolved in halophytes, it has been suggested that development of salt tolerant crops can be achieved through domestication of halophytes (Flowers, 2004), but many important limitations, such as variable germination and plant diseases of prospective crop halophytes and economic competition with conventional crops, have hindered this approach (Rozema & Flowers, 2008). Two ideas suggest that there could be practical applications to the evolutionary studies of salt tolerance for breeding salt tolerance. First, identifying close relatives of crop species with salt tolerance could help breed tolerance into crop species (Deinlein et al., 2014). Second, if we identify lineages that have more halophytes than expected, we could focus our search for new salt tolerant crops within these lineages, an approach that has been suggested for the search for medicinal plants (Saslis-Lagoudakis et al., 2012). Within those lineages, we could prioritise close relatives of known salt tolerant species for experimental screening of their tolerance (Joseph et al., 2013).

EVOLUTIONARY ASSOCIATIONS WITH OTHER STRESS-TOLERANCE TRAITS

In this chapter so far, we have pointed out that salt tolerance has had multiple, recent independent origins across many angiosperm families. This evolutionary pattern reveals a paradox: salt tolerance is a rare and complex trait, but it is an evolutionarily labile one, which appears to be gained often. One possible explanation for the evolutionary lability of salt tolerance is that it builds upon a more general suite of stress-tolerance traits that enable lineages to adapt to a wide range of environmental challenges (Chapter 9; Chapin et al., 1993). This would mean that salt tolerance is more easily acquired in lineages with particular background enabling traits as starting points (Bromham & Bennett, 2014), which might be linked to ecophysiological responses to aridity (Osborne & Freckleton, 2009; Rozema & Schat, 2013), and enhance the capacity of those lineages to evolve salt tolerance. One way to investigate this hypothesis is to employ a comparative phylogenetic approach (Christin et al., 2010; Hancock & Edwards, 2014). The strength of this approach is that it can place tolerance to several stresses within a unifying comparative framework. For any group of plants, we can plot the taxa with tolerance to different stresses on a phylogeny and then estimate the relatedness of taxa with resistance to these stresses. The detection of these associations on phylogenetic trees is the first step towards understanding their underlying causes. In the following paragraphs, we discuss the putative association of other types of stress tolerance with the evolution of salt tolerance. In particular, we focus on three ecophysiological traits: C_4 photosynthesis, heavy metal tolerance, and alkali-tolerance.

C_4 Photosynthesis

C_4 photosynthesis is a modified photosynthetic pathway that increases a plant's efficiency of carbon fixation, thus reducing photorespiration and allowing higher water-use efficiency. In turn, C_4 ecophysiology conveys advantages in harsh environmental conditions, including salinity (Christin et al., 2013; Sage, 2004; Sage & Monson, 1999; Sage et al., 2012). In particular, the greater water-use efficiency in C_4 plants can not only reduce the effect of osmotic stress, but also mitigate the effects of ionic stress of salinity, because C_4 plants utilise less water, and therefore intake less salt, per growth unit (Sage, 2001). Many C_4 plants are known to be halophytes (Aronson, 1989; Dajic, 2006; Sage & Monson, 1999) and plants with C_4 photosynthesis are often overrepresented in saline environments (Eallonardo et al., 2013; Feldman et al., 2008). For example, plants with C_4 photosynthesis are dominant in saline Argentinian grasslands (Feldman et al., 2008). Like salt tolerance, C_4 photosynthesis is surprisingly evolutionary labile for a complex trait: C_4 photosynthesis has evolved independently over 60 times in angiosperms (Sage et al., 2012), including several times within the grass family (Edwards & Smith, 2010; Grass Phylogeny Working Group II, 2012). C_4 plant species are found scattered across the phylogeny of angiosperms (Edwards & Ogburn, 2012; Sage et al., 2011).

Previous studies have pointed out the putative association between C_4 photosynthesis and salt tolerance, as part of a wider stress syndrome to aridity (Edwards & Smith, 2010; Sage, 2001), and it has been suggested that the evolution of C_4 in grasses enabled the colonisation of arid and saline habitats (Osborne & Freckleton, 2009). Other studies have explicitly tested for this association in a phylogenetic context. For example, in the chenopods, higher rates of gain of C_4 were inferred in salt tolerant lineages, which was attributed to shared adaptations between C_4 photosynthesis and salt tolerance as part of a wider drought tolerance syndrome (Kadereit et al., 2012). Also, salt tolerance and C_4 photosynthesis are significantly associated in the grass family. Using a phylogeny of the grass family, and mapping halophytes and C_4 plants on that phylogeny, we have recently shown that salt tolerance is more likely to occur in C_4 than C_3 grass lineages (Bromham & Bennett, 2014).

However, although these studies have found a significant correlation on the phylogeny between C_4 photosynthesis and salt tolerance, it is not clear whether these traits are directly or indirectly linked. This association may suggest that some traits involved in C_4 photosynthesis also directly convey tolerance to salinity or facilitate its evolution. On the other hand, it could be indirect factors that drive this association, such as biogeography. C_4 photosynthesis facilitated the expansion of certain lineages into arid and saline habitats (Osborne & Freckleton, 2009; Strömberg, 2011), and this exposure may have indirectly rendered these lineages more likely to evolve salt tolerance. Further, even if these two traits have a direct causal connection, inferring the direction of this connection is challenging (Bromham & Bennett, 2014).

Heavy Metal Tolerance

Another trait that has been suggested to be linked to salt tolerance is the ability to tolerate and accumulate heavy metals (Ghnaya et al., 2007; Manousaki & Kalogerakis, 2010; Rozema & Schat, 2013). Like salt, high concentrations of several heavy metals pose a lethal oxidative threat to most plants. However, some plant species—called heavy metal hyperaccumulators—are not only able to survive and reproduce in soils with high concentration of heavy metals, but are also able to accumulate them within their tissues. Both hyperaccumulators and halophytes face osmotic and oxidative stress and may use similar mechanisms to alleviate these stresses. One of the mechanisms employed by metal hyperaccumulators to deal with heavy metal stress is the production of compatible solutes, which serve as osmoprotectants, shielding cellular structures from damage and maintaining osmotic regulation (Lefevre et al., 2009; Schat et al., 1997; Sharma & Dietz, 2006). Halophytes also produce osmoprotectants (compatible solutes) to mitigate the effect of ion toxicity caused by salinity (Blumwald, 2000; Glenn et al., 1999; Munns & Tester, 2008). Some osmoprotectants, such as the compatible solute proline, are effective not only against salt (Flowers et al., 1977; Stewart & Lee, 1974), but can also protect plants living in soils with heavy metals like cadmium (Cd) and copper (Cu: Schat et al., 1997; Sharma & Dietz, 2006).

Because both salt and heavy metal tolerance require mechanisms to combat osmotic and oxidative stress, we might expect that species that can tolerate heavy metals might also be better at tolerating salt (Ghnaya et al., 2007; Jordan et al., 2002; Lefevre et al., 2009; Rozema & Schat, 2013). For example, some groups that contain halophytes (Table 1) are also known to contain metal hyperaccumulators [*e.g.,* Brassicaceae (Rascio & Navari-Izzo, 2011), Asteraceae (Prasad & de Olivera Freitas, 2003)]. Future phylogenetic investigations can test whether halophytes and hyperaccumulators are found in closely related lineages, even when species that convey both types of stress resistance are discounted from the analysis.

Alkali Tolerance

There are several reasons to believe that tolerance to salinity and alkalinity might be associated. First, salinity and sodicity often co-occur on the landscape (Rengasamy, 2006), so plants occupying environments that are saline and sodic need to adapt to both salinity and alkalinity (Bui, 2013). Second, these two stresses affect plants in similar ways. Both salinity and sodicity affect water use efficiency. Water uptake is hindered in sodic soils because of poor soil structure, and water loss is exacerbated because alkalinity interferes with stomatal closure through the accumulation of sodium ions (Bernstein, 1975). Also, both salinity and alkalinity adversely affect photosynthesis (Nishiuchi et al., 2010; Yang et al., 2008). Because of these common challenges that plants face in alkaline and saline environments, similar mechanisms might be employed in dealing with both stresses (Bromham et al., 2013).

We applied a macroevolutionary comparative approach to investigate the phylogenetic association between lineages that occur in high salinity and the ones found in high alkalinity. We collated occurrence data for Australian grasses and, using geochemical modelling, we estimated the likely levels of salinity and pH that species are exposed to in their natural distributions. Using these geochemical predictions, we identified species that are likely to occur in relatively higher salinity and alkalinity. Using a taxonomic randomisation, we found that there are significantly more species than expected by chance occurring in both high predicted salinity and alkalinity (Saslis-Lagoudakis et al., submitted). Additionally, we used a complete genus-level phylogeny of grasses (Bouchenak-Khelladi et al., 2010) and plotted species occurring in high predicted salinity and alkalinity conditions on that tree. We found a non-random phylogenetic association between lineages containing the sets of taxa that occur in high predicted salinity and those predicted in high predicted alkalinity. One possible explanation of the association of species found in high predicted salinity with those from high predicted alkalinity is that some lineages have pre-existing enabling traits that makes it easier for them to adapt to either saline or alkaline environments, or both. However there are other factors, such as biogeography and the spatial overlap of salinity and alkalinity (Rengasamy, 2006) that could drive this association. Further, it is likely that the scale of geochemical predictions in that study does not reflect the variation of geochemistry in the landscape.

FUTURE INVESTIGATIONS OF ASSOCIATIONS BETWEEN ECOPHYSIOLOGICAL STRATEGIES

The findings outlined in this section of the chapter suggest that salt tolerance is associated with other types of stress tolerance. However, further investigations can help us tease apart these associations. Mapping traits that potentially underlie environmental stress tolerance on phylogenies can help us explicitly test for the presence of shared mechanisms used in several abiotic tolerances (Fujita et al., 2006) within an evolutionary context. For example, investigating the evolution of a photosynthetic enzyme, a recent study showed that it was recruited in the evolution of C_4 and CAM metabolism within lineages of Caryophyllales (Christin et al., 2014). Because many of the mechanisms supporting salt tolerance are involved in osmoregulation, it would be logical to assume that these mechanisms could also be involved in other types of water-use efficiency stresses, like drought and flooding (Colmer & Voesenek, 2009; Munns & Tester, 2008; Osborne & Sack, 2012; Rozema & Schat, 2013). Therefore, further studies can explore specific mechanisms involved in different water-use efficiency stresses (drought, flooding, frost, salt tolerance) in a comparative phylogenetic framework. The presence of shared mechanisms recruited in the evolution of both salt tolerance and other types of water-use efficiency stresses might partially explain the association among them.

CONCLUSION

Salt tolerance is rare, as only 0.25% of all flowering plants are halophytes. It is a complex trait, involving several physiological, anatomical, and morphological adaptations. However, salt tolerance has been gained many times independently in several angiosperms lineages. We reveal at least 59 independent origins of salt tolerance at the family level, but the total number of origins in angiosperms is certainly much higher than this figure. Further, we find that the evolutionary dynamics of salt tolerance differ across lineages.

What could explain these different evolutionary dynamics in different lineages? Salt tolerance is associated with other ecophysiological strategies related to adaptations to harsh environments, such as C_4 photosynthesis, alkali, and heavy metal tolerance. One explanation for the association of salt tolerance with other ecophysiological strategies related to environmental stress is that salt tolerance builds upon a more general suite of stress-tolerance traits that provide starting points to environmental stress tolerance. Therefore, it is possible that salt tolerance is more easily acquired in lineages with particular enabling traits as starting points, and that these enabling traits also enhance the capacity of those lineages to evolve other ecophysiological strategies. However, other factors, such as biogeography, could also shape these associations. Further research is needed to explore these evolutionary patterns. Particularly the phylogenetic distribution of different anatomical and physiological traits can shed light on which traits can act as enabling traits in the evolution of salt tolerance and other types of environmental stress tolerance.

The availability of molecular phylogenies for many taxa allows us to test evolutionary hypotheses across several lineages in order to deduce general patterns of the evolution of environmental stress tolerance in angiosperms. Further, our knowledge of halophytes and plant species tolerant to other types of environmental stresses, although not complete, is richer nowadays than ever before. Additionally, methods such as geochemical modelling are being explored to test if they can predict which plants are tolerant to different types of environmental stress based on species distributions. Finally, the development of comparative phylogenetic methods in the last decade provides suitable tools to test for macroevolutionary hypotheses. All these factors make this is an exciting time to perform research in the evolution of salt tolerance, as well as other types of environmental stress.

REFERENCES

Aronson, J. (1989) *HALOPH: a data base of salt tolerant plants of the world*. Tucson, AZ: Office of Arid Land Studies, University of Arizona.

Bennett, T. H., Flowers, T. J. & Bromham, L. (2013) Repeated evolution of salt-tolerance in grasses. *Biology Letters 9*, 20130029.

Bernstein, L. (1975) Effects of salinity and sodicity on plant growth. *Annual Review of Phytopathology 13*, 295-312.

Blumwald, E. (2000) Sodium transport and salt tolerance in plants. *Current Opinion in Cell Biology 12*, 431-434.

Bouchenak-Khelladi, Y., Verboom, G. A., Savolainen, V. & Hodkinson, T. R. (2010) Biogeography of the grasses (Poaceae): A phylogenetic approach to reveal evolutionary history in geographical space and geological time. *Botanical Journal of the Linnean Society 162*, 543-557.

Boyle, B., Hopkins, N., Lu, Z., Raygoza Garay, J. A., Mozzherin, D., Rees, T., Matasci, N., Narro, M., Piel, W., Mckay, S., Lowry, S., Freeland, C., Peet, R. & Enquist, B. (2013) The taxonomic name resolution service: An online tool for automated standardization of plant names. *BMC Bioinformatics 14*, 16.

Bromham, L. & Bennett, T. H. (2014) Salt tolerance evolves more frequently in C4 grass lineages. *Journal of Evolutionary Biology 27*, 653-659.

Bromham, L., Saslis-Lagoudakis, C. H., Bennett, T. H. & Flowers, T. J. (2013) Soil alkalinity and salt tolerance: Adapting to multiple stresses. *Biology Letters 9*, 20130642.

Bui, E. N. (2013) Soil salinity: A neglected factor in plant ecology and biogeography. *Journal of Arid Environments 92*, 14-25.

Cayuela, L., Granzow-de la Cerda, Í., Albuquerque, F. S. & Golicher, D. J. (2012) taxonstand: An R package for species names standardisation in vegetation databases. *Methods in Ecology and Evolution 3*, 1078-1083.

Chapin, F. S. III, Autumn, K. & Pugnaire, F. (1993) Evolution of suites of traits in response to environmental stress. *The American Naturalist 142*, S78-S92.

Christin, P.-A., Freckleton, R. P. & Osborne, C. P. (2010) Can phylogenetics identify C4 origins and reversals? *Trends in Ecology & Evolution 25*, 403-409.

Christin, P.-A., Osborne, C. P., Chatelet, D. S., Columbus, J. T., Besnard, G., Hodkinson, T. R., Garrison, L. M., Vorontsova, M. S. & Edwards, E. J. (2013) Anatomical enablers and the evolution of C4 photosynthesis in grasses. *Proceedings of the National Academy of Sciences (USA) 110*, 1381-1386.

Christin, P.-A., Arakaki, M., Osborne, C. P., Bräutigam, A., Sage, R. F., Hibberd, J. M., Kelly, S., Covshoff, S., Wong, G. K.-S., Hancock, L. & Edwards, E. J. (2014) Shared origins of a key enzyme during the evolution of C4 and CAM metabolism. *Journal of Experimental Botany 65*, 3609-3621.

Colmer, T. D. & Voesenek, L. A. C. J. (2009) Flooding tolerance: Suites of plant traits in variable environments. *Functional Plant Biology 36*, 665-681.

Cuartero, J., Yeo, A. R. & Flowers, T. J. (1992) Selection of donors for salt-tolerance in tomato using physiological traits. *New Phytologist 121*, 63-69.

Dajic, Z. (2006) Salt stress. In: K. V. Madhava Rao, A. S. Raghavendra & K. Janardhan Reddy (Eds.). *Physiology and molecular biology of stress tolerance in plants* (pp. 41-82). Dordrecht: Springer.

Deinlein, U., Stephan, A. B., Horie, T., Luo, W., Xu, G. & Schroeder, J. I. (2014) Plant salt-tolerance mechanisms. *Trends in Plant Science 19*, 371–379.

Donoghue, M. J. (2005) Key innovations, convergence, and success: Macroevolutionary lessons from plant phylogeny. *Paleobiology 31*, 77-93.

Eallonardo, A. S., Leopold, D. J., Fridley, J. D. & Stella, J. C. (2013) Salinity tolerance and the decoupling of resource axis plant traits. *Journal of Vegetation Science 24*, 365-374.

Edwards, E. J. & Donoghue, M. J. (2013) Is it easy to move and easy to evolve? Evolutionary accessibility and adaptation. *Journal of Experimental Botany*, DOI:10.1093/jxb/ert220.

Edwards, E. J. & Ogburn, R. M. (2012) Angiosperm responses to a low-CO2 world: CAM and C4 photosynthesis as parallel evolutionary trajectories. *International Journal of Plant Sciences 173*, 724-733.

Edwards, E. J. & Smith, S. A. (2010) Phylogenetic analyses reveal the shady history of C4 grasses. *Proceedings of the National Academy of Sciences (USA) 107*, 2532-2537.

Feldman, S. R., Bisaro, V., Biani, N. B. & Prado, D. E. (2008) Soil salinity determines the relative abundance of C3/C4 species in Argentinean grasslands. *Global Ecology and Biogeography 17*, 708-714.

FitzJohn, R. G. (2012) Diversitree: Comparative phylogenetic analyses of diversification in R. *Methods in Ecology and Evolution 3*, 1084-1092.

Flowers, T. J. (2004) Improving crop salt tolerance. *Journal of Experimental Botany 55*, 307-319.

Flowers, T. J. & Colmer, T. D. (2008) Salinity tolerance in halophytes. *New Phytologist 179*, 945-963.

Flowers, T. J. & Yeo, A. R. (1995) Breeding for salinity resistance in crop plants: where next? *Australian Journal of Plant Physiology 22*, 875-884.

Flowers, T. J., Troke, P. F. & Yeo, A. R. (1977) The mechanism of salt tolerance in halophytes. *Annual Review of Plant Physiology 28*, 89-121.

Flowers, T. J., Galal, H. K. & Bromham, L. (2010) Evolution of halophytes: Multiple origins of salt tolerance in land plants. *Functional Plant Biology 37*, 604-612.

Fujita, M., Fujita, Y., Noutoshi, Y., Takahashi, F., Narusaka, Y., Yamaguchi-Shinozaki, K. & Shinozaki, K. (2006) Crosstalk between abiotic and biotic stress responses: A current view from the points of convergence in the stress signaling networks. *Current Opinion in Plant Biology 9*, 436-442.

Ghnaya, T., Slama, I., Messedi, D., Grignon, C., Ghorbel, M. H. & Abdelly, C. (2007) Effects of Cd2+ on K+, Ca2+ and N uptake in two halophytes *Sesuvium portulacastrum* and *Mesembryanthemum crystallinum*: Consequences on growth. *Chemosphere 67*, 72-79.

Glenn, E. P., Brown, J. J. & Blumwald, E. (1999) Salt tolerance and crop potential of halophytes. *Critical Reviews in Plant Sciences 18*, 227-255.

Grass Phylogeny Working Group II (2012) New grass phylogeny resolves deep evolutionary relationships and discovers C4 origins. *New Phytologist 193*, 304-312.

Hancock, L. & Edwards, E. J. (2014) Phylogeny and the inference of evolutionary trajectories. *Journal of Experimental Botany*, DOI: 10.1093/jxb/eru118

Jordan, F. L., Robin-Abbott, M., Maier, R. M. & Glenn, E. P. (2002) A comparison of chelator-facilitated metal uptake by a halophyte and a glycophyte. *Environmental Toxicology and Chemistry 21*, 2698-2704.

Joseph, S., Bhave, M., Miller, J. T. & Murphy, D. J. (2013) Rapid identification of *Acacia* species with potential salt tolerance by using nuclear ribosomal DNA markers. *Sustainable Agriculture Research 2*, 77-86.

Kadereit, G., Ackerly, D. & Pirie, M. D. (2012) A broader model for C4 photosynthesis evolution in plants inferred from the goosefoot family (Chenopodiaceae s.s.). *Proceedings of the Royal Society Series B: Biological Sciences 279*, 3304-3311.

Lefevre, I., Marchal, G., Meerts, P., Correal, E. & Lutts, S. (2009) Chloride salinity reduces cadmium accumulation by the Mediterranean halophyte species *Atriplex halimus* L. *Environmental and Experimental Botany 65*, 142–152.

Les, D., Cleland, M. & Waycott, M. (1997) Phylogenetic studies in Alismatidae II: Evolution of marine angiosperms (seagrasses) and hydrophily. *Systematic Botany 22*, 443-463.

Lowry, D. B., Hall, M. C., Salt, D. E. & Willis, J. H. (2009) Genetic and physiological basis of adaptive salt tolerance divergence between coastal and inland *Mimulus guttatus*. *New Phytologist 183*, 776-788.

Maddison, W. P. & Maddison, D. R. (2006) Mesquite: a modular system for evolutionary analysis. Available at http://mesquiteproject.org.

Manousaki, E. & Kalogerakis, N. (2010) Halophytes present new opportunities in phytoremediation of heavy metals and saline soils. *Industrial & Engineering Chemistry Research 50*, 656-660.

Mäser, P., Eckelman, B., Vaidyanathan, R., Horie, T., Fairbairn, D. J., Kubo, M., Yamagami, M., Yamaguchi, K., Nishimura, M., Uozumi, N., Robertson, W., Sussman, M. R. & Schroeder, J. I. (2002) Altered shoot/root Na+ distribution and bifurcating salt sensitivity in *Arabidopsis* by genetic disruption of the Na+ transporter AtHKT1. *FEBS Letters 531*, 157-161.

Menzel, U. & Lieth, H. (2003) Halophyte database version 2.0. In: H. Lieth & M. Mochtchenko (Eds.). *Cash crop halophytes: Recent studies* (pp. 221–250). Dordrecht: Kluwer Academic Publishers.

Munns, R. & Tester, M. (2008) Mechanisms of salinity tolerance. *Annual Review of Plant Biology 59*, 651-681.

Nguyen, N. T., Moghaieb, R. E. A., Saneoka, H. & Fujita, K. (2004) RAPD markers associated with salt tolerance in *Acacia auriculiformis* and *Acacia mangium*. *Plant Science 167*, 797-805.

Nguyen, N. T., Saneoka, H., Suwa, R. & Fujita, K. (2009) The interaction among provenances of *Melaleuca leucadendra* (Weeping Paperbark), salt, and aluminum. *Forest Science 55*, 443-454.

Nishiuchi, S., Fujihara, K., Liu, S. & Takano, T. (2010) Analysis of expressed sequence tags from a NaHCO3-treated alkali-tolerant plant, *Chloris virgata*. *Plant Physiology and Biochemistry 48*, 247-255.

Ogburn, R. M. & Edwards, E. J. (2012) Quantifying succulence: A rapid, physiologically meaningful metric of plant water storage. *Plant, Cell & Environment 35*, 1533-1542.

Ogburn, R. M. & Edwards, E. J. (2013) Repeated origin of three-dimensional leaf venation releases constraints on the evolution of succulence in plants. *Current Biology 23*, 722-726.

Osborne, C. P. & Freckleton, R. P. (2009) Ecological selection pressures for C4 photosynthesis in the grasses. *Proceedings of the Royal Society Series B: Biological Sciences 276*, 1753-1760.

Osborne, C. P. & Sack, L. (2012) Evolution of C4 plants: A new hypothesis for an interaction of CO2 and water relations mediated by plant hydraulics. *Philosophical Transactions of the Royal Society B: Biological Sciences 367*, 583-600.

Prasad, M. N. V. & de Oliveira Freitas, H. M. (2003) Metal hyperaccumulation in plants - Biodiversity prospecting for phytoremediation technology. *Electronic Journal of Biotechnology 6*, 285-321.

R Core Team (2014) *R: A language and environment for statistical computing*. Vienna: R Foundation for Statistical Computing.

Rajakaruna, N., Siddiqi, M. Y., Whitton, J., Bohm, B. A. & Glass, A. D. M. (2003) Differential responses to Na+/K+ and Ca2+/Mg2+ in two edaphic races of the *Lasthenia californica* (Asteraceae) complex: A case for parallel evolution of physiological traits. *New Phytologist 157*, 93-103.

Rascio, N. & Navari-Izzo, F. (2011) Heavy metal hyperaccumulating plants: How and why do they do it? And what makes them so interesting? *Plant Science 180*, 169-181.

Reich, P. B., Ellsworth, D. S., Walters, M. B., Vose, J. M., Gresham, C., Volin, J. C. & Bowman, W. D. (1999) Generality of leaf trait relationships: A test across six biomes. *Ecology 80*, 1955-1969.

Reimond, R. J. & Queen, W. H. (1974) *Ecology of halophytes*. New York, NY: Academic Press.

Rengasamy, P. (2006) World salinization with emphasis on Australia. *Journal of Experimental Botany 57*, 1017-1023.

Rozema, J. & Flowers, T. (2008) Crops for a salinized world. *Science 322*, 1478-1480.

Rozema, J. & Schat, H. (2013) Salt tolerance of halophytes, research questions reviewed in the perspective of saline agriculture. *Environmental and Experimental Botany 92*, 83-95.

Ruan, C.-J., da Silva, J. A. T., Mopper, S., Qin, P. & Lutts, S. (2010) Halophyte improvement for a salinized world. *Critical Reviews in Plant Sciences 29*, 329-359.

Sage, R. F. (2001) Environmental and evolutionary preconditions for the origin and diversification of the C4 photosynthetic syndrome. *Plant Biology 3*, 202-213.

Sage, R. F. (2004) The evolution of C4 photosynthesis. *New Phytologist 161*, 341-370.

Sage, R. F. & Monson, R. K. (1999) *C4 plant biology*. San Diego, CA: Academic Press.

Sage, R. F., Christin, P.-A. & Edwards, E. J. (2011) The C4 plant lineages of planet Earth. *Journal of Experimental Botany 62*, 3155-3169.

Sage, R. F., Sage, T. L. & Kocacinar, F. (2012) Photorespiration and the evolution of C4 Photosynthesis. *Annual Review of Plànt Biology 63*, 19-47.

Sahi, C., Singh, A., Blumwald, E. & Grover, A. (2006) Beyond osmolytes and transporters: novel plant salt-stress tolerance-related genes from transcriptional profiling data. *Physiologia Plantarum 127*, 1-9.

Saslis-Lagoudakis, C. H., Savolainen, V., Williamson, E. M., Forest, F., Wagstaff, S. J., Baral, S. R., Watson, M. F., Pendry, C. A. & Hawkins, J. A. (2012) Phylogenies reveal predictive power of traditional medicine in bioprospecting. *Proceedings of the National Academy of Sciences USA 109*, 15835–15840.

Saslis-Lagoudakis, C. H., Bui, E., Moray, C. & Bromham, L. (submitted) Evolutionary patterns of salinity and alkalinity tolerance in Australian grasses: A phylogenetic analysis combining distribution data and geochemical modelling. *Annals of Botany*.

Schat, H., Sharma, S. S. & Vooijs, R. (1997) Heavy metal-induced accumulation of free proline in a metal-tolerant and a nontolerant ecotype of *Silene vulgaris*. *Physiologia Plantarum 101*, 477-482.

Sharma, S. S. & Dietz, K.-J. (2006) The significance of amino acids and amino acid-derived molecules in plant responses and adaptation to heavy metal stress. *Journal of Experimental Botany 57*, 711-726.

Smith, S. A., Beaulieu, J. M., Stamatakis, A. & Donoghue, M. J. (2011) Understanding angiosperm diversification using small and large phylogenetic trees. *American Journal of Botany 98*, 404-414.

Stewart, G. R. & Lee, J. A. (1974) The role of proline accumulation in halophytes. *Planta 120*, 279-289.

Strömberg, C. A. E. (2011) Evolution of grasses and grassland ecosystems. *Annual Review of Earth and Planetary Sciences 39*, 517-544.

Teakle, N. L. & Tyerman, S. D. (2010) Mechanisms of Cl- transport contributing to salt tolerance. *Plant, Cell & Environment 33*, 566-589.

The Plant List (2010) http://www.theplantlist.com [database on the Internet].

Wright, I. J., Reich, P. B. & Westoby, M. (2003) Least-cost input mixtures of water and nitrogen for photosynthesis. *The American Naturalist 161*, 98-111.

Yang, C., Shi, D. & Wang, D. (2008) Comparative effects of salt and alkali stresses on growth, osmotic adjustment and ionic balance of an alkali-resistant halophyte *Suaeda glauca* (Bge.). *Plant Growth Regulation 56*, 179-190.

In: Plant Ecology and Evolution in Harsh Environments ISBN: 978-1-63321-955-7
Editors: N. Rajakaruna, R. S. Boyd and T. B. Harris

Chapter 5

THE ECOLOGY, ASSEMBLY AND EVOLUTION OF GYPSOPHILE FLORAS

Michael J. Moore[1*], Juan F. Mota[2], Norman A. Douglas[1], Hilda Flores Olvera[3] and Helga Ochoterena[3]

[1]Department of Biology, Oberlin College, Oberlin, OH, USA
[2]Departamento de Biología y Geología, Universidad de Almería, Almería, España
[3]Instituto de Biología, Universidad Nacional Autónoma de México, México, DF, México

ABSTRACT

Gypsum ($CaSO_4 \cdot 2H_2O$) exposures and gypseous soils occupy over 100 million ha worldwide, primarily in arid and semiarid regions, with particularly large areas of surface gypsum in southwestern Asia, the Mediterranean region, the Horn of Africa and southwestern North America. Each of these areas hosts a diverse assemblage of gypsum endemic plant taxa, known as gypsophiles. Although plant biologists have been interested in the causes of gypsophily for well over a century, it has only been over the past few decades that gypsophile floras have received sustained ecological and evolutionary study. Recent work, principally in Spain, has revealed that both physical (*e.g.*, gypsum crusts, soil porosity) and chemical (*e.g.*, high Ca and S, low cation exchange capacity) factors may control community structure on highly gypseous substrates. Plant-fungal interactions may also play a key role in plant establishment on gypsum, although few studies have examined this subject. Molecular systematic and population genetic studies over the past two decades have revealed several key similarities in the assembly and evolution of gypsophile floras and taxa. These studies imply that gypsophile lineages have frequently appeared multiple times within clades that are ancestrally tolerant of gypsum, that speciation has been common in the most widespread lineages of gypsophiles, and that most gypsophile lineages first appeared no earlier than the latest Miocene. Population genetic studies have revealed generally higher levels of among-population genetic differentiation and isolation-by-distance within gypsophile taxa, in line with expectations for taxa that are restricted to substrate archipelagoes such as gypsum. Despite these advances in our understanding of gypsophily, gypsum floras remain much more poorly

* Corresponding author: michael.moore@oberlin.edu.

studied compared to other important edaphic endemic communities, such as serpentine and halophilic floras, highlighting the need for additional work.

INTRODUCTION

Surface gypsum ($CaSO_4 \cdot 2H_2O$) deposits and gypsisols occur worldwide in arid and semi-arid regions, covering 100-207 million ha worldwide (Eswaran & Gong, 1991; Herrero, 2004; Herrero & Porta, 2000; Verheye & Boyadgiev, 1997). For example, large areas of exposed gypsum characterize parts of the Horn of Africa region (*e.g.*, Ethiopia, Somalia), North Africa (*e.g.*, Tunisia, Algeria), western Asia (*e.g.*, Iran, Iraq, Turkey), Australia, eastern Spain, and the Chihuahuan Desert region of North America (Escudero et al., 2014; Food and Agriculture Organization, 1998; Merlo et al., 2011). In contrast to most NaCl-rich soils, which are mainly concentrated along sea shores or less commonly in interior deserts and endorheic basins (Merlo et al., 2011), gypsum bedrock exists primarily in interior deposits and is derived from ancient, shallow hypersaline lagoons (Mota et al., 2011). Gypsum may also form in hot springs from volcanic vapors (Herrero et al., 2009) and can form pedogenically (Eswaran & Gong, 1991). Although subsurface gypsum deposits occur worldwide, the high solubility of gypsum means that it persists at the surface for evolutionarily meaningful times almost exclusively in arid and semiarid regions (Escudero et al., 2014; Parsons, 1976).

Gypsum outcrops can be relatively pure or may be combined with other salts, such as sodium chloride. Because of its high solubility, bedrock gypsum often becomes intermixed with surrounding soils, creating mosaics of soils with differing gypsum contents. Gypsum soils (or gypsisols) are characterized by gypsum contents > 5% and the presence of a gypsic horizon in which gypsum is accumulated (Food and Agriculture Organization, 1990). Gypsum outcrops can have different physical characteristics, as they can be exposed as massive gypsum evaporite bedrock, crystalline selenite, anhydrite, secondary evaporites or even sand dunes (Figure 1). Physical surface crusts commonly contain > 25% gypsum (Verheye & Boyadgiev, 1997).

Plants living on gypsum soils show varying degrees of fidelity to gypsum and employ a variety of survival strategies, both of which have been used as bases for ecological classification (*e.g.*, Davis et al., 1986; Duvigneaud & Denaeyer-de Smet, 1968). The vegetation of gypsum soils includes substrate generalist taxa that grow on and off of gypsum, taxa that grow mostly on gypsum, and taxa that are endemic to gypsum. In recent literature, these three groups of taxa have generally been referred to as gypsovags, gypsoclines, and gypsophiles, respectively (Meyer, 1986), although it is important to note that in older literature the word gypsophile had a much more variable meaning, often referring to any species commonly encountered on gypsum, regardless of its overall fidelity to the substrate (*e.g.*, Johnston, 1941; Powell & Turner, 1977). We follow Meyer's definitions for the purposes of this chapter.

Not coincidentally, the regions with the most extensive gypsum outcrops host the largest assemblages of gypsophiles. Particularly species-rich gypsophile floras exist in the Chihuahuan Desert (at least 200 species; *e.g.*, Johnston, 1941; Moore & Jansen, 2007; Powell & Turner, 1977), Somalia and Ethiopia (at least 50 species; Thulin, 1993; 1995; 1999; 2006), Turkey (at least 40 species; *e.g.*, Akpulat & Celik, 2005), and Spain (at least 40 species; Mota et al., 2009; 2011), with smaller gypsophile floras in Iran (*e.g.*, Akhani, 2004), North Africa

(Le Houérou, 1969), Australia (Symon, 2007), the Mojave Desert and Intermountain West of the United States (*e.g.*, Forbis de Queiroz et al., 2012; Meyer, 1986), Cyprus (Hadjikyriakou & Hand, 2011), and Yemen (Petrusson & Thulin, 1996).

Figure 1. Different physical characteristics of gypsum outcrops, as encountered in the Chihuahuan Desert of northern Mexico: A) rocks (Sierra Tlahualilo, Durango); B) crystals (Puerto de Lobos, Chihuahua); C) crusts (Sierra Roque, Chihuahua); D) sand dunes (Bolsón de Cuatro Ciénegas, Coahuila).

With the exception of the gypsum flora of Spain (Mota et al., 2011), gypsophiles have been poorly studied in most areas of the world, especially compared to serpentine and halophilic vegetation. For example, in most of the above regions, but particularly in western Asia and the Horn of Africa, gypsum habitats have been underexplored botanically, and it is likely that many more gypsophile taxa remain to be discovered and described. Even in the relatively well-botanized gypsum areas of Spain and the United States, more than a dozen new gypsophile taxa have been described in the past decade (*e.g.*, Atwood & Welsh, 2005; Erben & Arán, 2005; Sivinski & Howard, 2011). While great strides have been made in understanding the physiological and community ecology of gypsophile floras in Spain over the past 20 years, little or no corresponding research has been conducted in other gypsum environments, many of which have much different climates and/or rainfall regimes compared to Spain. Furthermore, it has only been over the 15 years that researchers have begun to assess the phylogenetic and population-level histories of gypsophiles. The present chapter reviews the current state of knowledge for gypsophile ecology and evolution, and identifies areas where additional research is needed to understand this globally important edaphic community.

GYPSOPHILE ECOLOGY

For well over a century, plant biologists have sought to understand the ecological controls on gypsum plant communities (*e.g.*, Contejan, 1881; Macchiati, 1888). Historically, ecologists have focused on physical (*e.g.*, Johnston, 1941; Meyer, 1986) and/or chemical (*e.g.*, Boukhris & Lossaint, 1970; Duvigneaud & Denaeyer-de Smet, 1968) causes for gypsophily, although more recent debates on the assembly of gypsophile floras have attempted to discriminate between two reference models: the specialist and refuge models (Escudero et al., 2014; Merlo et al., 1998; Palacio et al., 2007). These models closely link ecology with evolutionary processes and apply to other unusual geological substrates such as serpentine as well (Harrison & Rajakaruna, 2011). Below we review current understanding of the physical and chemical aspects of gypsophile ecology, as well as fungal-plant interactions.

Physical Soil Factors

Several physical characteristics have been posited to influence plant growth in soils with high gypsum content, including soil crusts, density and porosity, and associated phenomena such as water holding capacity. Among these factors, the crust that characterizes most gypsum soils has received perhaps the most attention as a physical soil attribute controlling gypsum endemism (*e.g.*, Cañadas et al., 2013; Romão & Escudero, 2005). In arid soils in general, traditionally two types of soil crusts, physical and biological, have been differentiated (but see Gil de Carrasco & Ramos, 2011). Reprecipitation of gypsum creates a physical crust in gypsum soils that contributes significantly to the formation of the structure of gypsic horizons (Daniells, 2012). However, gypsum soils are also frequently characterized by cryptogamic crusts that also influence soil chemistry and texture (Anderson et al., 1982; for more on cryptogamic crusts, see the section below). Hence it is not easy to separate the relative effects of physical vs. biotic crusts on germination and seedling establishment in gypsum environments. This is a clear example of how difficult it can be to separate the chemical, physical and biological factors when explaining gypsophily. Moreover, physical crusts are not exclusive to gypsum (*e.g.*, Anderson et al., 1982). A search in SCOPUS (10 April 2014) using the terms "soil crusts" and "arid" produced 388 results. Of these, only 39 included the word "gypsum."

While soil crusts have received the bulk of attention from ecologists, the hard upper soil horizons (gypsic and petrogypsic; Herrero & Porta, 2000) in highly gypseous soils also likely influence community composition. The gypsum content of soils influences porosity and root penetration capacity (Poch, 1998). Although gypsisols contain > 5% gypsum, much higher amounts of gypsum tend to characterize gypsophile floras. For example, Salmerón et al. (2014) found an average gypsum content approaching 60% in soils dominated by the gypsocline *Jurinea pinnata* in Spain, which qualifies such soils as hypergypsics (Herrero, 2004). Unfortunately, as Drohan & Merkler (2009) have noted, gypsum content of gypseous soils is rarely provided in most studies. Although there are not many field data, those that are available show that gypsum is a difficult environment for plant roots (Guerrero-Campo et al., 1999). Several studies, mainly of cultivated plants, have noted that gypsum contents > 25% hinder root development (*e.g.*, Boyadgiev, 1974; Mashali, 1996). Poch (1998) found that roots are seldom found in horizons with gypsum content > 60%, and when that percentage exceeds 80%, roots only grow through preexisting cracks or faunal channels and will otherwise form a mat above the upper boundary of these horizons. Poch (1998) also notes that gypsum soil pores may be irregular and discontinuous, which would seriously affect root development in plants whose roots are concentrated in shallow, highly gypseous horizons. Poch & Verplancke (1997) showed that gypsum content was positively correlated with penetration resistance, although they note that this alone does not explain the poor growth of roots in hypergypsic soils. Furthermore, the resistance of soil to root penetration may be increased upon drying, which may help explain why the effect of gypsum on plants appears much greater in arid and semiarid climates. Gibbens & Lenz (2001) reported that petrogypsid soils in the Chihuahuan Desert restricted rooting depth of shrubs to less than 1 m and thus contributed to vegetative sparseness. Nevertheless, some gypsophiles have been found to possess relatively deep roots (*e.g.*, Mota et al., 2011), and hence the effects of gypsum content on root penetration may not be universal.

Water holding capacity of gypsum soils is also likely to influence gypsum floras. However, here too, the data are contradictory. Several authors claim that gypsum soils have lower water holding capacity (*e.g.*, Meyer & García-Moya, 1989), whereas others have suggested the opposite (Hiouani, 2006). It has also been observed that gypsum soils are moist at depth even when surrounding soils dry completely (Meyer & García-Moya, 1989), and according to Hiouani (2006), moisture tends to increase as the percentage of gypsum increases. These apparent contradictions may be related to the irregular distribution of water in these soils, especially when their gypsum contents are very high (Food and Agriculture Organization, 1990). In these cases the pores in the gypsum may become plugged by the precipitation of leached gypsum (Poch, 1998), which may cause high mortality in the fine roots and limit their performance. Precipitation of gypsum and calcium carbonate around roots has also been reported to occur as a consequence of high calcium concentrations in the rhizosphere (Hinsinger et al., 2009). Further investigation of gypsum particle size and micromorphology, including their influence on soil matric potential, may reveal additional influences on community structure in gypsum soils.

Soil Chemistry

Although much of the ecological research into gypsophily recognizes that physical and chemical constraints may exist, chemical factors have largely been treated as secondary and have therefore been underexplored (Escudero et al., 2014; Romão & Escudero, 2005). The fact that many gypsophiles, and particularly those that are regionally dominant, seem to be characterized by certain nutritional or chemical profiles, strongly suggests that unusual soil chemistry of gypseous substrates has influenced the evolution of such taxa. Below we summarize the chemical aspects of gypsum soils that plants typically must contend with, with a focus on how gypsophiles deal with excess levels of calcium and sulfur.

In general, gypsum soils are characterized by alkaline pH, high content of carbonates, the dominance of Ca and Mg ions, low NaCl, and above all, reduced fertility (Salmerón-Sánchez et al., 2014). The pH of the gypsum soils varies between slightly and moderately alkaline (Drohan & Merkler, 2009) and is not very different from other calcareous soils (Salmerón-Sánchez et al., 2014). Gypsum does not significantly increase osmotic potential despite its high contents of certain salts and ions (Herrero et al., 2009). Electrical conductivity of these soils is usually below 3 dS m^{-1} (*e.g.*, Herrero et al., 2009; Salmerón-Sánchez et al., 2014). For Spanish gypsum soils, Gil de Carrasco & Ramos (2011) provide an average value of 2.76 dS m^{-1} and Herrero et al. (2009) provide a value of 2.25 dS m^{-1}.

Gypsum soils are characterized by their lowered fertility. Highly gypseous soils have very little organic matter and a low cation exchange capacity (CEC). The high pH and high concentrations of Ca promote rapid insolubilization of nutrients released by weathering (Gil de Carrasco & Ramos, 2011), and reduce the availability of key macro- and micronutrients such as Fe, K, Mg, Mn, P and Zn (Boscaiu et al., 2013; Oyonarte et al., 2002).

Plants growing in high Ca environments must also deal with the cytotoxicity of this element. Although Ca is an essential element for numerous biological functions, it is toxic at high concentrations in the cytoplasm (Hawkesford et al., 2012). Physiological mechanisms, such as sequestering Ca within cells or in the apoplast via oxalate crystallization, allow plants growing on calcium-rich soils to avoid this toxicity (*e.g.*, Fink, 1991; Franceschi & Nakata, 2005).

Plants tolerant of gypsum soils pose no exception. In their study of gypsovags from White Sands, New Mexico, USA, Borer et al. (2012) found that plants have different strategies that allow them to cope with the Ca excess, including the prevention of Ca uptake, the sequestration of foliar Ca in chemically unavailable forms (calcium oxalate), and the maintenance of foliar Ca in labile forms, which may allow it to be excreted from foliar salt glands. These mechanisms largely coincide with the four strategies that allow plants to deal with excess Ca and S found by Duvigneaud & Denaeyer-de Smet (1968; 1973) and Merlo et al. (1998; 2001) among plants growing on gypsum in Spain: the accumulator, the extruder, the assimilator, and the avoider. The first group includes species that accumulate large amounts of Ca, and often S and Mg; slight foliar succulence is characteristic of many of these plants (*e.g.*, *Gypsophila*, *Ononis tridentata*). The extruders contain species from primarily halophilic lineages that possess secretory glands, including *Frankenia* and some *Limonium* (Kleinkopf & Wallace, 1974). The assimilators include groups with S-rich secondary metabolites, including the many taxa of Brassicales (*e.g.*, the families Brassicaceae, Capparaceae, and Resedaceae) that are found on gypsum around the world (see below), all of which may be physiologically preadapted to gypsum. The avoiders are able to finely control ionic import and hence are able to survive on very poor and oligotrophic soils; Duvigneaud & Denaeyer-de Smet (1968) note that most avoider taxa on gypsum are gypsovags.

Since the seminal work by Duvigneaud & Denaeyer-de Smet (1966), gypsophiles have been viewed in general as Ca, Mg and S accumulators. This pattern is evident in Table 1, which summarizes foliar nutrient concentrations for various gypsophiles, gypsoclines, and gypsovags. Values for Ca concentration in the leaves of most plants typically range from 0.5-2.5% (Jones, 2012; Kalra, 1997; Parsons, 1976).

Among Spanish gypsophiles, highly elevated levels of Ca (> 5%) have been found in *Gypsophila struthium*, *G. hispanica*, *Ononis tridentata*, *Frankenia thymifolia*, and *Sedum gypsicola* (Table 1). All of these species have slightly succulent leaves (Merlo et al., 1998; 2001), with the exception of *F. thymifolia*, which is an extruder. Another group of Iberian gypsophiles also exhibit above average values (> 3%) of foliar Ca: *Helianthemum squamatum*, *Lepidium subulatum*, *Herniaria fruticosa*, *Coris hispanica*, and *Santolina viscosa* (Table 1). Two Spanish gypsovags, *Helianthemum syriacum* and *Sedum sediforme,* also have Ca levels above 3%, whereas locally endemic gypsophiles such as *Centaurea hyssopifolia*, *Thymus lacaitae* or *Teucrium turredanum* have lower values (Table 1).

The widely distributed Spanish gypsocline *Jurinea pinnata*, which grows on both gypsum and dolomite, also has relatively low levels of Ca (2.6%), although those values are higher on gypsum than on dolomites (Table 1). This behavior is very similar to that exhibited by the Iberian gypsovags *Rosmarinus officinalis*, *Linum suffruticosum* or *Salvia lavandulifolia* (Palacio et al., 2007). Although little nutrient data from other gypsophile floras are available, Ca concentrations above 5% were found in the Tunisian gypsoclines *Erodium glaucophyllum*, *Zygophyllum album,* and *Moricandia suffruticosa* (Boukhris & Lossaint, 1970; 1972). No data are available for the large and diverse gypsophile flora of the Chihuahuan Desert region, although almost all regionally dominant gypsophiles in that area have slightly succulent leaves (*e.g.* gypsophile species of *Dicranocarpus, Sartwellia, Acleisanthes, Nama, Tiquilia*, and *Nerisyrenia*), suggesting that these taxa are also likely accumulators.

Many of the gypsophile taxa with elevated Ca concentration in Table 1 also possess elevated S and Mg concentration, although the pattern is less consistent for Mg. For example, the Ca accumulators *Gypsophila struthium*, *G. hispanica*, and *Ononis tridentata* possess the highest known S contents of any plants growing on gypsum, and have elevated Mg concentrations as well (Table 1). Other Spanish gypsophiles, such as *Helianthemum squamatum* and *Lepidium subulatum*, have elevated S but much lower Mg concentrations. In contrast, the narrowly distributed gypsophile *Helianthemum conquense* has relatively low foliar concentrations of Ca, Mg, and S (Table 1).

Palacio et al. (2007) suggest that there are two broad categories of gypsophile species: those that are dominant on gypsum and broadly distributed geographically (the regionally dominant gypsophiles) and those that are narrowly distributed. The former group is composed of taxa that are typically succulent-leaved and often show a remarkable ability to accumulate Ca, Mg, and S, as well as the macronutrients that are scarce in gypsum soils such as N and P (Table 1). Whereas many narrowly distributed gypsophiles like *Helianthemum conquense* seem to behave more like gypsovags in terms of nutrient accumulation, some locally distributed gypsophiles such as *Coris hispanica* and *Santolina viscosa* behave similarly to the "stockpiling" regional dominants (Table 1). Even the gypsovag *Helianthemum syriacum* is difficult to separate from the latter two species based on Ca concentration (Table 1). Although not perfect, the relatively strong correlation between regional dominance, foliar succulence, and the strategy of accumulating certain nutrients suggests a syndrome of common adaptations to gypsum soil chemistry, which deserves much further physiological and ecological study. Indeed, as Merlo et al. (2011) have noted, foliar Ca, Mg, and S concentration, as well as Ca:Mg ratio, seem to be useful parameters for establishing differences in the nutritional behavior of plants growing on gypsum, dolomite, and serpentine.

Table 1. Community characteristics, succulence, and foliar nutrient content for selected gypsophiles, gypsoclines, and gypsovags. All nutrient values are mean percentages; dashes indicate that values were not available. Taxa in bold are gypsophiles; all other taxa are gypsovags, except for the gypsocline *Jurinea pinnata*. Average values for halophytes are provided at the bottom of the table. Key to references: (1) Drohan & Merkler (2009); (2) Duvigneaud & Denaeyer de Smet (1966); (3) Duvigneaud & Denaeyer de Smet (1968); (4) Escudero et al. (2014); (5) M. Merlo et al. (unpublished); (6) Salmerón-Sánchez et al. (2014)

Species	Population growing on gypsum?	Taxon dominant on gypsum?	Taxon widespread on gypsum?	Succulent?	Ca	Mg	S	Na	N	P	K	References
Arctomecon californica	yes	?	no	yes	3.83	2.47	0.33	0.29	–	0.06	1.88	[1]
Artemisia herba-alba	yes	no	yes	no	1.20	0.33	0.20	0.04	3.94	0.27	1.45	[4]
Centaurea hyssopifolia	yes	no	no	no	2.60	0.49	0.80	0.06	4.02	0.27	2.33	[4]
Coris hispanica	yes	no	no	no	3.72	0.13	–	< 0.10	1.65	0.02	0.68	[5]
Eriogonum corymbosum	yes	?	no	no	0.84	2.47	0.31	0.18	–	0.08	2.27	[1]
Frankenia thymifolia	yes	yes	yes	no	11.00	1.22	1.15	0.13	2.00	0.10	1.80	[2,3]
Frankenia thymifolia	yes	yes	yes	no	10.66	0.91	–	0.10	1.50	0.04	0.37	[5]
Gypsophila hispanica	yes	yes	yes	yes	7.83	2.23	4.99	0.03	1.75	0.10	0.93	[2,3]
Gypsophila hispanica	yes	yes	yes	yes	7.40	1.21	3.00	0.06	2.49	0.19	1.18	[4]
Gypsophila struthium	yes	yes	yes	yes	6.13	3.94	3.64	1.33	1.55	1.26	0.68	[2,3]
Gypsophila struthium	yes	yes	yes	yes	8.17	0.83	–	< 0.10	1.26	0.08	0.80	[5]

Species	Population growing on gypsum?	Taxon dominant on gypsum?	Taxon widespread on gypsum?	Succulent?	Ca	Mg	S	Na	N	P	K	References
Helianthemum alypoides	yes	yes	no	no	1.83	0.25	–	< 0.10	1.08	0.07	0.28	[5]
Helianthemum conquense	yes	no	yes	no	1.90	0.26	0.10	0.03	1.68	0.11	0.39	[4]
Helianthemum squamatum	yes	yes	yes	yes	3.43	0.65	2.90	0.08	1.65	0.12	0.62	[4]
Helianthemum squamatum	yes	yes	yes	yes	3.15	0.78	2.48	0.08	1.37	0.09	0.75	[2,3]
Helianthemum squamatum	yes	yes	yes	no	2.62	0.42	–	< 0.10	1.14	0.05	0.28	[5]
Helianthemum syriacum	yes	yes	yes	yes	3.10	0.50	1.30	0.02	1.10	0.07	0.70	[2,3]
Helianthemum syriacum	yes	yes	yes	no	3.00	0.31	1.00	0.02	1.76	0.11	0.50	[4]
Helianthemum syriacum	yes	yes	yes	yes	3.02	0.20	–	< 0.10	1.30	0.08	0.36	[5]
Herniaria fruticosa	yes	no	yes	no	2.90	0.77	1.10	0.05	2.53	0.11	0.92	[4]
Herniaria fruticosa	yes	no	yes	no	3.00	1.30	0.81	0.01	1.00	0.04	0.89	[2,3]
Jurinea pinnata (on dolomite)	no	no	yes	no	2.20	0.71	0.35	0.04	1.47	0.06	1.05	[6]
Jurinea pinnata (on gypsum)	yes	no	yes	no	2.62	0.46	0.51	0.06	2.12	0.04	1.37	[6]
Lepidium subulatum	yes	yes	yes	no	1.80	0.46	2.80	0.02	3.20	0.16	1.40	[2,3]
Lepidium subulatum	yes	yes	yes	no	2.70	0.38	2.30	0.06	5.12	0.25	0.97	[4]
Lepidium subulatum	yes	yes	yes	no	1.83	0.11	–	< 0.10	2.12	0.08	0.36	[5]

Table 1. (Continued)

Species	Population growing on gypsum?	Taxon dominant on gypsum?	Taxon widespread on gypsum?	Succulent?	Ca	Mg	S	Na	N	P	K	References
Linum suffruticosum	yes	no	yes	no	2.65	2.45	0.06	0.06	2.80	0.17	0.92	[4]
Linum suffruticosum	no	no	yes	no	2.70	0.33	0.08	0.06	2.31	0.14	0.73	[4]
Ononis tridentata	yes	yes	yes	yes	5.57	2.52	6.07	0.03	2.31	0.10	0.68	[2,3]
Ononis tridentata	yes	yes	yes	yes	5.75	1.84	4.50	0.09	2.42	0.12	0.28	[4]
Ononis tridentata	yes	yes	yes	yes	4.37	0.86	–	0.20	1.31	0.04	0.24	[5]
Rosmarinus officinalis	yes	no	yes	no	1.15	0.25	0.22	0.04	0.97	0.06	1.29	[2,3]
Rosmarinus officinalis	yes	no	yes	no	1.40	0.28	0.10	0.06	1.09	0.07	0.80	[4]
Rosmarinus officinalis	no	no	yes	no	1.20	0.17	0.10	0.06	1.17	0.09	1.01	[4]
Salvia lavandulifolia	yes	no	yes	no	1.95	0.33	0.10	0.05	1.77	0.10	0.52	[4]
Salvia lavandulifolia	no	no	yes	no	1.50	0.30	0.05	0.05	1.51	0.09	0.58	[4]
Santolina viscosa	yes	no	no	no	3.01	0.12	–	0.29	1.34	0.06	0.60	[5]
Sedum gypsicola	yes	no	yes	yes	8.18	0.18	–	< 0.10	2.76	0.05	0.49	[5]
Sedum sediforme	yes	no	yes	yes	4.31	0.10	–	< 0.10	0.69	0.04	0.55	[5]
Teucrium capitatum	yes	no	yes	no	1.90	0.24	0.06	0.04	2.65	0.13	0.72	[4]

Species	Population growing on gypsum?	Taxon dominant on gypsum?	Taxon widespread on gypsum?	Succulent?	Ca	Mg	S	Na	N	P	K	References
Teucrium capitatum	no	no	yes	no	1.80	0.26	0.05	0.05	2.30	0.15	0.63	[4]
Teucrium polium	yes	no	yes	no	2.00	0.61	0.60	0.07	1.67	0.07	0.76	[4]
Teucrium turredanum	yes	yes	no	no	1.37	0.22	–	< 0.10	0.99	0.03	0.60	[5]
Thymus lacaitae	yes	no	no	no	1.60	0.40	0.04	0.05	1.42	0.11	0.56	[4]
halophytes (several species)	no	no	yes	yes	1.10	1.64	2.36	9.37	2.28	0.20	1.98	[2,3]

Fungal-Plant Interactions

Mycorrhizal and endophytic fungal interactions with gypsophiles are poorly understood but may play an important role in structuring gypsophile plant communities. A handful of recent studies have begun to shed light on the community composition of these fungi in gypsum environments. In Spain, Alguacil et al. (2009a; b; 2012) have found an unusually diverse arbuscular mycorrhizal fungus (AMF) community on Spanish gypsum, comparable to that found on non-gypseous sites with much higher plant density. A total of 21 AMF types were found in association with four Spanish gypsophiles: *Gypsophila struthium*, *Teucrium libanitis*, *Helianthemum squamatum*, and *Ononis tridentata* (Alguacil et al., 2009b). As Alguacil et al. (2009a) note, this appears to be the first report of AMF in the genus *Gypsophila* (Wang & Qiu, 2006). Moreover, Alguacil et al. (2009a) found novel AMF sequences among roots of *G. struthium*, suggesting the presence of undescribed species. This new fungal type was found mainly in the less altered gypsum zone, raising the possibility that it could be associated with survival or proliferation of *G. struthium* on gypsum, which could be among the factors underlying the great colonizing power that this species exhibits in abandoned gypsum quarries, where it becomes almost monospecific (Mota et al., 2004). Alguacil et al. (2012) found a higher diversity of AMF in perennial gypsophiles and gypsovags vs. an annual gypsovag, and Porras-Alfaro et al. (2014) also found that regionally dominant gypsophile taxa in New Mexico have generally higher overall levels of AMF colonization than nearby non-gypseous grasslands. The same authors also found high levels of colonization by dark septate fungi and hyaline septate endophytic fungi.

All of these results accord with the a priori prediction of Palacio et al. (2012) that gypsophiles have a higher degree of mycorrhizal infection than gypsovags, although it is important to note that these authors did not find higher rates of AMF colonization in gypsophiles vs. gypsovags in their own study, nor did they find support for the hypothesis that AMF are responsible for the high levels of soil macronutrients that characterize such taxa. To explain both the high diversity of AMF on gypsophiles and the presence of potentially undescribed taxa, Alguacil et al. (2009a) postulate the existence of strong selective pressures that have been able to promote the specialization of symbiotic microorganisms, helping vascular gypsophiles to proliferate under heavy stress. This hypothesis adds a possible coevolutionary dimension to the mechanisms involved in gypsophily.

The Assembly and Evolution of Gypsophile Floras

A comparison of existing, albeit incomplete, checklists and other related literature concerning gypsophiles reveals several interesting patterns relevant to the assembly of gypsophile floras worldwide. First, it is clear that each of the major gypsophile floras evolved independently, drawing their constituent taxa from local plant lineages. For example, all of the common gypsophiles in the Chihuahuan Desert (*e.g.*, species of *Tiquilia, Acleisanthes, Nerisyrenia, Nama,* etc.; Figure 2), Spain (*e.g.*, species of *Helianthemum, Ononis, Teucrium, Limonium,* etc.; Figure 3) and Somalia (*e.g.*, species of *Commiphora, Euphorbia, Kleinia,* etc.) are members of larger genera or species groups with centers of diversity in the same region (Mota et al., 2011; Thulin, 1993; 1995; 1999; 2006; Turner & Powell, 1979).

Figure 2. Examples of Chihuahuan Desert gypsophiles: A) *Acleisanthes lanceolata* var. *megaphylla* (Nyctaginaceae); B) *Tiquilia hispidissima* (Ehretiaceae); C) *Sartwellia flaveriae* (Asteraceae); D) *Gaillardia henricksonii* (Asteraceae); E) *Fouquieria shrevei* (Fouquieriaceae); F) *Anulocaulis leiosolenus* var. *howardii* (Nyctaginaceae); G) *Nerisyrenia gracilis* (Brassicaceae); H) *Nama carnosum* (Hydrophyllaceae).

Figure 3. Examples of gypsum habitats and gypsophiles in Spain: A) Gypsum scarp with *Sedum gypsicola* (Crassulaceae) and the lichen *Parmelia pokorny* (Parmeliaceae); B) Gypsum scrubland at Venta de los Yesos, Almería; C) *Ononis tridentata* (Fabaceae); D) *Gypsophila struthium* subsp. *struthium* (Caryophyllaceae); E) *Chaenorhinum grandiflorum* (Plantaginaceae); F) *Helianthemum alypoides* (Cistaceae); G) *Frankenia thymifolia* (Frankeniaceae); H) *Teucrium lepicephalum* (Lamiaceae).

Although some individual gypsophile taxa may be widely distributed within a particular gypsum region (*e.g.*, *Dicranocarpus parviflorus* in the Chihuahuan Desert), there is no evidence of direct long-distance dispersal of gypsophiles among major gypsum regions, with the possible exception of *Campanula fastigiata*, which is found in both Spain and Cyprus (Hadjikyriakou & Hand, 2011; Mota et al., 2011). Even in larger cosmopolitan genera like *Euphorbia, Helianthemum*, and *Campanula*, which have different gypsophiles in multiple major gypsum regions of the world (*e.g.*, gypsophile taxa in *Euphorbia* exist in both Somalia and the Chihuahuan Desert, but these taxa are not shared between the two regions), it is clear based on morphological and/or molecular evidence that the gypsophiles within each genus are locally derived rather than the result of long-distance dispersal (Mota et al., 2011; Thulin, 1993; 1995; 1999; 2006; Turner & Powell, 1979).

A preliminary review of floristic literature also reveals that the overwhelming majority of gypsophiles fall within just a few major flowering plant clades. For example, of 44 Spanish taxa that ranked highest (a rating ≥ 4) on the lists of gypsum plant taxa from Mota et al. (2009) and Mota et al. (2011), and thus may be considered gypsophiles, 18 are asterids, 9 are Caryophyllales, 6 are Brassicales, and 11 belong to other groups (clade membership sensu Angiosperm Phylogeny Group, 2009). Although species lists are incomplete or absent for other regions of the world, patterns of clade membership appear similar to those seen in Spain. The same preponderance of asterids and Caryophyllales characterizes the gypsophile floras of the Chihuahuan Desert, Somalia and Australia, with key Brassicales groups in the former two regions as well [*e.g.*, *Nerisyrenia* (Brassicaceae) in the Chihuahuan Desert, and *Cleome* (Cleomaceae) and *Reseda* (Resedaceae) in Somalia] (Thulin, 1993; Turner & Powell, 1979). For example, 58% of the taxa listed as gypsophiles in Powell & Turner (1977) are asterids and 27% are Caryophyllales, while 9 of the 13 gypsophile taxa listed by Symon (2007) from southern Australia are asterids. This global bias toward clade membership in such groups as asterids, Caryophyllales and Brassicales likely reflects underlying predispositions for gypsum tolerance within these groups. To examine these clade membership patterns more rigorously requires a more thorough global checklist of gypsophiles, which we are currently assembling.

Recent phylogenetic studies that have included gypsophile taxa have also revealed several trends in the origin and evolution of gypsophiles. The overwhelming majority of such studies to date have examined Chihuahuan Desert gypsophiles, with several clear patterns having emerged from these studies. First, multiple origins of gypsophily are typical within plant lineages that appear to be ancestrally tolerant of gypsum. Excellent examples of this phenomenon have been documented in recent studies of regionally dominant gypsophile taxa in the Chihuahuan Desert: Marlowe & Hufford (2007) found three independent origins of gypsophily within *Gaillardia* (Asteraceae), Moore & Jansen (2007) found two origins of gypsophily in *Tiquilia* subg. *Eddya* (Ehretiaceae), Douglas & Manos (2007) found at least four origins of gypsophily in tribe Nyctagineae (Nyctaginaceae) [although not available to Douglas & Manos (2007), the inclusion of two more gypsophile species of Nyctagineae from Somalia (*Acleisanthes somalensis* and *Commicarpus reniformis*) raises the number of origins in this clade to at least six (Levin, 2000; M. Thulin, pers. comm.)], McKown et al. (2005) implied at least three origins of gypsophily in subtribe Flaveriinae (Asteraceae), Taylor (2012) found three origins of gypsophily in *Nama* (Hydrophyllaceae) and Schenk (2013) documented up to five origins of gypsophily in *Mentzelia* sect. *Bartonia* (Loasaceae). In all of these examples, the larger group containing the gypsophiles possesses numerous other taxa that are gypsovags. For example, all non-gypsophile taxa in *Tiquilia* subg. *Eddya* grow both on and off of gypsum (Moore & Jansen, 2007; Richardson, 1977), and numerous members of tribe Nyctagineae (*e.g.*, *Anulocaulis eriosolenus,* all non-gypsophile species of *Allionia* and *Cyphomeris* and many non-gypsophile taxa of *Acleisanthes, Boerhavia*, and *Mirabilis*), *Gaillardia* (*e.g.*, *G. pulchella, G. spathulata*, and *G. parryi*) and *Mentzelia* (*e.g.*, *M. nuda, M. mexicana*, and *M. saxicola*) are also gypsovags (Douglas & Manos, 2007; Schenk, 2013; Thompson & Powell, 1981; Turner & Watson, 2007). Although phylogenetic studies including gypsophiles from other regions of the world are scarcer, those that have been completed support the results from in the Chihuahuan Desert. For example, at least three origins of gypsophily have been confirmed or implied in Spanish *Helianthemum* (Cistaceae; leading to the gypsophiles *H. squamatum, H. alypoides*, and *H. conquense*) and Mediterranean *Campanula* (Campanulaceae; leading to the Spanish/Cypriot gypsophile *C. fastigiata*, the North African gypsophile *C. filicaulis* subsp. *reboudiana* and the Turkish gypsophile *C. pinnatifida* var. *germanicopolitana*) and can be expected in Spanish *Limonium* (Plumbaginaceae) (Mota et al., 2009; 2011; Parejo-Farnés et al., 2013; Roquet et al., 2008). Each of these genera is characterized by numerous other gypsovag taxa as well.

Within the gypsophile flora of the Chihuahuan Desert region, existing phylogenetic studies further suggest that speciation has occurred frequently after the acquisition of gypsophily, particularly in those lineages that comprise the regionally dominant taxa on gypsum. Clades of regionally dominant gypsophiles have been documented in phylogenetic studies of *Gaillardia* (which has two gypsophilic clades; Marlowe & Hufford, 2007), *Tiquilia* subg. *Eddya* (Moore & Jansen, 2007), *Acleisanthes* (Levin, 2000), *Nama* (Taylor, 2012), *Mentzelia* sect. *Bartonia* (Schenk & Hufford, 2011), *Leucophyllum* (Scrophulariaceae; Gándara & Sosa, 2013), and *Argemone* (Papaveraceae; Schwarzbach & Kadereit, 1999), and unpublished data in the senior author's lab suggest that clades of gypsophiles exist in *Sartwellia* (Asteraceae), *Haploësthes* (Asteraceae), *Nerisyrenia* and *Anulocaulis* (Nyctaginaceae). Although not typically dominant on gypsum, the gypsophile Cactaceae genera *Aztekium* (2 species) and *Geohintonia* (1 species) form a clade and have also speciated on gypsum (Hernández-Hernández et al., 2011). In several of these gypsophile clades, extensive speciation has occurred. For example, the Chihuahuan Desert gypsophile clade of *Acleisanthes* comprises six taxa; the principal clade of *Nama* gypsophiles comprises 10 taxa, of which 8 taxa are gypsophiles; while *Nerisyrenia* is composed almost entirely of gypsophiles, with all but one of 12 described taxa being gypsophiles (Bacon, 1978; Fowler & Turner, 1977; Taylor, 2012). Most of the gypsophile clades, and all such clades with the largest number of taxa, are broadly distributed across the Chihuahuan Desert, despite the island-like nature of gypsum exposures. Although these clades as a whole are broadly distributed, individual taxa within them generally occupy much narrower geographic ranges that are usually allopatric from one another, suggesting that allopatric speciation is typically responsible for taxon boundaries within these gypsophile lineages. A good example of this phenomenon is provided by the gypsophile clade of *Nama*, the distribution of which is illustrated in Figure 4.

While the lack of phylogenetic studies in other gypsophile floras prevents firm conclusions, it is possible that similar phylogenetic and biogeographic patterns may also characterize some of the other more broadly distributed gypsophile floras. For example, possible clades of gypsophiles may exist within *Ononis* (Fabaceae), *Teucrium* (Lamiaceae), *Orobanche* (Orobanchaceae) and *Chaenorhinum* (Plantaginaceae) in Spain, within *Psephellus*

(Asteraceae) in Turkey, and within *Pseudoblepharispermum* (Asteraceae) and *Xylocalyx* (Orobanchaceae) in the Horn of Africa region (Mota et al., 2011; Thulin, 2006; Wagenitz & Kandemir, 2008).

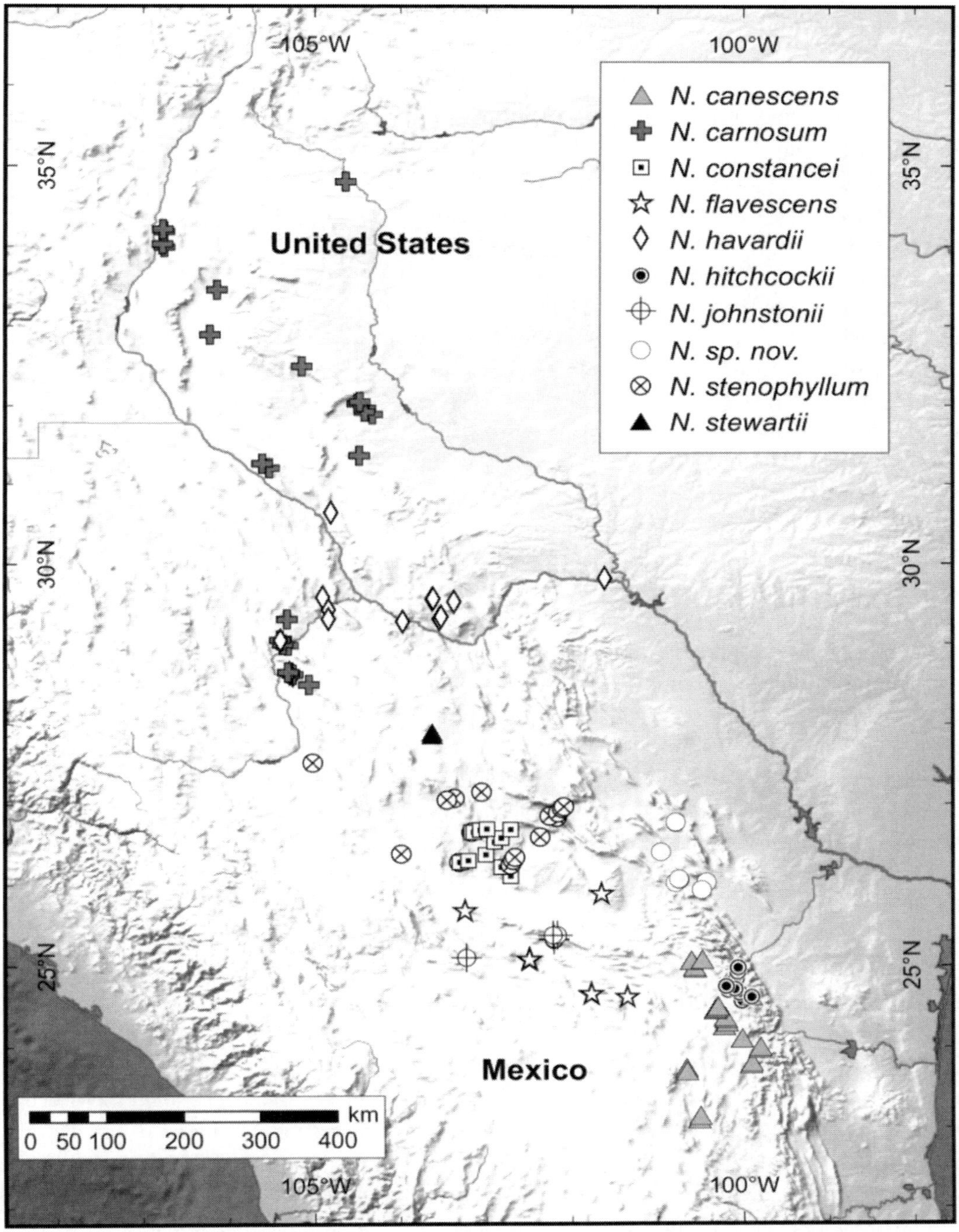

Figure 4. Distribution of taxa within the gypsophile clade of *Nama* (Hydrophyllaceae). All of these taxa are gypsophiles, with the exception of the gypsovags *N. johnstonii* and *N. havardii*.

Molecular evidence indicates that many gypsophile lineages around the globe may have appeared no earlier than the late Miocene (*ca.* 8-5.3 mya). Using molecular dating techniques, Moore & Jansen (2006; 2007) found that the two origins of gypsophily in *Tiquilia* subg. *Eddya* dated most likely to the early Pliocene and early-to-mid Pleistocene, respectively, with the earlier origin leading to the geographically widespread and regionally dominant *T. hispidissima* taxon complex, and the later origin leading to the geographically restricted clade of *T. turneri* and *T. tuberculata*. A late Miocene or early Pliocene divergence time was also favored for the split of the gypsophile (and morphologically quite distinctive) cactus genera *Aztekium* and *Geohintonia* (mean age = 5.67 mya), suggesting gypsophily is at least that old in that lineage (Hernández-Hernández et al., 2014).

In a molecular dating analysis of Cornales (which includes *Mentzelia* of the Loasaceae), Schenk & Hufford (2010) recovered a Pleistocene origin for *Mentzelia* sect. *Bartonia*, which includes numerous gypsophile taxa. Although they did not perform a separate dating analysis of these gypsophile lineages, it is clear from studies with more complete taxon sampling that the regionally dominant and geographically widespread Chihuahuan Desert gypsophile *Mentzelias* (*M. perennis, M. todiltoensis*, and *M. humilis*) diverged early in the history of the section, implying that they are older than the other gypsophile taxa of *Mentzelia*, which are all in more recently derived positions, have narrow distributions outside the Chihuahuan Desert, and represent distinct origins of gypsophily (Schenk, 2013; Schenk & Hufford, 2011). Gándara et al. (2014) recovered a late Miocene divergence time between the morphologically distinctive and monotypic gypsophile genus *Jaimehintonia* (Amaryllidaceae) and its nearest relative, suggesting that gypsophily arose in *Jaimehintonia* after that point. Wagstaff & Tate (2011) found a similar late Miocene divergence time between the Australian gypsophile *Lawrencia helmsii* (Malvaceae) and its congeners, again placing a late Miocene upper bound on the origin of gypsophily in this lineage.

In contrast, there are numerous other gypsophile lineages composed of single species that have restricted geographic ranges and are morphologically much more similar to their non-gypsophile relatives. Examples include *Tiquilia turneri, Mirabilis nesomii* (Nyctaginaceae), *Abronia nealleyi* (Nyctaginaceae), *Nama stevensii*, and *Gaillardia gypsophila,* all of which have been found to have very little phylogenetic distance separating them from morphologically very similar congeners, implying a very recent origin (Marlowe & Hufford, 2007; Moore & Jansen, 2007; Taylor, 2012; unpublished data). Presumably these taxa appeared in the Pleistocene, as was inferred for *Tiquilia turneri* in the molecular dating analyses of Moore & Jansen (2006; 2007). The existence of a mix of older and younger gypsophile lineages is important because it suggests that modern gypsophile floras have assembled gradually over the last several million years.

The post-Miocene assembly of gypsophile floras corresponds well with the current hypotheses concerning the spread of semi-arid and arid habitats during the Cenozoic. After a peak of global average temperature and precipitation in the late Paleocene and early Eocene, the Earth experienced several major episodes of cooling and drying, culminating in the most recent major episode during the latest Miocene and Pliocene (Graham, 2011; Zachos et al., 2008). Available paleoclimatic evidence suggests that it was not until this period that arid and semi-arid regions began to occupy relatively large portions of the Earth's surface (Arakaki et al., 2011; Axelrod, 1979; Graham, 2011; Hernández-Hernández et al., 2014; Salzmann et al., 2008). Given that all of the world's gypsophile plant assemblages occur in such habitats, it is unlikely that gypsum habitats dry enough and extensive enough to support gypsophile floras existed prior to the latest Miocene. Additional molecular dating analyses will be necessary to test this hypothesis further.

EVOLUTION AT THE POPULATION LEVEL

The population structure of gypsophiles should be largely determined by the island-like distribution of gypsum outcrops across the landscape. This edaphic restriction places an upper limit on the population size any gypsophile species can achieve. Once such a species has colonized a particular gypsum "island," however, it may remain on that outcrop indefinitely. Hence allele frequencies in gypsophiles should reach an equilibrium reflecting the combined effects of migration, mutation and genetic drift. The close correspondence of the island-like distributions of gypsophiles to the assumptions of well-studied theoretical models of population genetic structure (*e.g.*, the stepping-stone model; Kimura & Weiss, 1964) provides an opportunity to infer aspects of their demographic and evolutionary history from parameters commonly estimated in population genetic studies.

In gypsophiles, as with many desert and island taxa (Filner & Shmida, 1981), we would expect migration to be generally quite limited (and selection may actually favor reduced dispersibility; Schenk, 2013). Thus, isolation-by-distance should be evident. While low population size has no effect on mutation rates, at least one aspect of the biology of gypsophiles may serve to increase the effective population size, thus reducing the rate at which genetic diversity is lost due to genetic drift: almost all gypsophiles are perennials, with overlapping generations. Genetic diversity may also be maintained by outcrossing. Only a few gypsophiles are obvious selfers [though mixed mating systems may be common; for example *Acleisanthes* produces both cleistogamous and chasmogamous flowers (Douglas & Manos, 2007)]. On the other hand, biparental inbreeding in small populations may have the opposite effect. Finally, during the climatic oscillations of the Pleistocene, gypsophiles may not have been subjected to repeated genetic bottlenecks as severe as those suffered by plants in other habitats (*e.g.*, alpine taxa), because community composition on unusual substrates such as gypsum may be more stable than substrate generalist communities over a broader range of climates (Damschen et al., 2012; Harrison et al., 2009; Tapper et al., 2014).

In general, endemic taxa tend to have lower genetic diversity than widespread taxa, but measures of population structure do not seem to differ greatly between rare and common species (see reviews by Cole, 2003; Gitzendanner & Soltis, 2000; Hamrick & Godt, 1989). However, the number of migrants is typically much reduced in rare species (Cole, 2003) as compared to common ones.

To examine whether genetic variation and population structure in gypsophile taxa differ from that in "ordinary" endemics in predictable ways, we followed the example of these three reviews of genetic variation in plants with contrasting life histories (Cole, 2003; Gitzendanner & Soltis, 2000; Hamrick & Godt, 1989). We summarized available estimates of population genetic parameters from five studies that focused on gypsophile taxa; we also included five other studies of Spanish gypsoclines and a study of the gypsovag *Arctomecon californica*, a close congener of the gypsophile *A. humilis*, for comparison (Table 2).

Table 2. Population genetic parameters estimated for gypsophile (in bold) and selected gypsocline species, as well as the gypsovag *Arctomecon californica*. Abbreviations: Pops = number of populations, Inds = number of individuals, P = proportion of loci that are polymorphic, A = mean number of alleles per locus, H_o = observed heterozygosity, H_e = expected heterozygosity, H_t = Nei's gene diversity, A_e = effective number of alleles (calculated from H_o), H_s = mean within-population gene diversity, N_m = effective number of migrants, IBD = isolation-by-distance. For population differentiation, the following indicators apply: † = G_{st} or F_{st}; ‡ = among-population variance from AMOVA. Average parameter values calculated from diploid data only; # identifies polyploid taxa or haploid genomes excluded from parameter averages. Significant R^2 values identified by an asterisk (*). The final three entries report parameter averages from published reviews of population parameters in plants, for comparison

Citation	Taxon	Data Type	Pops	Inds	P	A	H_o	H_e	A_e	H_t	H_s	Population Differentiation	N_m	R^2 (IBD)
Allphin et al., 1998	***Arctomecon humilis***	isozyme	6	163	0.104	1.43	0.100	0.103	1.51	0.339		0.620†	0.16	
Hickerson & Wolf, 1998	*Arctomecon californica*	allozyme	16	480	0.554	1.71	0.158	0.163		0.239	0.163	0.320†	0.54	0.048*
Aguirre-Liguori et al., 2014	***Fouquieria shrevei***	cpDNA#	5	94								0.850†, 0.709‡		0.532*
Pérez-Collazos & Catalán, 2008	***Ferula loscosii***	allozyme	11	330	0.327	1.62	0.164	0.125	1.18	0.152	0.125	0.134†	1.62	0.839*
Pérez-Collazos et al., 2009	***Ferula loscosii***	AFLP	12	342	0.523						0.171	0.440‡	0.32	0.811*
Salmerón-Sánchez et al., 2014	*Jurinea pinnata*	AFLP	16	160								0.370‡	0.42	
Martínez-Nieto et al., 2013	***Gypsophila struthium* subsp. *hispanica***	AFLP	7	82	0.573				1.35	0.258	0.200	0.226†, 0.280‡	0.86	0.038
		cpDNA#								0.810	0.381	0.530†		
	Gypsophila struthium* subsp. *struthium	AFLP	16	185	0.562				1.29	0.224	0.160	0.286†, 0.334‡	0.62	0.128
		cpDNA#								0.827	0.292	0.647†		
Jiménez & Sánchez-Gómez, 2012	*Moricandia moricandioides* subsp. *pseudofoetida*	ISSR	1	50	0.817						0.213	0.533†, 0.665‡	0.19	
	Moricandia moricandioides subsp. *moricandioides*	ISSR	1	30	0.790						0.213			
Pérez-Collazos et al., 2008	*Boleum asperum*#	AFLP	10	240	0.913				3.91	0.744		0.202‡	0.99	0.443*

Table 2. (Continued)

Citation	Taxon	Data Type	Pops	Inds	P	A	H_o	H_e	A_e	H_t	H_s	Population Differentiation	N_m	R^2 (IBD)
Pérez-Collazos & Catalán, 2006	*Vella pseudocytisus* subsp. *paui*[#]	allozyme	6	162	0.636				2.23	0.552		0.200‡	1.00	
		AFLP	**6**	**162**	**0.625**				**2.39**	**0.581**		**0.219‡**	**0.89**	**0.560***
López-Pujol et al., 2004	*Thymus loscosii*[#]	allozyme	8	257	0.850	3.00	0.472	0.422	1.80	0.444	0.429	0.033†	7.33	
	Average				**0.531**	**1.59**	**0.141**	**0.130**	**1.33**	**0.242**	**0.169**	**0.423†, 0.466‡**	**0.589**	**0.425**
Hamrick & Godt, 1989	**Endemic**				0.400	1.80		0.096	1.15	0.263	0.163	0.248		
	Widespread				0.589	2.29		0.202	1.31	0.347	0.267	0.210		
Gitzendanner & Soltis, 2000	**Endemic**				0.367	1.94				0.219		0.206		
	Widespread				0.449	2.23				0.242		0.224		
Cole, 2003	**Rare**				0.407	1.74	0.100	0.113		0.142		0.212	1.190	
	Common				0.588	2.34	0.139	0.150		0.199		0.198	2.240	

We report the following statistics that reflect genetic diversity of these taxa: percentage of polymorphic loci, *P*; number of alleles per locus, *A*; Nei's total gene diversity, H_t; average genetic diversity within populations, H_s; effective number of alleles, A_e; and observed heterozygosity, H_o. We also tabulated estimates of population differentiation, including G_{st}, F_{st}, or among-population variance from analysis of molecular variance (AMOVA).

Some studies estimated the effective number of migrants, N_m. For the sake of comparison, we estimated N_m from F_{st} or G_{st} for the remaining studies where this was possible. Finally, we report the degree to which populations exhibited isolation-by-distance. Averages discussed in the following section exclude diversity parameters estimated from known polyploids, which typically have larger numbers of alleles, and from haploid chloroplast data.

Comparisons of parameter values to those obtained from the three reviews should be viewed qualitatively, since the small number of studies of gypsophiles precludes rigorous statistical analysis.

Genetic Diversity

The taxa in Table 2 tend to show levels of genetic variation similar to that expected for endemic (Hamrick & Godt, 1989) or rare (Cole, 2003; Gitzendanner & Soltis, 2000) plant species, although by some measures, they exceed the genetic diversity typical of widespread species. The percentage of polymorphic loci ranged from 0.10 to 0.82, with a mean of 0.53, intermediate to the averages for rare and widespread taxa in the three reviews cited above. The number of alleles per locus varied from 1.43 to 1.71 (mean 1.59). Though this was reported in only three enzyme studies of diploid taxa, the value lies slightly below averages for rare or endemic taxa in the three reviews. Nei's total gene diversity (H_t) averaged 0.24 (range 0.15-0.34), slightly below the estimated means of endemics in Hamrick & Godt (1989), but actually greater than later estimates for widespread species (Cole, 2003; Gitzendanner & Soltis, 2000). Average genetic diversity within populations is 0.17, slightly higher than found for endemics in general (0.16; Hamrick & Godt, 1989). Perhaps more significantly, the estimated effective number of alleles (Kimura & Crow, 1964), which we estimated as $1/(1- H_t)$, averaged 1.33 (1.18-1.51), while this statistic (which depends on total heterozygosity) averaged 1.31 among widespread species in Hamrick & Godt (1989). Thus, while gypsophiles tend to possess fewer alleles per locus than most rare or endemic plant species, they do not show obviously reduced heterozygosity. Finally, in the three studies that reported observed heterozygosity, H_o averaged 0.14, nearly equal to the figure reported for common taxa in Cole (2003). It is likely that the old ages of these populations, and their long-term stability, have allowed allele frequencies to reach equilibrium.

Population Differentiation

Measures of population differentiation (F_{st} or G_{st}: mean 0.42; or proportion of variation explained by differences between populations from AMOVA: 0.47) were in general higher than those found for rare or common species (Table 2) (Cole, 2003). This is not surprising given the static, island-like distribution of their habitats through time. In *Gypsophila struthium* subsp. *struthium* and *G. s.* subsp. *hispanica* (Martínez-Nieto et al., 2013), chloroplast haplotypes were used in addition to nuclear markers (AFLPs) to estimate population differentiation. The estimated values of among-population chloroplast haplotype differentiation were 0.65 and 0.53 in these two taxa respectively, compared to differentiations of 0.33 and 0.28, as estimated by AMOVA on the AFLP markers. A similarly high value (0.65) was obtained from an AMOVA of chloroplast haplotypes from *Fouquieria shrevei*, the only gypsophile taxon from the Chihuahuan Desert that has been studied at the population level (Aguirre-Liguori et al., 2014). The fact that chloroplast differentiation in *Gypsophila* is roughly twice that of the nuclear genome results from inherent differences in effective population size between genomic compartments, which for chloroplasts in hermaphroditic plants is expected to be ½ that of nuclear loci (Birky et al., 1989). Interestingly, there is no indication of greater differentiation in chloroplast data as one might expect if seed dispersal were more limited than pollen dispersal, or lesser differentiation, which could indicate additional nuclear gene flow via pollen dispersal. In the absence of nuclear data, there is no way to evaluate this in *Fouquieria shrevei*, but it is important to recognize that seemingly very high population differentiation values for chloroplast data do not necessarily imply that seed dispersal is necessarily more restricted than gene flow through pollen.

Migration and Isolation-by-Distance

Migration was estimated by some authors (Allphin et al., 1998; Hickerson & Wolf, 1998; López-Pujol et al., 2004; Pérez-Collazos & Catalán, 2006; Pérez-Collazos et al., 2009), who generally based their estimates on the value of F_{st}. While estimates of the number of effective migrants based on population differentiation must be viewed with extreme caution (Whitlock & McCauley, 1999), for the sake of comparison, we calculated values for the gypsophile species based on the reported among-population variation (Table 2). N_m values thus obtained averaged only 0.59, much lower than the mean value for rare species in Cole (2003). This may reflect the highly discontinuous nature of gypsum outcrops in Spain, where the majority of these population genetic surveys have been conducted. Additional reports from different areas may shed light on how much the patchiness of gypsum outcrops affects migration.

These studies often examined whether genetic distance was correlated with geographic distance, in other words, whether isolation-by-distance (IBD) was evident in their datasets. A stepping-stone model, in which gene flow is a function of geographic distance, is likely to produce such a pattern if populations are at equilibrium. However, if populations have recently expanded into new areas, such a pattern may not have had time to emerge (Slatkin, 1993). Moderate to strong IBD was manifest up to a distance of 24.8 km in the locally distributed gypsoclines *Vella pseudocytisus* subsp. *paui* (Pérez-Collazos & Catalán, 2006) and *Boleum asperum* (Pérez-Collazos et al., 2008). Isolation-by-distance is also evident in the gypsophiles *Gypsophila struthium* sensu lato (Martínez-Nieto et al., 2013) and *Ferula loscosii* (Pérez-Collazos & Catalán, 2008; Pérez-Collazos et al., 2009), which occur in multiple gypsum areas of Spain. In the Chihuahuan Desert, the regionally dominant gypsophile *Fouquieria shrevei* shows IBD as well (Aguirre-Liguori et al., 2014). IBD in these gypsophiles is obviously driven largely by the geographic separation of discrete populations with limited gene flow between them, rather than genetic structure within continuous habitat. In contrast, IBD appears to be weak within subspecies of *Gypsophila struthium* (Martínez-

Nieto et al., 2013) and in the gypsovag *Arctomecon californica* (Hickerson & Wolf, 1998), perhaps limited by sustained high gene flow in comparatively continuous habitat, or by recent population expansion.

General Phylogeographic Patterns

Few of these studies explicitly test a phylogeographic model; however, Pérez-Collazos et al. (2009) discerned a Pliocene colonization of the Iberian Peninsula from north Africa in *Ferula loscosii*, followed by south-to-north dispersal through the Pleistocene. In *Gypsophila struthium* (Martínez-Nieto et al., 2013), chloroplast data suggest that central and eastern Spain represents the ancestral range, which has expanded, and given rise to *G. struthium* subsp. *hispanica* in eastern and, more recently, northern Spain, specifically the Ebro Valley, which is home to several unique gypsophiles. Finally, Aguirre-Liguori et al. (2014) determined that chloroplast haplotypes in *Fouquieria shrevei* are invariant at low elevation sites in western Coahuila, Mexico, which were inundated during pluvials in the Pleistocene, whereas montane gypsum sites show greater diversity, consistent with longer residence of these populations *in situ*. As the number of phylogeographic investigations of gypsophiles increases, we will be able to better characterize the response of populations to historical climate fluctuations, which is key to understanding the diversification of gypsophile floras.

Conclusion

Although great progress has been made in understanding the ecology, assembly and evolution of gypsophile floras worldwide, much remains to be explored in this major but underappreciated edaphic community. The ecological mechanisms controlling the establishment of gypsophile floras deserve further study, especially with regard to interactions among physical, chemical, and biological factors operating in the rhizosphere. Ecological and floristic studies would be particularly welcome in areas with different climates than Spain, especially in places like the Chihuahuan Desert, Iran, and Somalia, all of which have important differences in rainfall amounts and seasonality compared to each other and to Spain, and which are likely to yield many additional gypsophile taxa. Finally, further phylogeographic and phylogenetic studies are needed in gypsum environments throughout the world to assess whether island biogeographic patterns are typical of gypsum archipelagoes, both at the community and genetic level, and to confirm whether different gypsophile communities share similar ages and assembly characteristics.

Acknowledgments

The authors thank Sara Palacio and two additional reviewers for helpful reviews that improved this manuscript, and Arianna Goodman for technical help with figures. This work was supported by grants from Oberlin College, the US National Science Foundation (DEB-1054539) and the National Geographic Society (8873-11) to MJM, and by grants from the

Consejería de Economía, Innovación y Ciencia of the Junta de Andalucía (co-financed with FEDER funds) to JFM.

References

Aguirre-Liguori, J. A., Scheinvar, E. & Eguiarte, L. E. (2014) Gypsum soil restriction drives genetic differentiation in *Fouquieria shrevei* (Fouquieriaceae). *American Journal of Botany 101*, 730–736.

Akhani, H. (2004) A new spiny, cushion-like *Euphorbia* (Euphorbiaceae) from south-west Iran with special reference to the phytogeographic importance of local endemic species. *Botanical Journal of the Linnean Society 146*, 107–121.

Akpulat, H. A. & Celik, N. (2005) Flora of gypsum areas in Sivas in the eastern part of Cappadocia in Central Anatolia, Turkey. *Journal of Arid Environments 61*, 27–46.

Alguacil, M. M., Roldán, A. & Torres, M. P. (2009a) Assessing the diversity of AM fungi in arid gypsophilous plant communities. *Environmental Microbiology 11*, 2649–2659.

Alguacil, M. M., Roldán, A. & Torres, M. P. (2009b) Complexity of semiarid gypsophilous shrub communities mediates the AMF biodiversity at the plant species level. *Microbial Ecology 57*, 718–727.

Alguacil, M. M., Torrecillas, E., Roldán, A., Díaz, G. & Torres, M. P. (2012) Perennial plant species from semiarid gypsum soils support higher AMF diversity in roots than the annual *Bromus rubens*. *Soil Biology and Biochemistry 49*, 132–138.

Allphin, L., Windham, M. D. & Harper, K. T. (1998) Genetic diversity and gene flow in the endangered dwarf bear poppy, *Arctomecon humilis* (Papaveraceae). *American Journal of Botany 85*, 1251–1261.

Anderson, D. C., Harper, K. T. & Holmgren, R. C. (1982) Factors influencing development of cryptogamic soil crusts in Utah deserts. *Journal of Range Management 35*, 180–185.

Angiosperm Phylogeny Group (2009) An update of the Angiosperm Phylogeny Group classification for the orders and families of flowering plants: APG III. *Botanical Journal of the Linnean Society 161*, 105–121.

Arakaki, M., Christin, P.-A., Nyffeler, R., Lendel, A., Eggli, U., Ogburn, R. M., Spriggs, E., Moore, M. J. & Edwards, E. J. (2011) Contemporaneous and recent radiations of the world's major succulent plant lineages. *Proceedings of the National Academy of Sciences, USA 108*, 8379–8384.

Atwood, N. D. & Welsh, S. L. (2005) New species of *Mentzelia* (Loasaceae) and *Phacelia* (Hydrophyllaceae) from New Mexico. *Western North American Naturalist 65*, 365–370.

Axelrod, D. I. (1979) Age and origin of Sonoran Desert vegetation. *Occasional Papers of the California Academy of Sciences, No. 132*.

Bacon, J. D. (1978) Taxonomy of *Nerisyrenia* (Cruciferae). *Rhodora 80*, 159–227.

Birky, C. W., Fuerst, P. & Maruyama, T. (1989) Organelle gene diversity under migration, mutation, and drift: equilibrium expectations, approach to equilibrium, effects of heteroplasmic cells, and comparison to nuclear genes. *Genetics 121*, 613–627.

Borer, C. H., Hamby, M. N. & Hutchinson, L. H. (2012) Plant tolerance of a high calcium environment via foliar partitioning and sequestration. *Journal of Arid Environments 85*, 128–131.

Boscaiu, M., Bautista, I., Lidón, A., Llinares, J., Lull, C., Donat, P., Mayoral, O. & Vicente, O. (2013) Environmental-dependent proline accumulation in plants living on gypsum soils. *Acta Physiologiae Plantarum 35*, 2193–2204.

Boukhris, M. & Lossaint, P. (1970) Sur la teneur en soufre de quelque plantes gypsophiles de Tunisie. *Oecologia Plantarum 5*, 345–354.

Boukhris, M. & Lossaint, P. (1972) Spécificité biogeochimique des plantes gypsophiles de Tunisie. *Oecologia Plantarum 7*, 45–68.

Boyadgiev, T. G. (1974) *Contribution to the knowledge of gypsiferous soils*. Rome: Food and Agriculture Organization of the United Nations.

Cañadas, E. M., Ballesteros, M., Valle, F. & Lorite, J. (2013) Does gypsum influence seed germination? *Turkish Journal of Botany 38*, 141–147.

Cole, C. T. (2003) Genetic variation in rare and common plants. *Annual Review of Ecology, Evolution, and Systematics 34*, 213–237.

Contejan, C. (1881) *Géographie botanique: Influence du terrain sur la végétation*. Paris: Librairie J. B. Baillere et fils.

Damschen, E. I., Harrison, S., Ackerly, D. D., Fernandez-Going, B. M. & Anacker, B. L. (2012) Endemic plant communities on special soils: Early victims or hardy survivors of climate change? *Journal of Ecology 100*, 1122–1130.

Daniells, I. G. (2012) Hardsetting soils: A review. *Soil Research 50*, 349–359.

Davis, D. S., Droop, S. J. M., Gregerson, P., Henson, L., Leon, C. J., Villa-Lobos, J. L., Synge, H. & Zantovska, J. (1986) *Plants in danger: What do we know?* Cambridge: International Union for Conservation of Nature and Natural Resources.

Douglas, N. A. & Manos, P. S. (2007) Molecular phylogeny of Nyctaginaceae: Taxonomy, biogeography, and characters associated with a radiation of xerophytic genera in North America. *American Journal of Botany 94*, 856–872.

Drohan, P. J. & Merkler, D. J. (2009) How do we find a true gypsophile? *Geoderma 150*, 96–105.

Duvigneaud, P. & Denaeyer-de Smet, S. (1966) Accumulation du soufre dans quelques espèces gypsophiles d'Espagne. *Bulletin de la Société Royale de Botanique de Belgique 99*, 263–269.

Duvigneaud, P. & Denaeyer-de Smet, S. (1968) Essai de classification chimique (éléments minéraux) des plantes gypsicoles du bassin de l'Ebre. *Bulletin de la Société Royale de Botanique de Belgique 101*, 279–291.

Duvigneaud, P. & Denaeyer-de Smet, S. (1973) Considerations sur l'ecologie de la nutrition minerale des tapis vegetaux naturels. *Oecologia Plantarum 8*, 219–246.

Erben, M. & Arán, V. J. (2005) *Limonium mateoi* (Plumbaginaceae), a new species from Central Spain. *Anales del Jardín Botánico de Madrid 62*, 3–7.

Escudero, A., Palacio, S., Maestre, F. T. & Luzuriaga, A. L. (2014) Plant life on gypsum: A review of its multiple facets. *Biological Reviews,* DOI: 10.1111/brv.12092

Eswaran, H. & Gong, Z.-T. (1991) Properties, genesis, classification, and distribution of soils with gypsum. In: W. D. Nettleton (Ed.). *Occurrence, characteristics, and genesis of carbonate, gypsum, and silica accumulations in soils* (pp. 89–119). SSSA Special Publication. Madison, WI: Soil Science Society of America.

Filner, S. & Shmida, A. (1981) Why are adaptations for long-range seed dispersal rare in desert plants? *Oecologia 51*, 133–144.

Fink, S. (1991) The micromorphological distribution of bound calcium in needles of Norway spruce [*Picea abies* (L.) Karst.]. *New Phytologist 119*, 33–40.

Food and Agriculture Organization (1990) *Management of gypsiferous soils (FAO Soils Bulletin 62)*. Rome: Food and Agriculture Organization of the United Nations.

Food and Agriculture Organization (1998) *World reference base for soil resources. World Soil Resources Report 84*. Rome: Food and Agriculture Organization of the United Nations.

Forbis de Queiroz, T., Baughman, C., Baughman, O., Gara, M. & Williams, N. (2012) Species distribution modeling for conservation of rare, edaphic endemic plants in White River Valley, Nevada. *Natural Areas Journal 32*, 149–158.

Fowler, B. A. & Turner, B. L. (1977) Taxonomy of *Selinocarpus* and *Ammocodon* (Nyctaginaceae). *Phytologia 37*, 177–208.

Franceschi, V. R. & Nakata, P. A. (2005) Calcium oxalate in plants: formation and function. *Annual Review of Plant Biology 56*, 41–71.

Gándara, E. & Sosa, V. (2013) Testing the monophyly and position of the North American shrubby desert genus *Leucophyllum* (Scrophulariaceae: Leucophylleae). *Botanical Journal of the Linnean Society 171*, 508–518.

Gándara, E., Specht, C. D. & Sosa, V. (2014) Origin and diversification of the *Milla* Clade (Brodiaeoideae, Asparagaceae): a Neotropical group of six geophytic genera. *Molecular Phylogenetics and Evolution 75,* 118-125.

Gibbens, R. P. & Lenz, J. M. (2001) Root systems of some Chihuahuan Desert plants. *Journal of Arid Environments 49*, 221–263.

Gil de Carrasco, C. & Ramos, J. J. (2011) Los suelos yesíferos (gypsisoles) en España. In: J. F. Mota Poveda, P. Sánchez Gómez & J. S. Guirado Romero (Eds.). *Diversidad vegetal de las Yeseras Ibéricas: El reto de los archipiélagos edáficos para la biología de la conservación* (pp. 33–50). Almería, Spain: ADIF-Mediterráneo Asesores Consultores.

Gitzendanner, M. A. & Soltis, P. S. (2000) Patterns of genetic variation in rare and widespread plant congeners. *American Journal of Botany 87*, 783–792.

Graham, A. (2011) The age and diversification of terrestrial New World ecosystems through Cretaceous and Cenozoic time. *American Journal of Botany 98*, 336–351.

Guerrero-Campo, J., Alberto, F., Hodgson, J., García-Ruiz, J. M. & Montserrat-Martí, G. (1999) Plant community patterns in a gypsum area of NE Spain. I. Interactions with topographic factors and soil erosion. *Journal of Arid Environments 41*, 401–410.

Hadjikyriakou, G. & Hand, R. (2011) *Teucrium salaminium* Hadjik. & Hand (Lamiaceae, *Teucrium* sect. *Polium*), a new species from Cyprus. *Candollea 66*, 341–351.

Hamrick, J. L. & Godt, M. J. W. (1989) Allozyme diversity in plant species. In: A. H. D. Brown, M. T. Clegg, A. L. Kahler & B. S. Weir (Eds.). *Plant population genetics, breeding, and genetic resources* (pp. 43–63). Sunderland, MA: Sinauer Associates.

Harrison, S., Damschen, E. & Going, B. M. (2009) Climate gradients, climate change, and special edaphic floras. *Northeastern Naturalist 16*, 121–130.

Harrison, S. & Rajakaruna, N. (Eds.) (2011) *Serpentine: The evolution and ecology of a model system*. Berkeley, CA: University of California Press.

Hawkesford, M., Horst, W., Kichey, T., Lambers, H., Schjoerring, J., Skrumsager Møller, I. & White, P. (2012) Functions of macronutrients. In: P. Marschner (Ed.). *Marschner's mineral nutrition of higher plants*. Third edition (pp. 135–190). London: Academic Press.

Hernández-Hernández, T., Hernández, H. M., De-Nova, J. A., Puente, R., Eguiarte, L. E. & Magallón, S. (2011) Phylogenetic relationships and evolution of growth form in Cactaceae (Caryophyllales, Eudicotyledoneae). *American Journal of Botany 98*, 44–61.

Hernández-Hernández, T., Brown, J. W., Schlumpberger, B. O., Eguiarte, L. E. & Magallón, S. (2014) Beyond aridification: Multiple explanations for the elevated diversification of cacti in the New World Succulent Biome. *New Phytologist 202,* 1382-1397.

Herrero, J. & Porta, J. (2000) The terminology and the concepts of gypsum-rich soils. *Geoderma 96*, 47–61.

Herrero, J. (2004) Revisiting the definitions of gypsic and petrogypsic horizons in Soil Taxonomy and World Reference Base for Soil Resources. *Geoderma 120*, 1–5.

Herrero, J., Artieda, O. & Hudnall, W. H. (2009) Gypsum, a tricky material. *Soil Science Society of America Journal 73*, 1757–1763.

Hickerson, L. L. & Wolf, P. G. (1998) Population genetic structure of *Arctomecon californica* Torrey & Frémont (Papaveraceae) in fragmented and unfragmented habitat. *Plant Species Biology 13*, 21–33.

Hinsinger, P., Bengough, A. G., Vetterlein, D. & Young, I. M. (2009) Rhizosphere: Biophysics, biogeochemistry and ecological relevance. *Plant and Soil 321*, 117–152.

Hiouani, F. (2006) *Influence de la teneur en gypse et de la taille de ses grains sur la capacité de retention en eau sur les sols de zone de Ain Benoui -Biskra.* MS Thesis: Algiers, Ministere de L'Enseignement Superieur et de la Recherche.

Jiménez, J. F. & Sánchez-Gómez, P. (2012) Molecular taxonomy and genetic diversity of *Moricandia moricandioides* subsp. *pseudofoetida* compared to wild relatives. *Plant Biosystems 146*, 99–105.

Johnston, I. M. (1941) Gypsophily among Mexican desert plants. *Journal of the Arnold Arboretum 22*, 145–170.

Jones, J. B. (2012) *Plant nutrition and soil fertility manual.* Second edition. Boca Raton, FL: CRC Press.

Kalra, Y. (1997) *Handbook of reference methods for plant analysis*. Boca Raton, FL: CRC Press.

Kimura, M. & Crow, J. F. (1964) The number of alleles that can be maintained in a finite population. *Genetics 49*, 725–738.

Kimura, M. & Weiss, G. H. (1964) The stepping stone model of population structure and the decrease of genetic correlation with distance. *Genetics 49*, 561–576.

Kleinkopf, G. E. & Wallace, A. (1974) Physiological basis for salt tolerance in *Tamarix ramosissima*. *Plant Science Letters 3*, 157–163.

Le Houérou, H. N. (1969) La végétation de la Tunisie steppique. *Annales de l'Institut National de la Recherche Agronomique de Tunisie 42*, 1–622.

Levin, R. A. (2000) Phylogenetic relationships within Nyctaginaceae tribe Nyctagineae: evidence from nuclear and chloroplast genomes. *Systematic Botany 25*, 738–750.

López-Pujol, J., Bosch, M., Simon, J. & Blanché, C. (2004) Allozyme diversity in the tetraploid endemic *Thymus loscosii* (Lamiaceae). *Annals of Botany 93*, 323–332.

Macchiati, L. (1888) Contribuzione alla flora del gesso. *Nuevo Giornale Botanico Italiano 20*, 418–422.

Marlowe, K. & Hufford, L. (2007) Taxonomy and biogeography of *Gaillardia* (Asteraceae): A phylogenetic analysis. *Systematic Botany 32*, 208–226.

Martínez-Nieto, M. I., Segarra-Moragues, J. G., Merlo, E., Martínez-Hernández, F. & Mota, J. F. (2013) Genetic diversity, genetic structure and phylogeography of the Iberian endemic *Gypsophila struthium* (Caryophyllaceae) as revealed by AFLP and plastid DNA sequences: Connecting habitat fragmentation and diversification. *Botanical Journal of the Linnean Society 173*, 654–675.

Mashali, A. M. (1996) Soil management practices for gypsiferous soils. In: R. M. Poch (Ed.). *Proceedings of the international symposium on soils with gypsum* (pp. 34–52). Lleida: Edicions Universitat de Lleida.

McKown, A. D., Moncalvo, J.-M. & Dengler, N. G. (2005) Phylogeny of *Flaveria* (Asteraceae) and inference of C4 photosynthesis evolution. *American Journal of Botany 92*, 1911–1928.

Merlo, M. E., Mota, J. F., Cabello, J. & Alemán, M. M. (1998) La gipsofilia en plantas: Un apasionante edafismo. *Investigación y Gestión 3*, 103–112.

Merlo, M. E., Rodríguez-Tamayo, M. L., Jiménez, M. L. & Mota, J. F. (2001) Recapitulación sobre el comportamiento biogeoquímico de algunos gipsófitos y halófítos ibéricos. *Monografías Flora Vegetación Béticas 12*, 97–106.

Merlo, M. E., Mota, J. F. & Sánchez-Gómez, P. (2011) Ecofisiología y adaptaciones de las plantas vasculares a las características físicas y químicas de sustratos especiales. In: J. F. Mota, P. Sánchez-Gómez & J. S. Guirado Romero (Eds.). *Diversidad vegetal de las Yeseras Ibéricas: El reto de los archipiélagos edáficos para la biología de la conservación* (pp. 53–73). Almería: ADIF-Mediterráneo Asesores Consultores.

Meyer, S. E. (1986) The ecology of gypsophile endemism in the eastern Mojave Desert. *Ecology 67*, 1303–1313.

Meyer, S. E. & García-Moya, E. (1989) Plant community patterns and soil moisture regime in gypsum grasslands of north central Mexico. *Journal of Arid Environments 16*, 147–155.

Moore, M. J. & Jansen, R. K. (2006) Molecular evidence for the age, origin, and evolutionary history of the American desert plant genus *Tiquilia* (Boraginaceae). *Molecular Phylogenetics and Evolution 39*, 668–687.

Moore, M. J. & Jansen, R. K. (2007) Origins and biogeography of gypsophily in the Chihuahuan Desert plant group *Tiquilia* subg. *Eddya* (Boraginaceae). *Systematic Botany 32*, 392–414.

Mota, J. F., Sola, A. J., Jiménez-Sánchez, M. L., Pérez-García, F. & Merlo, M. E. (2004) Gypsicolous flora, conservation and restoration of quarries in the southeast of the Iberian Peninsula. *Biodiversity and Conservation 13*, 1797–1808.

Mota, J. F., Sánchez Gómez, P., Merlo Calvente, M. E., Catalán Rodríguez, P., Laguna Lumbreras, E., De la Cruz Rot, M., Navarro Reyes, F. B., Marchal Gallardo, F., Bartolomé Esteban, C., Martínez Labarga, J. M., Sainz Ollero, H., Valle Tendero, F., Serra Laliga, L., Martínez Hernández, F., Garrido Becerra, J. A. & Pérez García, F. J. (2009) Aproximación a la checklist de los gipsófitos ibéricos. *Anales de Biología 31*, 71–80.

Mota, J. F., Sánchez Gómez, P. & Guirado Romero, J. S. (Eds.) (2011) *Diversidad vegetal de las Yeseras Ibéricas: El reto de los archipiélagos edáficos para la biología de la conservación*. Almería, Spain: ADIF-Mediterráneo Asesores Consultores.

Oyonarte, C., Sánchez, G., Urrestarazu, M. & Alvarado, J. J. (2002) A comparison of chemical properties between gypsophile and nongypsophile plant rhizospheres. *Arid Land Research and Management 16*, 47–54.

Palacio, S., Escudero, A., Montserrat-Martí, G., Maestro, M., Milla, R. & Albert, M. J. (2007) Plants living on gypsum: Beyond the specialist model. *Annals of Botany 99*, 333–343.

Palacio, S., Johnson, D., Escudero, A. & Montserrat-Martí, G. (2012) Root colonisation by AM fungi differs between gypsum specialist and non-specialist plants: Links to the gypsophile behaviour. *Journal of Arid Environments 76*, 128–132.

Parejo-Farnés, C., Albaladejo, R. G., Arroyo, J. & Aparicio, A. (2013) A phylogenetic hypothesis for *Helianthemum* (Cistaceae) in the Iberian Peninsula. *Botanica Complutensis 37*, 83–92.

Parsons, R. F. (1976) Gypsophily in plants—a review. *American Midland Naturalist 96*, 1–20.

Pérez-Collazos, E. & Catalán, P. (2006) Palaeopolyploidy, spatial structure and conservation genetics of the narrow steppe plant *Vella pseudocytisus* subsp. *paui* (Vellinae, Cruciferae). *Annals of Botany 97*, 635–647.

Pérez-Collazos, E. & Catalán, P. (2008) Conservation genetics of the endangered Iberian steppe plant *Ferula loscosii* (Apiaceae). *Plant Biology 10*, 492–501.

Pérez-Collazos, E., Segarra-Moragues, J. P. & Catalán, P. (2008) Two approaches for the selection of relevant genetic units for conservation in the narrow European endemic steppe plant *Boleum asperum* (Brassicaceae). *Biological journal of the Linnean Society 94*, 341–354.

Pérez-Collazos, E., Sánchez-Gómez, P., Jiménez, J. F. & Catalán, P. (2009) The phylogeographical history of the Iberian steppe plant *Ferula loscosii* (Apiaceae): A test of the abundant-centre hypothesis. *Molecular Ecology 18*, 848–861.

Petrusson, L. & Thulin, M. (1996) Taxonomy and biogeography of *Gymnocarpos* (Caryophyllaceae). *Edinburgh Journal of Botany 53*, 1–26.

Poch, R. M. (1998) Pore space characteristics as indicators of soil behaviour in gypsiferous soils. *Geoderma 87*, 87–109.

Poch, R. M. & Verplancke, H. (1997) Penetration resistance of gypsiferous horizons. *European Journal of Soil Science 48*, 535–543.

Porras-Alfaro, A., Raghavan, S., Garcia, M., Sinsabaugh, R. L., Natvig, D. O. & Lowrey, T. K. (2014) Endophytic fungal symbionts associated with gypsophilous plants. *Botany 92*, 295–301.

Powell, A. M. & Turner, B. L. (1977) Aspects of the plant biology of the gypsum outcrops of the Chihuahuan Desert. In: R. H. Wauer & D. H. Riskind (Eds.). *Transactions of the symposium on the biological resources of the Chihuahuan Desert region, United States and Mexico, Sul Ross State University, Alpine, Texas, 17-18 October 1974* (pp. 315–325). Washington, DC: U.S. Department of the Interior, National Park Service Transactions and Proceedings Series, Number 3.

Richardson, A. T. (1977) Monograph of the genus *Tiquilia* (*Coldenia*, sensu lato), Boraginaceae: Ehretioideae. *Rhodora 79*, 467–572.

Romão, R. L. & Escudero, A. (2005) Gypsum physical soil crusts and the existence of gypsophytes in semi-arid central Spain. *Plant Ecology 181*, 127–137.

Roquet, C., Sáez, L., Aldasoro, J. J., Susanna, A., Alarcón, M. L. & García-Jacas, N. (2008) Natural delineation, molecular phylogeny and floral evolution in *Campanula*. *Systematic Botany 33*, 203–217.

Salmerón-Sánchez, E., Martínez-Nieto, M. I., Martínez-Hernández, F., Garrido-Becerra, J. A., Mendoza-Fernández, A. J., de Carrasco, C. G., Ramos-Miras, J. J., Lozano, R.,

Merlo, M. E. & Mota, J. F. (2014) Ecology, genetic diversity and phylogeography of the Iberian endemic plant *Jurinea pinnata* (Lag.) DC. (Compositae) on two special edaphic substrates: Dolomite and gypsum. *Plant and Soil 374*, 233–250.

Salzmann, U., Haywood, A. M., Lunt, D. J., Valdes, P. J. & Hill, D. J. (2008) A new global biome reconstruction and data-model comparison for the middle Pliocene. *Global Ecology and Biogeography 17*, 432–447.

Schenk, J. J. (2013) Evolution of limited seed dispersal ability on gypsum islands. *American Journal of Botany 100*, 1811–1822.

Schenk, J. J. & Hufford, L. (2010) Effects of substitution models on divergence time estimates: simulations and an empirical study of model uncertainty using Cornales. *Systematic Botany 35*, 578–592.

Schenk, J. J. & Hufford, L. (2011) Phylogeny and taxonomy of *Mentzelia* section *Bartonia* (Loasaceae). *Systematic Botany 36*, 711–720.

Schwarzbach, A. E. & Kadereit, J. W. (1999) Phylogeny of prickly poppies, *Argemone* (Papaveraceae), and the evolution of morphological and alkaloid characters based on ITS nrDNA sequence variation. *Plant Systematics and Evolution 218*, 257–279.

Sivinski, R. & Howard, M. O. (2011) A new species of *Linum* (Linaceae) from the northern Chihuahuan Desert. *Phytoneuron 2011-33*, 1–7.

Slatkin, M. (1993) Isolation by distance in equilibrium and non-equilibrium populations. *Evolution 47*, 264–279.

Symon, D. E. (2007) Lists of gypsophilous plants from southern Australia. *Journal of the Adelaide Botanic Gardens 21*, 45–54.

Tapper, S.-L., Byrne, M., Yates, C. J., Keppel, G., Hopper, S. D., Van Niel, K., Schut, A. G. T., Mucina, L. & Wardell-Johnson, G. W. (2014) Prolonged isolation and persistence of a common endemic on granite outcrops in both mesic and semi-arid environments in south-western Australia. *Journal of Biogeography*, DOI: 10.1111/jbi.12343

Taylor, S. E. (2012) *Molecular systematics and the origins of gypsophily in* Nama *L. (Boraginaceae)*. PhD Dissertation. Austin, TX: University of Texas at Austin.

Thompson, H. J. & Powell, A. M. (1981) Loasaceae of the Chihuahuan Desert region. *Phytologia 49*, 16–32.

Thulin, M. (1993) *Flora of Somalia, Volume 1: Pteridophyta; Gymnospermae; Angiospermae (Annonaceae–Fabaceae)*. London: Royal Botanic Gardens, Kew.

Thulin, M. (1995) *Flora of Somalia, Volume 4: Angiospermae (Hydrocharitceae–Pandanaceae)*. London: Royal Botanic Gardens, Kew.

Thulin, M. (1999) *Flora of Somalia, Volume 2: Angiospermae (Tiliaceae–Apiaceae)*. London: Royal Botanic Gardens, Kew.

Thulin, M. (2006) *Flora of Somalia, Volume 3: Angiospermae (Ericaceae–Asteraceae)*. London: Royal Botanic Gardens, Kew.

Turner, B. L. & Powell, A. M. (1979) Deserts, gypsum, and endemism. In: J. R. Goodin & D. K. Northington (Eds.). *Arid land plant resources: Proceedings of the international arid lands conference on plant resources, Texas Tech University* (pp. 96–116). Lubbock, TX: International Center for Arid and Semi-Arid Land Studies.

Turner, B. L. & Watson, T. (2007) Taxonomic revision of *Gaillardia* (Asteraceae). *Phytologia Memoirs 13*. Gruver, TX; Texensis Publishing.

Verheye, W. H. & Boyadgiev, T. G. (1997) Evaluating the land use potential of gypsiferous soils from field pedogenic characteristics. *Soil Use and Management 13*, 97–103.

Wagenitz, G. & Kandemir, A. (2008) Two new species of the genus *Psephellus* (Compositae, Cardueae) from eastern Turkey. *Willdenowia 38*, 521–526.

Wagstaff, S. J. & Tate, J. A. (2011) Phylogeny and character evolution in the New Zealand endemic genus *Plagianthus* (Malveae, Malvaceae). *Systematic Botany 36*, 405–418.

Wang, B. & Qiu, Y.-L. (2006) Phylogenetic distribution and evolution of mycorrhizas in land plants. *Mycorrhiza 16*, 299–363.

Whitlock, M. C. & McCauley, D. E. (1999) Indirect measures of gene flow and migration: FST|[ne]|1/(4Nm+1). *Heredity 82*, 117–125.

Zachos, J. C., Dickens, G. R. & Zeebe, R. E. (2008) An early Cenozoic perspective on greenhouse warming and carbon-cycle dynamics. *Nature 451*, 279–283.

In: Plant Ecology and Evolution in Harsh Environments
Editors: N. Rajakaruna, R. S. Boyd and T. B. Harris
ISBN: 978-1-63321-955-7

Chapter 6

PHYSIOLOGICAL ADAPTATIONS OF PLANTS TO SERPENTINE SOIL

Emily R. Palm* and Elizabeth Van Volkenburgh
University of Washington, Department of Biology, Seattle, WA, USA

ABSTRACT

Serpentine soil is classically defined as having low calcium (Ca) availability coupled with high concentrations of magnesium (Mg) and toxic heavy metals. The distribution of plants onto serpentine soil is severally limited to only those species that are tolerant of these edaphic factors. The mechanisms for tolerance that have been historically explored include exclusion and tolerance of elevated Mg and heavy metals, and the ability of serpentine plants to acquire Ca despite its dramatically low availability. The uptake and accumulation of heavy metals has been extensively studied in regard to serpentine plants and has informed much of our understanding of heavy metal physiology in plants. The Ca and Mg aspect of serpentine physiology is not as well understood, but it is clear that the low Ca:Mg ratio severely limits the growth of plants not adapted to serpentine soil. This chapter describes our current knowledge of heavy metal, Ca, and Mg uptake, transport, and accumulation as it pertains to the adaptation of plants to serpentine soil. The techniques and approaches used to determine mechanisms of exclusion and sequestration of heavy metals may provide insight into the processes regulating Ca and Mg homeostasis in serpentine-adapted plants.

INTRODUCTION

Serpentine soil is a naturally occurring model system ideal for the study of the physiological responses of plants to edaphic factors. Numerous studies have found that serpentine soil is low in several of the macronutrients essential for plant growth, especially Ca, P, and K. Derived from ultramafic rock rich in the mineral serpentinite, these soils often exhibit extremely high concentrations of Mg (Kruckeberg, 2002; Proctor & Woodell, 1975).

* Corresponding author: eniniane@u.washington.edu.

The combined low availability of Ca with elevated concentrations of Mg creates an unusually low Ca:Mg ratio that is one of the defining characteristics of serpentine soil. The parent material from which serpentine soil originates is highly variable, subject to the local conditions (*e.g.*, temperature and the type of surrounding rock) at the time of its formation (Tyndall & Hull, 1999). As a result, there is a great deal of variation within serpentine soil with respect to the absolute concentrations of Ca and Mg and the presence and concentration of heavy metals. Within serpentine ecosystems, additional abiotic stresses are also often present, including drought, low nutrient cycling rates, and shallow soil depth. This is partly explained by the fact that serpentine soil frequently occurs on rocky, exposed slopes where soil stability is already low and prone to erosion (Jenny, 1980). This suite of abiotic factors and the sparse vegetation observed on serpentine soils has been described as the "serpentine syndrome" (Jenny, 1980). It is a physiologically challenging environment for most plant species and one that may be sufficient to isolate populations (Kay et al., 2011), thereby driving the development of unique physiological adaptations and high levels of endemism (Brooks, 1987; Kruckeberg, 1999). Avoidance of these conditions is a frequent way that plants cope, as exemplified by the seemingly barren landscapes (Tyndall & Hull, 1999), strongly defined ecotypes (Kruckeberg, 1951; Rajakaruna & Bohm, 1999; Rajakaruna et al., 2003), and unique species assemblages found in these geochemically distinct habitats (Koenigs et al., 1982a; b; Wood, 1984).

For the plants that do grow on serpentine soil, the physiological mechanisms by which they cope are not universal, nor are they clearly defined. To address this question, a number of hypotheses have been tested, including tolerance to low Ca:Mg, avoidance of Mg toxicity, an increased requirement for Mg, and tolerance or hyperaccumulation of heavy metals (reviewed by Brady et al., 2005; Kazakou et al., 2008). Despite recent efforts through soil amendment studies and analysis of the elemental composition of serpentine plant communities, a single common factor that limits plant growth on serpentine soil has yet to be identified (Lazarus et al.,2011; O'Dell & Rajakaruna, 2011). The adaptation of plants to serpentine soil is of particular interest to plant physiologists because many of these species tolerate heavy metals, through either exclusion or hyperaccumulation. Most of the known serpentine-endemic species exclude heavy metals; a smaller proportion are classified as hyperaccumulators (Kruckeberg, 1999). The research on heavy metal-hyperaccumulating species from serpentine soil systems is extensive (Baker, 1978; Lombini et al., 2003; Paliouris& Hutchinson, 1991; Proctor, 1970; Wenzel et al., 2003) and has provided excellent insight into the physiological mechanisms of ion uptake, transport and accumulation patterns by plants. This information has been valuable to, and perhaps motivated by, the emerging field of phytoremediation and the selection of plant species particularly well suited for this purpose. The effect of the low Ca and high Mg concentrations on plant growth remains unclear; comparisons of serpentine ecotypes have shown a range of responses to low Ca:Mg ratios, including indifference to elevated Mg (Asemaneh et al., 2007; Marrs & Proctor, 1976; Proctor, 1970; Rajakaruna et al., 2003), exclusion of Mg (Madhok, 1965; Madhok & Walker, 1969; Sambatti & Rice, 2007; Walker et al., 1955), greater acquisition of Ca despite low soil availability (Asemaneh et al., 2007; Tibbetts & Smith, 1993; Wallace et al., 1982) and a greater Mg requirement (Madhok, 1965; Madhok & Walker, 1969). This question is further complicated by evidence of indiscriminant cation uptake and translocation in plants (Hall & Williams, 2003; Shaul, 2002); however, some degree of selectivity may be achieved among serpentine-tolerant species as suggested by differences in the mineral composition of species growing in the same soil (Lazarus et al., 2011; Lyon et al., 1971; Pope et al., 2010).

Here we describe the current knowledge of Ca, Mg, and heavy metal uptake, transport, and accumulation as it pertains to the adaptation of plants to serpentine soil. We focus on the coping strategies found among serpentine-tolerant plants including exclusion, amelioration, and tolerance of nutrient stress. We address the outstanding questions in regard to Ca and Mg and suggest that the growing body of work on the known physiological mechanisms governing heavy metal ion transport and exclusion by plant tissues may provide insight into the processes regulating calcium and magnesium homeostasis in serpentine-adapted plants.

CALCIUM AND MAGNESIUM PHYSIOLOGY

The Role of Ca and Mg in Plant Physiology

The unifying feature of serpentine soils is the low Ca:Mg ratio, the result of decreased availability of Ca and excessive concentration of Mg. This low Ca:Mg ratio represents an edaphic stress to which most plants are intolerant—in general, plants grow best when the Ca:Mg of the soil is close to one or greater. Both Ca and Mg are macronutrients in that they are required at higher concentrations for normal plant growth, relative to the micronutrients. Calcium is a divalent cation that serves two main roles in plant physiology: as a bridging ion that increases the stability of cell walls and membranes and as a universal signaling molecule, facilitating responses to light, water availability, gravity, and abiotic stress (Marschner, 2011; White & Broadley, 2003). At deficient concentrations of Ca, one of the first symptoms observed is the cessation of growth near the meristematic regions. Calcium regulates the activity of polygalacuronase, an enzyme that facilitates the breakdown of pectates of the cell wall during growth (Konno et al., 1984). High Ca concentrations inhibit the activity of polygalacuronase, whereas Ca deficiency leads to the breakdown of cell walls and subsequent tissue instability. In nature, both extremes of calcium availability exist, and mechanisms for acquiring Ca under low availability (*e.g.*, calcifuge soils) and for limiting uptake under high concentrations (*e.g.*, calcareous soils) have been proposed (Lee, 1999). Magnesium, also a divalent cation, serves several vital roles in plant physiology: it is the central molecule of chlorophyll—the pigment responsible for a majority of the light capture in photosynthesis—and it is key for the activation of many enzymatic reactions, including the synthesis and activation of Rubisco—the enzyme that catalyzes the fixation of carbon in photosynthesis (Marschner, 2011). Whereas excessively high concentrations of Mg are not normally found in soils, Mg deficiencies are common, and as a result, they are better understood. In serpentine soils, Mg is not limited, but in fact, exists at concentrations found to be inhibitory to the growth and survival of most plant species; several common crop species and cultivars have been tested in comparison to serpentine-adapted plants and exhibit significant decreases in growth when grown on Mg-rich soils (Lazaroff & Pitman, 1966; Madhok, 1965; Madhok & Walker, 1969; Proctor, 1971). The fundamental role of both cations makes the physiological response and tolerance to low Ca and high Mg an interesting and complex aspect of serpentine-adapted species.

Regulation of Ca and Mg Uptake in Serpentine Plants

The ability to regulate uptake of Ca and Mg from the soil may be a key component of the tolerance observed among many serpentine-adapted plants. The rate of all nutrient acquisition is governed by the rate of transpiration, as the movement of a majority of ions is accomplished through bulk flow into the apoplast of roots (Yeo & Flowers, 2007). The low concentration of Ca in the soil requires that a plant have adaptations to acquire adequate concentrations for maintaining growth and other Ca-dependent processes such as cell signaling in response to stress. The elevated level of Mg in serpentine soil would be toxic if it were able to freely enter the plant. Among non-serpentine plants grown on serpentine soil, symptoms of Ca deficiency and Mg toxicity are indeed observed. The low availability of Ca in serpentine soil is inhibitory to the growth of most plants. Transplants of serpentine-adapted species onto non-serpentine and serpentine soils show little effect on their biomass.

In comparison, non-serpentine plants grown on serpentine soil show dramatic increases in biomass when the soil is amended with additional Ca. These results suggest that serpentine-adapted plants are somehow tolerant or indifferent to the low availability of Ca. Some serpentine-native species have been shown to absorb more Ca than Mg without a significant increase in their biomass, indicating that they are better able to acquire Ca from Ca-limited soil (Asemaneh et al., 2007; Walker et al., 1955). In a survey of several species growing in an area of serpentine soil in Northern California, Wallace et al. (1982) observed that four of the five species that were sampled maintained an internal Ca:Mg ratio that was close to or greater than 1, an surprising result given that the range of soil Ca:Mg was between 0.16 and 0.32. The ratio of Ca:Mg in plant tissue was higher than that of the soil due to high concentrations of Ca within the plant, not solely through exclusion of Mg. This suggests a level of ion selectivity.

The exclusion of Mg has been tested in a number of species at both the root and shoot levels (Madhok, 1965; Madhok & Walker, 1969; Palm et al., 2012; Rajakaruna et al., 2003; Sambatti & Rice, 2007; Walker et al., 1955). A serpentine ecotype of *Helianthus exilis* (Asteraceae) demonstrated Mg exclusion from its shoots that was not observed in a non-serpentine ecotype of *H. annuus* (Sambatti & Rice, 2007). However, in a comparison of serpentine and non-serpentine ecotypes of *Mimulus guttatus* (Phrymaceae), similar results were not found. Instead, the serpentine ecotype of *M. guttatus* had higher concentrations of Mg in the shoots than the non-serpentine ecotype, suggesting a difference in their respective ability to translocate Mg from the roots to the shoots (Palm et al., 2012). The soil chemistry of a serpentine soil outcrop in the Jasper Ridge Preserve in Northern California proved to be predictive of the presence of specific races of *Lasthenia californica* (Asteraceae) and the mineral composition of their tissues (Rajakaruna & Bohm, 1999). Race A plants occurred in areas of elevated Mg, Na, and pH levels, which correlated with increased internal concentrations of Mg and Na. In contrast, Race C plants had higher concentrations of Ca, K, and Ni and higher Ca:Mg ratios, all of which were a reflection of soil conditions at Race C sites. Cultivation of Race C plants in soil collected from Race A sites significantly decreased the development of leaf tissue and overall biomass. At the same time, Race A plants showed the same biomass and leaf number regardless of whether they were grown in soil from Race A or C sites. A mechanism based solely on exclusion may be too restrictive to describe the accumulation patterns observed among many serpentine-adapted species. Instead, a combination of mechanisms leading to differences in Mg accumulation patterns (*e.g.*, in roots versus shoots) and regulation of translocation should be considered, as has been shown for heavy metal accumulation.

Our understanding of the roles of Ca and Mg in serpentine physiology is complicated by the fact that both cations are divalent and that cation channels are often not ion-specific (Pineros & Tester, 1997; Shaul, 2002; White & Broadley, 2003). These factors may be among the reasons it has been difficult to determine whether the low availability of Ca or the excessive concentration of Mg is more problematic for plants not adapted to serpentine soil. Many studies documenting differences in the mineral composition of plant tissues from serpentine and non-serpentine populations have concluded that there must be some selective uptake mechanism; however, that mechanism has not been clearly defined for either Ca or Mg. As ions move from the soil and into the roots to the vasculature, there are a number of places where ion selectivity is likely to be important. As depicted in the root cross-section shown in Figure 1 (adapted from Shaul, 2002), the uptake of ions from the soil into the plant occurs via one of two pathways: symplastically, from cell-to-cell through the plasmodesmata, or apoplastically, between cells in the apoplastic space, driven by bulk flow (Marschner, 2011). Following the capillary-driven movement of water through the plant, ions will reach a barrier to their transport regardless of pathway. In the symplastic pathway, the plasma membrane of root epidermal cells exerts selective permeability achieved by ion channels to maintain ion homeostasis and permit ion movement through the cortex via plasmodesmata. For the apoplastic pathway, that barrier is the Casparian strip, a waxy layer that is embedded in the endodermis surrounding the vascular bundle. To move beyond the Casparian strip, ions in the apoplast must eventually enter the symplast through channels. The ability for serpentine-tolerant plants to acquire greater concentrations of Ca or to exclude Mg relative to non-tolerant plants in soils with low Ca:Mg ratios may be due to increased specificity of ion channels at the plasma membrane of the epidermal or cortical cells within the root. Within the vascular bundle, transport proteins in the xylem parenchyma cells regulate the loading and unloading of ions into and from the transpiration stream. Specificity of transport proteins responsible for the loading of ions into the xylem may reside in whether they efflux ions out of the cell or pump them into the vacuole (Shaul, 2002; Shaul et al., 1999). Thus far, no studies have demonstrated differences in membrane permeability or the degree of ion specificity of these transport proteins with regard to serpentine tolerance in plants.

Several nonselective channels have been shown to transport a wide range of divalent cations and some do have the potential to be selective under a certain set of conditions, such as depleted or elevated concentrations of a particular cation. For example, the *rca* channel transports Ca across the plasma membrane of root cells (Pineros & Tester, 1995; 1997; White et al., 2000). This channel is not specific to Ca; it has the ability to transport a wide range of ions, including Mg, in the absence of Ca. The specificity of this channel is based on its very low affinity for Mg, relative to Ca. Similar mechanisms have been observed in response to salt stress and the transport of Na and K. In *Arabidopsis thaliana* (Brassicaceae), a combination of high- and low-affinity K transporters is key to maintaining intracellular homeostasis, especially under elevated Na (Zhu, 2003). The affinities of Ca and Mg channels should be analyzed under concentrations representative of field conditions. A variety of electrophysiological techniques exist to measure flux rates in roots and across membranes, including microelectrodes to measure changes in membrane potentials in response to concentration gradients and the isolation of channels through patch clamp.

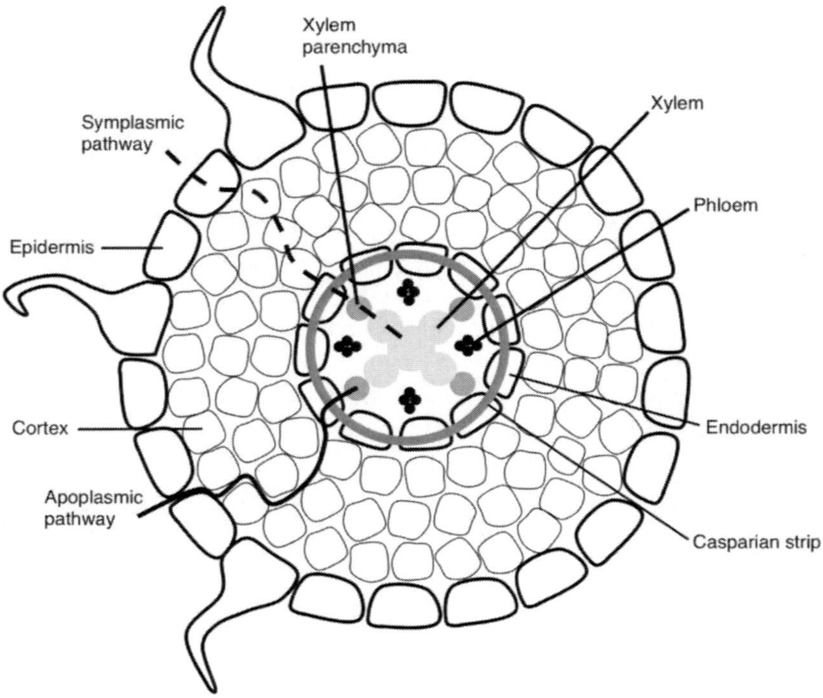

Figure 1. Schematic cross-section of a root. Similarly shaped and shaded cells indicate the same cell type as described by the labels provided. The symplasmic and apoplasmic ion pathways are indicated by dashed and solid lines, respectively.

Increased Mg Requirement

Among serpentine tolerant plants, there are some species that exhibit increased yield in response to elevated external concentrations of Mg. This observation leads to the suggestion that increased Mg in the growing condition is required for maintaining adequate internal conditions and the normal growth of these plants (Madhok, 1965; Madhok & Walker, 1969; Main, 1974; 1981). Madhok (1965) and Madhok & Walker (1969) showed that the serpentine endemic sunflower, *Helianthus bolanderi* subsp. *exilis* is one such example. When grown under increasing concentrations of Mg (0.005 mM to 2.0 mM) with a constant Ca concentration of 4.0 mM, *H. bolanderiexilis* showed symptoms of Mg deficiency under low Mg concentrations but this was reversed at higher concentrations of Mg. In contrast, the common cultivated sunflower, *H. annuus*, did not show Mg deficiency at any of the Mg concentrations used, but did demonstrate decreased yield with increasing Mg concentration. Similar results have been found in several studies, and the detailed chemical analyses of root and leaf tissues do show differences in the internal concentrations of Mg and Ca between serpentine-adapted and closely related non-adapted species. In the case of *H. bolanderii exilis*, the Mg concentration reached 10 mM before internal concentrations were comparable to the internal concentration of the non-serpentine-adapted *H. annuus* grown at 2 mM. In instances such as this, the slow absorption of Mg suggests a difference in the Mg-permeability of some serpentine species, with the result that high external concentrations of

Mg are required obtain adequate internal Mg concentrations and thereby escape Mg deficiency.

A wide range of genetic tools are available to identify quantitative trait loci (QTLs) that may pinpoint specific genes for Mg and Ca transporters. Such approaches have been used to identify candidate genes that may be involved in conferring tolerance to serpentine conditions. An induced loss of function mutation in the candidate gene *CAX1* in *A. thaliana* revealed individuals that were tolerant to a serpentine-like low Ca:Mg ratio (Bradshaw, 2005). *CAX1* encodes for a Ca-proton antiporter on the tonoplast that returns cytoplasmic Ca concentrations to within the normal range (10^{-7} mol^{-1}) after a Ca-dependent signaling event. When wild type *A. thaliana* is cultivated under conditions with low Ca: Mg ratios, the activity of the *CAX1* exchanger creates cytoplasmic concentrations of Ca that are significantly below the normal range. Bradshaw (2005) proposed that in combination with the influx of Mg through non-selective cation channels, ion homeostasis is disrupted, leading to plant death among wild type individuals. *Mimulus guttatus*, is a highly adaptable wildflower growing across the western United States. Populations grow on and off of serpentine soil and appear to be differentially adapted to their native soils (Palm et al., 2012). With its sequenced genome, *M. guttatus* has the potential to be a useful model for understanding Ca and Mg balance in terms of serpentine tolerance (Wu et al., 2008).

HEAVY METAL PHYSIOLOGY

Heavy Metals, Plants, and Serpentine Soil

Heavy metals found in serpentine soil represent both essential micronutrients (*e.g.*, Cu, Ni, Zn) that are required by plants in small quantities and non-essential ions (*e.g.*, Cd, Hg, Pb), both of which are toxic at high concentrations. Heavy metal concentrations in serpentine soil often far exceed those required for normal plant growth (Brooks, 1987). The heavy metal micronutrients (*i.e.*, Cu, Ni, Zn) play important roles in redox reactions and enzymatic activity, chloroplast development (Tripathy & Mohanthy, 1980), electron transfer in the light reactions (Baron et al., 1995;Mysliwa-Kurdziel et al., 2002), and lignification of the cell wall and protein synthesis (Marschner, 2011). Common symptoms of heavy metal toxicity include chlorosis of young leaves, inhibition of root growth, and destruction of membranes due to lipid peroxidation under elevated Cu concentrations (De Vos et al., 1989). A range of coping mechanisms have been observed, from exclusion by selective uptake at the roots to controlled translocation to the shoots and subsequent hyperaccumulation in the leaves.

Heavy Metal Exclusion and Phytochelators

While many hyperaccumulating species are endemic to serpentine soil, a majority of serpentine-adapted species and ecotypes are heavy metal excluders (Kruckeberg, 1999). Heavy metal excluders include species whose uptake and accumulation of excess heavy metal ions is restricted to the roots, consequently maintaining low concentrations of heavy metals in shoots relative to concentrations in the soil, and in some cases, concentrations in the roots

(Baker, 1981). Studies of plant communities growing on the same or adjacent serpentine soil sites have revealed that the shoot concentration of heavy metals can vary among species and growth form (*e.g.*, forb versus shrub; Lazarus et al., 2011; Nagy & Proctor, 1997; Wallace et al., 1982) and between ecotypes of the same species, (*e.g.*, *L. californica*; Rajakaruna & Bohm, 1999). The mechanism that prevents the uptake of heavy metals and subsequent translocation to the shoots of many serpentine-tolerant species is not known in each case. There is evidence that the addition of organic material to serpentine soil can significantly reduce the uptake of heavy metals (*e.g.*, Mn, Ni) by affecting soil pH and thereby altering the solubility of the metals (Fernandez et al., 1999; O'Dell & Claassen, 2006). Within the root, transporters may restrict the movement of heavy metals across the plasma membrane or the Casparian strip (Figure 1) based on their specificity (Hall & Williams, 2003). The root cell wall itself may act as a site of storage for some heavy metals, as shown in the heavy metal-indifferent fern, *Athyrium yokoscense* (Athyriaceae; Nishizono et al., 1987). This species only accumulates heavy metals when growing in the presence of high concentrations of Cd, Cu, and Zn. It will accumulate high concentrations of Zn and Cu in its roots, and Cd in its leaves without significant reduction in its growth. When exposed to increasing concentrations of Cd, Cu, and Zn, a greater proportion of the Zn found in the root cells was contained in the cytoplasm than was bound to the cell walls (Nishizono et al., 1987). The opposite result was observed for Cu, indicating that in *A. yokoscense*, the binding capacity of the root cell wall for Cu is greater than that for Zn. Therefore, the tolerance to Cu in this species of fern is due to the Cu ion-exchange capacity of the root cell walls. Because a greater proportion of the cellular Zn and Cd was found in the cytoplasm, a different tolerance mechanism is required for Zn and Cd, both of which have a lower cell wall ion-exchange capacity than does Cu. Through mechanisms such as sequestration by the root cell wall, a plant may limit the amount of the heavy metal that enters the shoot tissue.

Heavy Metal Hyperaccumulation

Translocation of ions from root to shoot tissue requires channels and transporters in root cortical cells and in the cells that aid in the loading of the xylem, the xylem parenchyma cells. This is a process that may vary in ion selectivity. *Noccaea caerulescens* (Brassicaceae) is a species that is commonly found growing in serpentine soils with elevated concentrations of Cd, Co, Ni, and Zn, with various ecotypes existing as Zn (Brown et al., 1995) and Cd (Lombi et al., 2000) hyperaccumulators. It has been shown to possess an altered ion transport mechanism allowing increased heavy metal influx and efficient translocation from the roots to the shoots (Papoyan & Kochian, 2004). The specific alteration in ion transport is due to an ATPase, coded for by the gene *TcHMA4* and a homolog of *AtHMA4* in *A. thaliana* that is responsible for mediating heavy metal efflux out of the cell (Hussain et al., 2004). It was determined through yeast complementation screening that *TcHMA4* is also a transporter responsible for heavy metal tolerance (Papayon & Kochian, 2004). When *TcHMA4* was expressed in yeast, it reduced accumulation of Cd. The length of the peptide that was used in the transformation vector proved to be important for the degree of tolerance that was conferred to the transformed yeast. The C-terminus cytoplasmic tail of the protein is thought to have a number of heavy metal binding domains. When portions of the full-length transporter are expressed, especially those that include the C-terminus tail, tolerance to increased concentration of Cd is conferred to the transformed yeast, coinciding with increased accumulation of Cd within the yeast cell. The full-length protein confers tolerance as well, but through efflux of Cd from the yeast cell, as observed from reduced accumulation of Cd. The partial peptides are thus able to bind the heavy metals and accumulate them within the transformed cell, whereas the full-length protein allows for transport of Cd out of the cell and avoidance of accumulation. From this, the authors propose that the role of *TcHMA4* in *N. caerulescens* is not directly involved in tolerance but rather in active xylem uploading of heavy metals for translocation to the shoot tissue (Papoyan & Kochian, 2004). Although complementation studies for *TcHMA4* were performed solely in response to Cd, the expression of the gene was shown to also be upregulated in *N. caerulescens* seedlings in response to elevated concentrations of Zn.

Hyperaccumulation of heavy metals into the shoot tissue requires transport through the xylem, a process that results in measurable changes in the xylem sap composition (Alves et al., 2011; Lasat et al., 1996; 1998). The xylem sap of the hyperaccumulator *N. caerulescens* has been observed to have an increased concentration of Zn as compared to a non-hyperaccumulating related species, *T. arvense* (Lasat et al., 1998). Lasat et al. (1996) showed that *N. caerulescens* has a higher influx of Zn in response to increasing Zn exposure than the non-hyperaccumulator, *T. arvense,* and greater accumulation of Zn in the shoots than in the roots, indicating a higher rate of Zn translocation to the shoot tissue. A similar process has been observed in *Alyssum serpyllifolium* subsp. *lusitanicum* (Brassicaceae), in which Ni is accumulated primarily in the epidermal cells of leaves (de la Fuente et al.,2007). The xylem sap concentrations of Ni in *A. serpyllifolium*, a known Ni hyperaccumulator, is higher than the concentrations observed in two other serpentine species growing in the same soil, *Cistus ladanifer* (Cistaceae; Lázaro et al., 2006) and *Quecus ilex* (Fagaceae; Nabais, 2000), neither of which is a Ni hyperaccumulator. Alves et al. (2011) concluded that the heavy metal concentration of the xylem sap may be a potential indicator of the heavy metal status of the aboveground tissues.

Heavy Metal Hyperaccumulation: Sequestration

The accumulation of heavy metals is made possible in many species by their compartmentalization to specific cell types, such as in the epidermal layer of leaves (de la Fuente et al., 2007; Küpper et al., 1999), the mesophyll cells (Küpper et al.,2000; Zhao et al.,2000), and the trichomes (Ghasemi & Ghaderian, 2009), and further by the subcellular sequestration of heavy metals into the vacuoles (Boyd & Martens 1992; Gabbrielli et al., 1991;Küpper et al., 2001). For hyperaccumulators, both the transport of heavy metals and the localization are important points of regulation for minimizing long-term damage. The cellular localization of Zn, Cd, and Ni has been extensively studied in heavy metal hyperaccumulators growing on serpentine soil (Cosio et al., 2005; Ghasemi & Ghaderian, 2009; Küpper et al., 1999; Zhao et al., 2000). In *N. caerulescens,* significant proportion of the Zn that is taken up by the plant is found in the epidermal cells of mature leaves, but the mechanism through which Zn is preferentially sequestered to these specific cells has yet to be identified. Whereas Zn is stored in the vacuoles of epidermal cells, it was concurrently observed that P was concentrated predominantly in the mesophyll cells of leaves (Küpper et al., 1999). Küpper et al. (1999) suggested that this separation of Zn and P into different cell types may prevent the co-precipitation of Zn with P that would otherwise lead to P deficiency. This particular pattern of Zn accumulation into the epidermal cells may be due to differences in selectivity between the interface of xylem cells, minor veins, mesophyll cells, and epidermal cells. The capacity of the vacuoles to store a higher quantity of a particular ion may also vary between cell types, leading to the accumulation of one ion over the other. Cosio et al. (2005) investigated the regulation of Cd and Zn accumulation patterns *Arabidopsis halleri* and two ecotypes of *N. caerulescens*, and showed that the plasma membranes of isolated mesophyll cells were not responsible for determining the differential final accumulation of Zn and Cd in the mesophyll cells of *A. halleri,* and Zn in the epidermal cells of one ecotype of *N. caerulescens*. They concluded that final accumulation of heavy metals must be directed prior to the plasma membrane, and is likely regulated by transport through the apoplast. The capacity of the cell wall to bind to particular ions should be further investigated and compared to the accumulation patterns of hyperaccumulating plants grown in the field.

Heavy Metal Hyperaccumulation: Detoxification

A second mechanism employed to avoid heavy metal toxicity by hyperaccumulators is that of detoxification. Several serpentine and heavy metal tolerant plants show increased concentrations of chelators, both phytochelatins and organic acids, that bind to heavy metals and reduce their ability to interact with proteins and membranes (Sanità di Toppi et al., 2003). Phytochelatins are metal-binding peptides that are involved in detoxification of heavy metals functioning mainly in the cytosol of cells. Synthesis of these peptides is strongly induced by exposure to Cd and, to a lesser degree, by Zn. Phytochelatin synthase, the enzyme responsible for phytochelatin synthesis was first identified and studied in *Silene vulgaris* (Caryophyllaceae; Grill et al., 1989), a serpentine-adapted species (Kruckeberg, 1984) with Cd-tolerant and Cd-sensitive ecotypes. Whereas the production of phytochelatins is induced only after exposure to heavy metals, phytochelatin synthase is constitutively produced, leading to the rapid production of metal binding peptides in response to heavy metal exposure. Despite the role that phytochelatins play in chelating heavy metals, a lower concentration of phytochelatins were found in the roots of the Cd-tolerant ecotype of *S. vulgaris* than in the Cd-sensitive ecotype when both were cultivated in a nutrient solution containing elevated levels of Cd (de Knecht et al., 1994). In Cd-tolerant *S. vulgaris,* phytochelatins are produced and may contribute to Cd detoxification, but they are not the sole mechanism responsible for conferring tolerance to Cd. In contrast, *N. caerulescens* ecotypes collected from a Zn-rich mine tailing site, a non-contaminated soil site, and a Zn-rich serpentine site showed elevated root and shoot concentrations of phytochelatins across all three ecotypes in response to Cd, Ni, and Zn (Schat et al., 2002). Increases in phytochelatins in response to the essential micronutrients Zn and Ni were small relative to the response to As, a non-essential heavy metal. Phytochelatin abundance increased dramatically in the three ecotypes with exposure to As, indicating that phytochelatins may be more important in the detoxification of non-essential metals relative to that of micronutrient metals. Similarly, the carboxyl groups of organic acids are negatively charged and may bind to heavy metal cations in the cytosol and vacuoles of cells. Citric, malic, oxalic, and malonic acids have been shown to play a role in heavy metal detoxification, long-distance transport, and hyperaccumulation in a number of species (Alves et al., 2011; Montargès-Pelletier et al., 2008). In *A.*

serpyllifolium, translocation of Ni is facilitated by citric acid, which, as the authors concluded, may be less costly to the plant than other potential, N-based metal ligands (Alves et al., 2011). In serpentine soils, where many macronutrients such as N are in limited supply, chelation of heavy metals by organic acids may provide a more efficient means of transporting heavy metals to the shoots.

Histidine, an amino acid, also acts as a heavy metal chelator. In *Alyssum lesbiacum*, a Ni hyperaccumulator, histidine levels increased in response to exposure of the plant to elevated concentrations of Ni (Krämer et al., 1996). In the closely related *Alyssum montanum,* a species not known for being a hyperaccumulator, no increase in histidine was observed. The exogenous application of histidine to the roots of *A. montanum* led to an increase in biomass, root elongation, and the rate of Ni transport through the xylem. Along with the increase in histidine, *A. lesbiacum* also exhibited an increase in biomass in response to Ni and greater accumulation of Ni in the shoots as compared to *A. montanum.* A lack of a measureable difference in the root concentrations between the two species suggests that histidine may bind to Ni and facilitate its translocation from the roots to the shoots in *A. lesbiacum.* A similar response to Ni exposure was found in the hyperaccumulators *Alyssum murale* and *A. bertolonii* (Krämer et al., 1996), and can be induced in the non-accumulator *Brassica juncea* (Brassicaceae) with application of exogenous histidine (Kerkeb & Krämer, 2003), indicating that the role of histidine as a metal chelator may be a common mechanism of Ni tolerance and xylem loading in plants, but it is not universal.

Hyperaccumulation of heavy metals and other non-essential ions may be beneficial to some plant species, facilitating the uptake of other essential macronutrients that may be found in limited quantities in the soil. Heavy metals may be useful as a way to reduce the inhibition of growth due to drought by acting as an osmoticum (Baker & Walker, 1989; Rajakaruna & Bohm, 1999), another abiotic stress commonly found in serpentine areas. Transport of heavy metals into the plant may be a by-product of attempting to accumulate adequate amounts of other essential nutrients from a source that is already naturally depleted. Heavy metals may also be hyperaccumulated in some plant species to provide additional resources necessary for the production of secondary metabolites, compounds that are an important protection against herbivory (Chapter 10). Therefore, sequestration is a solution to the potentially inhibitory effects of otherwise unregulated accumulation.

CONCLUSION

Two significant questions remain pertaining to the physiology of Ca and Mg in serpentine-adapted plants: 1) are serpentine plants more selective in their uptake of Ca and Mg from the soil, and if not, 2) are serpentine plants more tolerant of or acclimated to different internal concentrations of Ca and Mg relative to non-serpentine-adapted plants? Clearly, as multiple studies on the physiology of serpentine tolerance have shown, there is not a single answer for all serpentine-adapted plants. Differences in the mineral profiles among plants growing on the same site indicates that there is no hard and fast mechanism for serpentine tolerance, or one that is generally employed by all plants. As for serpentine physiology, we have extensive knowledge of the transport mechanisms for ions, sequestration, and localization of heavy metals, especially among hyperaccumulators

(Chapter 10). Similar approaches to those used for studying heavy metal regulation should be used to identify the mechanisms by which Ca and Mg homeostasis is achieved in serpentine-tolerant plants.

It is understood that nutrients may move into the roots of plants through the cortex by diffusion, governed by the flow of water regulated by the transpiration stream, but blocked by the significant regulation of the endodermis. Kuhn et al. (2000) showed that the rate of movement of Mg and Ca past the endodermis is up to two orders of magnitude slower than the observed movement through the root cortex. A majority of the nutrients absorbed from the soil may move freely through the root apoplastic space in this way, described as "bulk flow". The endodermis prevents the unregulated passage of ions into the vasculature and through the rest of the plant via the xylem tissue. Attempts have been made to analyze the entry of ions into roots and their subsequent translocation into the shoot tissue, but visualization and sample preparation techniques have proved problematic. Radioactively labeled Cu has been used to measure the uptake of Cu from the soil and has revealed differences in cell wall binding and sequestration by heavy metal-tolerant plants. Efforts to measure ion and water uptake have used fluorescent dyes or substitute ions such as Rb for Ca. If the goal of these studies is to elucidate the movement of specific ions through the root and across membranes, the use of fluorescent dyes and substitute ions may give misleading results due to differences in the size and charge of the tracer used and how that may be affected by the specificity of ion transporters. In a study that aimed to measure the uptake kinetics of Ca and Mg in mycorrhizal roots of *Picea abies* (Pinaceae), Kuhn et al. (2000) used stable isotopes of Ca and Mg. With the laser-microprobe-mass analyzer (LAMMA) technique, they were able to measure relative elemental concentrations of root cross-section samples after an incubation period in $^{25}Mg^{2+}$ and $^{44}Ca^{2+}$, evaluated over time. In general, there was no significant difference between the transport of Ca and Mg into the cortical cell walls or into the stele, but a greater amount of both tracers was observed in the stele at 22^{C} than at 6^{C}, suggesting a temperature dependence that may indicate protein-mediated transport across the plasma membrane. Transport proteins show temperature dependence, which is indicative of symplasmic transport of ions. Temperature could therefore be a relevant factor to consider when comparing the kinetics of ion uptake and the adaptive response of plant species to multiple abiotic factors in nature.

For serpentine-tolerant species that are not selective in the uptake of Ca and Mg, a mechanism for tolerance, or amelioration of internal concentrations must exist, especially for Mg. Two possibilities have been shown among heavy metal hyperaccumulators that may also play a role in Mg tolerance—chelation and sequestration. Tibbetts & Smith (1993) evaluated the vacuole contents of the leaves of *Sedum anglicum* (Crassulaceae), a serpentine-adapted species that grows in Mg-rich soils. The cell sap, derived from the vacuoles, decreased in Ca concentration and increased in Mg concentration with decreasing Ca:Mg ratio of the growing conditions, with a selective preference for Ca uptake over Mg at the root level. At the same time, the concentration of organic acids in the vacuole did not change in response to the Ca: Mg ratio and the proportion of metal-carboxylate to free cations remained constant. The majority of the Ca and Mg cations were bound to carboxylates in the vacuole with less than half the total concentration remaining as free ions regardless of the external Ca:Mg ratio. Thus, there was no preferential binding between the two divalent cations, and the carboxylates represent a flexible storage location for excess concentrations of either Ca or Mg.

Little is known about the effects of increased internal concentrations of Mg on plants in general and whether this is the main cause of reduced growth in non-serpentine plants grown under elevated Mg concentrations. Specifically, the effect of intracellular Mg imbalance on ion flux across the chloroplast envelope with respect to photosynthetic rate is unknown. Elevated concentrations of Mg in the cytosol decrease the photosynthetic rate, indicating that *in vivo*, the intracellular distribution of Mg is as tightly regulated as that of Ca. Cytosolic concentrations of Mg may alter the flux of K^+ and H^+ in and out of the stroma of the chloroplasts, respectively (Berkowitz & Wu, 1993). A change in stromal pH is required for the maintenance of proton gradients in the light reactions of photosynthesis. In the presence of high cytosolic concentrations of free Mg, cation-binding sites on the chloroplast membrane may be filled by Mg, changing the membrane potential and inhibiting the flux of H^+ out of the chloroplast stroma as the transport of the counterion K^+ is reduced (Gupta & Berkowitz, 1989). Under conditions of water stress, a decrease in photosynthetic rate has been attributed to increases in internal Mg concentrations in *Phaseolus vulgaris* (Fabaceae; Rao et al., 1987). It is not known whether reductions in growth are due to direct effects on growth or the photosynthetic process itself. As has been found in response to Cu and Zn, the activation of organic acid synthesis should be evaluated in response to Mg. A clear determination of the amount of free versus bound Mg needs to be made, as well as the potential for chelation as a mechanism for Mg detoxification.

The ability to translocate accumulated heavy metals from the roots to the shoots would demand that expression patterns of genes coding for putative heavy metal transporters coincide with specific cells involved in the sequestration and active uploading of ions into the vasculature (reviewed in Verbruggen et al., 2009). In *A. thaliana*, *AtHMA4* is expressed predominately in the vasculature of the roots, shoots, and leaves (Hussain et al., 2004). Whereas the cell-specific expression of *TcHMA4* was not measured in *N. caerulescens,* it was shown that *TcHMA4* is highly expressed in the roots, with increased expression in response to Cd exposure. Shaul et al. (1999) identified another cation transporter in *A. thaliana*, *AtMHX*, localized to the tonoplasts of xylem parenchyma cells, the cells immediately surrounding the xylem. The cell-specific expression pattern of *AtMHX* in the xylem parenchyma cells and the observed exchange of Mg ions with protons in patch clamp analysis makes it likely that this transporter is involved in the active uploading of ions from the roots (Shaul et al., 1999). *AtMHX* is mainly a Mg-transporter, but was shown also to actively transport Zn into the xylem of the roots (Shaul et al., 1999). Mg and H^+ exchangers identified in *Hevea brasiliense* (Euphorbiaceae) not only actively accumulate Mg in lactiferous vessels responsible for the synthesis of rubber but also show a high affinity for the transport of Cd and Zn (Amalou et al., 1994; Shaul, 2002).

The idea that a plant could accumulate a nutrient that would be, in most plants, completely inhibitory to its physiology and growth, and to do so at levels that far exceed the minimal requirements, is an attractive question for physiological research. The potential application of these plants to phytoremediation and the amelioration of soil made toxic through anthropogenic causes is further motivation to understand the physiological mechanisms of heavy metal tolerance in serpentine plants. However, there is a great deal of variation in the type and concentration of heavy metals present in serpentine soils and relatively few plant species can tolerate or hyperaccumulate heavy metals. Several key factors from the study of heavy metal physiology include the affinities of channels and transporters for specific ions, endodermal regulation of ion uptake, and the cellular and subcellular localization of heavy metal ions. Many of the techniques already used in the study of heavy metal transport and localization could be applied to the question of Ca and Mg homeostasis in serpentine-adapted plants. Recently developed genetic tools may broaden our understanding of the evolution of physiologically adaptive traits. Despite the extensive research on plant nutrition, what is clear from studies on serpentine physiology is that there is still much to learn about the effects of edaphic factors on plant growth and development.

REFERENCES

Alves, S., Nabais, C., Goncalves, M. L. S. & Correia dos Santos, M. M. (2011) Nickel speciation in the xylem sap of the hyperaccumulator *Alyssum serpyllifolium* ssp. *lusitanicum* growing on serpentine soils of northeast Portugal. *Journal of Plant Physiology 168,* 1715-1722.

Amalou, Z., Gibrat, R., Trouslot, P. & d'Auzac, J. (1994) Solubilization and reconstitution of the Mg^{2+}/H^{+}antiporter of the lutoid tonoplast from *Heveabrasiliensis* latex. *Plant Physiology 106*, 79-85.

Asemaneh, T.,Ghaderian, S. M. & Baker, A. J. M. (2007) Responses to Mg/Ca balance in an Iranian serpentine endemic plant, *Cleome heratensis* (Capparaceae), and a related non-serpentine species, *C. foliolosa*. *Plant and Soil 293*, 49-59.

Baker, A. J. M. (1978) Ecophysiological aspects of zinc tolerance in *Silenemaritima* With. *New Phytologist 80*, 635-642.

Baker, A. J. M. (1981) Accumulators and excluders – Strategies in the response of plants to heavy metals. *Journal of Plant Nutrition 3,* 643-654.

Baker, A. J. M. & Walker, P. L. (1989) Ecophysiology of metal uptake by tolerant plants. In: A. Shaw (Ed.). *Heavy metal tolerance in plants – Evolutionary aspects* (pp. 155-177). Boca Raton, FL: CRC Press.

Baron, M., Arellano, J. B. & Gorge, J. L. (1995) Copper and photosystem II: A controversial relationship. *Physiologia Plantarum 94*, 174-180.

Berkowitz, G. A. & Wu, W. (1993) Magnesium, potassium flux and photosynthesis. *Magnesium Research 6,* 257-265.

Boyd, R. S. & Martens, S. N. (1992) The raison d'être for metal hyperaccumulation by plants. In: A. J. M. Baker, J. Proctor& R. D. Reeves (Eds.).*The vegetation of ultramafic (serpentine) soils: Proceedings of the first international conference on serpentine ecology* (pp. 279-289). Andover, Hampshire: Intercept.

Bradshaw, H. D. Jr. (2005) Mutations in CAX1 produce phenotypes characteristic of plants tolerant to serpentine soils. *New Phytologist 167*, 81-88.

Brady, K. U., Kruckeberg, A. R. & Bradshaw, H. D. Jr. (2005) Evolutionary ecology of plant adaptation to serpentine soils. *Annual Review of Ecology, Evolution and Systematics 36*, 243-66.

Brooks, R. R. (1987) *Serpentine and its vegetation: A multidisciplinary approach*. Portland, OR: Dioscorides.

Brown, S., Chaney, R., Angle, J. & Baker, A. J. M. (1995) Zinc and cadmium uptake of *T. caerulescens* grown in nutrient solution. *Soil Science Society of America Journal 59*, 125-133.

Cosio, C., DeSantis, L., Frey, B., Diallo, S. & Keller, C. (2005) Distribution of cadmium in leaves of *Thlaspi caerulescens. Journal of Experimental Botany 56,* 765-775.

de Knecht, J. A., van Dillen, M., Koevoets, P. L. M., Schat, H., Verkleij, J. A. C. & Ernst, W. H. O. (1994) Phytochelatins in cadmium-sensitive and cadmium-tolerant *Silene vulgaris. Plant Physiology 104*, 255-261.

de la Fuente, V., Rodriguez, N., Diez-Garretas, B., Rufo, L., Asensi, A. &Amils, R. (2007) Nickel distribution in the hyperaccumulator *Alyssum serpyllifolium* Desf. spp. from the Iberian Peninsula. *Plant Biosystematics 141*, 170-180.

De Vos, C. H. R., Vooijs, R., Schat, H. & Ernst, W. H. O. (1989) Copper-induced damage to the permeability barrier in roots of *Silene cucubalus. Journal of Plant Physiology 135*, 165-169.

Fernandez, S., Seone, S. & Merino, A. (1999) Plant heavy metal concentrations and soil biological properties in agricultural serpentine soils. *Communications in Soil Science and Plant Analysis 30*, 1867-1884.

Gabbrielli, R., Mattioni, C. & Vergnano, O. (1991) Accumulation mechanisms and heavy metal tolerance of a nickel hyperaccumulator. *Journal of Plant Nutrition 14*, 1067-1080.

Ghasemi, R. & Ghaderian, S. M. (2009) Responses of two populations of an Iranian nickel-hyperaccumulating serpentine plant, *Alyssum inflatum* Nyar., to substrate Ca/Mg quotient and nickel. *Environmental and Experimental Botany 67*, 260-268.

Grill, E., Loeffler, S., Winnacker, E.-L. & Zenk, M. H. (1989) Phytochelatins, the heavy-metal binding peptides of plants are synthesized from glutathione by a specific gamma-glutamylcysteine dipeptidyl transpeptidase (phytochelatin synthase). *Proceedings of the National Academy of Sciences (USA) 86*, 6838-6842.

Gupta, A. S. & Berkowitz, G. A. (1989) Development and use of chlorotetracycline fluorescence as a measurement assay of chloroplast envelope-bound Mg. *Plant Physiology 89*, 753-761.

Hall, J. L. & Williams, L. E. (2003) Transition metal transporters in plants. *Journal of Experimental Botany 54*, 2601-2631.

Hussain, D., Haydon, M., Wang, Y., Wong, E., Sherson, S., Young, J., Camakaris, J., Harper, J. & Cobbett, C. (2004) P-type ATPase heavy metal transporters with roles in essential zinc homeostasis in *Arabidopsis. Plant Cell 16*, 1327-1339.

Jenny, H. (1980) *The soil resource: Origin and behavior*. New York, NY: Springer-Verlag.

Kay, K. M., Ward, K. L., Watt, L. R. &Schemske, D. W. (2011) Plant Speciation. In: S. H. Harrison & N. Rajakaruna (Eds.). *Serpentine: The evolution and ecology of a model system* (pp. 181-199). Berkeley, CA: University of California Press.

Kazakou, E., Dimitrakopoulos, P. G., Baker, A. J. M., Reeves, R. D. & Troumbis, A. Y. (2008) Hypotheses, mechanisms and trade-offs of tolerance and adaptation to serpentine soils: From species to ecosystem level. *Biological Reviews 83*, 495-508.

Kerkeb, L. & Krämer, U. (2003) The role of free histidine in xylem loading of nickel in *Alyssum lesbiacum* and *Brassica juncea. Plant Physiology 131*, 716-724.

Koenigs, R. L., Williams, W. A. & Jones, M. B. (1982a) Principal components analysis of vegetation data. *Hilgardia 50*, 1-14.

Koenigs, R. L., Williams, W. A., Jones, M. B. & Wallace, A. (1982b) Chemical composition of foliage and soil. *Hilgardia 50*, 15-26.

Konno, H., Yamaya, T., Yamasaki, Y. & Matsumoto, H. (1984) Pectic polysaccharide breakdown of cell walls in cucumber roots grown with calcium starvation. *Plant Physiology 76,* 633-637.

Krämer, U., Cotter-Howells, J. D., Charnock, J. M., Baker, A. J. M. & Smith, J. A. C. (1996) Free histidine as a metal chelator in plants that accumulate nickel. *Nature 379*, 635-638.

Kruckeberg, A. R. (1951) Intraspecific variability in response of certain native plant species to serpentine soil. *American Journal of Botany 38*, 408-419.

Kruckeberg, A. R. (1984) *California serpentines: Flora, vegetation, geology, soils, and management problems*. Berkeley, CA: University of California Press.

Kruckeberg, A. R. (1999) Serpentine barrens of western North America. In: R. C. Anderson, J. S. Fralish & J. M. Baskin (Eds.). *Savannas, barrens, and rock outcrop plant communities of North America* (pp.309-321). Cambridge, UK: Cambridge University Press.

Kruckeberg, A. R. (2002) *Geology and plant life: The effects of landforms and rock types on plants.* Seattle, WA: University of Washington Press.

Kuhn, A. J., Schroder, W. H. &Bauch, J. (2000) The kinetics of calcium and magnesium entry into mycorrhizal spruce roots. *Planta 210,* 488-496.

Küpper, H., Zhao, F. J. & McGrath, S. P. (1999) Cellular compartmentation of zinc in leaves of the hyperaccumulator *Thlaspi caeruslescens*. *Plant Physiology 119,* 305-311.

Küpper, H., Lombi, E., Zhao, F. J. & McGrath, S. P. (2000) Cellular compartmentation of cadmium and zinc in relation to other elements in the hyperaccumulator *Arabidopsis halleri. Planta 212,* 75–84.

Küpper, H., Lombi, E., Zhao, F. J., Wieshammer, G. & McGrath, S. P. (2001) Cellular compartmentation of nickel in the hyperaccumulators *Alyssum lesbiacum, Alyssum bertolonii,* and *Thlaspi goesingense. Journal of Experimental Botany 52*, 2291-2300.

Lasat, M. M., Baker, A. J. M. &Kochian, L.V. (1996) Physiological characterization of root Zn^{2+} absorption and translocation in Zn hyperaccumuator and nonaccumulator species of *Thlaspi*. *Plant Physiology 112*, 1715–1722.

Lasat, M. M., Baker, A. J. M. &Kochian, L. V. (1998) Altered zinc compartmentalization in the root symplasm and stimulated Zn^{2+} absorption into the leaf as mechanisms involved in zinc hyperaccumulation in *Thlaspi caerulescens*. *Plant Physiology 118,* 875-883.

Lazaroff, N. & Pitman, M. G. (1966) Calcium and magnesium uptake by barley seedlings. *Australian Journal of Biological Science 19*, 991-1005.

Lázaro, J. D., Kidd, P. S. & Martinez, C. M. (2006) A phytogeochemical study of the Tras-os-Montes region (NE Portugal): Possible species for plant-based soil remediation technologies. *Science of the Total Environment 354*, 265 – 277.

Lazarus, B. E., Richards, J. H., Claassen, V. P., O'Dell, R. E. & Ferrell, M. A. (2011) Species specific plant-soil interactions influence plant distribution on serpentine soils. *Plant and Soil 342,* 327-344.

Lee, J. A. (1999) The calcicole-calcifuge problem revisited. *Advances in Botanical Research 29*, 1-30.

Lombi, E., Zhao, F., Dunham, S. & McGrath, S. (2000) Cadmium accumulation in populations of *Thlaspi caerulescens* and *Thlaspi goesingense. New Phytologist 145,* 11-20.

Lombini, A., Llugany, M., Poschenrieder, C., Dinelli, E. & Barcelò, J. (2003) Influence of the Ca/Mg ratio on Cu resistance in three *Silene armeria* ecotypes adapted to calcareous soil or to different Ni- or Cu-enriched serpentine sites. *Journal of Plant Physiology 160*, 1451-1456.

Lyon, G. L., Peterson, P. J., Brooks, R. R. & Butler, G. W. (1971) Calcium, magnesium and trace elements in a New Zealand serpentine flora. *Journal of Ecology 59,* 421-429.

Madhok, O. P. (1965) Magnesium nutrition of *Helianthus annuus* L. and *Helianthus bolanderi* Gray subspecies *exilis* Heiser. PhD Dissertation. Seattle, WA: University of Washington.

Madhok, O. P. & Walker, R. B. (1969) Magnesium nutrition of two species of sunflower. *Plant Physiology 44*, 1016-1022.

Main, J. L. (1974) Differential responses to magnesium and calcium by native populations of *Agropyron spicatum. American Journal of Botany 61*, 931-937.

Main, J. L. (1981) Magnesium and calcium nutrition of a serpentine endemic grass. *American Midland Naturalist 105,* 196-199.

Marrs, R. H. & Proctor, J. (1976) The response of serpentine and nonserpentine *Agrosisstolonifera* L. to magnesium and calcium. *Journal of Ecology 64*, 953-964.

Marschner, H. (2011) *Mineral nutrition of plants.* Third edition. San Diego, CA: Academic Press.

Montargès-Pelletier, E., Chardot, V., Echevarria, G., Michot, L. J. & Morel, J. L. (2008) Identification of nickel chelators in three hyperaccumulating plants: An X-ray spectroscopic study. *Phytochemistry 69,* 1695-1709.

Mysliwa-Kurdziel, B., Prasad, M. N. V. & Strzalka, K. (2002) Heavy metal influence on the light phase of photosynthesis. In: M. N. V. Prasad & K. Strzalka (Eds.).*Physiology and biochemistry of metal toxicity and tolerance in plants* (pp. 229-255). Dordrecht: Kluwer Academic Publishers.

Nabais, C. (2000) *Seasonal transport, allocation and speciation of nickel in Quercus ilex trees grown on serpentine and nickel spiked soil.* PhD Dissertation. Coimbra: Universidade de Coimbra.

Nagy, L. & Proctor, J. (1997) Plant growth and reproduction on a toxic Alpine ultramafic soil: Adaptation to nutrient limitation. *New Phytologist 137*, 267-274.

Nishizono, H., Ichikawa, H., Suziki, S. & Ishii, F. (1987) The role of the root cell wall in the heavy metal tolerance of *Athyrium yokoscense. Plant and Soil 101*, 15-20.

O'Dell, R. E. & Claassen, V. P. (2006) Serpentine and non-serpentine *Achillea millefolium* accessions differ in serpentine substrate and response to organic and inorganic amendments. *Plant and Soil 279*, 253-269.

O'Dell, R. E. & Rajakaruna, N. (2011) Intraspecific variation, adaptation, and evolution. In: S. H. Harrison & N. Rajakaruna (Eds.). *Serpentine: The evolution and ecology of a model system* (pp. 97-137). Berkeley, CA: University of California Press.

Paliouris, G. & Hutchinson, T. C. (1991) Arsenic, cobalt, and nickel tolerances in two populations of *Silene vulgaris* (Moench) Garcke from Ontario, Canada. *New Phytologist 117*, 449-459.

Palm, E.,Brady, K. & Van Volkenburgh, E. (2012) Serpentine tolerance in *Mimulus guttatus* does not rely on exclusion of magnesium. *Functional Plant Biology 39*, 679-688.

Papoyan, A. & Kochian, L. V. (2004) Identification of *Thlaspicaerulescens* genes that may be involved in heavy metal hyperaccumulation and tolerance. Characterization of a novel heavy metal transporting ATPase. *Plant Physiology 136*, 3814-3823.

Pineros, M. & Tester, M. (1995) Characterization of a voltage-dependent Ca^{2+}-selective channel from wheat roots. *Planta 195*, 478-488.

Pineros, M. & Tester, M. (1997) Calcium channels in higher plants cells: Selectivity, regulation and pharmacology. *Journal of Experimental Botany 48*, 551-577.

Pope, T., Harris, T. B. & Rajakaruna, N. (2010) Vascular plants of adjacent serpentine and granite outcrops on Deer Isles, Maine USA. *Rhodora 112,* 105-141.

Proctor, J. (1970) Magnesium as a toxic element. *Nature 227*, 742-743.

Proctor, J. (1971) The plant ecology of serpentine: II. Plant response to serpentine soils. *Journal of Ecology 59*, 397-410.

Proctor, J. & Woodell, S. R. J. (1975) The ecology of serpentine soils. *Advances in Ecological Research 9*, 255-365.

Rajakaruna, N. & Bohm, B. A. (1999) The edaphic factor and patterns of variation in *Lasthenia californica* (Asteraceae). *American Journal of Botany 86*, 1576-1596.

Rajakaruna, N., Siddiqi, M. Y., Whitton, J., Bohm, B. A. & Glass, A. D. M. (2003) Differential responses to Na^+/K^+ and Ca^{2+}/Mg^{2+} in two edaphic races of the *Lasthenia californica* (Asteraceae) complex: A case for parallel evolution of physiological traits. *New Phytologist 157,* 93-103.

Rao, I. M., Sharp, R. E. & Boyer, J. S. (1987) Leaf magnesium alters photosynthetic response to low water potentials in sunflower. *Plant Physiology 84*, 1214-1219.

Sambatti, J. B. M. & Rice, K. J. (2007) Functional ecology of ecotypic differentiation in the Californian serpentine sunflower (*Helianthus exilis*). *New Phytologist 175*, 107-119.

Sanità di Toppi, L., Gremigni, P., Pawlik-Skowronska, B., Prasad, M. N. V. & Cobbett, C. S. (2003) Response to heavy metals in plants: A molecular approach. In: L. Sanità di Toppi & B. Pawlik-Skowronska (Eds.).*Abiotic stresses in plants* (pp.133-156). Dordrecht: Kluwer.

Schat, H., Llugany, M., Vooijs, R., Hartley-Whitaker, J. & Bleeker, P. M. (2002) The role of phytochelatins in constitutive and adaptive heavy metal tolerance in hyperaccumulator and non-accumulator metallophytes. *Journal of Experimental Botany 53,* 2381-2392.

Shaul, O. (2002) Magnesium transport and function in plants: The tip of the iceberg. *Biometals 15,* 309-323.

Shaul, O., Hilgemann, D. W., Almeida-Engler, J., Van Montagu, M., Inze, D. & Galili, G. (1999) Cloning and characterization of a novel Mg2+/H+ exchanger. *The EMBO Journal 18,* 3973-3980.

Tibbetts, R. A. & Smith, J. A. C. (1993) Vacuolar accumulation of calcium and its interaction with magnesium availability. In: A. J. M. Baker, J. Proctor & R. D. Reeves (Eds.).*The vegetation of ultramafic (serpentine) soils: Proceedings of the first international conference on serpentine ecology* (pp. 367-373). Andover, UK: Intercept.

Tripathy, B. C. & Mohanthy, P. (1980) Zinc-inhibited electron transport of photosynthesis in isolated barley chloroplasts. *Plant Physiology 66,* 1174-1178.

Tyndall, R. W. & Hull J. C. (1999) Vegetation, flora, and plant physiological ecology of serpentine barrens of eastern North America. In: R. C. Anderson, J. S. Fralish & J. M. Baskin (Eds.). *Savannas, barrens, and rock outcrop plant communities of North America* (pp. 67-82). Cambridge, UK: Cambridge University Press.

Verbruggen, N., Hermans, C. & Schat, H. (2009) Molecular mechanisms of metal hyperaccumulation in plants. *New Phytologist 181*, 759-776.

Walker, R., Walker, M. & Ashworth, P. R. (1955) Calcium-magnesium nutrition with special reference to serpentine soils. *Plant Physiology 30,* 214-221.

Wallace, A., Jones, M. B. & Alexander, G. V. (1982) Mineral composition of native woody plants growing on serpentine soil in California. *Soil Science 134*, 42-44.

Wenzel, W. W., Bunkowski, M., Puschenreiter, M. & Horak, O. (2003) Rhizosphere characteristics of indigenously growing nickel hyperaccumulator and excluder plants on serpentine soil. *Environmental Pollution 123,* 131-138.

White, P. J. & Broadley, M. R. (2003) Calcium in plants. *Annals of Botany 92,* 487-511.

White, P. J., Pineros, M., Tester, M. & Ridout, M. S. (2000) Cation permeability and selectivity of a root plasma membrane calcium channel. *Journal of Membrane Biology 174*, 71-83.

Wood, S. G. (1984) *Mineral element composition of forest communities and soils at Soldiers Delight, Maryland.* MS Thesis. Towson, MD: Towson State University.

Wu, C. A., Lowry, D. B., Cooley, A. M., Wright, K. M., Lee, Y. W. & Willis, J. H. (2008) *Mimulusguttatus* is an emerging model system for the integration of ecological and genomic studies. *Heredity 100,* 220-230.

Yeo, A. & Flowers, T. (2007) *Plant solute transport*. Hoboken, NJ: Wiley-Blackwell Press.

Zhao, F. J., Lombi, E., Breedon, T. & McGrath, S. P. (2000) Zinc hyperaccumulation and cellular distribution in *Arabidopsis halleri*. *Plant, Cell and Environment 23*, 507-514.

Zhu, J. K. (2003) Regulation of ion homeostasis under salt stress. *Current Opinion in Plant Biology 6*, 441-445.

In: Plant Ecology and Evolution in Harsh Environments ISBN: 978-1-63321-955-7
Editors: N. Rajakaruna, R. S. Boyd and T. B. Harris

Chapter 7

ECOLOGY AND EVOLUTION OF PLANTS IN ARCTIC AND ALPINE ENVIRONMENTS

***Amy L. Breen**[*,1], **David F. Murray**[2], **Martha K. Raynolds**[3], **Ina Timling**[3] **and Donald A. Walker**[3]*

[1]International Arctic Research Center, University of Alaska, Fairbanks, Alaska USA
[2]University of Alaska Museum of the North, University of Alaska, Fairbanks, Alaska USA
[3]Institute of Arctic Biology and Alaska Geobotany Center, University of Alaska, Fairbanks, Alaska USA

ABSTRACT

The main structuring element of a terrestrial biome lies in its vegetation. Hierarchical patterns, from the level of the plant community to the global biome, are at their core a reflection of the evolutionary response of plants to their environment. These processes provide the framework for our chapter on ecology and evolution of plants in arctic and alpine environments. Arctic and alpine plants grow above latitudinal and altitudinal treelines around the world. Short-statured shrubs, herbaceous plants, lichens, and mosses comprise the low vegetation of these regions that is collectively referred to as tundra. Arctic and alpine tundras are viewed as growing in uniformly and predictably harsh environments with low temperatures, even during the growing season. The harshness attributed to the tundra, however, vastly oversimplifies the limitations plants face in these environments. The Arctic is not spatially uniform at any scale; neither is the Alpine. The arctic flora in particular, with a history that exceeds two million years, developed through multiple glacial periods. There is ample evidence of major climatic changes over millennia through which tundra vegetation has persisted despite the perceived harshness. Components of the arctic flora may be ancient, but the modern flora is an amalgam of Tertiary, Quaternary, and Holocene contributions. Herein, we focus on recent insights into the ecology and evolution of arctic and alpine plants gained from molecular ecology, modeling, and remote-sensing studies. We review the history and evolution of arctic and alpine floras and discuss the current status of arctic and alpine plant biodiversity. We then

[*] Corresponding author: albreen@alaska.edu.

discuss the potential for arctic and alpine plants to adapt to a changing climate. We conclude with an overview of plant cross-kingdom interactions, with a focus on the plant-ectomycorrhizal fungi symbiosis in arctic and alpine environments.

INTRODUCTION

Arctic and alpine plants grow above latitudinal and altitudinal treelines around the world (Figure 1). Treeline is the limit of forest; beyond it conditions limit the growth, survival, and reproduction of trees. Short-stemmed shrubs, herbaceous plants, lichens, and mosses comprise the low vegetation of these regions. The vegetation of this treeless landscape is collectively referred to as tundra. Compact life forms are common, such as plants with dense basal rosettes or forming cushions, which protect vulnerable growing tissues from drying winds in summer and from blowing snow in winter.

Arctic and alpine tundras are viewed by some as uniformly and predictably harsh environments. Growth and productivity are constrained by the physical environment: timing of snowmelt, topography, moisture availability, exposure, and aspect. The vegetation is formed by species sufficiently tolerant of cold summer temperatures at any given location to survive freezing temperatures during the growing season, although frost-hardiness and frost-avoidance are not unique to arctic and alpine plants. The harshness attributed to the tundra, however, vastly oversimplifies the limitations plants face in these environments. To characterize tundra as harsh clearly represents our temperate zone bias (Murray, 1987). This bias makes it difficult to not view tundra plants as perilously close to the limits of life—which is simply not so. As Raup (1969) wrote:

> "…what we need is a first class Eskimo(sic) botanist—one who thinks of the tundra as a home, and a very good place to live. I think he would see the plants as they are, members of an ancient flora remarkably well adjusted to the habitat."

The Arctic is not spatially uniform at any scale; neither is the Alpine. Arctic and alpine environments are climatically variable from day to day, month to month, and year to year, yet they are predictable within limits. The Arctic flora in particular, with a history that exceeds two million years, developed through multiple glacial periods with contrasting demands imposed by the changing biological and physical environments over millennia through which tundra vegetation persisted, although the floristic composition of tundra varied over time—despite the perceived harshness.

Physiognomic similarities among the tundra regions can lead us to equate arctic and alpine environments. Dissimilarity among tundra types exists, however, notably in geographic distribution. Arctic tundra is beyond the latitudinal limit of trees in the northern hemisphere and comprises nearly 5% of the terrestrial surface of the Earth, or over 7 million km^2 (Walker et al., 2005). Approximately 5 million km^2 of the Arctic is covered by vegetation, and the remainder is covered by ice. In contrast, the Alpine is beyond the altitudinal limit of trees and comprises 3% of the terrestrial surface of the earth (Körner, 2003). Approximately 4 million km^2 of alpine tundra is scattered globally, with 82% occurring in the northern hemisphere. Plant species that occur in both the Arctic and Alpine, are designated as arctic-alpine taxa.

Our discussion here of alpine tundra is limited to the northern hemisphere as this is where most high altitude tundra occurs and where it is the most similar to the Arctic.

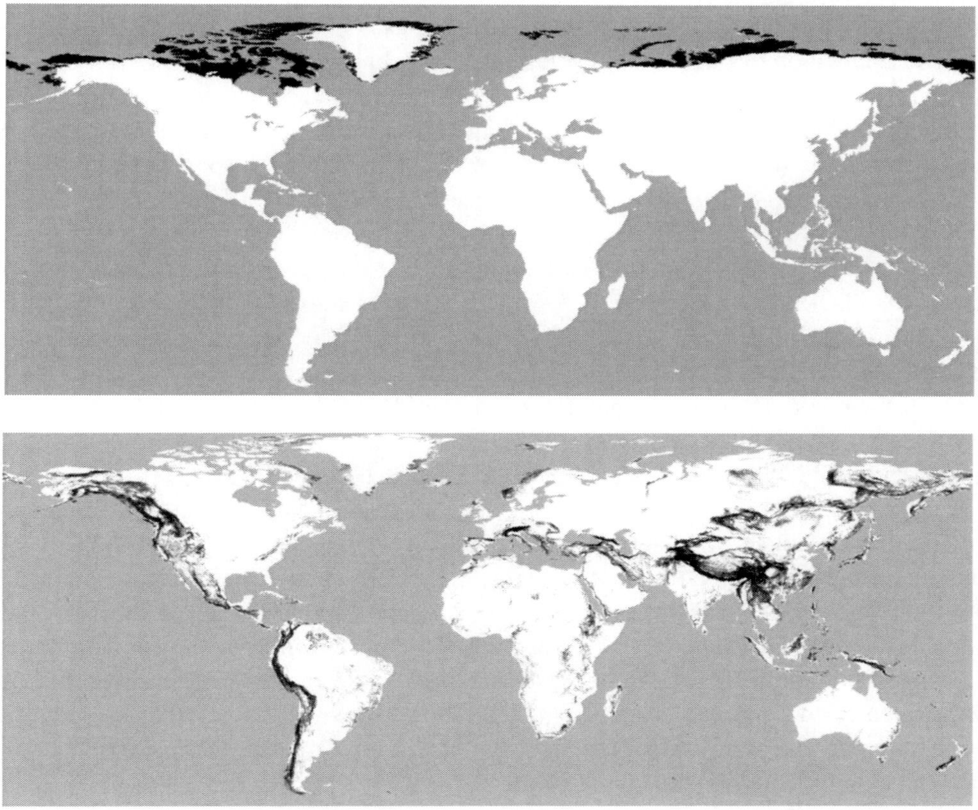

Figure 1. Global geographic distribution of arctic (top) and mountainous (bottom) regions. Alpine regions are fragmented and confined to above-treeline elevations within the mountainous regions, and thus difficult to depict at the global scale. The top panel is adapted from CAVM Team (2003), and the bottom panel is adapted from Körner et al. (2011).

Climate

The critical climatic attribute shared by arctic and alpine environments is low temperatures during the growing season. The arctic photoperiod is continuous during the growing season north of the Arctic Circle (latitude 66° 33' 44" N). However, during the continuous daylight of an arctic summer, the sun's angle remains low and the solar radiation is less intense than at lower latitudes. Mean July air temperature in the High Arctic is <6° C and can reach 10-12° C at its southern limit in the Low Arctic (Walker et al., 2005). There is a four-fold difference in length of the growing season across this gradient ranging from a few weeks to over three months.

A comparison of mean air temperature of the warmest month across alpine sites in the northern hemisphere shows a range from 5° C in the Austrian Central Alps to 8.5° C in the Rocky Mountains at Niwot Ridge in Colorado (Körner, 2003).

Plant Adaptations

The most profound limitation of the short growing season is its effect on plant reproduction. We discuss plant reproductive adaptations here and then specifically address the process of adaptation through natural selection in the *Adaptation and the Response of Arctic and Alpine Plants to Climate Change* section of this chapter. Plants must progress through anthesis, pollination, and seed set during a relatively short span of summer warmth. It is no surprise, therefore, that annuals are rare; the best example of an arctic annual being the arctic-alpine/bi-polar *Koenigia islandica* (Polygonaceae) (Jónsdóttir, 2011). The primary means to avoid this limitation is through vegetative reproduction; that is, by rhizomes (*e.g.*, graminoids), runners (*e.g.*, *Potentilla anserine* [Rosaceae], *Saxifraga flagellaris, Saxifraga platysepala* [Saxifragaceae]), bulbils (*e.g.*, *Bistorta vivipara* [Polygonaceae], *Saxifraga cernua, Saxifraga foliolosa,* viviparous grasses in *Festuca* and *Poa* [Poaceae]), or by producing seeds apomictically (*e.g.*, *Potentilla* spp.).

The majority of arctic and alpine plants can reproduce sexually, despite the prevalence of vegetative reproduction (cf. Murray, 1987), or if apomicts, may nevertheless require pollination. Self-incompatibility is rare and autogamy provides more assured seed set, but this can lead to genetic homogeneity and inbreeding depression. Mixed mating overcomes the many limitations imposed by arctic and alpine environments. Polyploidy buffers plants against the effects of inbreeding and genetic drift (Brochmann et al., 2004). An adaptation that is totally unexpected is heliotropism, the remarkable tracking of the sun by plants such as *Dryas* (Rosaceae) and *Papaver* (Papaveraceae). This occurs in conjunction with parabolic corollas (Kevin, 1972a, 1975; Wada, 1998) such that reflection of solar radiation from the inner surface of the corollas is focused on the reproductive structures thus warming them above ambient temperature and hastening development, as well as providing basking sites for insects (Hocking & Sharplin, 1965).

Advantages accrue to plants capable of producing pre-formed flower buds that overwinter surrounded by scales and leaves. These buds are developmentally advanced, in some cases up to and including meiosis, thus important steps of morphogenesis have already been completed when flowers open the following spring. Sørensen (1941) provided an excellent discussion in which he documented the wintering floral stages with photographs of meticulous dissections and cleared tissue.

Plants that are self-compatible and autogamous are more assured of seed set, although with some genetic cost through reduced recombination that accompanies inbreeding. Most outcrossers are self-compatible and through mixed-mating gain reproductive advantages. The outcrossing species are primarily wind- and insect-pollinated. Some are self-incompatible obligate outcrossers that require the mediation of insects (Kevin, 1972b). These plants offer both attraction and reward to potential pollinators. Attraction lies in flower shape and color, but "colors" not entirely within the spectra visible to humans. Among the white- and yellow-flowered taxa, so numerous in the flora, are ones with spectacular color elaborations in the ultra violet range, invisible to us but sensible to insects (Kevin, 1972c). Floral reward is typically in the form of pollen and nectar.

The relationship between plant and pollinator, attraction and reward, is so well established, co-evolved, that the loss of pollinators can limit the range of plants. Savile (1959) noted that the northern limit of Fabaceae in the islands of the eastern Canadian Arctic correlates well with the disappearance of the bumblebee. To someone accustomed to Low and

Middle Arctic floras (*sensu* Polunin, 1951), when on the ground, the absence of legumes in the High Arctic is noticed.

Vegetation Patterns

The response to summer temperatures is so consistent that the Arctic has been divided into five bioclimate subzones based largely on characteristic vegetation (CAVM Team, 2003). The bioclimate subzones are separated by approximately 2ºC in mean July temperature. Similar changes in temperatures occur with elevation in alpine areas in the Arctic, with elevation belts corresponding to the Arctic bioclimate subzones separated by approximately 333-m based on the adiabatic lapse rate of -6º C/1000 m (CAVM Team, 2003), although these subzones have been shown to shift upwards in Greenland due to a more continental climate with earlier snowmelt (Sieg & Daniëls, 2005).

In the Alpine, generally three biogeographical zones or alpine belts are recognized (Wielgolaski, 1997). The lower belt, with no trees and often with tall shrubs, is called the Low Alpine. The next belt, without shrub thickets and with a dominance of graminoids, is the Mid Alpine, although sometimes it is divided into two belts with the upper belt being referred to as Subnival. The belt of limited vegetation beyond the Mid Alpine that occurs on the highest peaks may be called either the High Alpine or Nival Belt. For a comparison of the biogeographical zones and belts of the Arctic and Alpine that pre-dates the Circumpolar Arctic Vegetation Map (CAVM Team, 2003) see Figure 1.1 in Wielgolaski (1997).

Each bioclimate subzone in the Arctic has characteristic plant growth forms. Bioclimate subzone A is the coldest part of the Arctic and includes mountain elevations closest to permanent snow cover. Most of the ground surface is barren, with only sparse vascular plant cover. What little vegetation is present grows mostly in soil cracks related to patterned ground or in sheltered areas provided by topography, where plants are protected from the wind and have a warmer microclimate. Nonvascular plants and biological soil crusts—consisting of a mixture of fungi, algae, and crustose lichens—are dominant, with a few scattered herbs (Vonlanthen et al., 2008). In bioclimate subzone B, there are a few more species of vascular plants and greater plant cover. Bare ground and biological soil crusts are still common, especially on ridges, dry hill slopes, and on the tops of hummocks. In bioclimate subzone C, the vegetation is still patchy, but covers most of the ground in flat, moist areas. Shrubs start to become an important component of the vegetation in sheltered sites. Bioclimate subzone D is mostly vegetated, with a mix of sedges, erect dwarf shrubs, forbs, lichens, and a thick layer of mosses (Kade et al., 2005). Bioclimate subzone E is adjacent to treeline and has the tallest shrubs and the most continuous vegetation cover.

Variation in plant communities at the sub-meter scale also occurs in relationship to patterned ground in most arctic and alpine areas (cf. Murray, 1997). Soil-frost processes create a range of patterned-ground features from 10- to 30-m diameter polygons with centers, rims, and troughs, to 1- to 5-m diameter frost circles and hummocks (Raynolds et al., 2008). Microhabitats associated with small differences in elevation above the water table, or differences in frost activity, are populated by different species. For example, in tussock tundra, shrubs grow on the warmer, well-drained areas, while mosses grow in the cooler, moister depressions.

Winter conditions affect plants mainly through snow and wind. Plants beneath the snow cover are protected from extreme temperatures, desiccation, abrasion, and herbivory, but can experience shorter growing seasons (Walker et al., 2001a). Taller vegetation is sometimes found in areas protected by moderately deep snow cover, but in deeper accumulations, snowbed plant productivity is strongly limited by the short snow-free period. Some evergreen species have developed the ability to photosynthesize beneath thin snow cover, giving them a head start in spring (Starr & Oberbauer, 2003). Plant communities specifically adapted to very short growing seasons are found in these snowbeds (Billings & Mooney, 1968).

Soil pH has a strong effect on arctic and alpine vegetation. Non-acidic areas in Arctic Subzones D and E are characterized by deeper thaw, non-tussock forming sedges and forbs, and frost circles with bare ground in their geomorphically active centers (Walker et al., 1998). In contrast, acidic areas have a deep moss layer, commonly including *Sphagnum* species (Sphagnaceae), which insulates the soil from summer warming. Plants growing in these moist, acidic soils include tussock sedges and ericaceous shrubs. This tussock tundra is found on old soils throughout Beringia, the vast region spanning from northeast Russia east across the Bering Land Bridge to northwest North America, that remained ice-free during Quaternary glaciations (see below). The effects of soil chemistry are especially obvious in areas of thin soil that are common in the Alpine, where plants are growing close to the source bedrock. Limestone bedrock weathers quickly and does not form soil as well as acidic bedrock, resulting in dry, calcium-rich soils, supporting vegetation that is often sparse, but forb-rich (Walker et al., 2001b).

Since the 1960s, many reviews on the topic of the ecology and evolution of plants of arctic and alpine environments have been published. In this chapter we provide a list of recommended readings by topic (Table 1) and summarize the insights gained from molecular ecology, modeling, and remote-sensing studies. We first provide an overview of the history and evolution of arctic and alpine floras and then discuss the biodiversity of arctic and alpine plants and their potential for adaptation to climate change. We conclude with an overview of plant cross-kingdom interactions, with a focus on the plant-ectomycorrhizal fungi symbiosis in arctic and alpine environments.

Table 1. Prominent reviews recommended by the authors on the topic of the ecology and evolution of arctic and alpine plants

Regional focus	Topic focus	Reference
Arctic	Adaptation	Savile (1972)
Arctic	Ecology	Chernov (1985)
Arctic and alpine	Biodiversity	Chapin & Körner (1995)
Arctic	Vegetation ecology	Bliss (2000)
Arctic	Phytogeography	Abbott & Brochmann (2003)
Alpine	Ecology	Körner (2003)
Arctic and alpine	Evolution	Abbott (2008)
Alpine	Vegetation ecology	Ellenberg (2009)
Arctic	Fungal ecology	Timling & Taylor (2012)
Arctic	Ecology & evolution	Brochmann et al. (2013)
Arctic	Biodiversity	Meltofte (2013)

HISTORY OF ARCTIC AND ALPINE FLORAS

There is a distinct arctic flora, one restricted to regions north of the latitudinal treeline, consisting of taxa that do not have ranges south of the Arctic, but including taxa with minor excursions into the northern boreal alpine zone. There are notable disjunctions south from the Arctic, as in North America, into the southern Rocky Mountains, and these comprise the arctic-alpine flora. There is also a distinct alpine flora that does not reach the Arctic, but is restricted to the southern Rocky Mountains and mountain ranges such as the Alps, Carpathians, Altai, and Caucasus.

The classic late 19th century model proposed a once widespread Tertiary arctic flora driven by advancing Pleistocene ice sheets south into high mountains, leaving nothing in their wake, the *tabula rasa* (clean slate) hypothesis (Nathorst, 1892). These southern migrants remained in the mountains and ascended to their summits when the post-glacial climate ameliorated and plants from south of the maximum extent of glaciation could migrate northward to repopulate the Arctic. Thus, the alpine flora was, by this reckoning, a Quaternary derivative of an early Tertiary arctic flora (cf. Darwin, 1859).

Weber (1965, 2003) has sought an explanation for the disjunctions of alpine plants in the Altai of south-central Siberia and in the southern Rocky Mountains of western North America. He presented abundant examples of taxa shared by both mountain systems and absent from the area between. To reconcile the huge geographic separation today, he envisioned (as did Darwin) a once more-or-less continuous blanket of these taxa at some time during late Tertiary and the subsequent destruction of these plants in the intervening area during the Quaternary. His proposal is logical and derives from inferences from long and detailed studies of floras; however, it must be said that this explanation is without empirical evidence.

Tolmachev (1960) proposed that the arctic flora had been derived from the alpine floras from the mountain ranges of Eurasia and North America. Although Hultén (1958) had earlier supposed a circumpolar arctic tundra at the onset of Quaternary glaciations, he accepted Tolmachev's hypothesis and put forward his own argument in favor of this account of history. He was aware of a common floristic core in mountain ranges surrounding the Arctic (W. A. Weber, pers. comm.). Which flora is the antecedent, arctic or alpine, is a question that remains unanswered.

Late Tertiary floras as reconstructed from plant remains at Lava Creek on the Seward Peninsula in Alaska USA (Hopkins et al., 1971) and at Kap København, 82° N latitude in Greenland (Bennike & Bøcher, 1990) do not provide evidence for a continuous late Tertiary arctic tundra. However, from that flora of Tertiary forests and forest-tundra, plants of bogs and similar cold sites, pre-adapted to conditions that would become widespread in the Quaternary, survived the shift from forest to tundra. Plants of pond margins and waterways faced little change in habitat as the cooling progressed. We can presume these plants persisted wherever riparian habitats remained extant (cf. Johnson & Packer, 1965). Macrofossils from the Tertiary Beaufort Formation of arctic Canada (Matthews & Ovenden, 1990) generally support this view. The occurrence of *Saxifraga oppositifolia* and *Dryas integrifolia* (Figure 2) in Canada and Greenland raise an important question: does the presence of quintessential tundra plants in today's world signify tundra in Late Tertiary?

Formation of the circumpolar arctic tundra we see today progressed throughout the nearly 2.5 to 3 million years of the Late Tertiary (Pliocene) and Quaternary. Arctic tundra reached its geographic extent, floristic richness, and current zonation (see Daniëls et al., 2013 for details) in post-glacial time. Those arctic areas wholly covered by ice sheets during the last glacial maximum, of course, were colonized as recently as 6,000 to 10,000 years ago. Components of the arctic flora may be ancient, but the modern flora is an amalgam of Tertiary, Quaternary, and Holocene contributions. As the vast continental ice sheets withdrew and eventually disappeared, plants moved onto the deglaciated terrain, a great many of them from south of the former ice margin, but patterns of plant distribution suggest also other sources, ones from within the area thought to be *tabula rasa*.

Fernald (1925) and Hultén (1937) drew our attention to areas of persistence, where plants adapted to harsh conditions avoided Quaternary glaciations in ice-free periglacial refugia when most of northern Eurasian and North America were otherwise ice covered. This meant *tabula rasa* but with special cases of plant survival. From centers of persistence, plants emerged and became geographically and ecologically sorted according to their dispersing ability and thresholds of tolerance to various abiotic and biotic limiting factors. Some plants moved faster and farther than others and established a circumpolar existence; some developed southern extensions along the Cordillera and formed the arctic-alpine flora. Others have continued to occupy restricted areas despite the millennia since their release from glacial conditions.

Whereas the boundaries of the huge Beringian refugium, as proposed by Hultén (1937), are now well documented by both geological and biological data, the extent and even the existence of smaller arctic and alpine refugia are still debated. Beringia is vast, but nunataks, used here in its broad sense as any non-glaciated area surrounded by glacier ice, are smaller in area, and discrete. Periglacial refugia have been used to explain numerous disjunct distributions, especially in alpine systems. Even so, questions remain: where did nunataks occur, when were they ice-free, when and how did the plants arrive at these locations, how did they survive there, and are they necessary to explain floristic novelties?

With the advent of molecular analysis of the genome in both plants and animals, and the rise of the field of phylogeography (Avise, 1994) there came an additional line of evidence by which to identify refugia, centers of phylogenetic and geographic origin, routes of migration, and instances of long distance dispersal from known sources.

More recently, information on plant cover has been gleaned from the bulk DNA extracted from frozen soil cores gathered at several sites in the Arctic: Russia, United States (Alaska), and Canada. Techniques have been developed that provide, for the most part, greater resolving power (*i.e.,* the ability to identify more taxa to species, than could be achieved through palynology alone). Importantly, these cores have been taken from exposures that date back to the last glacial maximum (cf. Willerslev et al., 2014).

Prior to molecular genetics, the thinking was that plants isolated for long periods of time in nunatak refugia would exist where, due to isolation, an influx of new genotypes was nil. Random fixation of genes by genetic drift and removal of less fit gene combinations through intense stabilizing selection acting upon these small populations would, theory predicts, result in a gene pool of low diversity but consisting of genotypes admirably adapted to the narrow constraints of a harsh, full glacial nunatak existence. This presumably left the survivors poorly equipped for post-glacial dispersal—except, perhaps, for the polyploid taxa.

A

B

Figure 2. A) *Dryas integrifolia* and B) *Saxifraga oppositifolia* were both a component of the Late Tertiary arctic flora as reconstructed from plant remains at Lava Creek on the Seward Peninsula in Alaska (Hopkins et al., 1971) and at Kap Køpenhavn in Greenland (Bennike & Bøcher, 1990) (photo credits: Martha Raynolds).

A tenet of phylogeography is that plant genomes undergo steady mutation in the neutral, or non-coding, regions of the genome. The longer populations are isolated, the longer the time for the fixation of unique gene combinations and rare alleles; hence, genetic identities form as long as interbreeding with other populations does not occur, as that would swamp any unique haplotypes. In phylogeography, the expectation is for greater genetic diversity as the signal of refugia persistence.

Disjunct occurrences in mountains were taken by some to be *prima facie* refugial survivors; the bicentric distribution pattern in the Scandinavian mountains is an example (Dahl, 1955). Although a thorough reconsideration by Brochmann et al. (2003) concluded that refugia were not necessary to account for both the disjunctions and endemics, a more recent study (Westergaard et al., 2011) has found examples explained by nunatak survival. Thus, these publications are a perfect illustration of the wisdom of Berg (1963):

"...most ... biogeographers explain the arctic-alpine disjunction in terms of glacial survival...It is my opinion that no single explanation can account for all the arctic-alpine disjunctions...a great deal of argumentation has resulted from a futile search for one universal cause."

The aggregate of disjunct occurrences of Rocky Mountain plants in eastern North America are what first led Fernald to propose his persistence theory (Fernald, 1925). What made some of his examples controversial was the absence of geological evidence for ice-free areas. Ives (1974), in his splendid review of biological refugia and the nunatak hypothesis, chastised those making claims for periglacial refugia without supporting evidence for full glacial, ice-free conditions, even in the face of strong geological evidence against such claims.

A counter-argument to refugial survival was that disjuncts were ecological specialists that arrived at their current position in post-glacial time. Why they remain today as small isolated populations was thought to be the result of drastically reduced ecotypes, the conservative species of Fernald (Fernald, 1925), the rigid species of Hultén (Hultén, 1937), and thus a genetically determined inability to disperse and compete elsewhere. An excellent review of Fernald and Hultén and the debate over refugial existence or post-glacial arrival is provided in Raup (1941, pt. 1).

Long distance dispersal has always been offered as a mechanism to explain disjunct species, but one which we are unlikely to confirm by direct evidence. Savile (1956, 1972), a great field biologist, believed in the efficacy of winter transport by strong winds over a landscape of ice and snow. However, for some geographic problems, greater distances must be traversed. Plant propagules are believed to have been carried across the Atlantic Ocean by migratory waterbirds such as those moving from western Europe to northeastern North America, contributing to the Amphiatlantic flora. The discussion has long gone back and forth, with reasons supporting both why long distance dispersal is probable and why it is not (Dahl, 1963; Löve, 1963).

Abbott & Brochmann (2003) have provided an excellent review of the molecular evidence for transatlantic dispersal. Since then, more examples have appeared: *Carex bigelowii* (Cyperaceae; Schönswetter et al., 2008) and *Saxifraga rivularis* (Westergaard et al., 2010). Moreover, in a remarkable study Alsos et al. (2007) demonstrated how Svalbard could be supplied with plants from elsewhere in post-glacial time, even from distant sources, without involving refugial populations—a suggestion that would have been in conflict with glacial geologists who have said that periglacial refugia did not exist there.

Mountains high enough to support alpine vegetation today were for the most part ice-covered during glacial maxima, certainly during the last glacial maximum, but alpine plants could have persisted in peripheral nunataks at the margins of an ice shield as Schönswetter et al. (2004) postulated for *Ranunculus glacialis* (Ranunculaceae) in the Alps. In the case of *Eritrichium* (Boraginaceae; Stehlik et al., 2002) at high elevations in the Alps, snow and ice would make refugia problematic. Similarly, Marr et al. (2008), having examined the genetics of *Oxyria digyna* (Polygonaceae) over much of the North America Cordillera and elsewhere, reported genetic diversity among disjunct occurrences that they interpreted as the consequence of periglacial refugia, albeit where geological evidence for ice thickness would appear to rule out ice-free areas. The implication is that genetic evidence trumps geological projections, yet the genetic diversity could be the result of post-glacial secondary contact and the rare alleles, at least in small populations, by fixed random processes.

How did the arctic species disperse southward down the Rocky Mounting chain, getting as far south as Montana, where there are about 100 arctic taxa found in the alpine zone (P. Lesica, pers. comm.)? There are even some arctic-alpine plants on the summits of the San Francisco Peaks of Arizona (Deaver Herbarium; www.nau.edu/deaver). We assume this is due to migrations southward from the Arctic, but it remains unclear when this would have occurred. During the glacial maxima, ice cover was nearly complete and thus, we presume, a barrier to dispersal. Prior to the final glacial advances and/or as glaciers receded in early post-glacial time, there would have been both the arctic environment and open corridors through which plants could have dispersed southward from Beringia (and some southern alpine plants northward). Thus a post-glacial process cannot be ruled out; in fact it seems likely. Despite numerous studies and discussions on the history and evolution of arctic and alpine floras for more than a century, there is still much to be learned.

BIODIVERSITY OF ARCTIC AND ALPINE PLANTS

Species richness of arctic and alpine plants tends to decline with increasing latitude and elevation. Low temperatures and a short growing season are environmental filters that are hypothesized to exclude species from increasingly more severe climates (Chapin & Körner 1995; Walker, 1995). There is no consensus, however, on a single explanation for the decline in biodiversity. Hypotheses fall into two groups, those based on ecological mechanisms of species co-occurrence and those based on evolutionary mechanisms governing rates of diversification and Earth history (Payer et al., 2013). These hypotheses are not necessarily mutually exclusive, as observed patterns may be due to interactions between both abiotic and biotic factors.

On a more regional scale, species richness of arctic and alpine plants is best explained by the ancestral stock of species, long-distance migration following deglaciation, evolution of new taxa, and proximity to a rich species pool as within and near Beringia (Chapin & Körner, 1995; Murray, 1995). Migration is essential for the assemblage of arctic and alpine floras, especially following glacial periods and associated extinctions. In the Arctic, the flora tends to intergrade continuously from a few centers of persistence. In contrast, alpine floras are more discrete due to their restricted habitat and geographic isolation, thereby leading to higher levels of endemism. Thus, mountain ranges in different regions tend to have disparate assemblages of alpine dominants, while the dominant plant species across the Arctic tend to have a circumpolar distribution.

Whereas it is often remarked that the flora of arctic and alpine regions is species-poor, even depauperate, the question arises: species-poor in relation to what? Summer climate is sufficiently cool and winds strong enough to preclude trees and tall shrubs, thus a major component of boreal and temperate vegetation is missing from tundra. But, are there niches unfilled? Are there families, genera, or species missing that we should expect? These questions have not been addressed, but are of interest as we discuss the flora of arctic and alpine regions in this section of the chapter.

Our knowledge of the arctic flora differs for each of the three main taxonomic groups of plants—vascular plants, bryophytes, and algae. Vascular plants are the best-known group. This is due in part to the recent publication of the *Checklist of Panarctic Flora (PAF) Vascular Plants* (Elven, 2011). The Panarctic flora includes 2,218 taxa, 91 families and 430 genera; which is less than 1% of the world flora (Daniëls et al., 2013). There are few gymnosperm taxa: 96% of the flora are angiosperms. Eight species-rich families account for more than 50% of the flora, of which the top three families are Asteraceae (254 taxa), Poaceae (224 taxa), and Cyperaceae (190 taxa). About 5%, or 106 taxa, are endemic. Most endemics are Beringian, occur arctic-wide, and are forbs. There are no endemic woody species.

As a whole, the arctic flora is viewed as taxonomically, ecologically, biologically, and genetically coherent with the many species having a circumpolar distribution. Biodiversity is low in comparison to temperate or tropical ecosystems. Trends in species richness are largely attributed to history, including glaciations, land-bridges, and north-south trending mountain ranges (Yurstev, 1994). Bryophytes are ubiquitous in the Arctic and contribute significantly to species richness, particularly in moist to wet habitats (Daniëls et al., 2013). There are an estimated 900 arctic bryophyte species and approximately 4,000 freshwater and marine algal species. The biodiversity of microalgae is still largely unknown. At present, there are few introduced species (101 taxa; Elven, 2011). The most widespread non-native stabilized introduced species are *Lepidotheca suaveolens* (Asteraceae, pineapple weed), *Plantago major* subsp. *major* (Plantaginaceae, common plantain), and *Trifolium pratense* (Fabaceae, red clover). Most of the introduced species are not invasive and are restricted to disturbed habitats. For example, hay brought in to protect disturbed slopes from erosion where the trans-Alaska oil pipeline passes through the Arctic created an influx of invasive species, but most were gone after the first winter. Although not currently a threat in the Arctic, invasive species are likely to increase due to increasing human activity coupled with climate change. For example, a recent study showed that visitors to Svalbard transport a minimum of four seeds on their shoes. Most of these seeds are from species known to be invasive elsewhere and over a quarter of these seeds were found to be capable of germination under current climatic conditions (Ware et al., 2012).

Plant species diversity of the world-wide alpine flora is much greater than in the Arctic. Körner (2003) estimates 8,000-10,000 vascular plants, comprising 100 families and about 2,000 genera, or nearly 4% of the world flora. The most common families in the Alpine are similar to those also common in the Arctic: Asteraceae, Poaceae, Brassicaceae, Caryophyllaceae, Cyperaceae, Rosaceae, and Ranunculaceae. Regional alpine floras, from the Teton Range in Wyoming to the Hokkaido alpine zone in Japan, typically include 200-280 species, with a mean diversity of 241 species from nine distinct mountain ranges (Körner, 2003). In contrast, in the Arctic, mean species richness of vascular plants from the 21 Panarctic floristic provinces (Elven, 2007) is estimated at 544 species (Daniëls et al., 2013). The most species rich floristic province is Western Alaska (825 species), and the least species rich region is Ellesmere Land-North Greenland (199 species). These data are not directly comparable to estimates of diversity for alpine floras as floristic provinces are not analogous to more regional mountain ranges. Within the alpine zone, total plant species richness within a given region declines by about 40 species of vascular plants per 100 m of elevation (Körner, 2002). Mosses (also see Chapter 12) and lichens (also see Chapter 3) deviate from this pattern as they often increase in abundance with increasing altitude, although their richness eventually decreases at the highest altitudes. Most alpine species occur at 1,000 m or lower, although a few species have been found as high as 5,900 m in the Tibetan Himalaya (Rongfu & Miehe, 1988) or 6,300 m on Mount Everest (Grabherr et al., 1995). Given the geographic isolation of mountains that often are functionally islands, endemism is high with the highest degree of endemism found at moderate, rather than at extreme altitudes.

There are several stressors to arctic biodiversity (Meltofte, 2013). These fall into two categories: anthropogenic and climatic stressors. Anthropogenic stressors include increased development, such as infrastructure associated with oil, gas, and other resource extraction. Further development will be made possible by increased opportunities for transportation including shipping lanes, road building, and regular air service to remote localities. There are also stressors from contaminants, such as persistent organic pollutants, and increased potential for oil spills.

Climatic stressors are the most serious threat to plant biodiversity in the Arctic and equally, or more so, to alpine environments. Climate warming is predicted to lead to migration of plants northward, altering the structure of vegetation through additions or even replacement from the sub-arctic to the low Arctic to the high Arctic. Terrestrial habitats in the Arctic are bounded to the north by a coastline so there is the potential that high arctic ecosystems may only survive in isolated refugia or in mountain habitats. A similar scenario is predicted for the Alpine, with expansion of treeline vegetation to higher elevations. Snowbed specialists, adapted to late snow melt and low soil temperatures are among the most threatened as both conditions are likely to be altered by climate change (Björk & Molau, 2007).

Many studies document changes in arctic and alpine plant distributions consistent with climate warming predictions. Re-sampling studies from over 100 mountains in Scandinavia and Europe, as well as on the arctic islands of Spitsbergen and Greenland, show that species richness on mountain summits has increased (Birks, 2013). This increase is predominantly an altitudinal ascent of grasses, dwarf shrubs, and low shrubs. In central Norway, Klanderud & Birks (2003) showed that changes in species richness from 1930 to 1998 varied by elevation belt. Total plant species richness in the lowest elevation belt (1,600-1,800 m) increased by 8-14 species, while in the mid-elevation belt (1,800-2,000 m) total plant species richness increased by 5-8 species. Above 2,000 m, little or no change in species richness was observed. No high-alpine species had gone extinct, although a few species had decreased in frequency since 1930. In Montana's Glacier National Park, arctic-alpine plant cover declined over two decades of study (1988-2011) with a concurrent increase in mean summer temperature (Lesica, 2014). Plants restricted to high elevations declined more so than those with a broader elevational distribution. In alpine areas of Europe, Gottfried et al. (2012) found increases in warm-adapted species and declines in cold-adapted species over a relatively short time period from 2001-2008. Warming experiments have shown an increase in shrubs in the Low Alpine in Europe (Cannone et al., 2007) and Asia (Klein et al., 2007) and from multiple sites across the Circumpolar Arctic (Elmendorf et al., 2011; Walker et al., 2006). Increases in satellite measures of greenness (related to aboveground plant biomass) have been observed (Epstein et al., 2012), as well as increases in shrub cover based on repeat photography in the warmest parts of the Arctic (*e.g.*, Tape et al., 2006), although grazing by reindeer, lemmings, and voles may limit shrub expansion (Olofsson et al., 2009). Studies in colder subzones of the Arctic have found increased vegetation cover and height, but little change in community composition (*e.g.,* Hudson & Henry, 2009), except in recently deglaciated areas where succession is occurring.

Equating biodiversity with species richness is one measure, but there is another level to be considered. From molecular studies, we now know that genetic diversity within Linnean, or biological, species can be high. The problem comes in assessing Linnean diversity, for there is often no parallel morphological differentiation to provide visible markers to genotypic boundaries. There is great genetic variation within the species (cf. Brochmann & Brysting, 2008). Reticulate evolution among arctic plants involves multiple genomes, secondary contact, hybridization, and polyploidization, all of which provide raw material for infraspecific variation and differentiation.

Some of the best information on biological species diversity comes from studies of *Draba* (Brassicaceae), initiated by Brochmann and continued by him with students and colleagues in Oslo. Grundt et al. (2006) conducted intraspecific crossing studies of three circumpolar diploid species in *Draba* and found, despite observations of limited morphological and genetic diversity, evidence for cryptic biological species, ones reproductively isolated from one another and thus evolutionarily independent. Hybrids from within populations were mostly fertile (63%), while those from within and among geographic regions (Alaska, Greenland, Svalbard, and Norway) were mostly infertile (8%). These results suggest that infraspecific diversity may be higher in the Arctic than previously realized.

Genetic diversity is essential to long-term persistence of arctic and alpine biodiversity as it provides opportunities for species to respond to changing environmental conditions. As abundance and geographic distributions of species shrink, genetic variability for selection to act upon is also often reduced. For most arctic and alpine plants, we lack information on how genetic variation, and therefore evolutionary potential, is generated and maintained. Species richness is often used as a surrogate for genetic diversity in conservation planning, although we are still learning how these two levels of biodiversity are related. To date, a few studies have addressed whether species and genetic levels of biodiversity co-vary in arctic and alpine plant communities.

Taberlet et al. (2012) showed that for the flora of the Alps and Carpathians, species richness and genetic diversity of high mountain vascular plants are not correlated. Their results showed that genetic diversity is instead associated with glacial history of a species, which in turn was linked with environmental and ecological characteristics of glacial refugia, range shifts, and associated demographic processes. In contrast, Eidesen et al. (2013) showed that patterns of genetic diversity across 17 vascular plant species are analogous to large-scale patterns of species diversity in the Arctic. Diversity was highest in Beringia and decreased gradually into more recently deglaciated regions. It should be noted that both of these studies assessed neutral genetic diversity, which is not under selection.

An aspect of genetic diversity in arctic plants was noted many decades ago as chromosome counts of northern plants were becoming known and diploids and polyploids were identified. It was further noted that there are more polyploids at higher latitudes than at low latitudes (Hagarup, 1928). The relationship between the frequency of polyploids and the northernmost regions became the preoccupation of many, for whom the underlying belief was that polyploidy *per se* gave the plants advantages in cold climates. The advantages of genetic diversity from multiple sets of chromosomes was presumed to endow polyploids with the ability to persist in the rigorous conditions, such as in glacial refugia (see above) and also to have the capacity to spread aggressively during deglaciation (Löve, 1959).

Johnson & Packer (1965, 1967) and Johnson et al. (1965) demonstrated, at Ogotoruk Creek in northwest arctic Alaska, a relationship between the frequency of polyploid taxa along gradients of soil texture, moisture and temperature, depth to permafrost, and degree of geomorphic disturbance. The diploids and low polyploids were found on more stable Tertiary surfaces, and the higher polyploids were found in habitats of the sort that became common and widespread during cold intervals of the Pleistocene, suggesting their more recent divergence.

Brochmann et al. (2004) examined the observations and explanations for polyploidy in arctic plants, particularly what can be concluded from recent molecular studies. Essentially, polyploidy is the means by which reticulate evolution proceeds and by which hybrids can gain fertility, stability, and independence. Research with hybrids showed there is *interspecific* gene flow across ploidy levels (Brochmann et al., 1992a), which demolishes the simplistic but long held belief in strong reproductive barriers between diploids and tetraploids and so-called abrupt speciation. Surprisingly, there can be two or three different parental species, all polyploids sharing parts of their genomes, which form polyphyletic hybrids. These hybrids attain fertility through polyploidization. Hence, taxa of different parental combinations, formed at different times and places, can exist within the same Linnean species (Brochmann et al., 1992b). Recent studies have shown that polyploidy has occurred at different times and places within *Vaccinium uliginosum* (Ericaceae; Eidesen et al., 2007) and that different ploidy levels overlap across the circumpolar distribution of *Saxifraga oppositifolia* (Müller et al., 2012).

Changes in biodiversity, driven by climate and other anthropogenic stressors, will provide new opportunities for recruitment and require adaptation and adjustment of arctic and alpine floras. Crawford (2008) argues that many widespread arctic and alpine plants occupy a range of different habitats, in terms of temperature and soil-moisture content for example, and are ecotypically diverse. If so, this should help buffer these species against extinction with increases in global temperatures. For other plants that are of recent origin or which are narrowly distributed, such ecotypic diversity does not exist. For species that may be outcompeted by more thermophilous species invading from the south, their survival depends on their ability to colonize newly deglaciated land at higher latitude or altitude where temperatures remain low. For alpine species that are already restricted in high altitude mountain ranges, there may be no new suitable habitat to exploit. If so, these species are likely to be among the most endangered in the future (Birks, 2008). In the next section of this chapter, we discuss adaptation and the response of arctic and alpine plants to climate change.

Adaptation and the Response of Arctic and Alpine Plants to Climate Change

Climate change in recent decades has led to changes in the composition and distribution of vegetation in arctic and alpine environments. These regions are changing, and as a consequence their biodiversity is also changing (Callaghan et al., 2004). Predicted increases in temperatures globally are 0.1°C per decade, which is amplified in the polar region compared to lower latitudes (ACIA, 2005).

In response to temperature increases, shrubs and trees are extending their limits both northwards and upwards. How will arctic and alpine plants be affected by climate change? Birks (2008) stated this question well:

> "Will arctic plants be pinched between advancing shrub tundra and forest and the rising sea-level in the low-land Arctic? Will alpine plants be squeezed off the tops of mountains?"

It is likely that some arctic and alpine plants will become extinct, particularly those with small endemic populations at the limit of plant life in the High Arctic or at high altitude. If we look to the past, however, to when the climate warmed in the early Holocene, temperatures were about 2°C warmer in the Arctic. Arctic and alpine plants persisted, and no arctic-alpine species with a fossil record is known to have gone extinct in the Quaternary (Birks, 2008). It therefore is likely that more ecotypically diverse species are resilient to climate change and will survive and adapt as long as some suitable habitat remains.

Ecotypes, variants within species, have long been recognized among temperate plants, where ecotypes show various morphological features adaptive to particular environmental conditions. A selective advantage may also accrue to ecotypes in their native site without a change in morphology as to be recognized as taxonomically distinct. There are many examples of ecotypes along latitudinal and altitudinal gradients, even along local gradients of microtopography where adaptations are less morphological and mainly physiological (Chapin & Chapin 1981; Shaver et al., 1979). Ecological amplitude in geographically wide-ranging species derives from the formation of entities with genetically fixed, adaptive properties. The effectiveness of this process is not fully appreciated. For ecotypes to undergo speciation there would first need to be sufficient genetic variation within them, and second, selection pressure to drive the process of differentiation. Absent one or both, further divergence does not occur; moreover, the infraspecific ecotypes allow for persistence across a range of environmental conditions. Raup (1969) evaluated the breadth of tolerance by species to gradients of soil moisture, plant cover, and geomorphic disturbance and found that some species exhibit great tolerances. This capacity of some tundra plants is a function either of phenotypic plasticity or of genetically fixed ecotypic differentiation, or a bit of both. It is likely that more ecotypically diverse species will have large ecological amplitudes, will be resilient to climate change, and will survive and adapt as long as the thresholds of tolerance to limiting factors are not exceeded.

Temperature, photoperiod, concentration of CO_2, and light intensity all affect photosynthesis and photosynthetic efficiency of plants. Species occurring in both arctic and alpine tundra provide examples of ecotypic differentiation for those environmental parameters. Ecotypes of these species are differentially adapted to the low light intensity and long photoperiod of the Arctic and to the high light intensity and short photoperiod of the Alpine. Even differences in the production of leaves, leaf width and thickness, and concentration of chlorophyll have been identified as part of ecotypic differentiation of physiological traits (cf. Mooney & Billings, 1961; Tieszen & Bonde, 1967).

Much of what we know about adaptation in arctic and alpine plants is based on common-garden studies as a means of identifying genetically controlled responses among plants grown in different adaptive norms. Work has ranged from the early reciprocal transplant studies of Clausen & Hiesey (1958) with *Potentilla glandulosa* and Clausen et al. (1948) with *Achillea lanulosa* (Asteraceae) across an elevational gradient in California, to work by Mooney & Billings (1961) with *Oxyria digyna* from a broad latitudinal range of arctic and alpine populations, to work by Tieszen & Bonde (1967) with *Deschampsia caespitosa* (Poaceae) and *Trisetum spicatum* (Poaceae) from arctic and alpine sites. The work of Clausen and colleagues revealed a sequence of climatic races. Mooney & Billings (1961) showed a clear differentiation of physiological traits in *Oxyria digyna* over a latitudinal gradient from northern Alaska south through the Rocky Mountains to Colorado. A more recent study returned to two separate reciprocal transplant experiments in Alaska 30 years later, *Dryas octopetala* subspecies along a snowbank gradient in the Alpine and *Eriophorum vaginatum* (Cyperaceae) along a latitudinal gradient in the Arctic (Bennington et al., 2012; McGraw & Antonovics, 1983; Shaver et al., 1986). For both species, differential survival in the ecotypes' native site provided strong evidence for local adaptation in these long-lived species. These findings show a broad range of ecotypes that would likely respond differently to climate change. Ultimately, the ecotypic differentiation revealed by these and other studies of arctic and alpine plants suggests extinction of wide-ranging species would be unlikely.

Just how the genes underlying genetic variation control ecotypic differentiation in arctic and alpine plants is unknown. Molecular evidence based on non-coding regions of the genome, so usefully applied in phylogeography is, however, neutral to the effects of selection. A focus on adaptive rather than neutral genetic variation will be needed for predicting responses to climate warming (Crawford, 2008). If we assume ecotypic diversity is a surrogate for adaptive genetic variation, it would seem, as discussed above, that species with high ecotypic diversity are likely to survive climate warming. To date, the genetics of adaptation have largely been studied in model organisms with short generation times and not for long-lived arctic and alpine plants.

We must note that an important distinction between arctic and alpine environments is both day length and light intensity. Phenology is often related to day length in plants. For example, arctic and alpine populations of *Oxyria digyna* show ecotypic differences in flower and rhizome production, and in growth responses, to temperature and day length (Mooney & Billings, 1961). Consequently, southern ecotypes cannot simply migrate northward to cooler temperatures in a warming climate, as day length varies from about 15 hours of solar radiation on the summer solstice at Niwot Ridge in Colorado (40° N) to continuous low intensity 24-hour solar radiation north of the Arctic Circle (>66° 33' 44" N). There are clearly limits to arctic and alpine plants escaping climate change by extending their ranges northwards and upwards.

Several recent global modeling studies have shed light on potential future states of vegetation in arctic and alpine environments. Alsos et al. (2012) analyzed range-wide genetic diversity of 27 northern plant species and used species distribution modeling to predict their future distributions and levels of genetic diversity through 2080. Their work predicts range reduction and loss of genetic diversity in nearly all species in their study, according to at least one scenario. Species that were more vulnerable to losses in genetic diversity lacked traits for long distance dispersal and had high levels of genetic differentiation among populations. In another study, Pearson et al. (2013) used ecological niche models, based on statistical associations between vegetation and climate, to predict the future distribution of arctic vegetation. Their study predicts that at least half of vegetated areas will shift to a different vegetation class, for example from graminoid tundra to shrub tundra, by 2050. Moreover, their model predicts woody plant cover, or shrub tundra and forest, will increase by as much as 52%. In contrast, Breen et al.'s (2014) regional modeling study for Alaska tundra predicts more modest shifts in woody plant cover. Their study used a state and transition model that is driven by both climate and fire dynamics. Treeline advance varies by the climate model used to drive the simulations. With greater tundra fire activity, 12% of tundra transitions to forest, and 11% of graminoid tundra transitions to shrub tundra, by 2100. In contrast, with more modest tundra fire activity, the amount of tundra that transitions to forest nearly doubles to 20%, but there is little change in the relative amounts of graminoid and shrub tundra.

Arctic and Alpine Plant Interactions with Other Organisms: The Ectomycorrhizal Symbiosis

Virtually every plant is full of endophytes (fungi, bacteria, viruses) that occur in all organs of plants. As in other ecosystems, plants in the Arctic and Alpine interact with organisms across kingdoms, including plants, animals (mammals, birds, insects), fungi, bacteria, archaea, and viruses. Many of these complex interactions, both direct and indirect, occur simultaneously. These interactions happen with different degrees of specificity and range from antagonistic to mutually beneficial. The outcome of such interactions depends in large part on the environment (Partida-Martinez & Heil, 2011), which in the Arctic and Alpine are dominated by low temperatures and a short growing season.

The use of molecular methods has not only revealed a great biodiversity of organisms in arctic and alpine environments, but also highlights the complex interactions of plants with other organisms, including fungi (Dahlberg et al., 2013; Gao & Yang, 2010; Timling & Taylor, 2012). Fungi are ubiquitous and benefit plants as mutualistic mycorrhizas (also see Chapters 2, 5) and saprotrophs by providing nutrients and water; they can harm plants as pathogens. We will illustrate how molecular methods have shed light on plant interactions with other organisms through the example of ectomycorrhizal fungi (EMF).

The ectomycorrhizal symbiosis is abundant throughout the Arctic and Alpine, where the fungi associate with shrubs, as well as a few sedges and forbs. Although EMF associate with only about 6% of the vascular plants in the Arctic, these plants are important components of plant communities that cover up to 69% of the ice-free Arctic (Walker et al., 2005). The symbiosis seems especially important in biomes with low nutrient availability, where the fungus provides nutrients and water to the plant and the plant provides carbohydrates to the fungus. In the Arctic, 61-86% of nitrogen in ectomycorrhizal plants is provided by their fungal symbionts while the plant provides 8-17% of photosynthetic carbon to the fungi (Hobbie & Hobbie, 2006).

In contrast to vascular plants in the Arctic, EMF associated with shrubs do not follow the classic pattern of species richness decline with latitude, which suggests that fungal species richness is not governed by temperature (Bjorbaekmo et al., 2010; Timling et al., 2012). The species-rich EMF communities that have been observed on host plants in the Arctic and Alpine are dominated by a few families that are especially species-rich (Thelephoraceae, Cortinariaceae, Inocybaceae) (Blaalid et al., 2014; Gao & Yang, 2010). Similarly, many plant communities are dominated by a few species-rich families (Asteraceae, Brassicaceae, Caryophyllaceae, Cyperaceae, Fabaceae, Poaceae, Ranunculaceae, and Rosaceae; Daniëls et al., 2013). This suggests that some plant and fungal families are especially adapted to arctic and alpine environments.

Furthermore, EMF communities in the Arctic appear to be dominated by generalist fungi with wide ecological amplitudes and which are excellent dispersers (Geml et al., 2012; Timling et al., 2012). In contrast to boreal (Taylor et al., 2010) and temperate forests (Ishida et al., 2007) and Mediterranean woodlands (Morris et al., 2008), the ectomycorrhizal symbiosis seems to have lower specificity in the Arctic and Alpine (Botnen et al., 2014; Gao & Yang 2010; Timling et al., 2012). While EMF communities in boreal, temperate and tropical climates show distinctive phylogeographic distribution patterns, with restrictions to continents and sub-continental regions (Geml et al., 2008; Talbot et al., 2014), the majority (73%) of EMF species observed in studies from Svalbard and across the entire bioclimatic gradient of North American Arctic have occurred also in other regions within and beyond the Arctic (Geml et al., 2012; Timling et al., 2012). Such wide distributions within the Arctic have been also observed for lichens (Geml et al., 2010) and vascular plants (Alsos et al., 2007). The wide distribution of fungi and lichens might be aided by the smaller size of their propagules. Finally the wide distribution suggests that terrestrial and trans-ocean long distance dispersal must be a common phenomenon in the wide open landscapes of the Arctic, aided by wind, snow, driftwood, sea ice, birds, and mammals (reviewed in Alsos et al., 2007).

Nevertheless, within the Arctic and Alpine, EMF communities show distribution patterns at the regional and local scale that often parallel those of plant communities found there. Ectomycorrhizal fungal communities associated with *Dryas integrifolia* and *Salix arctica* (Salicaceae) change gradually across the five bioclimatic subzones of the North American Arctic, corresponding with climate, plant productivity, glaciation history, geology, and soil factors (Timling et al., 2012). At a local scale, EMF communities often correlate with habitat, successional stage, plant community, and bedrock and edaphic factors such as pH, carbon, and nitrogen (Blaalid et al., 2014; Fujimura & Egger, 2012; Yao et al., 2013; Zinger et al., 2011).

Climatic changes in the Arctic have led to pan-arctic shrub expansion (Tape et al., 2006) and increases in plant productivity (Bhatt et al., 2010) and nutrient cycling (Rustad et al., 2001). Long-term warming experiments show not only changes in plant communities (Walker et al., 2006) but also changes in EMF community structure associated with *Betula nana* (Betulaceae), one of the shrubs most responsive to climate warming (Deslippe et al., 2011). The mutualistic nature of the ectomycorrhizal symbiosis, the low host specificity and the wide distribution support the idea that EMF may play critical roles in the expansion of shrubs in the tundra. Evidence from past climate changes comes from paleobotanical studies which show that plant and fungal communities changed with past glacial and interglacial cycles, with an increase in shrubs and trees and their ectomycorrhizal symbionts since the last glaciation (de Vernal & Hillaire-Marcel 2008; Lydolph et al., 2005). Soil analyses along a bioclimatic gradient in the North American Arctic show that subzone A, which is devoid of woody species, harbors EMF species, probably as spores, and that soil fungal communities in subzone E greatly overlap (74%) with communities of the boreal forests (Timling et al., 2014). Furthermore, a bioassay with soils from above treeline showed that these soils provide sufficient inoculum for a significant growth of conifers (Reithmeier & Kernaghan, 2013). The authors concluded that spores in the soils and shrubs above treeline could facilitate treeline expansion. With a warming climate one might expect changes of EMF community composition with a northward shift of some EMF taxa. Finally EMF may be critical in facilitating an establishment of woody species in subzone A and a treeline expansion into subzone E.

Conclusion

Despite considerable progress made in recent years, there remains much to learn about the ecology and evolution plants in arctic and alpine environments. Molecular ecology, modeling, and remote sensing studies, along with future fossil discoveries, will continue to build upon our knowledge of these biomes and improve our understanding of their potential response to future climate change. Brochmann et al. (2013) write that the species-poor arctic flora is likely to be adapted to environmental change, through selection for high mobility and buffering against inbreeding- and bottleneck-induced gene loss via polyploidy. However, today's flora of arctic and alpine environments will be challenged by a climate warmer than the Holocene and over a shorter period. There is a need to begin focusing on adaptive, rather than neutral genetic variation, to predict how arctic and alpine plants will respond to climate warming over the next century.

There is also a need to intensify biodiversity research on arctic and alpine floras, with an emphasis on vegetation classification, monitoring, and modeling (Daniëls et al., 2013). Efforts such as the Arctic Vegetation Archive (Walker et al., 2013) are underway to improve coordination and cooperation among arctic nations and to produce a pan-arctic vegetation classification. Furthermore, the archive will provide vegetation data from across the Circumpolar Arctic for use in biodiversity and ecosystem models. Jónsdóttir (2013) is also leading an initiative to develop a research framework on biodiversity-shaping forces that considers different spatial and temporal scales and identifies commonalities across biological hierarchies and organisms. This framework will provide for testing hypotheses about biodiversity trends in the face of climate change in the Arctic and Alpine.

Acknowledgments

We thank Inger Alsos, Teresa Hollingsworth, and two anonymous reviewers for their comments on an earlier version of the chapter that greatly improved its content. We also thank Christian Körner for fruitful discussions about the global distribution of arctic and alpine tundra and for sharing his map of mountainous regions of the world for adaptation and inclusion in the chapter.

References

Abbott, R. J. (Ed.). (2008) Special issue: History, evolution and future of arctic and alpine flora. *Plant Ecology & Diversity 1*, 129-349.

Abbott, R. J. & Brochmann, C. (2003) History and evolution of the arctic flora: In the footsteps of Eric Hultén. *Molecular Ecology 12*, 299-313.

ACIA (2005) *Arctic climate impact assessment*. Cambridge: Cambridge University Press.

Alsos, I. G., Eidesen, P. B., Ehrich D., Skrede, I., Westergaard, K., Jacobsen, G. H., Landvik, J. Y., Taberlet, P. & Brochmann, C. (2007) Frequent long-distance plant colonization in the changing Arctic. *Science 316*, 1606-1609.

Alsos, I. G., Ehrich, D., Thuiller, W., Eidesen, P. B., Tribsch, A., Schönswetter, P., Lagaye, C., Taberlet, P. & Brochmann, C. (2012) Genetic consequences of climate change for northern plants. Proceedings of the Royal Society Series B, DOI: 10.1098/rspb.2011.2363.

Avise, J. C. (1994) *Molecular markers, natural history and evolution*. New York, NY: Chapman & Hall.

Bennike, O. & Bøcher, J. (1990) Forest-tundra neighbouring the North Pole: Plant and insect remains from the Plio-Pleistocene Kap København Formation, North Greenland. *Arctic 43*, 332-338.

Bennington, C. C., Fetcher, N., Vavrek, M. C., Shaver, G. R., Cummings, K. J. & McGraw, J. B. (2012) Home site advantage in two long-lived arctic plant species: Results from two 30-yr reciprocal transplant studies. *Journal of Ecology 100*, 841-845.

Berg, R. (1963) Disjunctions in the Norwegian alpine flora and theories proposed for their explanation. *Blyttia 212*, 133-177, in Norwegian with an English summary (cited in Brochmann et al., 1996).

Bhatt, U. S., Walker, D. A., Raynolds, M. K., Comiso, J. C., Epstein, H. E., Jia, G., Gens, R., Pinzon, J. E., Tucker, C. J., Tweedie, C. E. & Webber, P. J. (2010) Circumpolar arctic tundra vegetation change is linked to sea ice decline. *Earth Interactions 14*, 1-20.

Billings, W. D. & Mooney, H. A. (1968) The ecology of arctic and alpine plants. *Biological Review 43*, 481-529.

Birks, H. H. (2008) The Late-Quaternary history of arctic and alpine plants. *Plant Ecology & Diversity 1*, 135-146.

Birks, H. J. B. (2013) Alpine plants and recent climate change. *Climatica.* Online: http://climatica.org.uk/alpine-plants-and-recent-climate-change. Accessed July 1, 2014.

Bjorbaekmo, M. F. M., Carlsen, T., Brysting, A., Vralstad, T., Hoiland, K., Ugland, K. I. & Geml, J., Schmacher, T. & Kauserud, H. (2010) High diversity of root associated fungi in both alpine and arctic *Dryas octopetala*. *BMC Plant Biology 10*, DOI:10.1186/1471-2229-10-244.

Björk, R. G. & Molau, U. (2007) Ecology of alpine snowbeds and the impact of global change. *Arctic, Antarctic and Alpine Research 39*, 34-43.

Blaalid, R., Davey, M. L., Kauserud, H., Carlsen, T., Halvorsen, R., Hoiland, K. & Eidesen, P. B. (2014) Arctic root-associated fungal community composition reflects environmental filtering. *Molecular Ecology23*, 649-659.

Bliss, L. C. (2000) Arctic tundra and the polar desert biome. In: M. G. Barbour & W. D. Billings (Eds.) *North American terrestrial vegetation.* Second edition (pp. 1-40). Cambridge, UK: Cambridge University Press.

Botnen, S., Vik, U., Carlsen, T., Eidesen, P. B., Davey, M. L. & Kauserud, H. (2014) Low host specificity of root-associated fungi at an Arctic site. *Molecular Ecology 23*, 975-985.

Breen, A. L., Bennett, A., Hewitt, R. E., Hollingsworth, T. N., Genet, H., Euskirchen, E. S., McGuire, A. D. & Rupp, T. S. (2014) Tundra fire and vegetation dynamics: Simulating the effect of climate change on fire regimes in Arctic ecosystems. In prep to submit to *Ecology Letters*.

Brochmann, C. & Brysting, A. K. (2008) The Arctic—An evolutionary freezer? *Plant Ecology & Diversity 1*, 181-195.

Brochmann, C., Stedje, B. & Borgen, L. (1992a) Gene flow across ploidal levels in *Draba* (Brassicaceae). *Evolutionary Trends in Plants 6*, 125–134.

Brochmann, C., Soltis, P. S. & Soltis, D. E. (1992b) Recurrent formation and polyphyly of Nordic polyploids in *Draba* (Brassicaceae). *American Journal of Botany 79*, 673-688.

Brochmann, C., Gabrielsen, T. M., Nordal, I., Landvik, J. Y. & Elven, R. (2003) Glacial survival or *tabula rasa*? The history of North Atlantic biota revisited. *Taxon 52*, 417-450.

Brochmann, C., Brysting, A. K., Alsos, I. G., Borgen, L., Grundt, H. H., Scheen A. C. & Elven, R. (2004) Polyploidy in arctic plants. *Biological Journal of the Linnean Society 82*, 521-536.

Brochmann, C., Edwards, M. & Alsos, I. G. (2013) The dynamic past and future of arctic vascular plants: Climate change, spatial variation and genetic diversity. In: K. Rhode (Ed.). *The balance of nature and human impact* (pp. 133-152). Cambridge, UK: Cambridge University Press.

CAVM Team (2003) *Circumpolar Arctic vegetation map, scale 1:7 500 000*. Conservation of Arctic Flora and Fauna (CAFF) map no. 1. Anchorage, AK: U.S. Fish and Wildlife Service.

Callaghan, T. V., Bjorn, L. O., Chernov, Y., Chapin, F. S., Christensen, T. R., Huntley, B., Ims, R. A., Johansson, M., Jolly, D., Jonasson, S., Matveyeva, N., Panikov, N., Oechel, W., Shaver, W., Elster, J., Henttonen, H., Laine, K., Taulavuori, K., Taulavuori, E. & Zockler, C. (2004) Biodiversity, distributions and adaptations of arctic species in the context of environmental change. *AMBIO 33*, 404-417.

Cannone, N., Sgorbayi, S. & Guglielmin, M. (2007) Unexpected impacts of climate change on alpine vegetation. *Frontiers in Ecology and Environment 5*, 360–364.

Chapin, F. S. III & Chapin, M. C. (1981) Ecotypic differentiation of growth processes in *Carex aquatilis* along latitudinal and local gradients. *Ecology 62*, 1000-1009.

Chapin, F. S. III & Körner, C. (Eds.) (1995) *Arctic and alpine biodiversity: Patterns, causes and ecosystem consequences*. Berlin: Spring-Verlag.

Chernov, Yu. I. (1985) *The living tundra*. Studies in Polar Research. Cambridge: Cambridge University Press.

Clausen, J. & Hiesey, W.M. (1958) *Experimental studies on the nature of species. IV. Genetic structure of ecological races*. Publication 615. Washington, DC: Carnegie Institution of Washington.

Clausen, J., Keck, D. D. & Hiesey, W. M. (1948) *Experimental studies on the nature of species. III: Environmental responses of climatic races of* Achillea. Publication 581. Washington, DC: Carnegie Institution of Washington.

Crawford, R. M. M. (2008) Cold climate plants in a warmer world. *Plant Ecology & Diversity 1*, 197-208.

Dahl, E. (1955) Biogeographic and geological indications of unglaciated areas in Scandinavia during the glacial ages. *Geological Society of America Bulletin 66*, 499-519.

Dahl, E. (1963) Plant migrations across the North Atlantic Ocean and their importance for the paleogeography of the region. In: A. Löve & D. Löve (Eds.). *North Atlantic biota and their history* (pp. 173-188). London: Pergamon Press.

Dahlberg, A., Bültmann, H., Cripps, C. L., Eyjólfsdóttir, G. G., Gulden, G., Kristinsson, H. & Zhurbenko, M. (2013) Fungi. In: H. Meltofte (Ed.). *Arctic biodiversity assessment: Status and trends in Arctic biodiversity* (pp. 354-371). Akureyri: Conservation of Arctic Flora and Fauna (CAFF).

Daniëls, F. A., Gillespie, L. J. & Poulin, M. (2013) Plants. In: H. Meltofte (Ed.). *Arctic biodiversity assessment: Status and trends in Arctic biodiversity* (pp. 310-353). Akureyri: Conservation of Arctic Flora and Fauna (CAFF).

Darwin, C. (1859) *On the origin of species by means of natural selection, or the preservation of favoured races in the struggle for life*. London: John Murray.

de Vernal, A. & Hillaire-Marcel, C. (2008) Natural variability of Greenland climate, vegetation, and ice volume during the past million years. *Science 320*, 1622-1625.

Deslippe, J. R., Hartmann, M., Mohn, W. W. & Simard, S. W. (2011) Long-term experimental manipulation of climate alters the ectomycorrhizal community of *Betula nana* in Arctic tundra. *Global Change Biology 17,* 1625-1636.

Eidesen, P. B., Alsos, I. G., Popp, M., Stensrud, Ø., Suda, J. & Brochmann, C. (2007) Nuclear vs. plastid data: Complex Pleistocene history of a circumpolar key species. *Molecular Ecology 16*, 3902-3925.

Eidesen, P. B., Ehrich, D., Bakkestuen, V., Alsos, I. G., Gilg, O., Taberlet, P. & Brochmann, C. (2013) Genetic roadmap of the Arctic: Plant dispersal highways, traffic barriers and capitals of diversity. *New Phytologist 200*, 898-910.

Ellenberg, H. H. (2009) The vegetation above the alpine treeline. In: H. H. Ellenberg. *Vegetation ecology of Central Europe*. Fourth edition (pp. 388-454). Cambridge: Cambridge University Press.

Elmendorf, S.C., Henry, G.H. R., Hollister, R. D., Björk, R. G., Boulanger-Lapointe, N., Cooper, E. J., Cornelissen, J.H. C., & Day, T. A., et al. (2012) Plot-scale evidence of tundra vegetation change and links to recent summer warming. *Nature Climate Change 2*, 453-457.

Elven, R. (2007) *Checklist of the panarctic flora (PAF): Vascular plants*. Oslo: University of Oslo, National Centre of Biosystematics, Natural History Museum.

Elven, R. (2011) *Annotated checklist of the panarctic flora (PAF): Vascular plants*. Oslo: University of Oslo, National Centre of Biosystematics, Natural History Museum.

Epstein, H. E., Raynolds, M. K., Walker, D. A., Bhatt, U. S., Tucker, C. J. & Pinzon, J. E. (2012) Dynamics of aboveground phytomass of the circumpolar arctic tundra over the past three decades. *Environmental Research Letters 7*, 015506. DOI:10.1088/1748-9326/7/1/015506.

Fernald, M. L. (1925) Persistence of plants in unglaciated areas of boreal America. *Memoirs of the American Academy of Arts and Sciences 15*, 237-342.

Fujimura, K. E. & Egger, K. N. (2012) Host plant and environment influence community assembly of High Arctic root-associated fungal communities. *Fungal Ecology 5*, 409-418.

Gao, Q. & Yang, Z. L. (2010) Ectomycorrhizal fungi associated with two species of *Kobresia* in an alpine meadow in the eastern Himalaya. *Mycorrhiza 20*, 281-287.

Geml, J., Laursen, G. A. & Taylor, D. L. (2008) Molecular diversity assessment of arctic and boreal *Agaricus* taxa. *Mycologia 100*, 577-589.

Geml, J., Kauff, F., Brochmann, C. & Taylor, D. L. (2010) Surviving climate changes: High genetic diversity and transoceanic gene flow in two arctic-alpine lichens, *Flavocetraria cucullata* and *Flavocetraria nivalis* (Parmeliaceae, Ascomycota). *Journal of Biogeography 37*, 1529-1542.

Geml, J., Timling, I., Robinson, C. H., Lennon, N., Nusbaum, C. H., Brochmann, C., Noordeloos, M. E. & Taylor, D. L. (2012) An arctic community of symbiotic fungi assembled by long-distance dispersers: Phylogenetic diversity of ectomycorrhizal basidiomycetes in Svalbard based on soil and sporocarp DNA. *Journal of Biogeography 39*, 74-88.

Gottfried, M., Pauli, H., Futschik, A., Akhalkatsi, M., Barančok, P., Benito Alonso, J. L., Coldea, G., Dick, J., Erschbamer, B., Fernández Calzado, M. R., Kazakis, G., Krajči, J., Larsson, P., Mallaun, M., Michelsen, O., Moiseev, D., Moiseev, P., Molau, U., Merzouki, A., Nagy, L., Nakhutsrishvili, G., Pedersen, B., Pelino, G., Puscas, M., Rossi, G., Stanisci, A., Theurillat, J.-P., Tomaselli, M., Villar, L., Vittoz, P., Vogiatzakis, I. & Grabherr, G. (2012) Continent-wide response of mountain vegetation to climate change. *Nature Climate Change 2*, 111-115.

Grabherr, G., Gottfried, M., Gruber, A. & Pauli, H. (1995) Patterns and current changes in alpine plant diversity. In: F. S. Chapin III & C. Körner (Eds.). *Arctic and alpine biodiversity: Patterns, causes and ecosystem consequences* (pp. 167-181). Berlin: Spring-Verlag.

Grundt, H. H., Kjølner, S., Borten, L., Rieseberg, L. H. & Brochmann, C. (2006) High biological species diversity in the arctic flora. *Proceedings of the National Academy of Sciences (USA) 103*, 972-975.

Hagarup, O. (1928) Morphological and cytological studies of *Bicornes*. *Dansk Botanisk Arkiv 6*, 1-27.

Hobbie, J. E. & Hobbie, E. A. (2006) N-15 in symbiotic fungi and plants estimates nitrogen and carbon flux rates in Arctic tundra. *Ecology 87*, 816-822.

Hocking, B. & Sharplin, C. D. (1965) Flower basking by arctic insects. *Nature 206*, 215

Hopkins, D. M, Matthews, J. V. Jr., Wolfe, J. A. & Silberman, M. L. (1971). A Pliocene flora and insect fauna from the Bering Strait region. *Paleogeography, Paleoclimatology & Paleoecology 9*, 111-231.

Hudson, J. M. G. & Henry, G. H. R. (2009) Increased plant biomass in a High Arctic heath from 1981–2008. *Ecology 90*, 2657–2663.

Hultén, E. (1937) *History of the arctic and boreal biota of the Quarternary* [sic] *Period*. Stockholm: Bokförlags Aktiebolaget.

Hultén, E. (1958) The Amphi-Atlantic Plants and their phytogeographical connections. *Kungliga Svenska Vetenskapsakademiens Handlingar 7*, 1-340.

Ishida, T. A., Nara, K. & Hogetsu, T. (2007) Host effects on ectomycorrhizal fungal communities: Insight from eight host species in mixed conifer-broadleaf forests. *New Phytologist 174*, 430-440.

Ives, J. (1974) Biological refugia and the nunatak hypothesis. In: J. Ives & R. Barry (Eds). *Arctic and alpine environments* (pp. 605-636). London: Methuen.

Johnson, A. W. & Packer, J. G. (1965) Polyploidy and environment in arctic Alaska. *Science 148*, 237-239.

Johnson, A. W. & Packer, J. G. (1967) Distribution, ecology, and cytology of the Ogororuk Creek flora and the history of Beringia. In: D. M. Hopkins (Ed.). *The Bering land bridge* (pp. 245-265). Stanford, CA: Stanford University Press.

Johnson, A. W., Packer, J. G. & Reese, G. (1965) Polyploidy, distribution, and environment. In: H. E. Wright, Jr. & D. G. Frey (Eds.). *The Quaternary of the United States* (pp. 497-507). Princeton, New Jersey: Princeton University Press.

Jónsdóttir, I. S. (2011) Diversity of plant life histories in the Arctic. *Preslia 83*, 281-300.

Jónsdóttir, I. S. (2013) Shaping forces of biodiversity in the Arctic – exploring the feasibility of a coherent research framework. *International tundra experiment 2013 meeting abstract.* Online: http://www.wsl.ch/epub/itex/sessions/detail_EN?id=147&session=37. Accessed July 1, 2014.

Kade, A. N., Walker, D. A. & Raynolds, M. K. (2005) Plant communities and soils in cryoturbated tundra along a climate gradient in the Low Arctic, Alaska. *Phytocoenologia 35*, 761–820.

Kevin, P. G. (1972a) Heliotropism in some arctic flowers. *Canadian Field Naturalist 86,* 41-44.

Kevin, P. G. (1972b) Insect pollination of high arctic flowers. *Journal of Ecology 60,* 831-847.

Kevin, P. G. (1972c) Floral colors in the high arctic with reference to insect-flower relations and pollination. *Canadian Journal of Botany 50*, 2289-2316.

Kevin, P. G. (1975) Sun-tracking solar furnaces in high arctic flowers: Significance for pollination and insects. *Science 189*, 723-726.

Klanderud, K. & Birks, H. J .B. (2003) Recent increases in species richness and shifts in altitudinal distributions of Norwegian mountain plants. *The Holocene 13*, 1-6.

Klein, J. A., Harte, J. & Zhao, X. (2007) Experimental warming, not grazing, decreases rangeland quality on the Tibetan Plateau. *Ecological Applications 17*, 541–557.

Körner, C. (2002) Mountain biodiversity, its causes and function: An overview. In: C. Körner & E. M. Spehm (Eds.). *Mountain biodiversity: A global assessment* (pp. 3-20). New York, NY: Parthenon.

Körner, C. (2003) *Alpine plant life: Functional plant ecology of high mountain ecosystems.* Second edition. Berlin: Spring-Verlag.

Körner C., Paulsen J. & Spehn E. M. (2011) A definition of mountains and their bioclimatic belts for global comparisons of biodiversity data. *Alpine Botany 121*, 73-78.

Lesica, P. (2014) Arctic-alpine plants decline over two decades in Glacier National Park, Montana, U.S.A. *Arctic, Antarctic and Alpine Research 46*, 327-332.

Löve, A. (1959) Origin of the arctic flora. Problems of the Pleistocene and Arctic. *Publications of the McGill University Museum 1*, 82-95.

Löve, D. (1963) Dispersal and survival of plants. In: A. Löve & D. Löve (Eds). *North Atlantic biota and their history* (pp. 189-205). London: Pergamon Press.

Lydolph, M. C., Jacobsen, J., Arctander, P., Gilbert, M. T., Gilichinsky, D. A., Hansen, A. J.,Willerslev, E.& Lange, L. (2005) Beringian paleoecology inferred from permafrost-preserved fungal DNA. *Applied Environmental Microbiology 71*, 1012-1017.

Marr, K. L., Allen, G. A. & Hebda, R. J. (2008) Refugia in the Cordilleran ice sheet of western North America: Chloroplast DNA diversity in the arctic-alpine plant *Oxyria digyna*. *Journal of Biogeography 35*, 1323-1334.

Matthews, J. V. Jr. & Ovenden, L. E. (1990) Late Tertiary plant macrofossils from localities in Arctic/Subarctic North America: A review of the data. *Arctic 43*, 364-392.

McGraw, J. B. & Antonovics, J. (1983) Experimental ecology of *Dryas octopetala* ecotypes. I. Ecotypic differentiation and life cycle stages of selection. *Journal of Ecology71*, 879–897.

Meltofte, H. (Ed.) (2013) *Arctic biodiversity assessment: Status and trends in Arctic biodiversity*. Akureyri: Conservation of Arctic Flora and Fauna (CAFF).

Mooney, H. A. & Billings, W. D. (1961) Comparative physiological ecology of arctic and alpine populations of *Oxyria digyna*. *Ecological Monographs 31*, 1-29.

Morris, M. H., Smith, M. E., Rizzo, D. M., Rejmánek, M. & Bledsoe, C. S. (2008) Contrasting ectomycorrhizal fungal communities on the roots of co-occurring oaks (*Quercus* spp.) in a California woodland. *New Phytologist 178*, 167-176.

Müller, E., Eidesen, P. B., Ehrich, D. & Alsos, I. G. (2012) Frequency of local, regional, and long-distance dispersal of diploid and tetraploid *Saxifraga oppositifolia* (Saxifragaceae) to Arctic glacier forelands. *American Journal of Botany 99*, 459-71.

Murray, D. F. (1987) Breeding systems in the vascular flora of Arctic North America. In: K. M. Urbanska (Ed.). *Differentiation patterns in higher plants* (pp. 239-262). Orlando, FL: Academic Press.

Murray, D. F. (1995) Causes of arctic plant diversity: origin and evolution. In: F. S. Chapin III & C. Körner (Eds.) *Arctic and alpine biodiversity: Patterns, causes and ecosystem consequences* (pp. 21-32). Berlin: Spring-Verlag.

Murray, D. F. (1997) Regional and local vascular plant diversity in the Arctic. *Opera Botanica 132*, 9-18.

Nathorst, A. G. (1892) Über den gegenwärtigen Stand unserer Kenntnis der Verbreitung fossiler Glazialpflanzen: Bihang til Kungliga Svenska Vetenskaps Akademiens. *Handlingar 17, Afd. III, 5*, 1–35.

Olofsson, J., Oksanen, L., Callaghan, T., Hulme, P. E., Oksanen, T. & Suominen, O. (2009) Herbivores inhibit climate-driven shrub expansion on the tundra. *Global Change Biology 15*, 2681-2693.

Partida-Martinez, L. P. & Heil, M. (2011) The microbe-free plant: Fact or artifact? *Frontiers in Plant Science 100,* 1-16.

Payer, D. C., Josefson, A. B. & Fjeldså, J. (2013) Species Diversity in the Arctic. In: H. Meltofte (Ed.). *Arctic biodiversity assessment: Status and trends in Arctic biodiversity* (pp. 66-77). Akureyri: Conservation of Arctic Flora and Fauna (CAFF).

Pearson, R. G., Phillips, S. J., Loranty, M. M., Beck, P. S., Damoulas, T., Knight, S. J. & Goetz, S. J. (2013) Shifts in Arctic vegetation and associated feedbacks under climate change. *Nature Climate Change* 3, 673-677.

Polunin, N. (1951) The real Arctic: Suggestions for its delimitation, subdivision, and characterization. *Journal of Ecology 39*, 308-315.

Raup, H. M. (1969) Studies of vegetation in northwest Greenland. In: K. N. H. Greenridge (Ed.). *Essays in plant geography and ecology*. Halifax: Nova Scotia Museum.

Raup, H. M. (1941) Botanical problems in boreal America. *Botanical Review Part 1, 7,* 147-208; *Part 2,7*, 209-248.

Raynolds, M. K., Walker, D. A., Munger, C. A., Vonlanthen, C. M. & Kade, A. N. (2008) A map analysis of patterned-ground along a North American Arctic transect. *Journal of Geophysical Research 113*, DOI:10.1029/2007JG000512.

Reithmeier, L. & Kernaghan, G. (2013) Availability of ectomycorrhizal fungi to black spruce above the present treeline in eastern Labrador. *PLoS One 8*, e77527. DOI:10.1371/journal.pone.0077527

Rongfu, H. & Miehe, G. (1988) An annotated list of plants from Southern Tibet. *Willdenowia 18*, 81-112.

Rustad, L. E., Campbell, J. L., Marion, G. M., Norby, R. J., Mitchell, M. J., Hartley, A. E., Cornelissen, J. H. C., Gurevitch, J. & CGTE-NEWS (2001) A meta-analysis of the response of soil respiration, net nitrogen mineralization, and aboveground plant growth to experimental ecosystem warming. *Oecologia 126*, 543-562.

Savile, D. B. O. (1956) Known dispersal rates and migratory potentials as clues to the origin of the North American biota. *American Midland Naturalist 56*, 434-453.

Savile, D. B. O. (1959) The botany of Somerset Island, District of Franklin. *Canadian Journal of Botany 37*, 959-1002.

Savile, D. B. O. (1972) *Arctic adaptations in plants.* Monograph No. 6. Ottawa: Research Branch, Agriculture Canada.

Schönswetter, P., Tribsch, A., Stehlik, I. & Niklfeld, H. (2004) Glacial history of the high alpine *Ranunculus glacialis* (Ranunculaceae) in the European Alps in a comparative phylogeographical context. *Biological Journal of the Linnean Society 81*, 183-195.

Schönswetter, P., Elven, R. & Brochmann, C. (2008) Trans-Atlantic dispersal and large-scale lack of genetic structure in the circumpolar, arctic-alpine sedge, *Carex bigelowii* s.l. (Cyperaceae). *American Journal of Botany 95*, 1006-1014.

Shaver, G. R., Chapin, F. S., & Billings, W. D. (1979) Ecotypic differentiation in *Carex aquatilus* as related to ice wedge polygonization in the Alaskan coastal tundra. *Journal of Ecology 67*, 1025-1046.

Shaver, G.R., Fetcher, N. & Chapin, F.S. (1986) Growth and flowering in *Eriophorum vaginatum*: annual and latitudinal variation. *Ecology 67*, 1524–1535.

Sieg, B. & Daniëls, F. J.A. (2005) Altitudinal zonation of vegetation in continental West Greenland with special reference to snowbeds. *Phytocoenologia 35,* 887-908.

Sørensen, T. (1941) Temperature relations and phenology of the northwest Greenland flowering plants. *Meddelelser om Grønland 125*, 1-305.

Starr, G. & Oberbauer, S. F. (2003) Photosynthesis of arctic evergreens under snow: Implications for tundra ecosystem carbon balance. *Ecology 84*, 1415-1420.

Stehlik, I., Blattner, F. R., Holderegger, R. & Bachmann, K. (2002) Nunatak survival of the high alpine plant *Eritrichium nanum* (L.) Gaudin in the central Alps during the ice ages. *Molecular Ecology 11*, 2027-2036.

Taberlet, P., Zimmermann, N. E., English, T., Tribsch, A. & Holderegger, R. et al. (2012) Genetic diversity in widespread species is not congruent with species richness in alpine plant communities. *Ecology Letters 15*, 1439-1448.

Talbot, J. M., Bruns, T. D. , Taylor, J. W., Smith, D. P., Branco, S., Glassman, S. I., Erlandson, S., Vilgalsy, R., Liao, H.-L., Smith, M. E.& Peay, K. G. (2014) Endemism and functional convergence across the North American soil mycobiome. *Proceedings of the National Academy of Sciences (USA)*, DOI: 10.1073/pnas.1402584111.

Tape, K., Sturm, M. & Racine, C. (2006) The evidence for shrub expansion in Northern Alaska and the Pan-Arctic. *Global Change Biology 12*, 686-702.

Taylor, D. L., Herriott, I. C., Stone, K. E., McFarland, J. W., Booth, M. G. & Leigh, M. B. (2010) Structure and resilience of fungal communities in Alaskan boreal forest soils. *Canadian Journal of Forest Research 40*, 1288-1301.

Tieszen, L. L. & Bonde, E. K. (1967) The influence of light intensity on growth and chlorophyll in arctic, subarctic and alpine populations of *Deschampsia caespitosa* and *Trisetum spicatum*. *University of Colorado Studies, Series in Biology 25*, 1-21.

Timling, I. & Taylor, D. L. (2012) Peeking through a frosty window: Molecular insights into the ecology of Arctic soil fungi. *Fungal Ecology 5*, 419-429.

Timling, I., Dahlberg, A., Walker, D. A., Gardes, M., Charcosset, J. Y., Welker, J. M. & Taylor, D. L. (2012) Distribution and drivers of ectomycorrhizal fungal communities across the North American Arctic. *Ecosphere 3*, DOI: 10.1890/ES12-00217.1.

Timling, I., Walker, D. A., Nusbaum, C., Lennon, N. J. & Taylor, D. L. (2014) Rich and cold: Diversity, distribution and drivers of fungal communities in patterned-ground ecosystems of the North American Arctic. *Molecular Ecology, 23,* 3258-3272.

Tolmachev, A. I. (1960) Der autochthone Grundstock der arktischen Flora und ihre Beziehungen zu den Hochgebirgsfloren Nord- und Zentralasiens. *Botanisk Tidsskrift 55*, 269-276.

Vonlanthen, C. M., Walker, D. A., Raynolds, M. K., Kade, A. N., Kuss, H. P., Daniëls, F. J. A. & Matveyeva, N. V. (2008) Patterned-ground plant communities along a bioclimate gradient in the High Arctic, Canada. *Phytocoenologia 38*, 23-63.

Wada, N. (1998) Sun-tracking flower movement and seed production of mountain avens, *Dryas octopetala* L. in the High Arctic, Ny Ålesund, Svalbard. *Proceedings of the NIPR Symposium on Polar Biology 11*, 128-136.

Walker, M. D. (1995) Patterns of arctic plant community diversity. In: Chapin, F. S. III & C. Körner (Eds.). *Arctic and alpine biodiversity: Patterns, causes and ecosystem consequences* (pp. 3-20). Berlin: Spring-Verlag.

Walker, D. A., Auerbach, N. A., Bockheim, J. G., Chapin, F. S. III, Eugster, W., King, J. Y., McFadden, J. P., Michaelson, G. J., Nelson, F. E., Oechel, W. C., Ping, C. L., Reeburg, W. S., Regli, S., Shiklomanov, N. I. & Vourlitis, G. L. (1998) Energy and trace-gas fluxes across a soil pH boundary in the Arctic. *Nature 394*, 469-472.

Walker, D. A., Billings, W. D. & De Molenaar, J. G. (2001a) Snow-vegetation interactions in tundra environments. In: H. G. Jones, J. W. Pomeroy, D. A. Walker & R. W. Hoham (Eds). *Snow ecology* (pp. 266-324). New York, NY: Cambridge University Press.

Walker, D. A., Bockheim J. G., Chapin, F. S. III, Eugster, W., Nelson, F. E. & Ping, C.-L. (2001b) Calcium-rich tundra, wildlife, and the "Mammoth Steppe." *Quaternary Science Reviews 20*, 149-163.

Walker, D. A., Raynolds, M. K., Daniëls, F. J. A., Einarsson, E. & Evlvebakk A., Gould, W. A., Katenin, A. E., Kholod, S. S., Markon, C. J., Evgeny, S., Moskalenko, N. G., Talbot, S. S., Yurtsev, B. A. & the other members of the CAVM Team. (2005) The circumpolar Arctic vegetation map. *Journal of Vegetation Science 16*, 267-282.

Walker, M. D., Wahren, C. H., Hollister, R. D., Henry, G. H. R., Ahlquist, L. E., Alatalo, J. M., Bret-Harte, M. S., Calef, M. P., Callaghan, T. V., Carroll, A. B., Epstein, H. E., Jónsdóttir, I. S., Klein, J. A., Magnusson, B., Molau, U., Oberbauer, S. F., Rewa, S. P., Robinson, C. H., Shaver, G. R., Suding, K. N., Thompson, C. C., Tolvanen, A., Totland, Ø., Turner, P. L., Tweedie, C. E., Webber, P. J. & Wookey, P. A. (2006) Plant community responses to experimental warming across the tundra biome. *Proceedings of the National Academy of Sciences (USA) 103*, 1342–1346.

Walker, D. A., Alsos, I. G., Bay, C., Boulanger-Lapointe, N., Breen, A. L., Bültmann, H., Christiansen, T., Damgaard, C., Daniëls, F. J. A., Hennekens, S., Reynolds, M. K., Le Roux, P. C., Luoto, M., Pellisier, L., Peet, R. K., Schmidt, N.-M., Stewart, L., Virtanen, R., Yoccoz, N. G. & Wisz, M. S. (2013) Rescuing valuable Arctic vegetation data for biodiversity models, ecosystem models and a Panarctic vegetation classification. *Arctic: Info North 66*, 133-137.

Ware, C., Bergstrom, D. M., Müller, E. & Alsos, I. G. (2012) Humans introduce viable seeds to the Arctic on footwear. *Biological Invasions 14*, 567-577.

Weber, W. A. (1965) Plant geography in the southern Rocky Mountains. In: H. E. Wright, Jr. & D. G. Frey (Eds.) *The Quaternary of the United States* (pp. 453-468). Princeton, NJ: Princeton University Press.

Weber, W. A. (2003) The Middle Asian element in the southern Rocky Mountain flora of the western United States: A critical biogeographical review. *Journal of Biogeography 30*, 649-685.

Westergaard, K. B., Jørgensen, M. H., Gabrielsen, T. M., Alsos, I. G. & Brochmann, C. (2010) The extreme Beringian/Atlantic disjunction in *Saxifraga rivularis* (Saxifragaceae) has formed at least twice. *Journal of Biogeography 37*, 1262-1276.

Westergaard, K. B., Alsos, I. G., Popp, M., Engelskjøn, T., Flatberg, K. I. & Brochmann, C. (2011) Glacial survival may matter after all: Nunatak signatures in the rare European populations of two west-arctic species. *Molecular Ecology 20*, 376-393.

Wielgolaski, F. E. (1997). *Polar and alpine tundra.* Ecosystems of the World 3. Amsterdam: Elsevier.

Willerslev, E., Davison, J., Moora, M., Zobel, M., Coissac, E., Edwards, M. E., Lorenzen, E. D., Vestergärd, M., Gussarova, G., Haile, J., Craine, J., Gielly, L., Boessenkool, S., Epp, L. S., Pearman, P. B., Cheddadi, R., Murray, D., Bräthen, K. A., Yoccoz, N., Binney, H., Cruaud, C., Wincker, P., Goslar, T., Alsos, I. G., Bellemain, E., Brysting, A. K., Elven, R., Sønstebø, J. H., Murton, J., Sher, A., Rasmussen, M., Rønn, R., Mourier, T., Cooper, A., Austin, J., Møller, P., Froese, D., Zazula, G., Pompanon, F., Rioux, D., Niderkorn, V., Tikhonov, A., Savvinov, G., Roberts, R. G., MacPhee, R. D., Gilbert, M. T., Kjær, K. H., Orlando, L., Brochmann, C. & Taberlet, P. (2014) Fifty thousand years of Arctic vegetation and megafaunal diet. *Nature 506*, 47-51.

Yao, F., Vik, U., Brysting, A. K., Carlsen, T., Halvorsen, R. & Kauserud, H. (2013) Substantial compositional turnover of fungal communities in an alpine ridge-to-snowbed gradient. *Molecular Ecology 19*, 5040-5042.

Yurstev, B. A. (1994) Floristic divisions of the Arctic. *Journal of Vegetation Science 5*, 765-776.

Zinger, L., Lejon, D. P. H., Baptist, F., Bouasria, A., Aubert, S., Geremia, R. A. & Choler, P. (2011) Contrasting diversity patterns of crenarchaeal, bacterial and fungal soil communities in an alpine landscape. *PLoS One 6*, DOI:10.1371/journal.pone.0019950.

In: Plant Ecology and Evolution in Harsh Environments
Editors: N. Rajakaruna, R. S. Boyd and T. B. Harris
ISBN: 978-1-63321-955-7

Chapter 8

DRIVERS OF DIVERSITY IN EVERGREEN WOODY PLANT LINEAGES EXPERIENCING CANOPY FIRE REGIMES IN MEDITERRANEAN-TYPE CLIMATE REGIONS

Michael C. Vasey* and V. Thomas Parker
Department of Biology, San Francisco State University, CA, USA

ABSTRACT

California has the most extreme Mediterranean-type climate (MTC) region on the planet. Between May and September there are almost no significant rain events. The topography of uplands tends to be rugged with many different soil types juxtaposed by tectonic activity into a complex landscape mosaic. Prolonged seasonal drought during warmer months favors shrubland over forest and sets the stage for relatively frequent, predictable, and intense canopy fires. Chaparral, a shrubland vegetation, is not only adapted to fire but many species are dependent upon fire regimes for recruitment of new generations. Geology, climate, and fire are intimately connected to the composition and distribution of chaparral. In this chapter, we hypothesize that it is the interactive nature of all three factors that drives chaparral ecology and, ultimately, the evolutionary processes that have shaped chaparral phylogenetic diversity as well. Ultimately, despite the extreme conditions that characterize MTC shrublands in California and elsewhere—or perhaps because of them—these MTC shrublands are renowned for their species diversity and, particularly, their abundance of local endemics. We use the genus *Arctostaphylos* (Ericaceae), a shrub lineage in California with the most taxa of all other shrubs (95) in this region, as an example to highlight this phenomenon. In particular, we focus on the relationship between *Arctostaphylos* and the summer marine layer along the California coast where almost half of its taxa (46) are narrow endemics (distribution < 1000 km^2) restricted to the narrow coastal zone. We further hypothesize that the more mild and moist climate regime along the coast, combined with greater rainfall on windward coastal uplands, sets the stage for a variety of interactions between geology, climate, and fire that ultimately drive the diversification of this genus. These interactions are enhanced by a

* Corresponding author: mvasey@sfsu.edu.

number of circumstances in these specialized habitats. In the coastal region, chaparral occurs in archipelagos of nutrient poor, shallow, rocky soils that are situated in island-like stands in a forest matrix. These coastal sites are subject to long fire return intervals associated with higher fuel moisture levels. This environmental context has facilitated a number of biological responses, including selection for obligate seeder species, hybridization among species in these isolated stands, and episodes of selection for optimal genotypes following each fire event which, ultimately, leads to new species of probable hybrid origin. Finally, along with these neoendemics, we argue that the coast also harbors many paleoendemics that have found refuge in these more mild and mesic habitats. Consequently, the extraordinary endemism of *Arctostaphylos* along the coast reflects both production of new species and preservation of species arising from past ages.

INTRODUCTION

Mediterranean-type climate (MTC) ecosystems are among the most biologically diverse ecosystems on the planet (Cowling et al., 1996; Myers et al., 2000; Rundel, 2004). A Mediterranean-type climate is characterized by mild, wet winters and dry, hot summers. California has the driest summer periods with a tenth or less summer rain compared to other MTC regions (average for June-August: 3-6 mm vs. 40-125 mm in other MTC regions) and with almost no summer rain days (<1 summer rain day vs. 10-20 days for other MTC regions). Extremes of climate, soil, and fire conditions vary among the MTC areas as a consequence of different features. California, the Mediterranean Basin, and Chile all experience considerable tectonic and volcanic activity and have relatively young soils, but the latter two have some summer rainfall. Summer rainfall also occurs in SW Australia and the Cape Region of SW South Africa, but these sites have old and weathered landscapes containing nutrient-poor soils. Although all five MTC regions are dominated by sclerophyllous shrublands, a convergence first noted by Schimper (1903), we can expect variation among these regions because of their different selective contexts and these differences are well described elsewhere (Keeley et al., 2012). Here we concentrate on patterns of plant diversity in California chaparral, a sclerophyllous shrub-dominated vegetation inhabiting an environment with the most pronounced version of reduced summer rain among the five MTC regions (Cowling et al., 2005).

While climate and soils act as a principal influence on vegetation patterns, all terrestrial systems also experience some type of fire regime (Pausas & Keeley, 2009). Shrub dominated ecosystems in MTC regions exhibit high intensity canopy-type fire regimes. Other significant dimensions of fire regimes are spatial extent, temporal frequency, and intensity, among others. The two extremes of fire type, surface fire and canopy fire regimes, generally introduce quite different types of environmental change and have very different selective influences on plant traits. Surface fire ecosystems, such as is characteristic of many conifer forests, exhibit mortality in small plants or young age classes and biomass removal only near the ground. Consequently, dominant plants experience little mortality and minimal loss of canopy, resulting in slow change in the community's composition and demographic dynamic even though surface fires may differentially impact species. Canopy fire ecosystems such as chaparral, in contrast, experience considerable change in postfire environmental conditions. With the removal of the leaf canopy and most above ground biomass, light levels are drastically greater following a canopy fire and the potential for water deficit or temperature extremes increases in more arid locations. While in both fire regime types newly establishing individuals are released from competition for light energy to some extent, seedlings of some species arising postfire in canopy fire vegetation experience a potentially harsher environment compared to seedlings in the understory prior to the fire due to this potential for temperature extremes and water deficit.

Fire acts as a peculiar type of disturbance regime because the spatial extent of fires is usually quite large, limiting long distance dispersal, and thus persisting organisms must survive all the different dimensions of the fire regime. In addition to opening space for recruitment dynamics at the local scale, fire interacts with aspects of plant longevity, persistence, or reproductive traits, and consequently plant composition. If the fire-return intervals and other dimensions of the fire regime are relatively predictable, fire leads to most plants exhibiting traits adaptive to fire. A particular fire regime will favor particular sets of fire-adaptive traits. The stronger the selection for fire-adaptive traits, often the less the vegetation seems to change postfire in composition within local communities. While there are arguments about whether traits have arisen *de novo* as a result of fire regimes (*e.g.*, Bradshaw et al., 2011; Keeley et al., 2011), fire-adaptive traits clearly characterize most dominant plants. Fire-adaptive traits may include dormant buds and underground storage organs that allow resprouting of surviving individuals. In many species other fire-adaptive traits mostly restrict seedling establishment to the first postfire growing season, for example, due to serotinous cones or fruits, or because seed dormancy is broken by the smoke or heat-pulse from a fire.

Fire Dependence

Chaparral vegetation best responds to fires at 30-150+ year intervals, generally longer than occurs in most other MTC regions. Within those timeframes, chaparral returns quickly to pre-fire conditions dominated by sclerophyllous shrubs because of the low mortality of resprouting individuals and the rapid reestablishment of shrub seedlings from fire-dependent seed banks (Hanes, 1971). Critically, however, chaparral is adapted to a particular fire regime of a range of frequency, intensity, and seasonal timing so that excluding fire or increasing the frequency of fire reduces the sustainability of this ecosystem (Keeley et al., 2005; Parker, 1990; Parker & Pickett, 1998; Zedler et al., 1983). Chaparral dynamics correspond to the cycle of wildfire, postfire recovery, and stand maturation.

All woody plants in chaparral display fire-adaptive traits and can be grouped into three general postfire life history categories based on combinations of postfire resprouting and seed dormancy characteristics (Keeley, 1987; Keeley et al., 2012; Parker & Kelly, 1989). Most genera contain species that survive fire as adults and, even though their aboveground stems are killed, the plants resprout from stem or root crowns. This first group is termed obligate resprouters because they persist by vigorous resprouting and contain transient seed banks and have no postfire seedling recruitment. A second cluster of species considered facultative seeders are found principally in three genera in California that resprout following fire, but these genera also have soil seed banks that generally depend upon fire-stimulated recruitment. The third life history pattern is obligate seeding shrubs and trees, species that typically are killed by fire but that produce seeds that are dormant at maturity and remain as persistent soil or aerial (canopy) seed banks until recruitment is stimulated after fire events. Similar to facultative seeders, their seeds are wholly or principally stimulated by wildfire and they typically germinate and establish in postfire stands. Populations of obligate seeders persist in chaparral exclusively because of the success of their seed banks and the flush of postfire seedling recruitment. Obligate seeders in California chaparral are generally found in *Arctostaphylos* and *Ceanothus* (Rhamnaceae) among shrubs and as serotinous lineages in *Pinus* (Pinaceae) and *Hesperocyparis* (Cupressaceae) among the trees.

Chaparral is a diverse ecosystem in the context of its woody species (Keeley & Keeley, 1988; Sawyer et al., 2009; Wells, 1962). The importance of fire in this system is clear when considering that species with fire-dependent reproduction represent about 80% of average stand dominance regardless of location (Vasey, 2012; Vasey et al., *In press*). Also, genera with a large number of chaparral species are either facultative or obligate seeders. While not considered in detail here, chaparral also contains a highly diverse component of annuals, herbaceous perennials, and suffrutescent shrubs that, with few exceptions, are found only in postfire stands of chaparral and are lacking in mature stands (Hanes, 1981; Sweeney, 1956). Dormant seed banks of these species are also fire-dependent and triggered by either intense heat shock or combustion products from smoke or charred wood (*e.g.*, Keeley, 1991; Keeley & Fotheringham, 2000). Composition of herbaceous species or suffrutescent shrubs varies among sites and their dominance depends on the rainfall and temperature pattern of the initial year following the fire event. Overall, chaparral typically has the highest plant compositional diversity in the first and second years after fire (Keeley et al., 2005; Sweeney, 1956) because of these herbaceous species in combination with the woody dominants.

DIVERSITY OF CHAPARRAL IN CALIFORNIA

The compositional diversity of chaparral communities must be distinguished from phylogenetic diversity of chaparral lineages, and particularly for sclerophyllous shrub genera. With regard to compositional diversity, Keeley et al. (2012) have argued that the interaction of multiple factors drives diversity rather more than any one particular factor (Figure 1). According to this conceptual framework, from a regional species pool (consisting of various lineages), an environmental template consisting primarily of climate factors, soil factors, and a fire regime filters the pool into different functional types (*e.g.*, fire-adapted sclerophyllous shrubs) that are then assembled into local communities. Observations of local community *alpha* diversity (number of species per plot) in California are generally not high in mature postfire chaparral which is characterized by closed canopies and stands dominated by relatively few species of mature shrubs with a sparse herbaceous understory (Vasey et al., *In press*). By contrast, low soil nutrient levels in mature fynbos and kwongan habitats are presumed to allow a more open canopy in these mature postfire shrublands and thus the coexistence of many more visible herbaceous and subshrub species situated in gaps between dominant shrubs (Keeley et al., 2012). Immediate postfire chaparral habitats often have about the same number of species per 0.1 h plot as immediate postfire fynbos and kwongan communities (Wisheu et al., 2000), but these higher levels of chaparral *alpha* diversity result from the expression of herbaceous species and subshrubs typically dormant within chaparral seed banks and not visible in mature stands. Since fire-return intervals are long, estimating the true *alpha* diversity in a particular chaparral stand is a challenge.

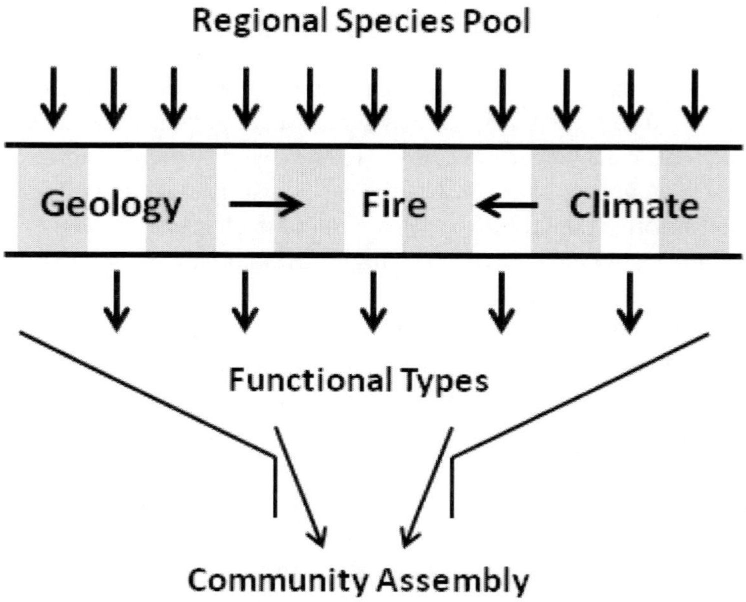

Figure 1. The "Geology-Fire-Climate" model of community assembly in chaparral modified with permission from Keeley et al. (2012).

Conversely, especially in coastal chaparral, the particular combination of shrub species can vary greatly among stands (also called species turnover)—a manifestation of *beta* diversity—or the distribution of species diversity at larger scales.

To a significant degree, *beta* diversity in mature stands of chaparral intersects with the phylogenetic diversity of dominant sclerophyllous shrub lineages in linking patterns of compositional diversity and phylogenetic diversity together. This is because high species turnover among mature stands of chaparral is primarily associated with high levels of local endemism of sclerophyllous shrub species found within those stands. Further, at these larger spatial scales, discernable patterns of higher versus lower *beta* diversity, as well as higher *gamma diversity* (overall diversity at regional scales), are associated with climatic gradients and the distribution of different soil types. For example, where chaparral is found in highly infertile serpentine "islands" distributed as a mosaic embedded in more fertile sandstone soils, higher levels of *beta* diversity were found on the serpentine substrates due to the greater presence of local endemic species on those substrates (Chapter 6; Harrison & Inouye, 2002).

In the fynbos, Ojeda (1998) found that the highest concentration of obligate seeder endemism in *Erica* (Ericaceae) occurs in the southwestern region of the Cape in an area with reliable summer rains and more mild summer drought conditions. Another example of this phenomenon relates to an under-appreciated link between high *beta* diversity in coastal chaparral, extreme local endemism of *Arctostaphylos* in this region, and to a lesser extent in *Ceanothus*, and its linkage to the influence of a summer marine layer (coastal fog and low cloud cover) that is a persistent climatic phenomenon along the California coast.

INFLUENCE OF SUMMER MARINE LAYER ON SHRUB ENDEMISM IN COASTAL CHAPARRAL

The diversity of woody plants in California chaparral has long been recognized (Cooper, 1922), particularly in *Arctostaphylos* and *Ceanothus* (Wells, 1969); however, only during the last forty years has the exceptional diversity of *Arctostaphylos* been fully appreciated (Parker et al., 2009, 2012). Recent treatments of *Arctostaphylos* recognize 104 taxa (67 species and 37 subspecies), 95 of which occur in California and most of which occur in chaparral (Table 1). Diversity in *Ceanothus* (61 taxa in California) is also better known today. *Ceanothus* is also centered in California, and most species are found in chaparral as well (Wilken, 2012). In the early 1980s, interest emerged in conserving natural communities rather than species *per se*, and two natural communities that stood out as being both exceptionally rich in local endemics and at-risk due to human activities were Central Maritime Chaparral and Northern Maritime Chaparral (Holland, 1986). The conservation value of maritime chaparral was further bolstered by recognition of the high rate of species turnover along the coast in *Arctostaphylos* and *Ceanothus* (Cody, 1986). Later, Keeley (1992) recognized coastal chaparral as the closest approximation to the endemic-rich fynbos in California, and he recognized that the mildness of the climate in the endemic rich Pacific coastal chaparral was analogous to that of the endemic-rich fynbos in the southwestern Cape region.

Despite the recognition that coastal chaparral harbors the highest concentration of local endemics compared to other chaparral in California, few hypotheses have been offered to explain this phenomenon other than mountainous terrain near the sea (Cody, 1986) or higher levels of precipitation in coastal uplands (Loarie et al., 2008; Richerson & Lum, 1980; but see Keeley, 1992). These hypotheses did not, however, account for the fact that most local endemics are located in coastal lowlands below 400 m elevation rather than on coastal uplands. The summer marine layer generally occurs below the 400 m elevation (Johnstone & Dawson, 2010), and in the late 1990s, it became recognized as an important source of moisture for coastal vegetation, particularly lowland vegetation (Dawson, 1998; Corbin et al., 2005; Fischer et al., 2009). These earlier studies focused principally on conifer forests. Vasey and colleagues (Vasey, 2012; Vasey et al., 2012; Vasey et al., *In press*) hypothesized that the summer marine layer may create significant differences in water relations of chaparral shrubs along a coast-inland gradient and that chaparral diversity patterns might be correlated with more favorable (*i.e.*, less harsh) water availability conditions along the coast.

These studies revealed that significant differences were found in late dry season water potentials among *Arctostaphylos* shrubs along the lowland coast (maritime), upland coast (transition), and interior gradient (Figure 2). Further, chaparral composition sampling revealed that in 0.1 h plots, coastal lowlands and coastal uplands had significantly higher levels of *beta* diversity than interior plots, as might have been predicted based on Cody (1986) (Figure 3). Correlations with dry season and wet season climate data showed that lowland coastal chaparral had a more favorable climate regime primarily due to the summer marine layer whereas upland coastal chaparral had a similarly favorable climate regime primarily due to significantly greater rainfall than the lowland coast or the interior (Vasey et al., *In press*).

Table 1. Summary of *Arctostaphylos* taxa (67 species, 37 subspecies, total 104), with clade identity (1=small, 2=large), ploidy level (D=diploid, T=tetraploid), life history (seeder=obligate seeder, sprouter=facultative seeder), distribution (C=coast, < 50 km from coast, I=interior, > 50 km from coast), range classes (0=1-10 Km^2, 1=10-100 Km^2, 2=100-1000 km^2, 3=1000-10000 km^2, 4=10000-100000 km^2, 5=100000-1000000 km^2, 6=>1000000 km^2), and locations (Cal=California, W=Western North America, Pac NW= Pacific Northwest, AZ=Arizona, Mex=Mexico, N Hem=Northern Hemisphere). Source of taxonomy is Parker et al. (2009; 2012) or International Plant Names Index (http://www.ipni.org)

Taxon	Clade	Ploidy	Life History	Distribution	Range Class	Location
A. andersonii	2	D	seeder	C	3	Cal
A. auriculata	2	D	seeder	C	2	Cal
A. australis	2	D	seeder	C	2	Baja
A. bakeri	2	T	seeder	C	1	Cal
A. b. subsp. *sublaevis*	2	T	seeder	C	1	Cal
A. bolensis	2	D[2]	seeder	C	0	Baja
A. canescens	2	D	seeder	I	4	Cal
A. c. subsp. *sonomensis*	2	D	seeder	I	4	Cal
A. catalinae	2	D[2]	seeder	C	2	Cal
A. columbiana	2	D	seeder	C	3	Cal, Pac NW
A. confertiflora	2	D	seeder	C	1	Cal
A. cruzensis	2	D	seeder	C	0	Cal
A. crustacea	2	T	sprouter	C	4	Cal
A. c. subsp. *crinita*	2	T	sprouter	C	3	Cal
A. c. subsp. *eastwoodiana*	2	T	sprouter	C	1	Cal
A. c. subsp. *insulicola*	2	T	sprouter	C	2	Cal
A. c. subsp. *rosei*	2	T	sprouter	C	2	Cal
A. c. subsp. *subcordata*	2	T	sprouter	C	2	Cal
A. densiflora	1	D	seeder	C	0	Cal
A. edmundsii	2	D	seeder	C	1	Cal

Table 1. (Continued)

Taxon	Clade	Ploidy	Life History	Distribution	Range Class	Location
A. franciscana	2[1]	D	seeder	C	0	Cal
A. gabilanensis	2	D	seeder	C	2	Cal
A. glandulosa	2	T	sprouter	I	4	Cal, Baja
A. g. subsp. *adamsii*	2	T	sprouter	I	3	Cal, Baja
A. g. subsp. *atumescens*	2	T	seeder	C	0	Baja
A. g. subsp. *crassifolia*	2	T	sprouter	C	3	Cal
A. g. subsp. *cushingiana*	2	T	sprouter	I	4	Cal
A. g. subsp. *erecta*	2	T	sprouter	I	1	Baja
A. g. subsp. *gabrielensis*	2	T	sprouter	I	2	Cal
A. g. subsp. *leucophylla*	2	T	sprouter	I	3	Cal, Baja
A. g. subsp. *mollis*	2	T	sprouter	I	4	Cal
A. glauca	2	D	seeder	I	4	Cal, Baja
A. glutinosa	2	D	seeder	C	1	Cal
A. hispidula	2	D[2]	seeder	I	4	Cal
A. hookeri	1	D	seeder	C	2	Cal
A. h. subsp. *hearstiorum*	1	D	seeder	C	0	Cal
A. hooveri	2	D	seeder	C	3	Cal
A. imbricata	2	D	seeder	C	0	Cal
A. incognita	2	D	sprouter	I	1	Baja
A. insularis	2	D	seeder	C	1	Cal
A. klamathensis	2	D[2]	seeder	I	2	Cal
A. luciana	2	D	seeder	C	1	Cal
A. manzanita	2	T	seeder	I	5	Cal
A. m. subsp. *elegans*	2	T	seeder	I	3	Cal
A. m. subsp. *glaucescens*	2	T	seeder	I	3	Cal
A. m. subsp. *laevigata*	2	T	seeder	C	1	Cal

Taxon	Clade	Ploidy	Life History	Distribution	Range Class	Location
A. m. subsp. *roofii*	2	D^2	sprouter	I	3	Cal
A. m. subsp. *wieslanderi*	2	D^2	seeder	I	3	Cal
A. malloryi	2	D^2	seeder	I	3	Cal
A. mewukka	2	T	sprouter	I	3	Cal
A. m. subsp. *truei*	2	T	seeder	I	2	Cal
A. montana	2	T	seeder	C	1	Cal
A. m. subsp. *ravenii*	2	T	seeder	C	0	Cal
A. montaraensis	2	D	seeder	C	2	Cal
A. montereyensis	2	D	seeder	C	2	Cal
A. moranii	2	T	sprouter	I	1	Baja
A. morroensis	2	D	seeder	C	0	Cal
A. myrtifolia	2	D	seeder	I	1	Cal
A. nevadensis	2	T	seeder	I	5	Cal
A. n. subsp. *knightii*	2	T	sprouter	I	2	Cal
A. nissenana	2	D	seeder	I	3	Cal
A. nortensis	2	D	seeder	I	2	Cal
A. nummularia	1	D	seeder	C	2	Cal
A. n. subsp. *mendocinoensis*	1	D	seeder	C	1	Cal
A. obispoensis	2	D	seeder	C	3	Cal
A. ohloneana	1	D	seeder	C	0	Cal
A. osoensis	2	D^2	seeder	C	0	Cal
A. otayensis	2	D	seeder	C	0	Cal
A. pacifica	1	T^2	seeder	C	0	Cal
A. pajaroensis	2	D	seeder	C	2	Cal
A. pallida	2	D	seeder	C	1	Cal
A. parryana	1^2	T	seeder	I	3	Cal
A. p. subsp. *deserticum*	1^2	T	sprouter	I	2	Cal

Table 1. (Continued)

Taxon	Clade	Ploidy	Life History	Distribution	Range Class	Location
A. p. subsp. *tumescens*	1[2]	T	sprouter	I	0	Cal
A. patula	1	D	sprouter[1]	I	5	Cal, Baja,W
A. pechoensis	2	D	seeder	C	1	Cal
A. peninsularis	2	D	sprouter	I	3	Baja
A. p. subsp. *juarezensis*	2	D	seeder	I	1	Baja
A. pilosula	2	D	seeder	C	3	Cal
A. pringlei	2	D	seeder	I	4	Cal, Baja, AZ
A. p. subsp. *drupacea*	2	D	seeder	I	4	Cal
A. pumila	2	D	seeder	C	2	Cal
A. pungens	2[1]	D	seeder	I	5	Cal, SW, Mex
A. purissima	2	D	seeder	C	3	Cal
A. rainbowensis	2	D	seeder	C	2	Cal
A. refugioensis	2	D	seeder	C	1	Cal
A. regismontana	2	D	seeder	C	2	Cal
A. rudis	1	D	seeder	C	3	Cal
A. sensitiva	1	D	seeder	C	3	Cal
A. silvicola	2	D[2]	seeder	C	1	Cal
A. stanfordiana	1	D	seeder	I	3	Cal
A. s. subsp. *decumbens*	1	D	seeder	I	0	Cal
A. s. subsp. *raichei*	1	D	seeder	I	1	Cal
A. tomentosa	2	T	sprouter	C	2	Cal
A. t. subsp. *bracteosa*	2	T	sprouter	C	1	Cal
A. t. subsp. *daciticola*	2	T	sprouter	C	0	Cal
A. t. subsp. *hebeclada*	2	T	sprouter	C	1	Cal

Taxon	Clade	Ploidy	Life History	Distribution	Range Class	Location
A. uva-ursi	1[1]	T	sprouter[1]	I	6	Cal, N Hem
A. u. subsp. *cratericola*	1	T	sprouter	I	2	Guatemala
A. virgata	2	D	seeder	C	1	Cal
A. viridissima	2	D	seeder	C	0	Cal
A. viscida	2	D	seeder	I	5	Cal
A. v. subsp. *mariposa*	2	D	seeder	I	4	Cal
A. v. subsp. *pulchella*	2	D[2]	seeder	I	4	Cal

[1]Both conditions are known to occur.
[2]Trait has been predicted, but not confirmed.

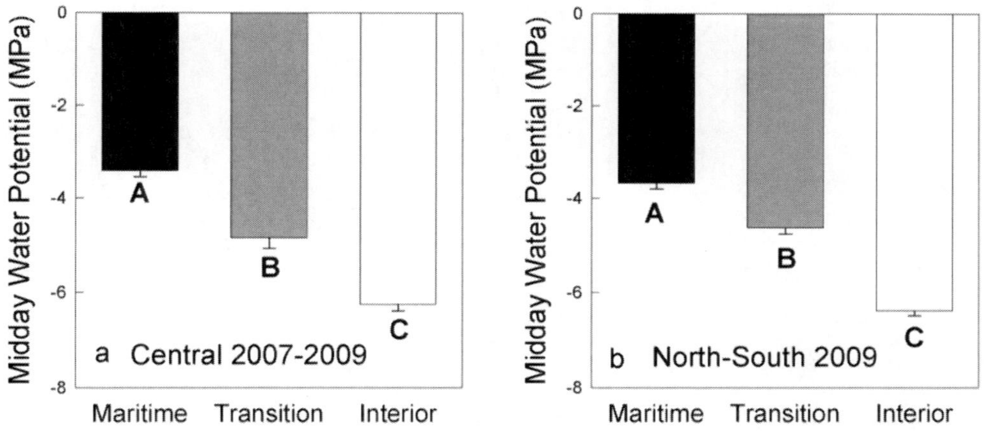

Figure 2. Water potential values from *Arctostaphylos* populations at the end of the dry season in maritime (lowland coast), transition (upland coast), and interior chaparral habitats (Vasey et al., 2012). Figure 2a shows three years of data from a transect in the Monterey Bay region; Figure 2b shows values from north, central, and south transects (San Francisco Bay region to Morro Bay, San Luis Obispo) that experience pronounced summer marine layer effects. Water potential gradient is thus consistent over time and space. Capital letters indicate significant differences at $p < 0.0001$ based on linear mixed model and Tukey HSD test, error bars represent standard error.

In summary, the hypothesis is supported that more favorable climatic conditions exist in coastal chaparral and that these conditions are positively correlated with high *beta* diversity in coastal chaparral. This higher level of *beta* diversity is a function of greater levels of local endemism in coastal chaparral, particularly in *Arctostaphylos, Ceanothus, Pinus,* and *Hesperocyparis*. Yet, this observation does not explain why more favorable climatic conditions along the coast might translate into a higher concentration of local chaparral endemics. Based on our studies of *Arctostaphylos*, we suggest a framework for future testing that ties together the ecological model suggested by Keeley et al. (2012) with an evolutionary process over time that might account for the hyperaccumulation of local *Arctostaphylos* endemics along the California coast. Potentially, this framework could apply to other analogous situations for other genera in California chaparral as well as shrublands in other MTC regions.

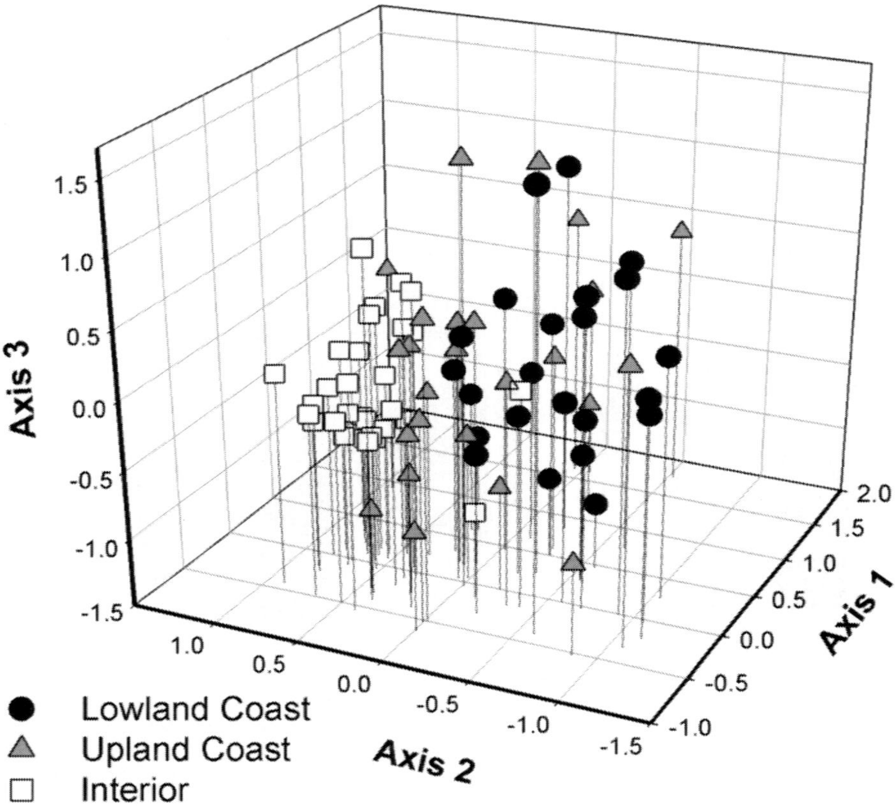

Figure 3. Non-metric multidimensional scaling analysis of maritime (lowland coast), transition (upland coast) and interior 0.1 h chaparral plots (n=87 for all) sampled at a regional scale from the San Francisco Bay region to northern Santa Barbara County (Vasey et al., *In press*). Dissimilarity among maritime and transition plots compared to interior plots suggests greater *beta* diversity for coastal chaparral. A multivariate analysis of dispersion (Anderson, 2006) confirmed that the maritime and transition plots were significantly different ($p < 0.0001$) than interior but not significantly different from each other.

LOCAL ENDEMISM IN ARCTOSTAPHYLOS ALONG THE CALIFORNIA COAST: AN EVOLUTIONARY PERSPECTIVE

Of the 95 taxa of *Arctostaphylos* recognized in California (Parker et al., 2012), over half (54%) are restricted to ranges within 50 km of the coast. Of these 57 species, 46 (81%) are local endemics, occupying distributions of less than 1000 km^2 (*i.e.*, an area approximately 30 km x 30 km) (Figure 4). Conversely, *Arctostaphylos* species and subspecies that primarily or exclusively occur in the interior constitute little more than a third of the taxa (36%) and only 10 of these taxa (26%) have a range of less than 1000 km^2; almost three quarters of the interior species (74%) are relatively widespread (Vasey & Parker, unpublished data). Local endemics along the coast are primarily concentrated in lowland gaps in otherwise relatively continuous Coast Range mountains (*e.g.*, Monterey Bay, San Francisco Bay, Morro Bay, and the Santa Maria plains).

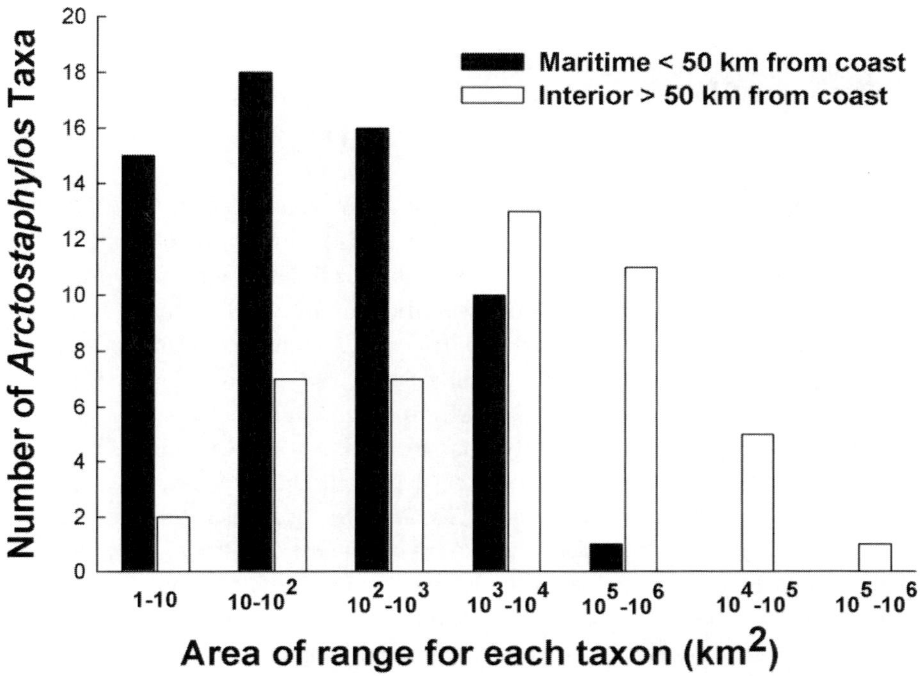

Figure 4. Comparison of range distributions for 104 *Arctostaphylos* taxa (species and subspecies) restricted to maritime habitats (< 50 km from the coast) or mostly distributed in the interior (> 50 km from the coast). Note that the great majority of coastal taxa are local endemics (ranges < 1000 km^2). Based upon Vasey & Parker (unpublished data).

Monterey Bay provides a good example. Within a 1000 km^2 area centered on Monterey Bay, a total of twelve species and subspecies of *Arctostaphylos* can be found in stands of coastal chaparral on a variety of different substrates from deep sand to granite to Monterey shale to Aromas sandstone. Of these twelve taxa, four species and two subspecies are local endemics restricted to this area while a seventh, *A. gabilanensis*, is a local endemic restricted to the Gabilan Range with one small population located slightly to the south of this artificial boundary. This pattern of local endemism first drew the attention of ecologists such as Griffin (1978) who coined the term "maritime chaparral", and later Holland (1986) and state agencies that implemented policies to protect lowland maritime chaparral (Sawyer et al., 2009).

Given the intrinsically harsh conditions associated with MTC regions, and particularly California (Cowling et al., 2005), it is somewhat of a conundrum that MTC regions in general are renowned for their high levels of biodiversity and, in particular, species richness (*alpha* diversity) and species turnover (*beta* diversity) associated with local endemism. The compelling question that arises from this observation is intrinsically both ecological *and* evolutionary; *i.e.*, what ecological factors over time would give rise to so many recognizably different, closely related lineages occupying such relatively small ranges within such close proximity and what biological factors would enhance such diversification? This question is all the more vexing given that *Arctostaphylos* and other dominants of MTC shrublands are relatively long-lived, woody plants. Below is a conceptual framework for considering how to address this question, specifically with regard to *Arctostaphylos* and also in the context of its pattern of local endemism along the California coast.

A 'Geology-Fire-Climate' Model for Diversification in *Arctostaphylos*

We propose that the "geology-fire-climate" model of Keeley et al. (2012) as a process shaping compositional diversity in MTC shrublands is also a good framework from which to build a phylogenetic diversity model. Variation in soils, climate, and fire regimes provide a context for examining long-term patterns of diversification in *Arctostaphylos* and particularly the concentration of local endemic *Arctostaphylos* taxa along the California coast. As a relatively "young" coast, California's coastal margin has a high level of topographic heterogeneity and tectonic activity along a transform fault system that has been present during the past thirty million years. Consequently, many different rock formations, and hence soil types, have been juxtaposed into a rich landscape mosaic characterized by archipelagos of different-sized "edaphic islands" (Kruckeberg, 2002). Although these soils are not as nutrient deficient as those in southwest South Africa or southwest Australia, they are relatively inhospitable compared to surrounding soils due to shallow, rocky conditions, deep sand, or challenging nutrient compositions (*e.g.*, serpentines). As previously discussed, coastal lowlands are cooler and more humid due to the summer marine layer whereas winter rainfall on coastal uplands is almost double that of coastal lowlands or interior mountains (Vasey et al., *In press*). These relatively moist and equable conditions consequently favor a penetration of conifer forest and mixed evergreen forest far down the coast while more fine-grained and deep clay soils favor coastal prairie and coastal scrub. Chaparral typically is restricted to isolated edaphic islands within these topographically heterogeneous landscapes in a mosaic with forest, grassland, or coastal scrub (Wells, 1962). Finally, chaparral along the coast has been found to have much longer fire return intervals than interior chaparral (Odion & Tyler, 2002). This could well be due, in part, to higher fuel moisture levels, especially through the summer dry season due to the marine layer (Figure 5), but also less frequent lightning strikes (Keeley, 1982). Cooler, moist conditions during the dry season and longer fire intervals allow more flammable biomass to accumulate in coastal chaparral stands and thus when canopy fires occur they tend to be high intensity. This constellation of environmental conditions characterizing the California coastal template is distinctive compared to conditions in chaparral around the rest of the state (Parisien & Moritz, 2009).

Within this environmental context, we begin the discussion of *Arctostaphylos* diversification by considering a classic paper by Stebbins & Major (1965) that identifies several centers of endemism in California. Most of these endemic areas were concentrated along the California coast. Subsequent work by Richerson & Lum (1980) and Loarie et al. (2008) have been consistent with this pioneering work in terms of recognizing the California coast as a hot spot for plant species diversity. Stebbins & Major (1965) focused on two alternative routes to local endemism: 1) the case of "paleoendemics" which evolved in the distant past under different climatic regimes, were once widespread, and have now become restricted to local refugia within their present range (*e.g.*, the Catalina ironwood, *Lyonothamnus floribundus* [Rosaceae]; Raven & Axelrod, 1978); and 2) the case of "neoendemics" which evolved in the recent past under climatic regimes similar to the present but are restricted to particular substrates or habitat conditions that inherently limit widespread dispersal and establishment (*e.g.*, *Stephanomeria malheurensis* [Asteraceae]; Gottlieb, 1978).

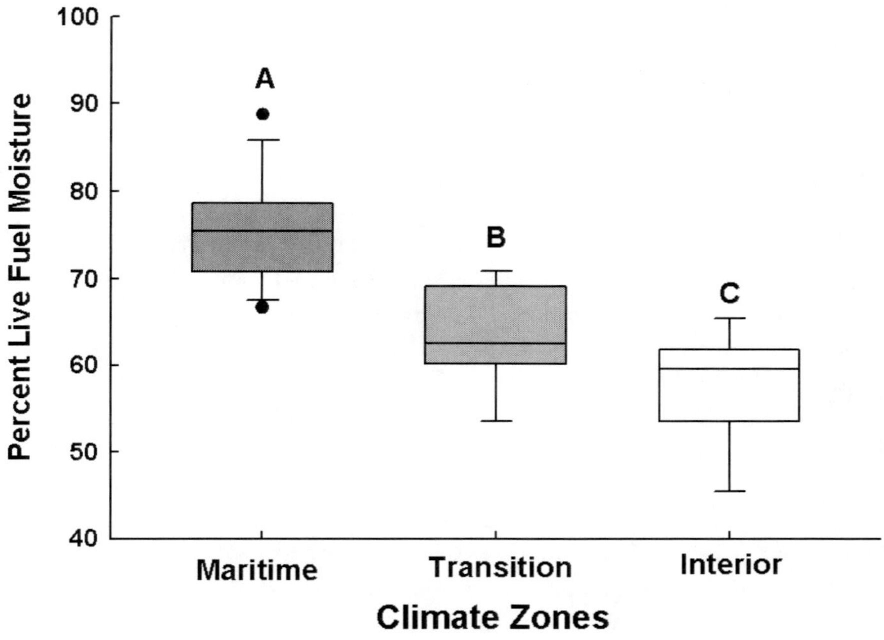

Figure 5. Live Fuel Moisture (LFM) percent calculated from leaves obtained from individual *Arctostaphylos* shrubs (n=10 per species) during sampling for water potentials and stable isotope analyses from maritime (n=100), transition (n=69), and interior (n=71) localities during the end of the dry season, 2009 (based on data collected in Vasey et al., 2012). Live Fuel Moisture is defined as the water content of live vegetation expressed as a percentage of the dry mass of vegetation (*md*) where LFM = *mw* – *md* / *md* and *mw* = is the mass of undried vegetation (Dennison et al., 2008). Box plots present the results of ANOVA analysis with Tukey HSD test for differences among means. Capital letters indicate significant differences at $p < 0.0001$, error bars represent 95% confidence intervals and "periods" represent outliers.

Since recognizable *Arctostaphylos* fossils are known from western Nevada as far back as 13.7 mya in the middle Miocene (Edwards, 2004), when climatic conditions were presumably more mesic (Axelrod, 1973), possibly several locally endemic *Arctostaphylos* species represent "paleoendemics" (*i.e.*, relict lineages) now restricted to relatively favorable refugial habitats. In fact, molecular genetic studies in *Arctostaphylos* have demonstrated strong evidence that there are two "deep" lineages within the genus (Boykin et al., 2005; Markos et al., 1998; Wahlert et al., 2009) that are likely to be a legacy of this long evolutionary history. *Arctostaphylos* species have retained a limited ability to hybridize between these two deep lineages but empirical observation has shown that hybridization between diploids from these two clades is minimal (typically <4 % in the field) (Parker & Vasey, unpublished). Tetraploid species in *Arctostaphylos* in some cases may represent stabilized allopolyploid hybrids between the two deep clades (Scheirenbeck et al., 1992). Particularly because coastal chaparral is relatively more mesic, contains edaphic islands that resist colonization by surrounding forest, and tends to have longer fire intervals, several locally endemic *Arctostaphylos* species may be examples of paleoendemics. Yet, this is probably not the only answer to our question of *why so many local coastal endemics in Arctostaphylos* because a large number of these species have essentially the same ITS molecular sequence despite presenting widely divergent morphologies (Boykin et al., 2005, Wahlert et al., 2009). This pattern is consistent with a more recent diversification of these *Arctostaphylos* taxa, similar to the pattern of more recent diversification of several Cape *Erica* clades (Pirie et al., 2011). Therefore, we turn to possible mechanisms that may have driven the production of these more "neoendemic" species.

Of the 46 local *Arctostaphylos* endemics along the California coast, 33 are diploid species that belong to the larger clade (Wahlert et al., 2009). These 33 species represent over half of the 62 species found in the state (Table 1). Further, most of these species are in the group that is morphologically distinct but possess virtually identical ribosomal ITS sequences. Allopatric speciation or founder effects would provide a prominent mechanism for ecological selection to promote the diversification of species among the different soil types. Additionally, diploid species within each clade appear to be able to freely hybridize, so it is possible—if not probable—that many of these species arose via homoploid hybridization followed by ecological selection. Some of these species clearly contain traits that link them to two morphologically different species (*e.g.*, Parker & Vasey, 2004; Wahlert et al., 2009).

Another biological trait that all the endemic species have in common is persistent soil seed banks (Parker & Kelly, 1989). Seedlings generally occur only in postfire habitats and are subjected to strong ecological selection pressures at that stage. Most of the endemic species are obligate seeders. Due to a higher turnover rate of generations, obligate seeding may permit a higher rate of speciation over time (Wells, 1969; Wisheu et al., 2000). Obligate seeding species not only depend upon fire to stimulate germination, but pre-existing adults in the population are killed by intense canopy fires. Thus, new individuals that establish from the postfire seed bank are influenced by current, rather than past, climatic and edaphic conditions. New gene combinations of individuals within the population are relatively free from the swamping effects of a large population of pre-existing adults. Further, virtually all of the obligate seeding species occur on distinctive soil types, suggesting that this life history trait is effective at fine-tuning surviving genotypes to their particular *in situ* conditions over multiple fire events. Fewer coastal endemics are facultative seeders (resprouters), and while ecological selection would act on their seedlings as well, surviving adults may be expected to slow the genetic transformation of those populations through gene flow. Overall, these dynamics indicate some of the biological dimensions to the "geology-fire-climate" evolutionary model that may help to account for this relatively recent proliferation of local endemics along the California coast.

The key to this model is the *interaction* of all three factors—geology, fire, and climate—facilitated by biological features like persistent soil seed banks and obligate seeding that are critical to overcoming the harsh environmental conditions that characterize the California MTC region and that lead to the exceptional pattern of local endemism described herein along the California coast. Anacker et al. (2011) demonstrated that an evolutionary trait, specific leaf area, is linked to the interaction of soil fertility, a coast-interior climate gradient, and fire history. They cautioned that it might be impossible to tease apart any one of these factors in explaining specific leaf area differences among chaparral shrubs. This may well be because of the interactive nature of this relationship.

Another key factor may be the coastal mosaic of vegetation restricting maritime chaparral to separate "islands". The isolation of chaparral patches by forest or coastal scrub would limit gene flow among populations and permit localized ecological specialization as described above. Coastal forests invade chaparral only slowly because soils are shallow or nutrient poor (Dunne & Parker, 1999; Horton et al., 1999). Adjacent chaparral communities persist at the site because their persistence is reinforced by more intense fires than the invading forest can tolerate (Odion et al., 2010).

To summarize, the geological template is a "slow variable" that provides a relatively fixed environmental setting in which different plant species move over the landscape in response to different climate regimes that characterize a given region on millennial time scales. As in all MTC regions, fires are both predictable and important disturbance events that operate on different lineages over time to select for different fire adaptive traits and, in extreme conditions (such as chaparral), fire dependent recruitment in postfire environments. Among obligate seeders, these fire events create lottery-style filters for surviving genotypes and set the stage for natural selection to shift populations in one direction or another over relatively long time scales after multiple fire events. In the coastal region, relaxation of the extremes of the summer dry period allow the establishment of multiple vegetation types, isolating most chaparral stands. Allopatric speciation, founder effects, and hybridization all contribute to the diversification seen along the coast of California.

CONCLUSION

California is the most extreme of all MTC regions in terms of its summer dry period (Cowling et al., 2005) yet it makes an important contribution along with other MTC regions to approximately 15-20% of the world's total vascular plant diversity (Cowling et al., 1996). In California chaparral, the greater diversity of shrub lineages in coastal regions is associated with higher moisture availability due to different combinations of the summer marine layer, more rainfall at higher elevations, and more mild summer temperatures and reduced evapotranspiration near the coast.

The relaxation of California's otherwise harsh MTC environment permits other vegetation, such as forests, to coexist and restrict coastal chaparral to isolated areas characterized by azonal soil conditions. This diversity of soil types, isolation of chaparral stands, and prolonged but intense fire regimes drives the dynamics of fire dependent lineages such as *Arctostaphylos* in this coastal region, and it also is most likely responsible for the preservation of paleoendemic species isolated in coastal refugia. Obligate seeders contribute to chaparral diversity by their faster generation times, permitting relatively rapid speciation and ecological specialization.

In the context of an intrinsically harsh climate regime like the extreme MTC summer drought characterizing California, the relaxed end of the gradient of this harsh environment seems to express the greatest species diversity.

ACKNOWLEDGMENTS

We thank Dr. N. Rajakaruna for inviting us to contribute to this book. This chapter is the culmination of decades of thinking about the dynamics of *Arctostaphylos* and how it relates, in particular, to patterns of chaparral diversity along the coast of California. Dr. Jon Keeley, an expert in MTC vegetation, has been an important source of inspiration and a great "foil" to brainstorm ideas that relate to this topic. We particularly appreciate his helpful comments regarding this manuscript and his seminal ideas recognizing the importance of geology, climate, and fire to the ecology of chaparral. Dr. Karen Holl, Dr. Michael Loik, Brett Hall, U.C. Santa Cruz, and many other students and colleagues have provided invaluable research and insights that helped to inform this long-standing inquiry. An anonymous reviewer also provided some valuable feedback concerning ideas reflected in this chapter. Lastly, San Francisco State University has long supported our work in various ways and we appreciate its commitment to pushing the boundaries of scientific knowledge concerning the natural history, ecology, and evolutionary biology that distinguishes California and its treasure trove of unique native species.

REFERENCES

Anacker, B., Rajakaruna, N., Ackerly, D., Harrison, S, Keeley, J. & Vasey, M. (2011) Ecological strategies in California chaparral: Interacting effects of soils, climate, and fire on specific leaf area. *Plant Ecology & Diversity 4*, 179-188.

Anderson, M. J., Ellington, K. E. & McArdle, B. H. (2006) Multivariate dispersion as a measure of beta diversity. *Ecology Letters 9*, 683-693.

Axelrod, D. I. (1973) History of the Mediterranean ecosystem in California. In: F. di Castri & H. A. Mooney (Eds.). *Mediterranean ecosystems: Origin and structure* (pp. 225-277). New York, NY: Springer.

Boykin, L. M., Vasey, M. C., Parker, V. T. & Patterson, R. (2005) Two lineages of *Arctostaphylos* (Ericaceae) identified using the Internal Transcribed Spacer (ITS) of the nuclear genome. *Madroño 52*, 139-147.

Bradshaw, S. D., Dixon, K. W., Hopper, S. D., Lambers, H. & Turner, S. R. (2011) Little evidence for fire-adapted plant traits in Mediterranean climate regions. *Trends in Plant Science 16*, 69-76.

Cody, M. L. (1986) Diversity, rarity, and conservation in Mediterranean-climate regions. In: M. E. Soulé (Ed.). *Conservation biology: The science of scarcity and diversity* (pp. 122-152). Sunderland, MA: Sinauer and Associates.

Cooper, W. S. (1922) *The broad-sclerophyll vegetation of California: An ecological study of chaparral and its related communities.* Publication 319. Washington, DC: Carnegie Institution of Washington.

Corbin, J. D., Thomsen, M. A., Dawson, T. E. & D'Antonio, C. M. (2005) Summer water use by California coastal prairie grasses: fog, drought, and community composition. *Oecologia 145*, 511-521.

Cowling, R. M., Rundel, P. W., Lamont, B. B., Arroyo, M. K. & Arianoutsou, M. (1996) Plant diversity in Mediterranean-climate regions. *Trends in Ecology and Evolution 11*, 362-366.

Cowling, R. M., Ojeda, F., Lamont, B. B., Rundel, P. W. & Lechmere-Oertel, R. (2005) Rainfall reliability a neglected factor in explaining convergence and divergence of plant traits in fire-prone Mediterranean-climate ecosystems. *Global Ecology and Biogeography 14*, 509-519.

Dawson, T. E. (1998) Fog in the California redwood forest: ecosystem inputs and use by plants. *Oecologia 117*, 476–485.

Dunne, J. A. & Parker, V. T. (1999) Seasonal soil water potential patterns and establishment of *Pseudotsuga menziesii* seedlings in chaparral. *Oecologia 119*, 36-45.

Edwards, S.W. (2004) Paleobotany of California. *Four Seasons 12,* 3-75.

Fischer, D. T., Still, C. J. & Williams A. P. (2009) Significance of summer fog and overcast for drought stress and ecological functioning of coastal California endemic plant species. *Journal of Biogeography 36*, 783–799.

Gottlieb, L. D. (1978) *Stephanomeria malheurensis* (Compositae), a new species from Oregon. *Madroño 25,* 44-46.

Griffin J. R. (1978) Maritime chaparral and endemic shrubs of the Monterey Bay region, California. *Madroño 25*, 65-81.

Hanes, T. L. (1971) Succession after fire in the chaparral of southern California. *Ecological Monographs 41*, 27-52.

Hanes, T. L. (1981) California chaparral. In: F. di Castri, D. W. Goodall & R. L. Specht (Eds.). *Ecosystems of the World. 11. Mediterranean-type shrublands* (pp. 139-174). New York, NY: Elsevier.

Harrison, S. & Inouye, B. D. (2002) High ß diversity in the flora of Californian serpentine "islands." *Biodiversity and Conservation 11,* 1869-1876.

Holland, R. F. (1986) *Preliminary descriptions of the terrestrial natural communities of California.* Unpublished report. Sacramento, CA: California Department of Fish and Game, Natural Heritage Division.

Horton, T. R., Bruns, T. & Parker, V. T. (1999) Ectomycorrhizal fungi in *Arctostaphylos* patches contribute to the establishment of *Pseudotsuga menziesii. Canadian Journal of Botany 77*, 93-102.

Johnstone, J. A. & Dawson, T. E. (2010) Climatic context and ecological implications of summer fog decline in the coast redwood region. *Proceedings of the National Academy of Sciences (USA) 107*, 533-4538.

Keeley, J. E. (1982) Distribution of lightning and man-caused wildfires in California. In: C. E. Conrad & W. C. Oechel (Eds.). *Proceedings of the international symposium on the dynamics and management of Mediterranean type ecosystems* (pp. 431-437). USDA Forest Service, General Technical Report PSW-58.

Keeley, J. E. (1987) Ten years of change in seed banks of the chaparral shrubs, *Arctostaphylos glauca* and *A. glandulosa. American Midland Naturalist 117*, 446–448.

Keeley, J. E. (1991) Seed germination and life history syndromes in the California chaparral. *Botanical Review 57*, 81-116.

Keeley, J. E. (1992) A Californian's view of fynbos. In: R. M. Cowling (Ed). *The ecology of fynbos: Nutrients, fire, and diversity* (pp. 372-378). Oxford, UK: Oxford University Press.

Keeley, J. E. & Fotheringham, C. J. (2000) Role of fire in regeneration from seed. In M. Fenner (Ed). *Seeds: The ecology of regeneration in plant communities.* Second edition (pp. 311-330). Oxon: CAB International.

Keeley, J. E. & Keeley, S. C. (1988) Temporal and spatial variation in fruit production by chaparral shrubs. In: F. di Castri, Ch. Floret, S. Rambal & J. Roy (Eds.). *Time-scales and water stress: Proceedings of the 5th International Conference on Mediterranean Ecosystems* (pp. 457-463). Paris: International Union of Biological Sciences.

Keeley, J. E., Fotheringham, C. J. & Baer-Keeley, M. (2005) Factors affecting plant diversity during postfire recovery and succession of Mediterranean-climate shrublands in California, USA. *Diversity and Distributions 11*, 535-537.

Keeley, J. E., Pausas, J. G., Rundel, P. W., Bond, W. J. & R. A. Bradstock. (2011) Fire as an evolutionary pressure shaping plant traits. *Trends in Plant Science 16*, 406-411.

Keeley, J. E., Bond, W. J., Bradstock, R. A., Pausas, G. & Rundel, P. W. (2012) *Fire in Mediterranean ecosystems: Ecology, evolution, and management.* Cambridge: Cambridge University Press.

Kruckeberg, A. R. (2002) *Geology and plant life: The effects of landforms and rock types on plants.* Seattle, WA: University of Washington Press.

Loarie, S. R., Carter, B. E., Hayhoe, K., McMahon, S., Moe, R., Knight, C. A. & Ackerly, D. D. (2008) Climate change and the future of California's endemic flora. *PLoS One 3*, e2502.

Markos, S. M., Hileman, L. M., Vasey, M. C. & Parker, V. T. (1998) Phylogeny of the *Arctostaphylos hookeri* complex (Ericaceae) based on nrDNA data. *Madroño 45*, 187–199.

Myers, N., Mittermeier, R. A., Mittermeier, C. G., da Fonseca, G. A. B. & Kent, J. (2000) Biodiversity hotspots for conservation priorities. *Nature 403*, 853-858.

Odion, D. C. & Tyler, C. (2002) Are long fire-free periods needed to maintain the endangered, fire-recruiting shrub *Arctostaphylos morroensis* (Ericaceae)? *Conservation Ecology 6*, 4. URL: http://www.consecol.org/vol6/iss2/art4.

Odion, D. C., Moritz, M. A. & DellaSala, D. A. (2010) Alternative community states maintained by fire in the Klamath Mountains, USA. *Journal of Ecology 98*, 96-105.

Ojeda, F. (1998) Biogeography of seeder and resprouter *Erica* species in the Cape Floristic Region—Where are the resprouters? *Biological Journal of the Linnaean Society 63*, 331-347.

Parker, V. T. (1990) Problems encountered while mimicking nature in vegetation management: an example from a fire-prone vegetation. In: R. S. Mitchell, C. J. Sheviak & D. J. Leopold (Eds.). *Ecosystem management: Rare species and significant habitats. Proceedings of the 15th Annual Natural Areas Conference* (pp. 231-234). New York State Museum Bulletin 471. New York, NY: New York State Museum.

Parker, V. T. & Kelly, V. R. (1989) Seed banks in California chaparral and other Mediterranean climate shrublands. In: M. A. Leck, V. T. Parker & R. L. Simpson (Eds.). *Ecology of soil seed banks* (pp. 231-255). New York, NY: Academic Press.

Parker, V. T. & Pickett, S. T. A. (1998) Historical contingency and multiple scales of dynamics in plant communities. In: D. L. Peterson and V. T. Parker (Eds.). *Ecological scale: Theory and applications* (pp. 171-191). New York, NY: Columbia University Press.

Parker, V. T. & Vasey, M. C. (2004) *Arctostaphylos gabilanensis*, a newly described auriculate-leaved manzanita from the Gabilan Mountains, California. *Madroño 51*, 322-325.

Parker, V. T., Vasey, M. C. & Keeley, J. E. (2009) *Arctostaphylos. Flora of North America, North of Mexico*. Volume 8, Magnoliophyta: Paoniaceae to Ericaceae (pp. 406-445). New York, NY: Oxford University Press.

Parker, V. T., Vasey, M. C. & Keeley, J. E. (2012) *Arctostaphylos*. In: B. G. Baldwin et al. (Eds.). *The Jepson manual: Vascular plants of California*. Second edition (pp. 686-699). Berkeley, CA: University of California Press.

Parisien, M. A. & Moritz, M. A. (2009) Environmental controls on the distribution of wildfires at multiple spatial scales. *Ecological Monographs 79*, 127-154.

Pausas, J. G. & Keeley, J. E. (2009) A burning story: The role of fire in the history of life. *BioScience 59,* 593-601.

Pirie, M. D., Oliver, E. G. H., Bells, D. U. & Bellsted, D. U. (2011) A densely sampled ITS phylogeny of the Cape flagship genus *Erica* L. suggests numerous shifts in floral macro-morphology. *Molecular Phylogenetics and Evolution 61,* 593-601.

Raven, P. H. & Axelrod, D. I. (1978) Origin and relationships of the California flora. *University of California Publications in Botany 72*, 1-134.

Richerson, P. J. & Lum, K. (1980) Patterns of species diversity in California: Relation to weather and topography. *The American Naturalist 116,* 504-536.

Rundel, P. W. (2004) Mediterranean-climate ecosystems: defining their extent and community dominance. In: M. Arianoutsou & V. P. Panastais (Eds.). *Ecology, conservation, and management of Mediterranean climate ecosystems* (pp. 1-12). Amsterdam: IOS Press.

Sawyer, J. O., Keeler-Wolf, T. & Evens, J. M. (2009) *A manual of California vegetation.* Second edition. Sacramento, CA: California Native Plant Society.

Schierenbeck, K. R., Stebbins, G. L. & Patterson, R. W. (1992) Morphological and cytological evidence for polyphyletic allopolyploidy in *Arctostaphylos mewukka* (Ericaceae). *Plant Systematics and Evolution 179,* 187-205.

Schimper, A. F. W. (1903) *Plant geography upon a physiological basis.* Oxford: Clarendon Press.

Stebbins, G. L. & Major, J. (1965) Endemism and speciation in the California flora. *Ecological Monographs 35*, 1–35.

Sweeney, R. (1956) Responses of vegetation to fire: A study of the herbaceous vegetation following chaparral fires. *University of California Publications in Botany 29*, 143-216.

Vasey, M. C. (2012) *Influence of the summer marine layer on maritime chaparral and implications for conservation policy in the California coastal zone.* PhD Dissertation. Santa Cruz, CA: University of California, Santa Cruz.

Vasey, M. C., Loik, M. E. & Parker, V. T. (2012) Influence of summer marine fog and low cloud stratus on water relations of evergreen woody shrubs (*Arctostaphylos*, Ericaceae) in the chaparral of central California. *Oecologia 170,* 325-337.

Vasey, M. C., Parker, V. T., Holl, K. D., Loik, M. E. & Hiatt, S. (*In press*) Maritime climate influence on chaparral composition and diversity in the coast range of Central California. *Ecology and Evolution.*

Wahlert, G., Parker, V. T. & Vasey, M. C. (2009) A phylogeny of *Arctostaphylos* (Ericaceae) inferred from nuclear ribosomal ITS sequences. *Journal of the Botanical Research Institute of Texas 3*, 672-683.

Wells, P. V. (1962) Vegetation in relation to geological substratum and fire in the San Luis Obispo quadrangle, California. *Ecological Monographs 32*, 79-103.

Wells, P. V. (1969) The relation between mode of reproduction and the extent of speciation in woody genera of the California chaparral. *Evolution 23,* 264-267.

Wilken, D. H. (2012) *Ceanothus*. In: B. G. Baldwin et al. (Eds.). *The Jepson manual: Vascular plants of California*. Second Edition (pp. 686-699). Berkeley, CA: University of California Press.

Wisheu, I. C., Rosenweig, M. L., Olsvig-Whittaker, L. & Shmida, A. (2000) What makes nutrient-poor Mediterranean heathlands so rich in plant diversity? *Evolutionary Ecology Research 2*, 935-955.

Zedler, P. H. (1995) Fire frequency in southern California shrublands: Biological effects and management options. In: J. E. Keeley & T. Scott (Eds.). *Brushfires in California: Ecology and resource management* (pp. 101-112). Fairfield, WA: International Association of Wildland Fire Publications.

Zedler, P. H., Gautier, C. R. & McMaster, G. S. (1983) Vegetation change in response to extreme events: The effect of a short interval between fires in California chaparral and coastal scrub. *Ecology 64*, 809-818.

In: Plant Ecology and Evolution in Harsh Environments ISBN: 978-1-63321-955-7
Editors: N. Rajakaruna, R. S. Boyd and T. B. Harris

Chapter 9

THE EVOLUTIONARY ECOLOGY AND GENETICS OF STRESS RESISTANCE SYNDROME (SRS) TRAITS: REVISITING CHAPIN, AUTUMN AND PUGNAIRE (1993)

Eric J. B. von Wettberg[1,•], Jayanti Ray-Mukherjee[1,2], Nathan D'Adesky[1], Damian Nesbeth[1,3] and Seeta Sistla[4]

[1]Department of Biological Sciences, Florida International University, Miami, FL, USA

[2]School of Agricultural, Earth, and Environmental Sciences, University of KwaZulu-Natal, Durban, South Africa

[3]University of Technology, Kingston, Jamaica

[4]Department of Ecology and Evolutionary Biology, University of California, Irvine, CA, USA

ABSTRACT

The classic paper, "Evolution of Suites of Traits in Response to Environmental Stress" (Chapin, Autumn & Pugnaire. *The American Naturalist* 1993, v 142, s78-s92), examines ways in which Stress Resistance Syndrome (SRS) may evolve. The authors argued that plant traits that confer tolerance of low productivity environments are correlated with slow growth and tissue turnover. Chapin et al. also suggested that there may be an underlying simple genetic switch that allows organisms to evolve suites of slow growth traits adapting them to low productivity environments. More than two decades after it was initially posed, we revisit this hypothesis to explore whether support for their argument has emerged. Despite the explosive growth in ecological genomics over the past decade, we identify a significant paucity in our understanding of non-model SRS-exhibiting taxa. We suggest several lines for future work, from phylogenomics to population genomics, developmental genetics, and comparative biology of non-model plants displaying SRS traits.

• E-mail: Eric.vonwettberg@fiu.edu.

INTRODUCTION

It has long been observed that plants found in stressful habitats tend to share a suite of traits, or a Stress Resistance Syndrome (SRS), that provide broad adaptation across a range of low productivity, harsh, or edaphically toxic habitats (*e.g.,* Chapin, 1980; Grime, 1977; Grime, 1988; Grime & Hunt, 1975; Leshem & Kuiper, 1996; Odum & Pinkerton, 1955). As sessile organisms, terrestrial plants are prone to experience environmental stress and nutrient limitation. When grown under a broad range of stressful conditions, from water and nutrient limitation to the presence of toxins, pollutants, or excesses of required nutrients, plants display a range of traits that confer stress resistance, including low growth rates and tissue turnover coupled with slow rates of nutrient uptake and relatively highly-defended, carbon-rich tissue. Chapin et al. (1993) posited that if SRS traits are in fact geographically and taxonomically widespread, they likely could evolve via a simple genetic basis. If the genetic basis of the SRS is simple, and if similarities at the genetic level do exist, it should have profound impacts on the evolution of stress tolerance and patterns of biodiversity across stressful habitats. Chapin et al.'s exploration of how the mechanisms by which SRS may evolve from simple responses to stressful environments sharing a common genetic basis has profoundly shaped thinking about plant ecological genomics. In this chapter, we revisit Chapin and colleagues' hypotheses and review emerging genetic evidence that could support this observation.

Many of the Earth's terrestrial habitats are stressful to plants, as reflected by low primary productivity. High-stress environments tend to lack abundant available water, either through a shortage of water (*e.g.,* deserts, forest canopy habitats for epiphytes), toxicity of available water (*e.g.*, saline habitats; Chapters 2, 4, 11, 14), heavy metal-rich habitats (Chapters 2, 3, 6, 10, 11, 14, 15), rapid loss of water (*e.g.*, sandy soils), or extreme seasonality of available water (*e.g.*, some deserts, alpine and tundra habitats; Chapter 7). This makes water limitation a fundamental component of plant life in most harsh environments (Chapter 12). Harsh environments also often share a shortage of essential nutrients, such as N and P (Chapter 6), are based on newly formed or recently disturbed soils, or are characterized by climatic extremes that retard productivity (*e.g.,* alpine and high latitude habitats; Chapter 7). Some osmotic stresses, such as drought can compound nutrient limitation by making nutrients less available (*e.g.,* Chaves et al., 2003). These habitat similarities are paralleled by physiological and (potentially) genotypic characteristics of the stress resistant plants that occupy them.

Comparisons of plant traits across stressful environments clearly show that many SRS traits are shared (*e.g.,* Chapin, 1980; Grime, 1977, 1988). Classic SRS traits include low growth rates, low specific leaf area (SLA; *e.g.*, pine needles have a low SLA relative to maple leaves), slow tissue turnover, low photosynthetic and nutrient uptake rates, and high investment in roots and in secondary metabolites (such as defense compounds) relative to photosynthetically active tissues (Herms & Mattson, 1992). One example provided by Grime (1977) of SRS taxa is arctic shrubs that have low rates of growth and tissue turnover, high investment in roots, and high levels of secondary compounds relative to congeners in temperate habitats (*e.g.,* Chapter 7). Another example are sclerophyllous trees like oaks and olives of some seasonally arid regions (*e.g.*, Mediterranean climates; Chapter 8). These trees have small, thick leaves with low turnover rates, as well as low growth rates. Correlations among these suites of traits have long been observed within species and communities and across habitats (*e.g.,* Chapin, 1980; Grime, 1977; Grime & Hunt, 1975; Odum & Pinkerton, 1955), supporting the hypothesis that there is a fundamental trade-off between power output (growth rate) and efficiency of resource use (Chapin, 1980; Lambers & Pinkerton, 1955). For example, there is an intrinsic tradeoff between seed size vs. seed number, which on one hand displays correlation among traits such as larger seeds, low SLA, slow individual growth rate, and low rates of nutrient uptake, which leads to higher stress resistance (Adler et al., 2014). On the other hand, greater seed number is associated with high SLA and high tissue turnover rates, leading to faster growth and decreased stress resistance (Reich et al., 2003; Westoby, 1998; Wright et al., 2004, 2005).

Strauss & Agrawal (1999) drew a distinction between resistance and tolerance to herbivores or pathogens. Resistance is the ability to ward off an herbivore, whereas tolerance is the capacity to maintain reproductive output despite herbivory. Although both of these aspects of plant responses to their pests may come from the same experiment, how they are measured is also distinct. Resistance is measured by the amount of tissue lost, whereas tolerance is reproduction despite any damage from the pest. This distinction has not been universally drawn for abiotic stresses. Rarely are they reported as distinct measures of plant performance. Grime (1977) defined stress as any factor reducing plant productivity, but conflates tolerance and resistance. Much of the subsequent literature continues to conflate the two. However, a distinction between resistance (or tolerance) and avoidance or escape has been more widely drawn. For stresses that are seasonal and somewhat predictable—such as drought, salinity, or cold stress—seasonal growth (*e.g.*, in annuals or through seasonal senescence) can allow a taxon to escape the stress in a mostly dormant form (*e.g.*, as a seed in an annual or by dropping leaves in a deciduous tree; Baskin & Baskin, 2001; Stanton et al., 2000). Avoidance of a stress such as drought can occur through growth of deeper roots, providing a plant access to otherwise unavailable water. Escape and avoidance is common in Mediterranean and temperate floras, as well as many seasonal tropical floras. Seasonal growth can also be favored for stresses such as heavy metal contamination in soils, even if they do not have substantial seasonal variability, because exacerbating factors such as drought or temperature or possibly N availability can favor seasonal growth during the more moderate parts of the year. Although many taxa displaying SRS traits have seasonal growth (*e.g.,* nearly all temperate SRS taxa), in many taxa, avoidance of stress is not the primary or most salient way of handling the stress. Instead, avoidance is part of a suite of traits that characterizes non-SRS taxa.

Despite the phylogenetic diversity of plants displaying SRS traits, the similarity of the suite of stress resistance traits expressed across biomes suggests that adaptations to stressful habitats share common physiological mechanisms and potentially a common molecular basis. A common molecular basis to SRS traits could allow stress responses to be re-deployed when a new stress is faced, facilitating the evolution of stress tolerance in rapidly changing environments (*e.g.*, soil acidification due to increased N deposition). Developing our understanding of the molecular basis underlying SRS traits is pivotal, especially given that unpredictable climatic and anthropogenic conditions are giving rise to novel stressful conditions.

In this chapter, we revisit Chapin et al. (1993) to explore progress in our understanding of the evolution and underlying genetics of SRS traits. Chapin and colleagues suggested that all land plants have a "simple genetic switch" that can be turned on, shifting either a taxonomic group (*i.e.*, an evolutionary shift) or an individual (*i.e.*, a developmental shift) between a low resource-low growth and a high resource-high growth strategy. Ecological genomics has rapidly expanded in the past decade, allowing researchers tractable methodologies to begin to explore SRS genetics. Paralleling the order and reasoning of their argument, we review what has been learned in the two decades since these hypotheses were posed and highlight areas of ongoing investigation into the genomic underpinnings of SRS traits.

Proposed Evolutionary Mechanism of the Stress Resistance Syndrome

Chapin et al. (1993) suggested that SRS traits could rapidly evolve with a change in a "genetic switch or trigger". They observed that plants grown in low-resource environments express many of the same SRS traits, whether or not they express them in their home habitat. The traits include low growth rates, which correlate with low rates of photosynthesis and nutrient uptake, low rates of tissue turnover, and high investment in roots and in secondary metabolites relative to photosynthetically active tissues. They noted, therefore, that SRS-response can be phenotypically plastic, such that it is selectively expressed under stressful conditions. Notably, plants endemic to relatively high-resource environments tend to be more responsive to increased nutrient conditions and to have broader reaction norms of response between high- and low-nutrient conditions. Chapin and colleagues suggest that the phenotypic and genetic underpinnings of SRS plasticity are parallel, leading to the hypothesis that a "simple genetic switch or trigger" underlies shifts in SRS expression. Evidence of these linkages, however, remains sparse, despite great advances in molecular biology and plant physiology, and the emergence of new fields like genomics, systems biology, and evolutionary developmental genetics.

The nature of the hypothetical switch depends on the number of genes involved in evolution of suites of traits; fewer requisite genes will simplify the evolution of the switch. Chapin et al. (1993) cite a long history of debate about the number and effect size of genes underlying SRS traits. The effect sizes of adaptive alleles, such as those that might become fixed in an evolutionary shift toward the SRS, are thought have an exponential distribution, with few of large effect and many of small effect (Orr, 1998). Chapin et al. (1993) argue that changes either in a few or many genes could lead to some form of genetic switch or trigger. Our emerging understanding of the developmental genetics of plants, as well as the development of fields such as evolutionary developmental genetics, genomics, and systems biology, sheds some insight on the possibility of genetic switches underlying shifts in SRS behavior both plastically during ontology and genetically among clades differing in SRS expression.

For example, a combination of molecular approaches have identified most of the genes involved in the control of flowering time, which is one of the best understood plant developmental pathways (Chapter 11; Andrés & Coupland, 2012; Ehrenreich et al., 2009). Much of the molecular work on flowering time has been done in the model plant *Arabidopsis thaliana* (Brassicaceae); comparative work shows that much of the pathway is conserved across angiosperms, albeit with some important shifts (Ballerini & Kramer, 2011). Whereas many environmental cues influence flowering time, from photoperiod to temperature to resource availability and via several developmental and hormonal pathways, a relatively small number of genes actually integrates these cues and cause the shift to flowering (*e.g.*, Ehrenreich et al., 2009). Environmental cues can phenotypically shift flowering via these actors, and mutations in these integrators or the pathways leading to them can alter responsiveness to environmental and developmental cues (*e.g.*, Franklin & Whitelam, 2004). For example, cold temperature exposure in some temperate plants with vernalization sensitivity decreases expression of a critical gene called *FLC,* thus stopping flowering inhibition (*e.g.,* Andrés & Coupland, 2012; Ehrenreich et al., 2009). Given the centrality of flowering in the plant life cycle, shifts in flowering time invariably have pleiotropic effects on other phenotypes, such as size at flowering and growth rate, and can therefore affect SRS expression.The plasticity of flowering time and its underlying genetic regulation suggests that triggers for SRS expression are developmentally possible. However, our more limited understanding of other aspects of development is less supportive of a central shift in leaf turnover, seed size, wood density, or other aspects of the SRS. Unlike flowering time, many components of plant growth appear to be genetically decentralized. For example, patterns of leaf development that follow a generic developmental plan are also influenced by environmental cues, resource availability, and hormonal signals. However, they do not appear to have the same sort of simple developmental switch that can shift a plant between forms (as reviewed by Efroni et al., 2010). Furthermore, variation in important aspects of leaf physiology, such as leaf size, thickness, longevity, and stomatal density can be achieved by several different developmental mechanisms, such as differences in cell number or cell expansion (*e.g.,* Aguirrezabal et al., 2006; Tsukaya, 2014), complicating the concept of a central genetic control mechanism. Aspects of investment into stems relative to roots are likely under similar decentralized control, although are still poorly understood (*e.g.*, Kellermeier et al., 2014). Responses of roots to nitrate are multifaceted and lack the sort of central developmental control that characterizes the transition to flowering (Medici & Krouk, 2014). Consequently, although the idea of a simple genetic trigger that can switch plants between SRS forms is still plausible, the hypothesis remains equivocal due to our lack of understanding of how environmental cues mediate the developmental genetics of key SRS traits.

Physiological Link between Growth and the Stress Response Syndrome, and the Physiological Basis of the Stress Response Syndrome

Although many questions remain regarding the ecological and evolutionary development of SRS traits, our knowledge of the genetic and mechanistic basis of stress responses and the suite of SRS traits has rapidly grown during the past decade. Using forward and reverse genetic screens using the model plant *Arabidopsis*, crop plants (*e.g.,* rice, maize, soybean), and "emerging" models (*e.g.*, *Medicago* [Fabaceae], *Mimulus* [Phrymaceae], *Populus* [Salicaceae], and *Vitus* [Vitaceae], plant species for which we have a well annotated genome and many other resources such as germplasm, mutants, and transformation procedures that are present in *Arabidopsis*, but where the extent and ease of use of the resources is not as well developed as *Arabidopsis*), several key SRS traits have been linked to candidate genes (Table 1).

Table 1. Selected papers identifying candidate genes for stress resistance syndrome (SRS) traits

SRS Trait	Paper title	Citation	Summary and comments
Relative growth rate	Increased chilling tolerance following transfer of a betA gene enhancing glycinebetaine synthesis in cotton (*Gossypium hirsutum* L.)	Zhang et al. (2012)	An osmolant provided by genetic transformation and overexpression can provide chilling tolerance and affect growth rate.
	Effect of leghemoglobin A gene expression from soybean on tobacco plant growth and antioxidant state under damaging action of cadmium	Dmitryukova et al. (2011)	Leghomoglobin provided by genetic transformation and overexpression can affect growth rate and heavy metal tolerance.
	Homologs of FT, CEN and FD respond to developmental and environmental signals affecting growth and flowering in the perennial vine kiwifruit	Varkonyi-Gasic et al. (2013)	Conserved central players in the flowering time pathway, such as FT, have the capacity to exert widespread effects on plant growth rate.
	Gibberellin-associated cisgenes modify growth, stature and wood properties in Populus	Han et al. (2011)	Transgenic modification of Populus gibberellin metabolism has wide ranging effects on growth rate
	Using knockout mutants to reveal the growth costs of defensive traits	Züst et al. (2011)	Loss of function mutations in genes controlling plant defenses illustrate the costs of such genes to growth rate in the absence of herbivores
	A deficiency in the flavoprotein of *Arabidopsis* mitochondrial complex II results in elevated photosynthesis and better growth in nitrogen-limiting conditions	Fuentes et al., (2011)	Mutations to genes in the tricarboxylic acid cycle can increase growth rate under certain conditions
	How plants cope with water stress in the field? Photosynthesis and growth	Chaves et al. (2002)	Multiple variations are found in plants that are under water stress compared to well watered controls. Changes in root:shoot ratio, activity of enzymes essential for carbon metabolism, photosynthesis, leaf area, dry weight, leaf conductance, and stomatal closure are analyzed.
Photosynthetic rates	5-Aminolevulinic acid enhances photosynthetic gas exchange, chlorophyll fluorescence and antioxidant system in oilseed rape under drought stress	Liu et al. (2013)	Foliar application of a tetrapyrol precursor can increase drought stress and raise photosynthetic rates in Brassica crops.
	Antisense-mediated depletion of GMPase gene expression in tobacco decreases plant tolerance to temperature stresses and alters plant development	Wang et al. (2012)	Decreasing plant anti-oxidant activity decreased stress tolerance and reduced photosynthetic rate.

SRS Trait	Paper title	Citation	Summary and comments
	Improvement of water use efficiency in rice by expression of HARDY, an *Arabidopsis* drought and salt tolerance gene	Karaba et al. (2007)	Expression of HARDY (HRD) gene in *Arabidopsis* is linked to improved water efficiency by improving photosynthetic assimilation and decreasing transpiration, which ultimately leads to greater biomass gained per water used. HRD overexpression was found to have affected plants by causing greater leaf biomass and more bundle sheath cells which are thought to assist in the overall photosynthetic increase.
	Enhanced accumulation of BiP in transgenic plants confers tolerance to water stress	Alvim et al. (2001)	Increased BiP in plant cells leads to increased turgidity and better ability to maintain water levels in progressive drought conditions. Under drought conditions BiP overexpression was shown to lead to increased photosynthesis, stomatal conductance and transpiration than when compared to the control which had a significant reduction of CO_2 assimilation induced by drought.
	Relationship between the heat tolerance of photosynthesis and the thermal stability of Rubisco activase in plants from contrasting thermal environments	Salvucci & Crafts-Brandner (2004)	A decrease in photosynthesis during heat stress is thought to result from Rubisco activase's inability to change Rubisco to its active form. This research analyses photosynthetic rate and Rubisco activase under heat stress to determine if Rubisco activase is indeed responsible for the decrease in photosynthetic rate.
	The two senescence-related markers, GS1 (cytosolic glutamine synthetase) and GDH (glutamate dehydrogenase), involved in nitrogen mobilization, are differentially regulated during pathogen attack and by stress hormones and reactive oxygen species in *Nicotiana tabacum* L. leaves	Pageau et al. (2006)	Expression of genes involved in nitrogen assimilation were analyzed to explore their role in pathogen attack, and phytohormone application. Some of the genes responded generally to all stresses while other genes were more specialized and expressed only under certain stress conditions. These genes may play a role in natural leaf senescence and stress-related leaf senescence as well.
Tissue N concentration	Cadmium toxicity induced changes in nitrogen management in *Lycopersicon esculentum* leading to a metabolic safeguard through an amino acid storage strategy	Chaffei et al. (2004)	Analyzes the effect of cadmium toxicity on nitrogen mobilization and change. Cadmium was found to have inhibitory effects on a number of enzymes and accumulation of ammonium and amino acids, along with a decrease in protein. Results suggest that cadmium exposure leads to a shift in N storage and accumulation of N in roots.
	Gene expression profiles during the initial phase of salt stress in rice	Kawasaki et al. (2001)	Plants under high salt reduce their photosynthetic rate and start to synthesize stress inducible transcription factors and proteins, and then later produce defensive related factors.
Transpiration rate	Drought, abscisic acid and transpiration rate effects on the regulation of PIP aquaporin gene expression and abundance in Phaseolus vulgaris plants	Aroca et al. (2006)	The effects of ABA, MTW (inhibitor of stomatal opening), and drought were analyzed by measuring transpiration rate, leaf water, ABA abundance in tissue, as well as gene expression and protein abundance. The three treatments did not change leaf water or leaf ABA abundance, but instead led to a reduction in transpiration rate and increased expression in PVPIP2 and PIP1 (a plamsa membrane intrinsic protien) in leaves.

Table 1. (Continued)

SRS Trait	Paper title	Citation	Summary and comments
	Tolerance of pea (*Pisum sativum* L.) to long-term salt stress is associated with induction of antioxidant defences	Hernández et al. (2000)	Compares salt-tolerant to salt-sensitive pea plants. Salt-tolerant plants have increased expression of certain enzymes which induce antioxidative properties which are likely part of the changes that induce salt-tolerance and affect transpiration rate
	Important roles of drought- and cold-inducible genes for galactinol synthase in stress tolerance in *Arabidopsis thaliana*	Taji et al. (2002)	Profiles expression of genes related to raffinose and galactinol which are found to be accumulated in plants under environmental stresses. Different genes are induced by high salinity or drought stress than are for cold stress. Overexpression of AtGolS2 is shown to increase raffinose and galactinol which lead to drought tolerance by lowering transpiration levels.
	Stress-inducible expression of At DREB1A in transgenic peanut (*Arachis hypogaea* L.) increases transpiration efficiency under water-limiting conditions	Bhatnagar-Mathur et al. (2007)	Tested the transpiration efficiency of transgenic peanut plants that expressed stress inducible transcription factor DREB1A. One of the selected transgenic events had a 40% percent reduction rate in transpiration efficiency.
Root:shoot ratio	Locating genes associated with root morphology and drought avoidance in rice via linkage to molecular markers	Champou et al. (1995)	An early QTL study that identified chromosomal regions associated with drought tolerance. Many QTL's found were related to root thickness, root/shoot ratio, dry root weight, and overall root morphology.
Carbon-based defenses	Coordinated activation of metabolic pathways for antioxidants and defence compounds by jasmonates and their roles in stress tolerance in *Arabidopsis*	Sasaki-Sekimoto et al. (2005)	The signaling molecule jasmonate has been linked to response to oxidative stress, such as pathogen attack. Mutants confirm that jasmonate deficient plants lack induction of antioxidant genes and that mutants were more sensitive to O_3 exposure.
Leaf turnover rate	Delayed leaf senescence induces extreme drought tolerance in a flowering plant	Rivero et al. (2007)	Suppression of leaf senescence in transgenic plants induces stress tolerance under drought conditions.
	The ER luminal binding protein (BiP) mediates an increase in drought tolerance in soybean and delays drought-induced leaf senescence in soybean and tobacco	Valente et al. (2009)	Lumen binding protein in the ER, BiP, was overexpressed in soybean and tobacco plants. Overexpression of this gene caused drought tolerance when compared to the wild type. In soybean there was less wilting, less stomatal closure and photosynthetic rates and transpiration rates were less inhibited than in wild types. During drought stress in tobacco and soybean plants, leaf senescence was delayed.
	Rootstock-mediated changes in xylem ionic and hormonal status are correlated with delayed leaf senescence, and increased leaf area and crop productivity in salinized tomato	Albacete et al. (2009)	Grafted tomato with a rootstock from a salt tolerant relative (a *Solanum lycopersicum* × *S. cheesmaniae* cross) was studied under salinity stress. Concentration of ions and phytohormones were analyzed. Root factors had possible effects on leaf growth and senescence.

SRS Trait	Paper title	Citation	Summary and comments
	The delayed leaf senescence mutants of *Arabidopsis*, ore1, ore3, and ore9 are tolerant to oxidative stress	Woo et al. (2004)	Mutants that displayed tolerance to oxidative stress and delayed leaf senescence in the plant *Arabidopsis* thaliana differed in leaf turnover rates. Interestingly, antioxidant activity was similar or lower in the mutants, which cancels out the possibility of antioxidant activity being the factor contributing to oxidative tolerance.
Root phosphatase	WRKY75 transcription factor is a modulator of phosphate acquisition and root development in *Arabidopsis*	Devaiah et al. (2007)	WRKY75 is a transcription factor triggered under phosphate stress. RNAi was used to suppress this transcription factor and high affinity Pi transporters, phosphatases and Mt4/TPS1-like genes decreased: root morphology also changed. This shows that the WRKY75 transcription factor is a regulator of both phosphorous acquisition and root development.
Immobilization of heavy metals	The *Arabidopsis* heavy metal P-type ATPase HMA5 interacts with metallochaperones and functions in copper detoxification of roots	Andrés-Colás et al. (2006)	HMA5 (heavy metal ATPase) is involved in Cu uptake and storage in plants. The Cu transporting gene was identified and inserted in mutants, making them hypersensitvie to Cu, and more Cu accumulated in the roots when compared to the wild type
	Coordinated responses of phytochelatin synthase and metallothionein genes in black mangrove, *Avicennia germinans* exposed to cadmium and copper	Gonzalez-Mendoza et al. (2007)	Phytochelatins and metallothioneins are involved in metal detoxification in the black mangrove. An increase in expression of AvPCS may contribute to Cu^{2+} and Cd^{2+} detoxification. Overexpression of AvMt2 andAvPCS may contribute to a joint effort to detoxify non-essential metals.
	Analysis of transgenic Indian mustard plants for phytoremediation of metal-contaminated mine tailings	Bennett et al. (2003)	Transgenic plants that produced γ-glutamylcysteine synthetase (ECS) or glutathione synthetase (GS) in excess were analyzed for their metal concentration. Transgenics had greater metal accumulation than the wild type and more metal was removed from soil than by the wild type.
	The cytosolic O-acetylserine(thiol)lyase gene is regulated by heavy metals and can function in cadmium tolerance	Dominguez-Solis et al. (2001)	The studied gene is activated by heavy metal stress. *Arabidopsis* plants overexpressing this gene have greater tolerance to cadmium.
Leaf abscisic acid	*Arabidopsis* AtMYC2 (bHLH) and AtMYB2 (MYB) function as transcriptional activators in abscisic acid signaling	Abe et al. (2003)	Transgenic plants that overexpress AtMYC2 and AtMYB2 are ABA sensitive. This shows that the proteins AtMYC2 and AtMYB2 act as transcriptional regulators of ABA during drought.
	Regulation of drought tolerance by gene manipulation of 9-cis-epoxycarotenoid dioxygenase, a key enzyme in abscisic acid biosynthesis in *Arabidopsis*	Iuchi et al. (2001)	Manipulation of a NCED gene in *Arabidopsis* called AtNCED3 can promote drought tolerance. Overexpressors of AtNCED3 have higher ABA levels, which led to closing of stomata and lower transpiration rates that enabled drought tolerance.

Table 1. (Continued)

SRS Trait	Paper title	Citation	Summary and comments
	Arabidopsis basic leucine zipper proteins that mediate stress-responsive abscisic acid signaling	Kang et al. (2002)	ABA sensitivity was observed in transgenic plants overexpressing ABF3 and ABF4 (which are ABA responsive element binding factors). The transgenic plants had reduced transpiration and greater drought tolerance than controls.
	Disease resistance and abiotic stress tolerance in rice are inversely modulated by an abscisic acid–inducible mitogen-activated protein kinase	Xiong & Yang (2003)	Disease resistance and abiotic stress resistance are inversely affected by OsMAPK5 expression. Suppression of the gene increases resistance to fungi and bacteria, but decreases resistance to cold, drought and salt stresses.
Cytokinins	Cytokinin oxidase gene expression in maize is localized to the vasculature, and is induced by cytokinins, abscisic acid, and abiotic stress	Brugière et al. (2003)	Cytokinin oxidase can help control cytokinin levels, which play a role in plant development under abiotic stress conditions.
	Cytokinin-mediated source/sink modifications improve drought tolerance and increase grain yield in rice under water-stress	Peleg et al. (2011)	Modifications in cytokinin levels in transgenic plants provide higher yield and increased stress tolerance.
	Enhanced cytokinin synthesis in tobacco plants expressing PSARK::IPT prevents the degradation of photosynthetic protein complexes during drought	Rivero et al. (2010)	Transgenic plants expressing PSARK::IPT had increased cytokinin production and Brassinosteroid synthesis and reduced ABA drought stress response. Transgenic plants had less damage to photosynthetic proteins and better maintenance of photosynthesis during prolonged water stress.
	Root-synthesized cytokinins improve shoot growth and fruit yield in salinized tomato (*Solanum lycopersicum* L.) plants	Ghanem et al. (2011)	Over expression of root IPT genes (which are central in cytokinin production) gives better yield and growth to tomato plants under salt stress. IPT expressing lines showed increased CK production, delayed leaf senescence and stomatal closure, and greater growth than the wild type.
	Arabidopsis cytokinin receptor mutants reveal functions in shoot growth, leaf senescence, seed size, germination, root development, and cytokinin metabolism	Riefler et al. (2006)	Loss of function mutants of three histidine kinases, AHK2, AHK3 and CRE1/AHK4 (all known as cytokinin receptors), shows that triple mutants had seed sizes twice as big as the wild type. Leaves of AHK2 and AHK3 mutants had a significant decrease in chlorophyll content and lacked inhibition of dark induced chlorophyll loss. Double AHK2 and AHK3 mutants showed an enhanced root system with increased branching.

Comparisons of genes expressed during exposure of model plants and crops to different stresses have shown striking similarities among responses to different stresses, such as drought, salinity and cold. This phenomenon is particularly intriguing because crop plants have been selected for fast growth (*e.g.*, producing nutrient-rich, high SLA leaves) and elevated fruit production, traits antithetical to classic SRS physiology. The similarity of gene expression response to stressors among these phylogenetically distinct plant species suggests that the SRS may be highly (but not entirely) conserved at the genetic level (*e.g.*, Nakano et al., 2006).

Chapin and colleagues made two key points about the connections between physiology and the SRS. The first is that slow growth on its own confers some aspects of SRS expression because of its linkage to low rates of nutrient uptake, slow tissue turnover, etc. This stems in part from the closure of stomata and slowing of photosynthesis and leaf initiation that marks responses to drought and other osmotic stresses in plants (Chaves et al., 2002). Nutrient limitation, inherent in many stressful environments, will further necessitate slow growth (Chaves & Oliveira, 2004). Although we have identified many genes and pathways that mediate plant responses to drought and to some extent other responses to stress (see below, and topical reviews such as Bray, 2004; Chinnusamy et al., 2007; Munns & Tester, 2008; Shinozaki & Yamaguchi-Shinozaki, 2007), we know much less about the genetic basis of differences in growth rate or responsiveness to resources across populations or species, highlighting an area where ongoing research is needed.

The second observation made by Chapin et al. (1993) is that tight integration among traits in the SRS means that small changes in one trait could have pleiotropic effects on other aspects of the SRS. This hypothesis is based on the observation that plant responses to a variety of stresses are largely mediated by a few phytohormones that can integrate changes in the nutrient, water, and carbon balances of plants (Chaves & Oliveira, 2004). As we show below, our understanding of the developmental genetics of hormonal signaling in plants has grown remarkably since the 1990s, with a large number of studies identifying genes involved in the metabolism of key plant hormones.

Advances in Gene Expression and Systems Biology: Insights into Stress Response Syndrome Mechanisms

Since the early 1990s several technologies and approaches have emerged that have radically altered studies of plant physiology. Much of the growth in our ability to link genomics to plant hormonal expression comes from increased capacity to characterize the expression of genes activated by exposure to different conditions in which phytohormone expression is involved. Sequencing of expressed genes and whole genomes now parallels analysis of the entirety of expressed genes, proteins, and metabolites in any tissue of interest (Cramer et al., 2011; Obata & Fernie, 2012), and utilization of mutants and transgenics in a growing number of plant taxa is now possible. Reviewing these approaches is beyond our scope (but see reviews such as Glenn, 2011; Wright & von Wettberg, 2009); however, these techniques have drastically altered the amount and type of data available for understanding how environmental stresses are relayed through receptors, hormonal signaling, and gene expression into a phenotypic response. For example, research on mutants or transgenic constructs of particular genes and integrative approaches strengthen our understanding of the emergent properties of cellular signaling, metabolic networks, and other complex interactions within biological systems in a holistic fashion (*e.g.*, Systems Biology, Guffanti, 2002).

The power of molecular systems biology is the capacity to find and characterize all of the genes, proteins, hormones, and metabolites that are expressed or present in a particular tissue, or even cell, exposed to any environmental condition at a particular point in time. This can give unprecedented detail into the scope and timing of plant responses to stress. However, the majority of SRS-related studies focus on *Arabidopsis thaliana*, crop plants (*e.g.*, rice, maize, and soybean), and a handful of emerging model plants (*e.g.*, *Populus*, *Mimulus*, and *Medicago*). All of these are fast growing plants, with the possible exception of *Mimulus,* which can vary between annual and biennial phenologies. None are taxa with notable SRS expression. Because most work characterizing patterns of gene expression make use of rapidly growing plants, it remains unclear to what extent it applies to SRS-expressing plants, which are inherently slow growing.

Despite this limitation, molecular studies of these model organisms have greatly expanded our understanding of the physiological basis of the SRS. Researchers have identified many genes and a number of gene networks that are part of the SRS (*e.g.,* Table 1). Among stresses, the responses to drought, salinity, and cold are particularly similar to the extent that they involve responses to fundamental water shortage, which affects turgor, cell expansion and division, respiration, and (indirectly) growth rate (Shinozaki & Yamaguchi-Shinozaki, 2007; Xiong et al., 2002; Zhu, 2002). These osmotic stressors also stimulate responses to contain the potential for damage to the cell via antioxidant expression (*e.g.*, Munns & Tester, 2008). Responses to these stresses converge on limiting water loss, using water more efficiently, and protecting cells from free radical damage. Many stresses also elicit similar hormonal responses (*e.g.*, ABA expression, Cutler et al., 2010). Furthermore, there is extensive crosstalk among hormonal networks, suggesting that a single hormone alone is rarely responsible for the response to a single stress (*e.g.*, Cramer et al., 2011; Kiba et al., 2011). Similarly, responses to many stressors, such as different types of nutrient limitation, converge on sugar metabolism, ultimately affecting downstream gene networks that impact plant allocation to reproduction and further investment in shoots and roots (Eveland & Jackson, 2012; Gazzarrini & McCourt, 2001).

A key component of response to stress is hormonal signaling. Molecular work on hormone signaling has shown it to have more intricacies than were once appreciated. Auxin signals are perceived by the membrane associated auxin binding protein (*ABP1*), and then affect gene expression through a large family of auxin response factors (of which there are over 20 in *Arabidopsis*) (*e.g.*, Bargmann & Estelle, 2014; Baumann, 2014). Multiple response factors allow subtle gradients in auxin level to lead to tissue-specific expression and the multi-faceted roles of auxin (*e.g.,* Sablowski, 2013), and also create some redundancy through overlapping expression and function (Bargmann & Estelle, 2014). The large number of factors makes it less clear how a few mutations would lead to a simple trigger. Although auxin response factors are particularly numerous, other phytohormones have some level of redundancy in how hormonal signaling is transduced into gene expression (*e.g.,* Cutler et al., 2010; Miransari, 2012).

Notably, the genetic basis of differences among ecotypes in stress responsiveness has rarely been identified, with little information beyond *Arabidopsis* (*e.g.*, Aguirrezabal et al., 2006). Current work suggests that small changes in traits like sensitivity to hormones could have broad implications on plant performance, yet we lack examples in SRS-exhibiting taxa. Work in SRS-exhibiting taxa could conclusively demonstrate that small genetic changes can have cascading effects on physiology leading to shifts in SRS expression (*i.e.*, provide better support for a simple genetic switch to a SRS state). We need studies of taxa other than *Arabidopsis thaliana.* In particular, we need a thorough examination of taxa that vary in the range of SRS traits expressed. Studies of the mechanistic and molecular basis of SRS expression in multiple taxa with clades occurring on a variety of substrates or across a range of conditions varying the degree of stress they impose could be particularly insightful.

Stress Response Syndrome Evolution: A Molecular Perspective

Chapin et al. (1993) suggested that the SRS could evolve rapidly with changes in relatively few genes. Beyond the physiological and phenotypic commonalities among SRS expressing plants, including consistent traits of stress tolerant lineages in different stressful habitats, the existence of single gene mutants (*e.g.*, Imber & Tal, 1970; Quarrie, 1982) that affect several components of SRS pleiotropically, and the potential for species to undergo rapid evolution of resistance to novel stresses, such as heavy metal-rich soils (*e.g.*, Brady et al., 2005), suggest that rapid SRS evolution is possible. The intervening 20 years since Chapin et al.'s original work have provided only limited evidence for evaluating the evolution of SRS traits.

Mutants were emphasized by Chapin et al. (1993) as a target for studying the evolution of SRS traits because an artificial mutation affecting SRS expression would indicate that a naturally occurring variation might have similar cascading effects on SRS traits. For example, transcriptional regulators controlling phytohormones could have large-scale pleiotropic effects on suites of traits and quickly convert a plant between a stress tolerant and intolerant form. However, the mutants described by Chapin and colleagues have severe negative fitness effects (Imber & Tal, 1970; Quarrie, 1982), suggesting that they would likely be purged from any population. There are examples of such genes of very large effect segregating among populations. In crop plants, very strong selection pressure on a few loci has dramatically altered plant growth habit, phenology, and seed size (Doebley et al., 2006; Gottlieb, 1986; Gross & Olsen, 2010). Similar examples from uncultivated populations are present in species adapted to heavy metal-rich soils (Courbot et al., 2007; Hanikenne et al., 2008; Mcnair, 1992; Willems et al., 2007) and saline soils (Baxter et al., 2010; Rus et al., 2006). In the latter case, the locus underlying heavy metal toxicity and osmotic stress is a variant membrane transport protein (*e.g.*, *HMA4* or *HKT1*), which confers specific adaptations to excess heavy metals or sodium. Highlighting the physiological tradeoffs in SRS traits, alleles of these transporters are costly in that they reduce growth rates in mesic conditions and in the absence of excess metal ions. Similarly, mutants of other transporters, such as *CAX1*, confer tolerance of low Ca:Mg ratios and impact growth on Ca:Mg balanced media (Bradshaw, 2005). It seems unlikely that altered transporters alone underlie the SRS because they should be specific to ionic stresses only. However, if mechanistic work should conclude that variant transporters have broader effects—perhaps through their impacts on water balance or basic metabolism—then perhaps they could in fact act as the hypothesized SRS switches.

There are other mutations that could generate the immediate effects leading to SRS trait expression. One is a chromosomal inversion that reduces gene flow between ecotypes that vary in their SRS characteristics. For example, an inversion separates annual and perennial ecotypes of *Mimulus guttatus* (Lowry & Willis, 2010). Chromosomal inversions can create incompatibilities among ecotypes and may allow many loci that confer divergent adaptations to contrasting environments to be inherited as a supergene. Inversions have been shown to be important in adaptive divergence, such as the parallel clines in *Drosophila melanogaster* (Drosophilidae) on different continents (Reinhardt et al., 2014). They are also likely involved in some cases of speciation (*e.g.*, *Cicer reticulatum* and *C. echinospermum* [Fabaceae]; Ladzinsky & Adler, 1976). However, we have too few examples to do more than speculate about inversions as a mechanism for a single mutation to confer a large effect.

Chapin et al. (1993) also noted the potential for rapid evolution of stress tolerance, such as heavy metal tolerance or loss of defensive compounds of *Betula nana* colonizing Iceland in the absence of mammalian herbivores (Bryant et al., 1989). These are fascinating examples of rapid evolution, and they do suggest that aspects of the SRS can shift in tens of generations rather than hundreds or thousands. However, the taxa in which adaptation to a stress such as serpentine soils is most likely to evolve are those that already have stress tolerance (Chapin et al., 1993). If that is so, many of these groups should already have suitable genetic variation for expressing SRS traits. Therefore, phylogenetic analysis of taxa tolerating stresses such as heavy metal-rich soils is essential for strengthening our understanding of SRS evolution and expression.

A very informative study from Anacker & Harrison (2012) shows that in the California Floristic Province, most serpentine tolerators are tip taxa recently derived from non-serpentine taxa. Serpentine taxa display the classic SRS phenotype, but may have evolved from taxa from other habitats but that also displayed SRS traits. The source habitat of all the taxa, however, is missing, but in most cases the likely original habitat for ancestral taxa is a harsh environment. In one serpentine-tolerating group (the genus *Streptanthus* and allies [Brassicaceae]), however, serpentine endemism is not a mere "dead-end," with speciation continuing in the group within serpentine habitats (Ivalú-Cacho et al., 2013), highlighting serpentine-tolerating taxa that have radiated onto other habitats. Nevertheless, these habitats are also generally stressful. Similar work in a broader range of habitats is needed to determine the extent to which the SRS is part of adaptation to stressful environments and the extent to which adaptation to one stressful environment pre-adapts taxa to another stressful environment. Such work is essential to clarifying how SRS expression is gained and potentially lost.

Losses of SRS traits are equally important to gains for understanding the evolution of the SRS, although they are fundamentally different. Chapin et al. (1993) provided the example of *Betula* in Iceland, in which a simple loss of defensive compounds has occurred. Secondary compounds, such as the papyrific acid made by *Betula,* have a range of functions (from defense to antioxidant suppression) and are also physiologically expensive, and thus may be lost in some clades like *Betula.* Without the cost of expressing secondary metabolites, the Icelandic birch trees may grow more quickly, although complete evidence documenting the nature and extent of this cost had not been obtained (Bryant et al., 1989). Nevertheless, it is very plausible that the loss of secondary compounds generally relieves a physiological cost and reduces expression of the SRS. Although a defensive compound and the cost of making it can be eliminated with a single mutation, in many plants the opposite transition, making a new toxin for example, is less likely to evolve by way of a single mutation. Phylogenetics has shown that some reversal of trait loss does happen (*i.e.,* that Dollo's Law does not always hold; Collin & Miglietta, 2008), but losses of a trait can happen with fewer mutations and in a greater variety of ways than the acquisition of a complicated trait.

Some examples of plant populations from differing soil types do not have single genes of strong effect but appear to be multigenic. For example, a large number of differences exist between populations of *Arabidopsis lyrata* from serpentine and non-serpentine soils (Turner et al., 2010). A similarly large number of genetic differences separate saline and non-saline origin populations of *Medicago truncatula* (Friesen et al., 2010). In both of these examples, the divergence of populations on different soil types may have developed over thousands of years, even if genome wide F_{st} estimates (a measure of the extent of population differentiation) are quite low. Although not immediate, these are still potentially relatively rapid shifts in SRS expression. But whether a trigger underlies these changes remains unknown. In *Medicago truncatula*, flowering time appears to be an important component of differential salt tolerance, primarily by conferring avoidance of the times of year when soil salinity is highest (Friesen et al., in review). This shift, controlled by a smaller number of genes, could in fact be a shift that occurred in a few generations. However, avoidance of salinity through phenological adjustment is the opposite strategy from the SRS, as it revolves around high growth rates, higher nutrient uptake, and faster tissue turnover during the mesic conditions of the growing season.

These relatively slower shifts of plant taxa among habitats are consistent with our growing understanding of niche conservatism (Harvey & Pagel, 1991; Holt & Gaines, 1992; Peterson et al., 1999; Pearman et al., 2008; Prinzing et al., 2001; Wiens & Graham, 2005; Weins et al., 2010). Most plants grow in environments where they are well adapted, leading to genotype-environmental correlations (*e.g.,* Ackerly, 2003). These correlations should lead to stabilizing selection on most traits and the retention of niche-related ecological traits over time (*i.e.*, phylogenetic niche conservatism; Wiens et al., 2010). In only a few circumstances (*e.g.*, on habitat islands or the trailing edges of species ranges when they are migrating poleward or upslope with climate change) do we expect to find segregating variation in the suite of SRS traits (Ackerly, 2003). Elsewhere, niche conservatism predicts we should not observe variation in the SRS. Consequently, evolution of SRS traits is predicted to occur only under relatively rare conditions.

If harsh environments are in fact similar, SRS capacity should allow pre-adaptation among these habitats, and we predict that phylogenetic surveys of such habitats would reveal this. However, there are very few studies of this sort. The phylogenetic work mentioned above from Anacker & Harrison (2012) provides a model of what such work might look like. Rapid changes in sequencing technology mean that work of this nature can be scaled to phylogenomics (*e.g.,* Chan & Ragan, 2013). If harsh habitats vary in the degree of their stressfulness, relatively less stressful habitats may serve as launching pads for colonization of other more extreme habitats. For example, tolerance of mafic soils such as diabase or dolerite may be an entryway for colonization of soils derived from ultramafic rocks such as serpentinic soils. Diabase soils are derived from igneous rocks with large components of feldspar and pyroxene and have a high pH as well as high levels of Mg and Ca. Plants adapted to these soils may be pre-adapted to the low Ca:Mg, heavy metals, and low nutrient availability of ultramafic soils. In the mid-Atlantic region of the United States, diabase and ultramafic substrates co-occur within 50 km of one another. In situations such as this, the less stressful diabase may serve as a habitat where pre-adaptation to the more stressful serpentine can occur. If this is the case, diabase- and ultramafic-adapted species would be useful tools for examining the genetic basis of serpentine tolerance (Chapters 6, 11). As of yet, however, this system remains unexamined from a genomics perspective.

CLIMATE CHANGE, RESTORATION AND CONSERVATION

The influence of Chapin et al.'s (1993) SRS hypothesis rests partly on its consideration of the implications of SRS for responses to climate change and agriculture. If stress tolerant lineages have colonized multiple stressful habitats, we can consider this in protecting and restoring populations in low productivity habitats. In some restoration cases we must consider moving plants great distances. For example, sea level rise will inundate low-level areas and sometimes the nearest upland habitat may be far away. In the Florida Keys, for example, the nearest (relatively) high habitat within the United States is hundreds of kilometers north in the Lake Wales ridge, with a much more temperate climate and a very different substrate (sand vs. Karstic coral deposits; Maschinski et al., 2011). We have little sense of how most endangered plant groups in the Florida Keys would respond to the Lake Wales ridge. In more diverse island groups, such as those which have radiated across disparate habitats, the challenge of conservation in the face of climate change and habitat loss is even more severe because multiple habitats must be considered. A more detailed understanding of the genetic basis of adaptation to stressful habitats in adaptive radiations, such as *Scheidea* (Caryophyllaceae; Kapralov et al., 2013) or *Argyroxiphium* (Asteraceae; Baldwin & Sanderson, 1998), could be particularly useful. These groups occur across a range of habitats that vary in aspects of stress and exhibit just such a broad range of forms. This ecological genomic information could be used in anticipating shifts in habitats due to climate change (Chapter 13), or mitigating ongoing effects of the loss, degradation, and fragmentation of habitats (Ackerly, 2000).

IMPLICATIONS IN AGRICULTURE

The basic premise of the green revolution is to convert agricultural habitats into maximally productive habitats through extensive fertilization and irrigation programs, and then breed plants for the consequences of this management (Carroll et al., 1990). This has led breeders to largely select on investment into harvestable organs (*e.g.*, fruit) and on nutrient uptake. While this strategy has been successful from a production standpoint, it is likely not sustainable in a rapidly changing world with a growing population (Carroll et al., 1990). Both limited water availability and rising costs of mineral fertilizer will impose limits on expanding green revolution practices. Furthermore, agricultural practices have degraded vast areas of formerly productive agricultural land, through loss of soil, salinization, pollution, and other forms of damage. Agriculture on degraded and marginal lands without the benefit of green revolution inputs is likely going to be essential to feed a growing population of 9 billion or more. Degraded and marginal agricultural habitats will require crops bred for stress resistance. To do this effectively, understanding the nature of the SRS is critical. Perhaps most essentially, we need to understand the nature of the trade-off between stress tolerance and yield in crops.

Wild relatives of crops may be an effective place to look at the trade-off between stress tolerance and agricultural yield. Wild relatives of many crops often occur on stressful soils, and frequently express aspects of the SRS. This is true of maize, rice, and wheat, the three most widely grown cereals. It is also true in chickpea, the crop wild relative with which we are most familiar. The wild relatives of chickpea can grow in rock crevices across either basaltic or calcareous boulder fields and mountainsides in a Mediterranean climate. When grown under agricultural conditions, the chickpea wild relatives show the reduced growth rates, delayed flowering, and low responsiveness to added N that one would expect from an SRS-exhibiting species (von Wettberg et al., unpublished). We believe that adaptations to stressful habitats characteristic of wild crop relatives could confer greater stress tolerance to novel crop varietals. However, we are unsure of the extent to which crops with large amounts of introgressed wild genetic background will retain the full range of SRS traits from the wild.

CONCLUSION

Shortcomings of Existing Model Organisms

Although our understanding of ecological genomics has rapidly grown over the past two decades, the extent of evolutionary similarities and underlying genetic mechanisms that regulate SRS trait expression remains unclear. While the power of molecular approaches is growing, we still lack detailed community-wide comparisons of evolutionary trajectories in communities of plants in stressful habitats. Furthermore, we still have a paucity of information on the underlying molecular basis of adaptations across multiple stressful habitats. One major concern is that our model organisms are not ideal for the study of the genetic basis of the SRS. The taxa that have been chosen as genetic model organisms are all fast growing, weedy species. As such, genomic model taxa are by definition not SRS-exhibiting species, and their responses to stress are often phenological ones that allow them to avoid stress rather than tolerate it. Most "emerging models," such as *Medicago*, *Vitus*, or *Mimulus* (see Chapter 11) are similar, although *Mimulus* perhaps to a lesser extent because of variation in lifespan among populations.

Ecological genomics relies on rapid growth rate as an essential trait in a model plant, but it leaves researchers studying the SRS ill equipped to understand the genetic basis of slow growing plants. This is a shortcoming that can never be fully corrected with ongoing acceleration in DNA sequencing technology. The intensive work needed to do genetics in long-lived plants inevitably requires an investment in long-term studies. However, several stress tolerant models are now emerging, such as other species of *Arabidopsis*, *Thellungiella* (Brassicaceae), and a handful of other plants. These new model taxa may provide the framework necessary to deeply explore SRS evolution and genomics.

Notably, some caution is warranted with these new models. Many potential "SRS models," such as *Thellungiella* or *Steptanthus*, are in groups in which all taxa are stress tolerant. For example, *Streptanthus* has radiated across serpentinic and extremely dry habitats in California and the desserts of the North American southwest (Ivalú-Cacho et al., 2013).

The extent of the stress tolerance of *Streptanthus* makes it a fascinating group, but complicates certain approaches to uncovering the genetic basis of SRS evolution and expression. For example one cannot set up crosses for quantitative trait loci (QTL) mapping or develop association mapping panels in groups where all individuals exhibit the SRS (*e.g.*, Stinchcombe & Hoekstra, 2008). An ideal complement to taxa such as these would be taxa in which an SRS model is sister to, or even interfertile with, a non-SRS taxon. There are taxa like this, such as the genus *Collinsia* that has taxa that tolerate serpentine soil and in many ways behave like classic "SRS-expressing" species, and other, sometimes interfertile, taxa that are restricted to mesic soils. Even *Arabidopsis,* with its rapid cycling model *A. thaliana*, and more SRS-like taxa such as *A. halleri* or *A. lyrata*, has potential as a model genus that has not been fully exploited. We look forward to seeing more work in non-model plants from many habitats and with a variety of life forms, and most importantly with a range of SRS traits.

Moving beyond model plants, the emergence of powerful tools in phylogenomics and population genomics (*e.g.*, Chapters 2, 4, 5, 11) provides novel opportunities to identify patterns of similarities and differences across communities of stress tolerant plants. Clarifying the phylogenetics and phylogenomics of entire communities of plants expressing stress resistance syndromes is an obvious step forward in increasing our understanding of the evolution and underlying genetic mechanisms of the SRS. The habitat preferences of SRS sister taxa are understood only in a few plant communities (*e.g.*, serpentine communities of California to a limited extent); whether there are predictable patterns of shifting from one stressful habitat to another (as would be suggested by shared underlying SRS loci) remains unclear. Population genomic studies of SRS taxa to understand the genetic basis of their stress resistance are now possible. These studies can be used for inferring the demographic history of stress resistant taxa and to understand the demographic processes involved in thriving in low productivity habitats.

ACKNOWLEDGMENT

We thank Jessica Wright for helpful comments on earlier drafts of this manuscript. Eric von Wettberg's work on agricultural ecology is supported by NSF-PGRP 133934, HHMI award #52006924 to FIU's discipline-based education research group, and USDA-NIFA-Hispanic Serving Institutions Grant 2011-38422-30804 to FIU's agroecology program.

REFERENCES

Abe, H., Urao, T., Ito, T., Seki, M., Shinozaki, K. & Yamaguchi-Shinozaki, K. (2003) *Arabidopsis* AtMYC2 (bHLH) and AtMYB2 (MYB) function as transcriptional activators in abscisic acid signaling. *The Plant Cell Online 15*, 63-78.

Ackerly, D. D. (2000) Taxon sampling, correlated evolution, and independent contrasts. *Evolution 54*, 1480-1492.

Ackerly, D. D. (2003) Community assembly, niche conservatism, and adaptive evolution in changing environments. *International Journal of Plant Sciences 164*(S3), S165-S184.

Adler, P. B., Salguero-Gómez, R., Compagnoni, A., Hsu, J., Ray-Mukherjee, J., Mbeau-Ache, C. & Franco, M. (2014) Functional traits explain variation in plant life history strategies. *Proceedings of the National Academy of Sciences (USA) 111,* 740-745.

Aguirrezabal, L., Bouchier-Combaud, S., Radziejwoski, A., Dauzat, M., Cookson, S. J. & Granier, C. (2006) Plasticity to soil water deficit in *Arabidopsis thaliana*: Dissection of leaf development into underlying growth dynamic and cellular variables reveals invisible phenotypes. *Plant, Cell and Environment 29*, 2216-2227.

Albacete, A., Martinez-Anduljar, C., Ghanem, M. E., Acosta, M., Sanchez-Bravo, J., Asins, M. J. & Perez-Alfoca, F. (2009) Rootstock-mediated changes in xylem ionic and hormonal status are correlated with delayed leaf senescence, and increased leaf area and crop productivity in salinized tomato. *Plant, Cell and Environment 32*, 928-938.

Alvim, F. C., Carolino, S. M., Cascardo, J. C., Nunes, C. C., Martinez, C. A., Otoni, W. C. & Fontes, E. P. (2001) Enhanced accumulation of BiP in transgenic plants confers tolerance to water stress. *Plant Physiology* 126, 1042-1054.

Anacker, B. L. & Harrison, S. P. (2012) Climate and the evolution of serpentine endemism in California. *Evolutionary Ecology 26*, 1011-1023.

Andrés, F. & Coupland, G. (2012) The genetic basis of flowering responses to seasonal cues. *Nature Reviews Genetics 13*, 627-639.

Andrés-Colás, N., Sancenón, V., Rodríguez-Navarro, S., Mayo, S., Thiele, D. J., Ecker, J. R. & Peñarrubia, L. (2006) The *Arabidopsis* heavy metal P-type ATPase HMA5 interacts with metallochaperones and functions in copper detoxification of roots. *The Plant Journal 45*, 225-236.

Aroca, R., Ferrante, A., Vernieri, P. & Chrispeels, M. J. (2006) Drought, abscisic acid and transpiration rate effects on the regulation of PIP aquaporin gene expression and abundance in *Phaseolus vulgaris* plants. *Annals of Botany 98*, 1301-1310.

Baldwin, B. G. & Sanderson, M. J. (1998) Age and rate of diversification of the Hawaiian silversword alliance (Compositae). *Proceedings of the National Academy of Sciences (USA) 95*, 9402-9406.

Ballerini, E. S. & Kramer, E. M. (2011) In the light of evolution: a reevaluation of conservation in the CO–FT regulon and its role in photoperiodic regulation of flowering time. *Frontiers in Plant Science 2,* DOI: 10.3389/fpls.2011.00081.

Bargmann, B. O. & Estelle, M. (2014) Auxin perception: In the IAA of the beholder. *Physiologia Plantarum 151*, 52-61.

Baskin, C. C. & Baskin, J. M. (2001) *Seeds: Ecology, biogeography, and evolution of dormancy and germination*. Amsterdam: Elsevier.

Baumann, K. (2014) Plant cell biology: Auxin signalling: ABP1 finds its pair. *Nature Reviews Molecular Cell Biology 15*, 221-221.

Baxter, I., Brazelton, J. N., Yu, D., Huang, Y. S., Lahner, B., Yakubova, E. & Salt, D. E. (2010) A coastal cline in sodium accumulation in *Arabidopsis thaliana* is driven by natural variation of the sodium transporter AtHKT1. *PLoS Genetics 6*, e1001193.

Bennett, L. E., Burkhead, J. L., Hale, K. L., Terry, N., Pilon, M. & Pilon-Smits, E. A. (2003) Analysis of transgenic Indian mustard plants for phytoremediation of metal-contaminated mine tailings. *Journal of Environmental Quality 32*, 432-440.

Bhatnagar-Mathur, P., Devi, M. J., Reddy, D. S., Lavanya, M., Vadez, V., Serraj, R. & Sharma, K. K. (2007) Stress-inducible expression of At DREB1A in transgenic peanut

(*Arachis hypogaea* L.) increases transpiration efficiency under water-limiting conditions. *Plant Cell Reports 26*, 2071-2082.

Bradshaw, H. D. (2005) Mutations in CAX1 produce phenotypes characteristic of plants tolerant to serpentine soils. *New Phytologist 167*, 81-88.

Brady, K. U., Kruckeberg, A. R. & Bradshaw, H. D. Jr. (2005) Evolutionary ecology of plant adaptation to serpentine soils. *Annual Review of Ecology, Evolution, and Systematics 36*, 243-266.

Bray, E. A. (2004) Genes commonly regulated by water-deficit stress in *Arabidopsis thaliana*. *Journal of Experimental Botany 55*, 2331-2341.

Brugière, N., Jiao, S., Hantke, S., Zinselmeier, C., Roessler, J. A., Niu, X. & Habben, J. E. (2003) Cytokinin oxidase gene expression in maize is localized to the vasculature, and is induced by cytokinins, abscisic acid, and abiotic stress. *Plant Physiology 132*, 1228-1240.

Bryant, J. P, Tahvanainen, J., Sulkinoja, M., Julkunen-Tiitto, R., Reichardt, P. & Green, T. (1989) *The American Naturalist 134*, 20-34.

Carroll, C. R., Vandermeer, J. H. & Rosset, P. M. (1990) Agroecology. New York, NY: McGraw-Hill.

Chaffei, C., Pageau, K., Suzuki, A., Gouia, H., Ghorbel, M. H. & Masclaux-Daubresse, C. (2004) Cadmium toxicity induced changes in nitrogen management in *Lycopersicon esculentum* leading to a metabolic safeguard through an amino acid storage strategy. *Plant and Cell Physiology 45*, 1681-1693.

Champoux, M. C., Wang, G., Sarkarung, S., Mackill, D. J., O'Toole, J. C., Huang, N. & McCouch, S. R. (1995) Locating genes associated with root morphology and drought avoidance in rice via linkage to molecular markers. *Theoretical and Applied Genetics 90*, 969-981.

Chan, C. X. & Ragan, M. A. (2013) Next-generation phylogenomics. *Biology Direct 8, 3.* DOI: 10.1186/1745-6150-8-3.

Chapin, F. S. III. (1980) Nutrient allocation and responses to defoliation in tundra plants. *Arctic and Alpine Research 12*, 553-563.

Chapin, F. S. III, Autumn, K. & Pugnaire, F. (1993) Evolution of suites of traits in response to environmental stress. *The American Naturalist 142*, S78-S92.

Chaves, M. M. & Oliveira, M. M. (2004) Mechanisms underlying plant resilience to water deficits: Prospects for water-saving agriculture. *Journal of Experimental Botany 55*, 2365-2384.

Chaves, M. M., Pereira, J. S., Maroco, J., Rodrigues, M. L., Ricardo, C. P. P., Osório, M. L. & Pinheiro, C. (2002) How plants cope with water stress in the field? Photosynthesis and growth. *Annals of Botany 89*, 907-916.

Chaves, M. M., Maroco, J. P. & Pereira, J. S. (2003) Understanding plant responses to drought—from genes to the whole plant. *Functional Plant Biology,* 30, 239-264.

Chinnusamy, V., Zhu, J. & Zhu, J. K. (2007) Cold stress regulation of gene expression in plants. *Trends in Plant Science 12*, 444-451.

Collin, R. & Miglietta, M. P. (2008) Reversing opinions on Dollo's Law. *Trends in Ecology and Evolution 23*, 602-609.

Courbot, M., Willems, G., Motte, P., Arvidsson, S., Roosens, N., Saumitou-Laprade, P. & Verbruggen, N. (2007) A major quantitative trait locus for cadmium tolerance in *Arabidopsis halleri* colocalizes with HMA4, a gene encoding a heavy metal ATPase. *Plant Physiology 144*, 1052-1065.

Cramer, G. R., Urano, K., Delrot, S., Pezzotti, M. & Shinozaki, K. (2011) Effects of abiotic stress on plants: A systems biology perspective. *BMC Plant Biology 11*, 163.

Cutler, S. R., Rodriguez, P. L., Finkelstein, R. R. & Abrams, S. R. (2010) Abscisic acid: Emergence of a core signaling network. *Annual Review of Plant Biology,* 61, 651-679.

Devaiah, B. N., Karthikeyan, A. S. & Raghothama, K. G. (2007) WRKY75 transcription factor is a modulator of phosphate acquisition and root development in *Arabidopsis*. *Plant Physiology 143*, 1789-1801.

Dmitryukova, M. Y., Baimiev, A. K., Fedyaev, V. V. & Rakhmankulova, Z. F. (2011) Effect of leghemoglobin A gene expression from soybean on tobacco plant growth and antioxidant state under damaging action of cadmium. *Russian Journal of Plant Physiology 58*, 1055-1061.

Doebley, J. F., Gaut, B. S. & Smith, B. D. (2006) The molecular genetics of crop domestication. *Cell 127*, 1309-1321.

Domínguez-Solís, J. R., Gutiérrez-Alcalá, G., Romero, L. C. & Gotor, C. (2001) The cytosolic O-acetylserine (thiol) lyase gene is regulated by heavy metals and can function in cadmium tolerance. *Journal of Biological Chemistry 276*, 9297-9302.

Efroni, I., Eshed, Y. & Lifschitz, E. (2010) Morphogenesis of simple and compound leaves: A critical review. *The Plant Cell Online 22*, 1019-1032.

Ehrenreich, I. M., Hanzawa, Y., Chou, L., Roe, J. L., Kover, P. X. & Purugganan, M. D. (2009) Candidate gene association mapping of *Arabidopsis* flowering time. *Genetics 183*, 325-335.

Eveland, A. L. & Jackson, D. P. (2012) Sugars, signalling, and plant development. *Journal of Experimental Botany 63*, 3367-3377.

Franklin, K. A. & Whitelam, G. C. (2004) Light signals, phytochromes and cross-talk with other environmental cues. *Journal of Experimental Botany 55*, 271-276.

Friesen, M. L., Cordeiro, M. A., Penmetsa, R. V., Badri, M., Huguet, T., Aouani, M. E., Cook, D. R. & Nuzhdin, S. V. (2010) Population genomic analysis of Tunisian *Medicago truncatula* reveals candidates for local adaptation. *The Plant Journal 63*, 623-635.

Fuentes, D., Meneses, M., Nunes-Nesi, A., Araújo, W. L., Tapia, R., Gómez, I. & Jordana, X. (2011) A deficiency in the flavoprotein of *Arabidopsis* mitochondrial complex II results in elevated photosynthesis and better growth in nitrogen-limiting conditions. *Plant Physiology 157*, 1114-1127.

Gazzarrini, S. & McCourt, P. (2001) Genetic interactions between ABA, ethylene and sugar signaling pathways. *Current Opinion in Plant Biology 4*, 387-391.

Ghanem, M. E., Albacete, A., Smigocki, A. C., Frébort, I., Pospíšilová, H., Martínez-Andújar, C. & Pérez-Alfocea, F. (2011) Root-synthesized cytokinins improve shoot growth and fruit yield in salinized tomato (*Solanum lycopersicum* L.) plants. *Journal of Experimental Botany 62*, 125-140.

Glenn, T. C. (2011) Field guide to next-generation DNA sequencers. *Molecular Ecology Resources 11,* 759-769.

Gottlieb, L. D. (1986) The genetic basis of plant form. *Philosophical Transactions of the Royal Society of London Series B, Biological Sciences 313*, 197-208.

Gonzalez-Mendoza, D., Moreno, A. Q. & Zapata-Perez, O. (2007) Coordinated responses of phytochelatin synthase and metallothionein genes in black mangrove, *Avicennia germinans*, exposed to cadmium and copper. *Aquatic Toxicology 83*, 306-314.

Grime, J. P. (1977) Evidence for the existence of three primary strategies in plants and its relevance to ecological and evolutionary theory. *American Naturalist* 111, 1169-1194.

Grime, J. P. (1988) The CSR model of primary plant strategies—origins, implications and tests. In: L. D. Gottlieb & S. Jain (Eds.). *Plant Evolutionary Biology* (pp. 371-393). London: Chapman and Hall.

Grime, J. P. & Hunt, R. (1975) Relative growth-rate: Its range and adaptive significance in a local flora. *The Journal of Ecology 63*, 393-422.

Gross, B. L. & Olsen, K. M. (2010) Genetic perspectives on crop domestication. *Trends in Plant Science 15*, 529-537.

Guffanti, A. (2002) Modeling molecular networks: A systems biology approach to gene function. *Genome Biology 3*, 4031-1.

Han, K. M., Dharmawardhana, P., Arias, R. S., Ma, C., Busov, V. & Strauss, S. H. (2011) Gibberellin-associated cisgenes modify growth, stature and wood properties in *Populus*. *Plant Biotechnology Journal 9*, 162-178.

Hanikenne, M., Talke, I. N., Haydon, M. J., Lanz, C., Nolte, A., Motte, P. & Krämer, U. (2008) Evolution of metal hyperaccumulation required cis-regulatory changes and triplication of HMA4. *Nature 453*, 391-395.

Harvey, P. H. & Pagel, M. D. (1991) The comparative method in evolutionary biology (Vol. 239). Oxford: Oxford University Press.

Herms, D. A. & Mattson, W. J. (1992) The dilemma of plants: To grow or defend. *Quarterly Review of Biology 67*, 283-335.

Hernández, J. A., Jimenez, A., Mullineaux, P. & Sevilia, F. (2000) Tolerance of pea (*Pisum sativum* L.) to long-term salt stress is associated with induction of antioxidant defences. *Plant, Cell and Environment 23*, 853-862.

Holt, R. D. & Gaines, M. S. (1992) Analysis of adaptation in heterogeneous landscapes: Implications for the evolution of fundamental niches. *Evolutionary Ecology,* 6 433–447.

Imber, D. & Tal, M. (1970) Phenotypic reversion of flacca, a wilty mutant of tomato, by abscisic acid. *Science 169*, 592-593.

Iuchi, S., Kobayashi, M., Taji, T., Naramoto, M., Seki, M., Kato, T. & Shinozaki, K. (2001) Regulation of drought tolerance by gene manipulation of 9-cis-epoxycarotenoid dioxygenase, a key enzyme in abscisic acid biosynthesis in *Arabidopsis*. *The Plant Journal 27*, 325-333.

Ivalú-Cacho, N., Burrell, M., Pepper, A. E. & Strauss, S. Y. (2013) Novel nuclear markers inform the systematics and the evolution of serpentine use in *Streptanthus* and allies (Thelypodieae, Brassicaceae). *Molecular Phylogenetics and Evolution 72*, 71-81.

Kang, J. Y., Choi, H. I., Im, M. Y. & Kim, S. Y. (2002) *Arabidopsis* basic leucine zipper proteins that mediate stress-responsive abscisic acid signaling. *The Plant Cell Online 14*, 343-357.

Kapralov, M. V., Votintseva, A. A. & Filatov, D. A. (2013) Molecular adaptation during a rapid adaptive radiation. *Molecular Biology and Evolution 30*, 1051–1059.

Karaba, A., Dixit, S., Greco, R., Aharoni, A., Trijatmiko, K. R., Marsch-Martinez, N. & Pereira, A. (2007) Improvement of water use efficiency in rice by expression of HARDY, an *Arabidopsis* drought and salt tolerance gene. *Proceedings of the National Academy of Sciences (USA) 104*, 15270-15275.

Kawasaki, S., Borchert, C., Deyholos, M., Wang, H., Brazille, S., Kawai, K. & Bohnert, H. J. (2001) Gene expression profiles during the initial phase of salt stress in rice. *The Plant Cell Online 13*, 889-905.

Kellermeier, F., Armengaud, P., Seditas, T. J., Danku, J., Salt, D. E. & Amtmann, A. (2014) Analysis of the root system architecture of *Arabidopsis* provides a quantitative readout of crosstalk between nutritional signals. *The Plant Cell Online 26*, 1480-1496.

Kiba, T., Kudo, T., Kojima, M. & Sakakibara, H. (2011) Hormonal control of nitrogen acquisition: Roles of auxin, abscisic acid, and cytokinin. *Journal of Experimental Botany 62*, 1399-1409.

Ladzinsky, G. & Adler, A. (1976) The origin of chickpea, *Cicer arietinum* L. *Euphytica 25*, 211-217.

Lambers, H. & Poorter, H. (1992) Inherent variation in growth rate between higher plants: A search for physiological causes and ecological consequences. *Advances in Ecological Research 23*, 187–261.

Leshem, Y. Y. & Kuiper, P. J. C. (1996) Is there a GAS (general adaptation syndrome) response to various types of environmental stress? *Biologia Plantarum 38*, 1-18.

Liu, D., Wu, L., Naeem, M. S., Liu, H., Deng, X., Xu, L. & Zhou, W. (2013) 5-Aminolevulinic acid enhances photosynthetic gas exchange, chlorophyll fluorescence and antioxidant system in oilseed rape under drought stress. *Acta Physiologiae Plantarum 35*, 2747-2759.

Lowry, D. B. & Willis, J. H. (2010) A widespread chromosomal inversion polymorphism contributes to a major life-history transition, local adaptation, and reproductive isolation. *PLoS Biology 8*, e1000500. DOI: 10.1371/journal.pbio.1000500.

Maschinski, J., Ross, M. S., Liu, H., O'Brien, J., von Wettberg, E. J. & Haskins, K. E. (2011) Sinking ships: Conservation options for endemic taxa threatened by sea level rise. *Climatic Change 107*, 147-167.

Medici, A. & Krouk, G. (2014) The Primary Nitrate Response: A multifaceted signalling pathway. *Journal of Experimental Botany* DOI: 10.1093/jxb/eru245.

Mcnair, M. (1992) Preliminary studies on the genetics and evolution of the serpentine endemic *Mimulus nudatus* Curran. In: A. J. M. Baker, J. Proctor & R. D. Reeves (Eds.). The vegetation of ultramafic (serpentine) soils (pp. 409-419). Andover: Intercept Limited.

Miransari, M. (2012) Role of phytohormone signaling during stress. In: P. Ahmad & M. N. V. Prasad (Eds.). Environmental adaptations and stress tolerance of plants in the era of climate change, (pp. 381-393). New York, NY: Springer.

Munns, R. & Tester, M. (2008) Mechanisms of salinity tolerance. *Annual Review of Plant Biology 59*, 651-681.

Nakano, T., Suzuki, K., Fujimura, T. & Shinshi, H. (2006) Genome-wide analysis of the ERF gene family in *Arabidopsis* and rice. *Plant Physiology 140*, 411-432.

Obata, T. & Fernie, A. R. (2012) The use of metabolomics to dissect plant responses to abiotic stresses. *Cellular and Molecular Life Sciences 69*, 3225-3243.

Odum, H. T. & Pinkerton, R. C. (1955) Time's speed regulator: The optimum efficiency for maximum power output in physical and biological systems. *American Scientist 43*, 331-343.

Orr, H. A. (1998) The population genetics of adaptation: The distribution of factors fixed during adaptive evolution. *Evolution 52*, 935–949.

Pageau, K., Reisdorf-Cren, M., Morot-Gaudry, J. F. & Masclaux-Daubresse, C. (2006) The two senescence-related markers, GS1 (cytosolic glutamine synthetase) and GDH (glutamate dehydrogenase), involved in nitrogen mobilization, are differentially regulated during pathogen attack and by stress hormones and reactive oxygen species in *Nicotiana tabacum* L. leaves. *Journal of Experimental Botany 57*, 547-557.

Pearman, P. B., Guisan, A., Broennimann, O. & Rand, C. F. (2008) Niche dynamics in space and time. *Trends in Ecology and Evolution 23*, 149–158.

Peleg, Z., Reguera, M., Tumimbang, E., Walia, H. & Blumwald, E. (2011) Cytokinin-mediated source/sink modifications improve drought tolerance and increase grain yield in rice under water-stress. *Plant Biotechnology Journal 9*, 747-758.

Peterson, A. T., Soberón, J. & Sánchez-Cordero, V. (1999) Conservatism of ecological niches in evolutionary time. *Science 285*, 1265–1267.

Prinzing, A., Durka, W. and Klotz, S. (2001) The niche of higher plants: Evidence for phylogenetic conservatism. *Proceedings of the Royal Society London Series B 268*, 2383–2389.

Quarrie, S. A. (1982) Droopy: A wilty mutant of potato deficient in abscisic acid. *Plant, Cell and Environment 5*, 23-26.

Reich, P. B., Wright, I. J., Cavender-Bares, J., Craine, J. M., Oleksyn, J., Westoby, M. & Walters, M. B. (2003) The evolution of plant functional variation: Traits, spectra, and strategies. *International Journal of Plant Sciences 164*(S3), S143-S164.

Reinhardt, J. A., Kolaczkowski, B., Jones, C. D., Begun, D. J. & Kern, A. D. (2014) Parallel geographic variation in *Drosophila melanogaster*. *Genetics 197*, 361-373.

Riefler, M., Novak, O., Strnad, M. & Schmülling, T. (2006) *Arabidopsis* cytokinin receptor mutants reveal functions in shoot growth, leaf senescence, seed size, germination, root development, and cytokinin metabolism. *The Plant Cell Online 18*, 40-54.

Rivero, R. M., Kojima, M., Gepstein, A., Sakakibara, H., Mittler, R., Gepstein, S. & Blumwald, E. (2007) Delayed leaf senescence induces extreme drought tolerance in a flowering plant. *Proceedings of the National Academy of Sciences (USA) 104*, 19631-19636.

Rivero, R. M., Gimeno, J., Van Deynze, A., Walia, H. & Blumwald, E. (2010) Enhanced cytokinin synthesis in tobacco plants expressing PSARK: IPT prevents the degradation of photosynthetic protein complexes during drought. *Plant and Cell Physiology 51*, 1929-1941.

Rus, A., Baxter, I., Muthukumar, B., Gustin, J., Lahner, B., Yakubova, E. & Salt, D. E. (2006) Natural variants of AtHKT1 enhance Na+ accumulation in two wild populations of *Arabidopsis*. *PLoS Genetics 2*(12), e210.

Sablowski, R. (2013) Roots of beauty. *Nature Reviews Molecular Cell Biology 14*, 268-268.

Salvucci, M. E. & Crafts-Brandner, S. J. (2004) Relationship between the heat tolerance of photosynthesis and the thermal stability of Rubisco activase in plants from contrasting thermal environments. *Plant Physiology 134*, 1460-1470.

Sasaki-Sekimoto, Y., Taki, N., Obayashi, T., Aono, M., Matsumoto, F., Sakurai, N. & Ohta, H. (2005) Coordinated activation of metabolic pathways for antioxidants and defence compounds by jasmonates and their roles in stress tolerance in *Arabidopsis*. *The Plant Journal 44*, 653-668.

Shinozaki, K. and Yamaguchi-Shinozaki, K. (2007) Gene networks involved in drought stress response and tolerance. *Journal of Experimental Botany 58*, 221-227.

Stanton, M. L., Roy, B. A. & Thiede, D. A. (2000) Evolution in stressful environments. I. Phenotypic variability, phenotypic selection, and response to selection in five distinct environmental stresses. *Evolution 54*, 93-111.

Stinchcombe, J. R. & Hoekstra, H. E. (2008) Combining population genomics and quantitative genetics: finding the genes underlying ecologically important traits. *Heredity 100*, 158-170.

Strauss, S. Y. & Agrawal, A. A. (1999) The ecology and evolution of plant tolerance to herbivory. *Trends in Ecology and Evolution 14*, 179-185.

Taji, T., Ohsumi, C., Iuchi, S., Seki, M., Kasuga, M., Kobayashi, M. & Shinozaki, K. (2002) Important roles of drought-and cold-inducible genes for galactinol synthase in stress tolerance in *Arabidopsis thaliana*. *The Plant Journal 29*, 417-426.

Tsukaya, H. (2014) Comparative leaf development in angiosperms. *Current Opinion in Plant Biology 17*, 103-109.

Turner, T. L., Bourne, E. C., von Wettberg, E. J., Hu, T. T. & Nuzhdin, S. V. (2010) Population resequencing reveals local adaptation of *Arabidopsis lyrata* to serpentine soils. *Nature Genetics 42*, 260-263.

Valente, M. A. S., Faria, J. A., Soares-Ramos, J. R., Reis, P. A., Pinheiro, G. L., Piovesan, N. D. & Fontes, E. P. (2009) The ER luminal binding protein (BiP) mediates an increase in drought tolerance in soybean and delays drought-induced leaf senescence in soybean and tobacco. *Journal of Experimental Botany 60*, 533-546.

Varkonyi-Gasic, E., Moss, S., Voogd, C., Wang, T., Putterill, J. & Hellens, R. P. (2013) Homologs of FT, CEN and FD respond to developmental and environmental signals affecting growth and flowering in the perennial vine kiwifruit. *New Phytologist 198*, 732-746.

Wang, H. S., Zhu, Z. J., Feng, Z., Zhang, S. G. & Yu, C. (2012) Antisense-mediated depletion of GMPase gene expression in tobacco decreases plant tolerance to temperature stresses and alters plant development. *Molecular Biology Reports 39*, 10413-10420.

Westoby, M. (1998) A leaf-height-seed (LHS) plant ecology strategy scheme. *Plant and Soil 199*, 213-227.

Wiens, J. J. & Graham, C. H. (2005) Niche conservatism: Integrating evolution, ecology, and conservation biology. *Annual Review of Ecology, Evolution and Systematics 36*, 519–539.

Wiens, J. J., Ackerly, D. D., Allen, A. P., Anacker, B. L., Buckley, L. B., Cornell, H. V. & Stephens, P. R. (2010) Niche conservatism as an emerging principle in ecology and conservation biology. *Ecology Letters 13*, 1310-1324.

Willems, G., Dräger, D. B., Courbot, M., Godé, C., Verbruggen, N. & Saumitou-Laprade, P. (2007) The genetic basis of zinc tolerance in the metallophyte *Arabidopsis halleri* ssp. *halleri* (Brassicaceae): An analysis of quantitative trait loci. *Genetics 176*, 659-674.

Woo, H. R., Kim, J. H., Nam, H. G. & Lim, P. O. (2004) The delayed leaf senescence mutants of *Arabidopsis*, ore1, ore3, and ore9 are tolerant to oxidative stress. *Plant and Cell Physiology 45*, 923-932.

Wright, I. J., Reich, P. B., Cornelissen, J. H., Falster, D. S., Groom, P. K., Hikosaka, K. & Westoby, M. (2005) Modulation of leaf economic traits and trait relationships by climate. *Global Ecology and Biogeography 14*, 411-421.

Wright, I. J., Reich, P. B., Westoby, M., Ackerly, D. D., Baruch, Z., Bongers, F. & Villar, R. (2004) The worldwide leaf economics spectrum. *Nature 428*, 821-827.

Wright, J. W. & von Wettberg, E. V. (2009) "Serpentinomics"-An emerging new field of study. *Northeastern Naturalist 16*, 285-296.

Xiong, L. & Yang, Y. (2003) Disease resistance and abiotic stress tolerance in rice are inversely modulated by an abscisic acid–inducible mitogen-activated protein kinase. *The Plant Cell Online 15*, 745-759.

Xiong, L., Schumaker, K. S. & Zhu, J. K. (2002) Cell signaling during cold, drought, and salt stress. *The Plant Cell Online 14* (suppl. 1), S165-S183.

Zhang, K., Wang, J., Lian, L., Fan, W., Guo, N. & Lv, S. (2012) Increased chilling tolerance following transfer of a betA gene enhancing glycinebetaine synthesis in cotton (*Gossypium hirsutum* L.). *Plant Molecular Biology Reporter 30*, 1158-1171.

Zhu, J. K. (2002) Salt and drought stress signal transduction in plants. *Annual Review of Plant Biology 53*, 247-273.

Züst, T., Joseph, B., Shimizu, K. K., Kliebenstein, D. J. & Turnbull, L. A. (2011) Using knockout mutants to reveal the growth costs of defensive traits. *Proceedings of the Royal Society Series B 278*, 2598-2603.

In: Plant Ecology and Evolution in Harsh Environments ISBN: 978-1-63321-955-7
Editors: N. Rajakaruna, R. S. Boyd and T. B. Harris

Chapter 10

ECOLOGY AND EVOLUTION OF METAL-HYPERACCUMULATING PLANTS

Robert S. Boyd [•]
Department of Biological Sciences,
Auburn University, Auburn, AL, USA

ABSTRACT

Hyperaccumulator plants are often found in harsh environments, particularly on serpentine soils. At least 10 elements (metals and metalloids) have been described as being hyperaccumulated by plants. Hyperaccumulation has been hypothesized to be an adaptation to stress and several stresses have been suggested as driving evolution of this trait. Defense against herbivores and pathogens (enemies) is a leading hypothesis, as there is considerable evidence connecting high tissue metal levels with plant resistance to enemies. Tolerance of elevated soil metal levels has also been suggested as an evolutionary driver of hyperaccumulation. Sequestration of metals in particular tissues and cellular locations has been documented, but whether this is an adaptation that increases plant metal tolerance is unclear. Drought tolerance, another potential function of hyperaccumulation, has rarely been examined and experiments to date have provided mixed (but mainly negative) results. Other ecological phenomena (such as elemental allelopathy) also may be adaptive explanations for hyperaccumulation but, except for Se hyperaccumulation, little experimental evidence is available. Evolution of metal hyperaccumulation (particularly of Ni) has occurred multiple times in plants, suggesting both its adaptive value and that the specific adaptive value may vary depending on the metal, plant lineage, and specific habitat features involved in any particular case. Hyperaccumulator plants provide many opportunities for further research into plant adaptation to harsh habitats.

• E-mail: boydrob@auburn.edu.

INTRODUCTION

Plants contain many elements. Some, such as C, H, O, and N, are found in abundance whereas many others are present in low or extremely low amounts (Kirkby, 2012). Investigations of plant composition have revealed wide variation among species in the concentrations of many elements. In particular, some species contain two or three orders of magnitude more of some elements when compared to other species growing in the same habitat. The term hyperaccumulator was coined by Jaffré et al. (1976) to describe extraordinary concentrations of an element, and hyperaccumulation has since been developed as an important phenomenon in plant ecology, plant physiology, and plant evolutionary biology (Rascio & Navari-Izzo, 2011). Exact criteria for defining hyperaccumulation are difficult to establish (van der Ent et al., 2013) but the hyperaccumulation concept has been applied to a number of elements, including both metals and metalloids. The recent review by van der Ent et al. (2013) states that more than 500 hyperaccumulator plants have been described in the literature and that hyperaccumulated elements include As, Cd, Co, Cu, Mn, Ni, Pb, Se, Tl, and Zn.

Although many elements are hyperaccumulated by at least some plants, the vast majority of hyperaccumulator species accumulate Ni. According to van der Ent et al. (2013), approximately 78% of hyperaccumulators are Ni hyperaccumulators. Most if not all of these Ni hyperaccumulators are found on serpentine soils. Hyperaccumulators of other elements may also be associated with chemically unusual soils. As examples, most Cu and Co hyperaccumulators are associated with soils enriched in those elements in southern Africa (van der Ent et al., 2013) and Se hyperaccumulators are found on high Se soils in the western U.S. and Australia (Reeves & Baker, 2000). This chapter will emphasize Ni hyperaccumulators, but also will refer to hyperaccumulators of other elements to illustrate particular facets of the ecology and evolution of hyperaccumulation.

As mentioned above, the vast majority of hyperaccumulator species are Ni hyperaccumulators that are found in serpentine habitats. Serpentine soils are renowned for a suite of harsh features comprising the "serpentine syndrome" (Chapter 6; Brady et al., 2005; Brooks, 1987; Kazakou et al., 2008; Proctor & Woodell, 1975), including low Ca:Mg ratio, low levels of important plant nutrients, high drought stress, and elevated concentrations of some heavy metals (*e.g.*, Ni, Cr), although it is important to realize that there is considerable variation in these features among serpentine habitats. The vegetation of serpentine habitats is often more sparse than on adjacent soils (Brooks, 1987), leading to greater exposure to potential stressors such as solar radiation and wind, and potentially greater exposure to herbivores. This chapter explores the ecology and evolution of hyperaccumulation as a possible adaptation to these harsh habitat features.

ECOLOGY AND EVOLUTION OF METAL-HYPERACCUMULATING PLANTS

An early review of the literature regarding metal hyperaccumulation (Boyd & Martens, 1992) gathered adaptive explanations for this trait into four general categories. These were: 1) defense against herbivores and pathogens (defense hypothesis), 2) drought tolerance, 3) metal tolerance/disposal, and 4) elemental allelopathy. In addition, the inadvertent uptake hypothesis suggested that hyperaccumulation might be an incidental consequence of enhanced nutrient uptake abilities. In this last case, hyperaccumulation of metals might not be directly adaptive but instead be a consequence a nutrient-scavenging adaptation. Each of the adaptive explanations, and the research regarding them, is explored in more detail below.

Defense Hypothesis

The defense hypothesis has been investigated relatively intensively compared with the other adaptive explanations (Boyd, 2012). Both herbivores and pathogens can be important selective features for plants and in this chapter I refer to them simply as "enemies" when considering them in general. It also is important to consider the meaning of "harsh." In the context of herbivory, habitats of metal hyperaccumulator plants may be considered harsh if enemy attack is more likely there or if the extent of damage inflicted there is greater. In addition, if the fitness impact of enemy attack is greater in those habitats, then they can be considered harsh.

Some authors suggest that herbivore pressure is relatively low on serpentine habitats. For example, Hobbs & Mooney (1991) reported that exclusion of aboveground herbivores did not influence composition of a serpentine grassland. But direct experimental comparisons of herbivore pressure between serpentine and other habitats are relatively few. Not surprisingly for a broad question in which specific plant species are selected for scrutiny, results of direct comparisons that have been conducted are mixed. For example, Strauss & Ivalú Cacho (2013) documented greater herbivore attack in serpentine habitats (compared to adjacent non-serpentine grassland) and showed this was in part due to greater plant apparency on the serpentine outcrops. On the other hand, Lau et al. (2008) documented lower levels of herbivory on *Collinsia sparsifolia* (Plantaginaceae) plants placed into serpentine habitats. Similarly, Meindl et al. (2013) found less florivore damage to *Mimulus guttatus* (Phrymaceae; also see Chapter 11) flowers in serpentine habitats. It is likely that the specific ecological setting will determine whether herbivore pressure is more or less in a serpentine habitat.

The above studies of herbivore pressure used non-hyperaccumulator species. To my knowledge, there has been no experimental test of herbivore pressure on a hyperaccumulator species planted in both serpentine and adjacent non-serpentine habitats. Furthermore, to determine the importance of hyperaccumulation to plant defense such an experiment should control for the effects of differing soil metal concentrations in serpentine versus non-serpentine soils. To my knowledge this has not been attempted. However, there is one recent study in which a hyperaccumulator was planted across a gradient in soil composition: Che-Castaldo and Inouye (2014) planted the European Zn/Cd hyperaccumulator *Noccaea caerulescens* (Brassicaceae) along transects from metal-polluted to non-polluted soils. The study focused on metal mine tailings, and seeds and seedlings were planted on both mine tailing soil and non-polluted (low metal) adjacent soil. The authors found slow plant growth in all locations and did not report herbivore damage as an important ecological factor, despite the fact that metal concentrations of plants varied significantly between tailing soil (where some plants reached hyperaccumulator concentrations of Zn and Cd) and adjacent soil. This species is not native to the study area (Colorado, USA), however, so that the ecological relevance of this experiment to natural populations of hyperaccumulators is unclear.

Whether or not serpentine communities are more prone to herbivory than other habitat types, damage by herbivores and pathogens is an important ecological feature that can drive plant adaptation (Karban & Agrawal, 2002). The defense hypothesis suggests that hyperaccumulation is a resistance trait that can decrease enemy damage. For herbivores, this can be accomplished by deterrence (reducing or preventing feeding) or by toxic effects on an herbivore after ingesting plant material. There is considerable evidence that hyperaccumulated elements can defend plants against both herbivores (reviewed by Boyd, 2007) and pathogens (*e.g.*, Fones et al., 2010; reviewed by Hörger et al., 2013). For herbivores, deterrence, toxicity, or both have been shown to occur in some cases (Boyd, 2007). Most defense studies have been laboratory examinations but there have been some field studies (*e.g.*, Martens & Boyd, 2002; Noret et al., 2007): more of these are needed to place defense into the full ecological context of those settings.

Perhaps the best exploration of hyperaccumulator ecology to date is for Se hyperaccumulator plants in Colorado, USA. As summarized by El Mehdawi & Pilon-Smits (2012), an extensive series of field and laboratory studies has been used to explore defense as well as other ecological phenomena there (see sections below). But a possibly unique feature of Se hyperaccumulation is that Se can be volatilized by plants (El Mehdawi & Pilon-Smits, 2012): both by hyperaccumulators (as dimethyldiselenide) and non-hyperaccumulators (as dimethylselenide). This air-borne pathway for release of Se may provide an opportunity for phenomena such as detection of Se hyperaccumulator plants by herbivores/pathogens and pollinators, signaling between hyperaccumulator plants and plant neighbors, transfer of Se (perhaps onto the soil surface), etc. Comparative study of heavy metal hyperaccumulator systems is needed to determine if the air-borne pathway of Se enables ecological interactions that are not possible for non-volatile heavy metals.

Studies of herbivory often measure levels of damage to plants, as these are relatively easily measured and are likely to correlate with fitness impact. But these measures may not encompass the entire ecological impact of herbivory because that impact may differ in differing habitats. For example, because environmental stress (such as low levels of nutrients or high levels of drought stress) may be greater in serpentine habitats (Strauss & Boyd, 2011), the ability of serpentine plants to tolerate herbivore damage may be less because stress may make replacement of damaged tissues more expensive physiologically. Thus the ecological impact of similar amounts of herbivore damage may be greater for serpentine soil plants. Most research regarding the defense hypothesis has focused on hyperaccumulation as a resistance trait (where presence of hyperaccumulation can reduce levels of damage). To my knowledge, the connection of hyperaccumulation and tolerance of herbivory has only been explored once. Palomino et al. (2007) compared growth of the Ni hyperaccumulator *Noccaea fendleri* subsp. *glauca* (Brassicaceae) when grown in high and low metal soils and artificially damaged to differing extents (by cutting leaves with scissors). They reported a significant interaction between the treatments, with hyperaccumulating plants being more tolerant than non-hyperaccumulating plants when exposed to the greatest level of damage, illustrating that hyperaccumulation correlated with increased herbivory tolerance. Thus hyperaccumulation may contribute to plant defense through both resistance and tolerance effects, but much more research into its contribution to herbivory tolerance is needed.

Considerable evidence shows hyperaccumulation can be a defensive trait. But organisms often adapt to surmount the adaptations of others, and this also has been shown for hyperaccumulation. Surveys of insect faunas associated with Ni hyperaccumulators (Boyd et al. 2006a; b; Mesjasz-Przybylowicz & Przybylowicz, 2001; Wall & Boyd, 2002) have discovered a number of insects that contain high whole-body concentrations of Ni. These "high-Ni insects" (Boyd, 2009) are mainly herbivores, but one carnivore has been noted from South Africa. It is suspected that these herbivores are specialist feeders, as has been demonstrated in the few cases that have been investigated in detail. For example, the beetle *Chrysolina pardalina* (Coleoptera: Chrysomelidae) feeds exclusively on leaves of the South African Ni hyperaccumulator *Berkheya coddii* (Asteraceae; Augustyniak et al., 2002). In an additional example, *Melanotrichus boydi* (Hemiptera: Miridae) has been shown to be a specialist on the Ni hyperaccumulator *Streptanthus polygaloides* (Brassicaceae; Wall & Boyd, 2006). It should be noted that studies of insects associated with Se hyperaccumulators also have discovered Se tolerant species or races (Freeman et al., 2006; 2012; Galeas et al., 2008). To my knowledge, resistant pathogens have yet to be discovered from hyperaccumulators but these also would be expected to have evolved. Recent discoveries of metal tolerant endophytes inhabiting tissues of hyperaccumulators (*e.g.,* Barzanti et al., 2007) support this expectation.

Drought Tolerance

Serpentine soils are often considered more likely to generate drought stress in plants (Chapter 6; Brady et al., 2005; Kruckeberg, 2002; Proctor & Woodell, 1975). This is generally attributed to their more open vegetation structure, leading to greater insolation, and because serpentine soils may be more shallow and/or rocky. As evidence of the importance of drought in this habitat type, some authors have pointed to the greater frequency of xeromorphism (considering traits such as leaf size, sclerophylly, root:shoot ratio, and plant stature) displayed by serpentine soil plant species (*e.g.*, Borhidi, 1996; Brooks, 1987; Kruckeberg, 1984). It is important to note, however, that not all authors agree that elevated drought stress is a general feature of these soils (*e.g.*, Alexander et al., 2007). Furthermore, many hyperaccumulator plants grow in tropical habitats with relatively high levels of precipitation (Reeves, 2003) and thus enhanced drought stress is not a universal feature of serpentine localities.

Relatively few studies have experimentally examined drought tolerance by hyperaccumulator species. Whiting et al. (2003) used the Zn hyperaccumulator *Noccaea caerulescens* and the Ni hyperaccumulator *Alyssum murale* (along with non-accumulators *Alyssum montanum* and *Lepidium heterophyllum* as controls; all Brassicaceae), growing plants in metal-amended or unamended solutions and then stressing plants with polyethylene glycol solutions. They found no effect of plant metal status on growth, or changes in metal accumulation, by hyperaccumulator plants that were stressed or unstressed. They also found little change in rate of evapotranspiration, plant relative water content, or leaf sap osmolality. Bhatia et al. (2005) grew field-collected plants of the Ni hyperaccumulator *Stackhousia tryonii* (Celastraceae) in their native soils under different levels of drought stress. They documented increased Ni concentrations in drought stressed plants, and calculated that the increased Ni could account for a large proportion of the osmoregulation that occurred in the drought stressed plants. They also concluded, however, that a drought protective effect of Ni was likely not the major function of hyperaccumulated Ni because its use as an osmoticum was not very flexible (it was unable to change rapidly as plant water status fluctuated).

Finally, Kachenko et al. (2011) conducted an experiment with the Ni hyperaccumulator *Hybanthus floribundus* subsp. *floribundus* (Violaceae) that exposed plants to drought stress and monitored their response. They reported that plants were relatively drought tolerant but that Ni concentration did not change significantly with water stress, implying that Ni did not play a role in osmotic adjustment in this species.

The studies discussed above focus on the effect of metal supply on plant drought resistance and thus attempt to directly address the drought tolerance hypothesis. Surprisingly, few studies document morphological responses of hyperaccumulators to manipulation of soil metal concentrations. Bhatia et al. (2005) reported hyperaccumulator (*Stackhousia tryonii*) plants under the greatest drought stress were stunted and produced relatively small leaves but they did not measure stomatal density/distribution, stomatal behavior, cuticle features, etc. If plant metal concentration influences those features, then plant drought tolerance in turn may be affected.

Metal Tolerance/Disposal

Serpentine soils often have relatively high levels of Mg, Ni, or Cr (Brady et al., 2005). Because Ni is needed by plants in only tiny amounts (Polacco et al., 2013), and Cr is not an essential element, plants that hyperaccumulate one of these metals must be able to tolerate high tissue concentrations of that metal. This feature has led to suggestions that metal hyperaccumulation is an adaptation that allows a metal hyperaccumulator to tolerate a relatively high metal habitat. Increased ability to tolerate high levels of a metal could occur either through sequestration of that metal into locations in the plant body that would separate it from sensitive physiological processes (tolerance: Boyd & Martens, 1998), or by putting the metal into plant parts that then could be shed as a mechanism of metal disposal (disposal: Boyd & Martens, 1998). To my knowledge the disposal concept has not received much attention and, given that decomposition of shed plant parts may only serve to return a metal to soil and thus not remove it for long, does not seem to be an attractive hypothesis.

Considerable research has been devoted to understanding the mechanisms involved in hyperaccumulation (see reviews by Jaffré et al., 2013; Krämer 2010; Maestri et al., 2010; Verbruggen et al., 2009). Studies of the inheritance of hyperaccumulation and metal tolerance suggest that the two traits are inherited independently from one another (Pollard et al., 2002). This in turn suggests that metal tolerance is a necessary precondition for the evolution of hyperaccumulation ability, but tolerance may or may not be its adaptive function. Hyperaccumulator species comprise only a small percentage of the flora of serpentine habitats and so other species tolerate the habitat without relying on hyperaccumulation (van der Ent et al., 2013). Because all hyperaccumulators must be metal tolerant, the difficulty is determining if the mechanisms resulting in hyperaccumulation increase metal tolerance of hyperaccumulator plants. Metal hyperaccumulation might contribute to metal tolerance if metal is sequestered into a location (within a cell or within a tissue) where it will be isolated from sensitive physiological processes (Boyd & Martens, 1998).

Studies of cellular or tissue locations of metals in hyperaccumulators provide evidence that metals are deposited in such isolated locations. For example, at the tissue level metals are often at highest concentrations in epidermal tissues while, at the cellular level, cell walls or vacuoles have been reported as locations containing high metal concentrations in hyperaccumulator plants (Küpper & Leitenmaier, 2013; Rascio & Navari-Izzo, 2011). While these locations may isolate metals from important cellular compartments and tissues, it is difficult to design an experiment to test for a direct connection between metal storage location and its contribution to plant metal tolerance. Perhaps the best approach may be to create a mutant strain of a hyperaccumulator in which key genes leading to typical metal storage are disabled. The response of mutated and unmutated plants to high and low soil metal conditions would compare plants that do not properly sequester metal to plants that sequester that metal in the usual fashion. Non-hyperaccumulator strains of hyperaccumulator species have been created and are being used in experimental work (*e.g.*, Kazemi-Dinan et al. (2014), using the Zn hyperaccumulator *Arabidopsis halleri*; Brassicaceae) so this technical tool is becoming available. Strauss et al. (2002) point out, however, that creation of mutant strains brings up important experimental design questions regarding how to control for other effects of genetic manipulation besides variation of the trait in question. Keeping those issues in mind, future experiments with mutant lines of hyperaccumulators may be valuable in teasing apart the adaptive function(s) of hyperaccumulation, both in tests of the tolerance hypothesis and in tests of the other hypotheses.

Elemental Allelopathy/Facilitation

The suggestion that metals are disposed from the body of a hyperaccumulator plant implies that they are consequently liberated from fallen plant litter through decomposition. This movement of a metal from within the soil to the surface may enrich surface soil in that metal, which in turn may have ecological consequences for co-occurring species. Originally this concentration of a metal was suggested to affect the bacterial flora of serpentine soils, leading to evolution of Ni tolerant bacterial strains (Schlegel et al., 1991). Negative effects on associated plant species which are less metal tolerant could result in elemental allelopathy, a term coined by Boyd and Martens (1998). On the other hand, if the elevated soil concentration has a positive effect on a co-occurring species, the interaction would be considered elemental facilitation.

A review of elemental allelopathy studies (Morris et al., 2009) found little direct experimental evidence to support that phenomenon, but elemental allelopathy has not been well studied for metal hyperaccumulator plants. In fact, at the time of the review by Morris et al. (2009), only Zhang et al. (2007) had conducted a manipulative experiment (and it was a pot test that took place in growth chambers). Zhang et al. (2007) concluded that elemental allelopathy by Ni was unlikely because Ni in hyperaccumulator tissues was bound onto soil particles after release by decomposition and thus did not affect germination of other plant species. However, more recent work with Se hyperaccumulators has provided evidence for elemental allelopathy in that experimental system. El Mehdawi et al. (2011a) showed that the Se hyperaccumulator species *Astragalus bisulcatus* (Fabaceae) and *Stanleya pinnata* (Brassicaceae) enriched surface soils with Se. Those Se-enriched soils significantly decreased germination and growth of a Se-sensitive plant species (*Arabidopsis thaliana*), demonstrating the potential for elemental allelopathy in field situations.

In contrast to elemental allelopathy, elemental facilitation occurs when elevated soil element levels produce positive ecological effects on plant species associated with hyperaccumulators. I know of no examples involving hyperaccumulated metals, but research with Se hyperaccumulators has provided a fascinating case study. El Mehdawi et al. (2011b) examined the response of Se-tolerant neighbors of two Se hyperaccumulator species (*Astragalus bisulcatus* and *Stanleya pinnata*). They reported a positive growth effect, which they attributed to increased Se concentrations in the Se-tolerant (but non-hyperaccumulator) neighbors that decreased herbivory damage and in turn increased growth. El Mehdawi et al. (2011b) suggested that, by creating patches of high Se concentration soil that through both allelopathic and facilitative effects modify local plant community makeup, hyperaccumulator plants may act as ecosystem engineers.

It is clear that metal hyperaccumulators may change localized soil metal distribution and this in turn may have ecological consequences. From the standpoint of the hyperaccumulator species, these consequences range from positive to negative and provide an excellent opportunity for future study. Past studies of serpentine ecology often have emphasized the large-scale (landscape scale) patchiness of these habitats (*e.g*., Harrison et al., 2006), but finer-scale investigations of elemental distribution and its ecological effects seem warranted. At these finer scales, important ecological impacts of hyperaccumulator plants on their communities (and *vice versa*) may be discovered.

Evolution of Metal Hyperaccumulation

Early counts of hyperaccumulator taxa showed that the trait was spread among a number of plant lineages, so that multiple independent origins of hyperaccumulation were likely. Borhidi (2001) made an initial analysis of the phylogenetic distribution of Ni hyperaccumulation and pointed out some general trends, such as that Ni hyperaccumulation was rare among plants with a climbing growth form, was lacking in alkaloid-producing taxa, and was rare among plant families specializing on nutrient-poor soils. Krämer (2010) reported that Ni hyperaccumulation evolved at least six times, and Zn hyperaccumulation three times, within the Brassicaceae, a family that is unusually rich in hyperaccumulator species. The phylogenetic tree (focusing on plant orders) presented by Jaffré et al. (2013) showed Ni hyperaccumulation occurring on 24 branches (in 40 plant families): while it is not known if each represents an independent evolutionary event it does show a broad distribution of the trait across plant groups. Another recent analysis (Cappa & Pilon-Smits, 2014) concurred that hyperaccumulation (of Ni as well as other elements) evolved multiple times and further suggested that the trait can vary between populations or among individuals within species and so is constantly evolving. Additional studies targeting the evolution of hyperaccumulation in particular genera (*e.g*., Burge & Barker 2010; Cecchi et al. 2010; Mengoni et al. 2003) have added to our knowledge of the fine-grained evolutionary pattern of hyperaccumulation and also support the conclusion that it has evolved independently multiple times. As pointed out above, studies of the mechanisms underlying hyperaccumulation show that it is a complex trait that involves specialized membrane transporters and metal-binding ligands (*e.g*., Rascio & Navari-Izzo, 2011). These features are believed to be metabolically costly (Maestri et al., 2010) so that their evolution should require a benefit great enough to justify this expense. To my knowledge we do not yet have enough knowledge of these mechanisms to enable us to estimate the metabolic cost of hyperaccumulation. It is hoped that current progress in studies of the physiologic mechanisms of hyperaccumulation will allow us to do so.

However, there probably are other costs to hyperaccumulation besides metabolic ones. As an illustration, Strauss et al. (2002) reviewed costs of herbivore resistance traits and found that there are both direct costs (resource-based tradeoffs) as well as ecological costs (in which interactions with other organisms are affected by evolution of a defensive trait). Ecological costs include effects such as deterrence of mutualists, increased parasite load or impact of enemies, reduced tolerance of enemy attack, and reduced competitive ability. Interestingly, extensive studies of Se hyperaccumulators to date have not revealed ecological costs (El Mehdawi & Pilon-Smits, 2012). In the case of metal hyperaccumulators, several recent studies suggest evolutionary connections between metal tolerance and defense. For example, Fones et al. (2013) suggested that the ability of a Zn hyperaccumulator (*Noccaea caerulescens*) to tolerate high Zn levels involved its ability to tolerate high levels of Reactive Oxygen Species (ROS). But high levels of ROS are an important part of the signaling pathway that activates plant defenses against pathogen attack, so that there was a tradeoff between this plant's being able to tolerate high levels of Zn and its becoming susceptible to pathogen attack. Fones et al. (2013) suggested that the hyperaccumulator evolved Zn hyperaccumulation to replace the deactivated pathogen defense, so that Zn tolerance eventually led to the evolution of hyperaccumulation. Similarly, Freeman et al. (2005) suggested that salicylic acid (SA) was an important signal molecule for pathogen defense in species of *Thlaspi* (*Noccaea*). Several Ni hyperaccumulator species had high constitutive levels of SA and these were associated with enhanced tolerance of high tissue Ni concentrations. When grown in low Ni soil, the hyperaccumulator *T. goesingense* was susceptible to pathogen infection but this susceptibility did not occur when plants were provided Ni. The authors suggested that in order to tolerate Ni, the pathway connecting SA to pathogen resistance was decoupled in the hyperaccumulator plants. The decoupling was feasible because Ni hyperaccumulation replaced the defense provided by the SA-connected resistance pathway and this protected hyperaccumulators from pathogen attack. One consequence of this evolutionary pathway is that the plant would be restricted to high-metal soils because of enhanced susceptibility to pathogens when growing on low metal substrates.

The idea that defense has propelled the evolution of hyperaccumulation was termed the "defensive enhancement hypothesis" by Boyd (2012), who produced a conceptual model for how hyperaccumulation might have evolved in this manner. Critical to this model is the concentration of an element in plant tissues that provides a fitness benefit to a plant that is high enough to overcome the costs of elemental accumulation. Boyd suggested that, if this threshold concentration is relatively low and if an increase will result in increased protective benefit, then natural selection should favor evolution of greater tissue element concentration. This hypothesis has been explicitly extended to defense by pathogens by Hörger et al. (2013). Experimental studies of defense using artificial growth media (*e.g.*, Boyd & Shaw, 2004; Cheruiyot et al., 2013; Coleman et al., 2005; Jhee et al., 2006) or whole plants (*e.g.*, Scheirs et al., 2006) have reported some defensive benefits from element concentrations below hyperaccumulator levels. This suggests that further exploration is warranted. It should be pointed out that the question of effective threshold concentration pertains to the other adaptive hypotheses: if relatively low tissue levels of metals provide drought, tolerance, or allelopathic benefits, then selection may act to increase those benefits and drive the evolution of hyperaccumulation.

Complementing the defensive enhancement hypothesis, Boyd (2012) suggested that one mechanism whereby relatively low concentrations of an element may become defensively valuable is through joint effects with other plant chemical constituents. This "joint effects hypothesis" states that positive effects (additivity or synergy) of chemicals when they occur in combination can lower the threshold concentration at which an element may be defensively effective to below hyperaccumulator levels. If this occurs, then evolution of greater element concentrations can occur through defensive enhancement (as described in the defensive enhancement scenario above).

Boyd (2012) posited that joint effects might occur between different elements accumulated by a plant, or between an element and an organic plant defense chemical (such as glucosinolates, alkaloids, etc.). Some investigation of joint effects has occurred and, to date, additive effects (Hörger et al., 2013; Jhee et al., 2005; Kazemi-Dinan et al., 2014) are reported more often than synergistic ones (Cheruiyot, 2012). This hypothesis also needs more experimental investigation.

As evident above, evolution of hyperaccumulation has often been framed in terms of the defense hypothesis. It is important to realize, however, that the hypothesized benefits of hyperaccumulation are not mutually exclusive. More than one hypothesis may explain the evolution of hyperaccumulation in any particular case or differing hypotheses may apply to different hyperaccumulated elements (Boyd, 2002). For example, Nguyen et al. (2014) reported that evolution of high constitutive levels of plant defensins in *Arabidopsis halleri* provided two benefits: elevated Zn tolerance and elevated pathogen defense.

It also should be pointed out that some species accumulate more than one element and this feature expands the number of potential evolutionary explanations. As a recent example, Tang et al. (2009) reported that *Arabis paniculata* (Brassicaceae) could simultaneously hyperaccumulate Cd, Pb, and Zn. These multi-element accumulators raise interesting questions, such as: 1) do different elements have different adaptive functions? For example, one element may function primarily in pathogen defense and another via elemental allelopathy; 2) do joint effects promote the adaptive function(s) of each element? For example, one element may synergistically enhance the effect of a second while that second may or may not affect the function of the first; or 3) do different elements produce different joint effects with the same plant secondary compound? Exploration of these and other questions provides many opportunities for future experiments.

Conclusion

Hyperaccumulation has evolved multiple times and probably has multiple adaptive functions. These include defense against herbivores and pathogens, drought tolerance, elemental allelopathy, and possibly metal tolerance. In any particular case, hyperaccumulation also probably has multiple ecological effects on associated organisms. Teasing apart those ecological connections, and interactions between connections, is a complicated task that will require concerted experimental efforts. Research to date has established a set of adaptive hypotheses that provide a framework for future research. Exploration of those hypotheses has the potential to explain the evolution of this fascinating plant trait and how it helps plants survive harsh environmental conditions.

Acknowledgment

I thank three anonymous reviewers for comments and suggestions on earlier versions of this manuscript. This work was supported by Alabama Agricultural Experiment Station project ALA021-109008.

References

Alexander, E. B., Coleman, R. G., Keeler-Wolf, T. & Harrison, S. P. (2007) Serpentine geoecology of western North America: Geology, soils and vegetation. New York, NY: Oxford University Press.

Augustyniak, M., Mesjasz-Przybylowicz, J., Miroslaw, N., Dybowska, M., Przybylowicz, W. & Migula, P. (2002) Food relations between *Chrysolina pardalina* and *Berkheya coddii*, a nickel hyperaccumulator from South African ultramafic outcrops. *Fresenius Environmental Bulletin 11*, 85-90.

Barzanti, R., Ozino, F., Bazzicalupo, M., Gabbrielli, R., Galardi, F., Gonelli, C. & Mengoni, A. (2007) Isolation and characterization of endophytic bacteria from the nickel hyperaccumulator plant *Alyssum bertolonii*. *Microbial Ecology 53,* 306-316.

Bhatia, N. P., Baker, A. J. B., Walsh, K. B. & Midmore, D. J. (2005) A role for nickel in osmotic adjustment in drought-stressed plants of the nickel hyperaccumulator *Stackhousia tryonii* Bailey. *Planta 223,* 134-139.

Borhidi, A. (1996) Phytogeography and vegetation ecology of Cuba. Budapest, Hungary: Akadémiai Kiadó.

Borhidi, A. (2001) Phylogenetic trends in Ni-accumulating plants. *South African Journal of Science 97,* 544-547.

Boyd, R. S. (2002) Commentary–Ecology of metal hyperaccumulation. *New Phytologist 162*, 563-567.

Boyd, R. S. (2007) The defense hypothesis of elemental hyperaccumulation: Status, challenges and new directions. *Plant and Soil 293*, 153-176.

Boyd, R. S. (2009) High-nickel insects and nickel hyperaccumulator plants: A review. *Insect Science 16, 19*-31.

Boyd, R. S. (2012) Plant defense using toxic inorganic ions: Conceptual models of the Defensive Enhancement and Joint Effects Hypotheses. *Plant Science 195*, 88-95.

Boyd, R. S. & Martens, S. N. (1992) The raison d'être for metal hyperaccumulation by plants. In: A. J. M. Baker, J. Proctor & R. D. Reeves (Eds.). The vegetation of ultramafic (serpentine) soils (pp. 279-289). Andover: Intercept Limited.

Boyd, R. S. & Martens, S. N. (1998) The significance of metal hyperaccumulation for biotic interactions. *Chemoecology 8,* 1-7.

Boyd, R. S. & Shaw, J. J. (2004) Response of *Xanthomonas campestris* to metals: Implications for hyperaccumulation as a pathogen defense. In: R. S. Boyd, A. J. M. Baker & J. Proctor (Eds.). Ultramafic rocks: Their soils, vegetation and fauna (pp. 279-282). St Albans: Science Reviews 2000 Ltd.

Boyd, R. S., Davis, M. A., Wall, M. A. & Balkwill, K. (2006a) Metal concentrations of insects associated with the South African Ni hyperaccumulator *Berkheya coddii* (Asteraceae). *Insect Science 13,* 85-102.

Boyd, R. S., Wall, M. A. & Jaffré, T. (2006b) Nickel levels in arthropods associated with Ni hyperaccumulator plants from an ultramafic site in New Caledonia. *Insect Science 13*, 271-277.

Brady, K. U., Kruckeberg, A. R. & Bradshaw, H. D. Jr. (2005) Evolutionary ecology of plant adaptation to serpentine soils. *Annual Review of Ecology, Evolution and Systematics 36*, 243-266.

Brooks, R. R. (1987) Serpentine and its vegetation: A multidisciplinary approach. Portland, OR: Dioscorides Press.

Burge, D. O. & Barker, W. R. (2010) Evolution of nickel hyperaccumulation by *Stackhousia tyronii* (Celastraceae), a serpentinite-endemic plant from Queensland, Australia. *Australian Systematic Botany 23,* 415-430.

Cappa, J. J. & Pilon-Smits, E. A. H. (2014) Evolutionary aspects of elemental hyperaccumulation. *Planta 239,* 267-275.

Cecchi, L., Gabbrielli, R., Arnetoli, M., Gonnelli, C., Hasko, A. & Selvi, F. (2010) Evolutionary lineages of nickel hyperaccumulation and systematics in European Alysseae (Brassicaceae): Evidence from nrDNA sequence data. *Annals of Botany 106*, 751-767.

Che-Castaldo, J. & Inouye, D. W. (2014) Field germination and survival of experimentally introduced metal hyperaccumulator *Noccaea caerulescens* (Brassicaceae) across a soil metal gradient. *American Midland Naturalist 171*, 229-245.

Cheruiyot, D. J. (2012) Investigation of elemental defense and trophic transfer of metals using beet armyworm, *Spodoptera exigua,* Hübner. PhD dissertation. Auburn, AL: Auburn University.

Cheruiyot, D. J., Boyd, R. S. & Moar, W. J. (2013) Exploring lower limits of plant elemental defense by cobalt, copper, nickel and zinc. *Journal of Chemical Ecology 39*, 666-674.

Coleman, C. M., Boyd, R. S. & Eubanks, M. D. (2005) Extending the elemental defense hypothesis: Dietary metal concentrations below hyperaccumulator levels could harm herbivores. *Journal of Chemical Ecology 31,* 1669-1681.

El Mehdawi, A. F. & Pilon-Smits, E. A. H. (2012) Ecological aspects of plant selenium hyperaccumulation. *Plant Biology 14*, 1-10.

El Mehdawi, A. F., Quinn, C. F. & Pilon-Smits, E. A. H. (2011a) Effects of selenium hyperaccumulation on plant-plant interactions: Evidence for elemental allelopathy? *New Phytologist 191*, 120-131.

El Mehdawi, A. F., Quinn, C. F. & Pilon-Smits, E. A. H. (2011b) Selenium hyperaccumulators facilitate selenium-tolerant neighbors via phytoenrichment and reduced herbivory. *Current Biology 21*, 1440-1449.

Fones, H., Davis, C. A. R., Rico, A., Fang, F., Smith, J. A. C. & Preston, G. M. (2010) Metal hyperaccumulation armors plants against disease. *PLoS Pathogens 6,* e1001093. DOI: 10.1371/journal.ppat.1001093.

Fones, H. N., Eyles, C. J., Bennett, M. H., Smith, J. A. C. & Preston, G. M. (2013) Uncoupling of reactive oxygen species accumulation and defence signalling in the metal hyperaccumulator plant *Noccaea caerulescens*. *New Phytologist 199*, 916-924.

Freeman, J. L., Garcia, D., Kim, D., Hopf, A. & Salt, D.E. (2005) Constitutively elevated salicylic acid signals glutathione-mediated nickel tolerance in *Thlaspi* nickel hyperaccumulators. *Plant Physiology 137,* 1082-1091.

Freeman, J. L., Quinn, C. F., Marcus, M. A., Fakra, S. & Pilon-Smits, E. A. H. (2006) Selenium-tolerant diamondback moth disarms hyperaccumulator plant defense. *Current Biology 16,* 2181-2192.

Freeman, J. L., Marcus, M. A., Fakra, S. C., Devonshire, J., McGrath, S. P., Quinn, C. F. & Pilon-Smits, E. A. H. (2012) Selenium hyperaccumulator plants *Stanleya pinnata* and *Astragalus bisulcatus* are colonized by Se-resistant, Se-excluding wasp and beetle seed herbivores. *PLOS One 7*(12), e50516. DOI: 10.1371/journal.pone.0050516.

Galeas, M. L., Klamper, E. M., Bennett, L. E., Freeman, J. L., Kondratieff, B. C., Quinn, C. F. & Pilon-Smits, E. A. H. (2008) Selenium hyperaccumulation reduces plant arthropod loads in the field. *New Phytologist 177*, 715-724.

Harrison, S., Safford, H. D., Grace, J. B., Viers, J. H. & Davies, K. F. (2006) Regional and local species richness in an insular environment: Serpentine plants in California. *Ecological Monographs 76*, 41-56.

Hobbs, R. J. & Mooney, H. A. (1991) Effects of rainfall variability and gopher disturbance on serpentine grassland dynamics. *Ecology 72*, 59-68.

Hörger, A. C., Fones, H. N. & Preston, G. M. (2013) The current status of the elemental defense hypothesis in relation to pathogens. *Frontiers in Plant Science,* DOI: 10.3389/fpls.2013.00395.

Jaffré, T., Brooks, R. R., Lee, J. & Reeves, R. D. (1976). *Sebertia acuminata*: A hyperaccumulator of nickel from New Caledonia. *Science 193*, 579-580.

Jaffré, T., Pillon, Y., Thomine, S. & Merlot, S. (2013) The metal hyperaccumulators from New Caledonia can broaden our understanding of nickel accumulation in plants. *Frontiers in Plant Science 4, 279*, DOI: 10.3389/fpls.2013.00279.

Jhee, E. M., Boyd, R. S. & Eubanks, M. D. (2006) Effectiveness of metal-metal and metal-organic compound combinations against *Plutella xylostella* (Lepidoptera: Plutellidae): Implications for plant elemental defense. *Journal of Chemical Ecology 32*, 239-259.

Kachenko, A. G., Bhatia, N. & Singh, B. (2011) Influence of drought stress on the nickel-hyperaccumulating shrub *Hybanthus floribundus* (Lindl.) F. Muell. subsp. *floribundus*. *International Journal of Plant Sciences 172*, 315-322.

Karban, R. & Agrawal, A. A. (2002) Herbivore offense. *Annual Review of Ecology and Systematics 33*, 641-664.

Kazakou, E., Dimitrakopoulos, P. G., Baker, A. J. M., Reeves, R. D. & Troumbis, A. Y. (2008) Hypotheses, mechanisms and trade-offs of tolerance and adaptation to serpentine soils: From species to ecosystem level. *Biological Reviews 83*, 495-508.

Kazemi-Dinan, A., Thomaschky, S., Stein, R. J., Krämer, U. & Müller, C. (2014) Zinc and cadmium hyperaccumulation act as deterrents towards specialist herbivores and impede the performance of a generalist herbivore. *New Phytologist,* DOI: 10.1111/nph.12663.

Kirkby, E. (2012) Introduction, definition and classification of nutrients. In: P. Marschner (Ed.). Marschner's mineral nutrition of higher plants. Third edition (pp. 3-5). Amsterdam: Academic Press.

Krämer, U. (2010) Metal hyperaccumulation in plants. *Annual Reviews in Plant Biology 61,* 517-534.

Kruckeberg, A. R. (1984) California serpentines: Flora, vegetation, geology, soils, and management problems. Berkeley, CA: University of California Press.

Kruckeberg, A. R. (2002) Geology and plant life: The effects of landforms and rock types on plants. Seattle, WA: University of Washington Press.

Küpper, H. & Leitenmaier, B. (2013) Cadmium-accumulating plants. In: A. Sigel, H. Sigel & R. K. O. Sigel (Eds.). Cadmium: From toxicity to essentiality (pp. 373-393). Metal Ions in Life Sciences 11. Dordrecht: Springer.

Lau, J. A., McCall, A. C., Davies, K. F., McKay, J. K. & Wright, J. W. (2008) Herbivores and edaphic factors constrain the realized niche of a native plant. *Ecology 89*, 754-762.

Maestri, E., Marmiroli, M., Visioli, G. & Marmiroli, N. (2010) Metal tolerance and hyperaccumulation: Costs and trade-offs between traits and environment. *Environmental and Experimental Botany 68*, 1-13.

Martens, S. N. & Boyd, R. S. (2002) The defensive role of Ni hyperaccumulation by plants: A field experiment. *American Journal of Botany 89*, 998-1003.

Meindl, G. A., Bain, D. J. A. & Ashman, T.-L. (2013) Edaphic factors and plant-insect interactions: Direct and indirect effects of serpentine soil on florivores and pollinators. *Oecologia 173,* 1355-1366.

Mengoni, A., Baker, A. J. M., Bazzicalupo, M., Reeves, R. D., Adigüzel, N., Chianni, E., Galardi, F., Gabbrielli, R. & Gonnelli, C. (2003) Evolutionary dynamics of nickel hyperaccumulation in *Alyssum* revealed by ITS nrDNA analysis. *New Phytologist 159,* 691-699.

Mesjasz-Przybylowicz, J. & Przybylowicz, W. J. (2001) Phytophagous insects associated with the Ni-hyperaccumulating plant *Berkheya coddii* (Asteraceae) in Mpumalanga, South Africa. *South African Journal of Science 97*, 596-598.

Morris, C., Grossl, P. R. & Call, C. A. (2009) Elemental allelopathy: Processes, progress and pitfalls. *Plant Ecology 202*, 1-11.

Nguyen, N. N., Ranwez, V., Vile, D., Soulié, M.-C., Dellagi, A., Expert, D. & Gosti, F. (2014) Evolutionary tinkering of the expression of *PDF1*s suggests their joint effect on zinc tolerance and the response to pathogen attack. *Frontiers in Plant Science,* DOI: 10.3389/fpls.2014.00070.

Noret, N., Meerts, P., Vanhaelen, M., Dos Santos, A. & Escarré, J. (2007) Do metal-rich plants deter herbivores? A field test of the defence hypothesis. *Oecologia 152,* 92-100.

Palomino, M., Kennedy, P. G. & Simms, E. L. (2007) Nickel hyperaccumulation as an anti-herbivore trait: Considering the role of tolerance to damage. *Plant and Soil 293*, 189-195.

Polacco, J. C., Mazzafera, P. & Tezotto, T. (2013) Opinion–Nickel and urease in plants: Still many knowledge gaps. *Plant Science 199-200,* 79-90.

Pollard, A. J., Powell, K. D., Harper, F. A. & Smith, J. A. C. (2002) The genetic basis of hyperaccumulation in plants. *Critical Reviews in Plant Sciences 21*, 539-566.

Proctor, J. & Woodell, S. R. (1975) The ecology of serpentine soils. *Advances in Ecological Research 9,* 255-365.

Rascio, N. & Navari-Izzo, F. (2011) Heavy metal accumulating plants: How and why do they do it? And what makes them so interesting? *Plant Science 180*, 169–181.

Reeves, R. D. (2003) Tropical hyperaccumulators of metals and their potential for phytoextraction. *Plant and Soil 249*, 57-65.

Reeves, R. D. & Baker, A. J. M. (2000) Metal-accumulating plants. In: I. Raskin & B. D. Ensley (Eds.). Phytoremediation of toxic metals: Using plants to clean up the environment (pp. 193-229). Hoboken, NJ: John Wiley.

Scheirs, J., Vandevyvere, I., Wollaert, K., Blust, R. & De Bruyn, L. (2006) Plant-mediated effects of heavy metal pollution on host choice of a grass miner. *Environmental Pollution 143*, 138-145.

Schlegel, H. G., Cosson, J.-P. & Baker, A. J .M. (1991). Nickel-hyperaccumulating plants provide a niche for nickel-resistant bacteria. *Botanica Acta 104,* 18-25.

Strauss, S. Y. & Boyd, R. S. (2011) Herbivory and other cross-kingdom interactions on harsh soils. In: S. H. Harrison & N. Rajakaruna (Eds.). Serpentine: The evolution and ecology of a model system (pp. 181-199). Berkeley, CA: University of California Press.

Strauss, S. Y. & Ivalú Cacho, N. (2013) Nowhere to run, nowhere to hide: The importance of enemies and apparency in adaptation to harsh soil environments. *The American Naturalist 182*, E1-E14.

Strauss, S. Y., Rudgers, J. A., Lau, J. A. & Irwin, R. E. (2002) Direct and ecological costs of resistance to herbivory. *TRENDS in Ecology and Evolution 17*, 278-285.

Tang, Y.-T., Qiu, R.-L., Zeng, X.-W., Ying, R.-R., Yu, F.-M. & Zhou, X.-Y. (2009) Lead, zinc, cadmium hyperaccumulation and growth stimulation in *Arabis paniculata* Franch. *Environmental and Experimental Botany 66*, 126-134.

van der Ent, A., Baker, A. J. M., Reeves, R. D., Pollard, A. J. & Schat, H. (2013) Hyperaccumulators of metal and metalloid trace elements: Facts and fiction. *Plant and Soil 362*, 319-334.

Verbruggen, N., Hermans, C. & Schat, H. (2009) Molecular mechanisms of metal hyperaccumulation in plants. *New Phytologist 181,* 759-776.

Wall, M. A. & Boyd, R. S. (2002) Nickel accumulation in serpentine arthropods from the Red Hills, California. *Pan-Pacfic Entomologist 78*, 168–176.

Wall, M. A. & Boyd, R. S. (2006) *Melanotrichus boydi* (Hemiptera: Miridae) is a specialist on the nickel hyperaccumulator *Streptanthus polygaloides* (Brassicaceae). *Southwestern Naturalist 51,* 481-489.

Whiting, S. N., Neumann, P. M. & Baker, A. J. M. (2003) Nickel and zinc hyperaccumulation by *Alyssum murale* and *Thlaspi caerulescens* (Brassicaceae) do not enhance survival and whole-plant growth under drought stress. *Plant, Cell and Environment 26*, 351-360.

Zhang, L., Angle, J. S. & Chaney, R. L. 2007. Do high-nickel leaves shed by the nickel hyperaccumulator *Alyssum murale* inhibit seed germination of competing plants? *New Phytologist 173,* 509-516.

In: Plant Ecology and Evolution in Harsh Environments ISBN: 978-1-63321-955-7
Editors: N. Rajakaruna, R. S. Boyd and T. B. Harris

Chapter 11

METHODS AND DISCOVERIES IN THE PURSUIT OF UNDERSTANDING THE GENETIC BASIS OF ADAPTATION TO HARSH ENVIRONMENTS IN *MIMULUS*

Jessica P. Selby[1,*], Annie L. Jeong[1,2], Katherine Toll[1], Kevin M. Wright[3] and David B. Lowry[4]

[1]Department of Biology, Duke University, Durham, North Carolina, USA
[2]University Program in Genetics and Genomics, Duke University Medical Center, Durham, North Carolina, USA
[3]Department of Organismic and Evolutionary Biology, Harvard University, Cambridge, Massachusetts, USA
[4]Department of Plant Biology, Michigan State University, East Lansing, MI, USA

ABSTRACT

The *Mimulus guttatus* D.C. species complex (Phrymaceae) is a model system for understanding the genetic basis of adaptation to variable environments. Recent studies in this system on the evolution of drought escape via shifts in flowering time as well as tolerance to serpentine, copper mine, and saline soils have provided new insights into the mechanisms of adaptation and speciation. Determining the genetic basis of plant adaptation to such harsh environmental conditions is of fundamental importance to biology and has many applied benefits. Here, we review research on adaptation to extreme habitats in *Mimulus* and describe how recent developments in high-throughput phenotyping and the use of genomic approaches are driving further advances in understanding the genetics of adaptation and speciation.

---

* Corresponding author: jessica.selby@duke.edu.

INTRODUCTION

Identifying the genetic and physiological basis of adaptations to complex environmental conditions is a major challenge in ecological genomics. Wild plant species offer particularly attractive systems for addressing these questions, as they often exhibit local adaptation of populations to different habitats across their geographic ranges (Hereford, 2009; Leimu & Fischer, 2008). Environmental heterogeneity results in selective pressures that differ between habitats which can promote population differentiation and maintain genetic variation (Clausen, 1951; Gillespie & Turelli, 1989; Hedrick, 1986; Kawecki & Ebert, 2004; Levene, 1953). Over time, adaptation to different habitats can lead to reproductive isolation (RI), either directly through the evolution of traits involved in local adaptation or indirectly if reproductive isolating barriers hitchhike along with adaptations (Coyne & Orr, 2004; Rundle & Nosil, 2005; Schluter & Conte, 2009; Wright et al., 2013). Plants that live in habitats characterized by harsh abiotic conditions—for example, drought, toxic soils, salinity, and thermal extremes—often provide particularly vivid examples of how habitat-mediated divergent selection creates biological diversity. Plants that are able to tolerate harsh environments are well suited for investigating the genetic and physiological basis of adaptation because selection in these habitats can be quite strong and, in some cases, the abiotic stress is known and can be manipulated in lab and/or field studies (Brady et al., 2005; Macnair, 1987). Furthermore, plants have repeatedly adapted to many of these stressful habitat types, providing opportunities to investigate the degree of parallel trait evolution and whether it is due to parallel changes at the genetic level.

Plants that are able to thrive in harsh habitats provide well-known, classic examples of adaptation (Antonovics & Bradshaw, 1970; Bradshaw, 1991; Kruckeberg, 1951; Macnair, 1981). Numerous field and lab-based studies have demonstrated that populations are often locally adapted to harsh environmental conditions (Hereford, 2009; Leimu & Fischer, 2008; O'Dell & Rajakaruna, 2011). Recent molecular work has offered insights into potential mechanisms of adaptation to harsh habitats. For example, the flowering time pathway is well characterized (Kobayashi & Weigel, 2007) and differences in flowering time often contribute to adaptive escape from drought (Franks, 2011; Hall & Willis, 2006; Ludlow, 1989; McKay et al., 2003) and cold (Mendez-Vigo et al., 2011). There has also been significant progress in understanding the molecular basis of plant ion homeostasis and metal tolerance (Baxter et al., 2010; Clemens, 2001; Colangelo & Guerinot, 2006; Hanikenne et al., 2008), which are likely important mechanisms for coping with extreme soil habitats. Notably lacking, however, are studies that have identified naturally segregating variants controlling these traits and characterized the fitness effects of these variants in native habitats.

In order to understand plant adaptation to harsh environments, there is a vital need for studies that integrate the cellular and molecular control of traits with an understanding of the ecological context of such traits. To elucidate how selection is operating on specific traits, researchers should test the relationship between phenotypic variation and fitness using manipulative field experiments whenever feasible. The best test of local adaptation is the classical reciprocal transplant experiment (Clausen et al., 1940). The genetic basis of adaptive differences can then be characterized using Quantitative Trait Locus (QTL) mapping, association mapping, or genome scan studies. To confirm that loci identified via these approaches actually contribute to adaptation these loci should be tested for their fitness effects in native field habitats, as elegantly demonstrated by several recent studies (Ågren et al., 2013; Leinonen et al., 2013; Lexer et al., 2003; Lowry & Willis, 2010; Prasad et al., 2012; Verhoeven et al., 2008).

In this chapter, we discuss the genetic basis of adaptation to harsh environments in the *Mimulus guttatus* species complex (yellow monkeyflower) and the approaches that have enabled these studies. We begin by briefly summarizing what is currently known about species in the *M. guttatus* complex that are able to tolerate serpentine soils (Chapter 6), copper mine tailings (Chapter 14), saline habitats, or water-limited environments. We then focus on QTL mapping and population genomic approaches and what they have revealed about the genetic basis of adaptation to several of these habitats. As sequencing becomes faster and cheaper, phenotyping has become a limiting step for forward genetic studies. Here we describe both field and lab-based, high-throughput phenotyping methods that have been successfully used in *Mimulus* to assay tolerance to several different abiotic stresses.

Mimulus Guttatus is a Model for the Genetics of Adaptation to Harsh Environments

The *Mimulus* genus contains approximately 160 species, which display an incredible degree of ecological variation including adaptation to numerous stressful habitat types (Table 1; Beardsley & Olmstead, 2002; Vickery, 1978; Wu et al., 2008). The center of diversity of the *M. guttatus* species complex is located in western North America. Members of the complex are broadly interfertile (Wu et al., 2008). The complex includes some species that have highly restricted ranges and are often associated with a specific marginal or harsh environment, including several edaphic endemic species (Macnair, 1989; Macnair & Gardner, 1998). Other species, in particular *M. guttatus,* are wide-ranging with populations occurring in countless different habitats (*calflora.org*; Vickery, 1964; Wu et al., 2008). Since the pioneering work of Robert Vickery (Clausen & Hiesey, 1958; Vickery, 1952), many of the studies on *Mimulus* have focused on elucidating the genetic basis of traits that contribute to RI and ecological divergence. With the development of genetic resources, *Mimulus* has become a model system for evolutionary and ecological genetics (Chapters 6, 9; Hellsten et al., 2013; Wu et al., 2008).

Mimulus guttatus, a focal member of the genus, combines incredible ecological diversity with the attributes of a true genetic model system. *Mimulus guttatus* is easily maintained in the laboratory with a short generation time (2-3 months), small size, high fecundity (100-400 seeds per cross), and reproductive flexibility (clonal propagation and self-fertile). The sequenced genome of *M. guttatus* (~430 Mbp) has been publicly available since 2010; the most recent annotated version (v2.0) is available on *www.phytozome.net* (Goodstein et al., 2012). In addition, there is extensive EST and RNA-seq data, over 1,000 highly polymorphic PCR gene-based markers, fingerprinted BAC libraries, and integrated genetic and physical maps (available on mimulusevolution.org; Wu et al., 2008). Gene-based (exon-primed intron-spanning) markers have been used successfully in widespread *M. guttatus* populations, as well as in distantly related species such as *M. aurantiacus, M. ringens,* and *M. primuloides* (Cooley et al., 2011; Griffin, 2010; Streisfeld et al., 2013). Finally, stable transformation protocols have been developed for *Mimulus* (Susič et al., 2014; Yuan et al., 2013) enabling critical functional tests of candidate genes identified via forward genetic approaches. The wealth of genomic resources coupled with the ecological variability of the *M. guttatus* species complex make it a powerful system for studying adaptation to harsh environments.

Table 1. Common harsh environments of *Mimulus*

Habitat	*Mimulus* species	Stressors	Reference
Coastal	*guttatus*	Soil salinity, wind, salt spray	Lowry et al., 2009
Copper mine tailings	*cupriphilus, guttatus*	Heavy metal (Cu) toxicity, early seasonal drought	Allen & Sheppard, 1971; Macnair 1981; 1989; Macnair & Christie, 1983; Wright et al., 2013
Geothermal soils	*guttatus*	High soil temperature, seasonal drought	Bunn & Zabinski, 2003; Delmer, 1974; Lekberg et al., 2012
Granite outcrops	*laciniatus*	Seasonal drought	Peterson et al., 2013
High elevation	*guttatus, laciniatus, mephiticus, primuloides, tilingii*	Cold temperatures, reduced growing season, seasonal drought, UV radiation	Douglas, 1981; Ferris et al., (*In press*).
Serpentine	*congdonii, douglasii, floribundus, glaucescens, guttatus, kellogii, layneae, mephiticus, nudatus, pardalis*	Low Ca:Mg and other nutrients, heavy metals, early seasonal drought	Consortium of California Herbaria, 2014; Macnair & Gardner, 1999; Nesom, 2012; Palm et al., 2012; Tilstone & Macnair, 1997; Hughes et al., 2001; Gardner & Macnair, 2000; Murren et al., 2006; Meindl et al., 2013.

NATURAL HISTORY OF ADAPTATIONS TO HARSH ENVIRONMENTS IN *MIMULUS*

Species in the *M. guttatus* complex have adapted to various abiotic stressors such as drought, high salinity, and soils with toxic metal concentrations and low essential nutrients. In this section, we describe what is known about the natural history of adaptation to harsh environmental conditions in the *M. guttatus* species complex. This natural history lays the foundation for the remainder of the chapter, which describes efforts to understand the physiological and genetic bases of these adaptations.

Serpentine Adaptations

Serpentine soils, derived from the weathering of ultramafic rocks, are characterized by a unique suite of edaphic variables: extremely low levels of Ca and high levels of Mg; deficiency in the major macronutrients N, P, and K; high concentrations of heavy metals such as Ni, Co, and Cr; and low water holding capacity (Chapter 6; Alexander et al., 2007). Many plant species are unable to grow in serpentine habitats because they cannot tolerate the chemical and physical properties of these soils. However, several species within the *M. guttatus* complex have adapted to these harsh soils. The widespread *M. guttatus* can be found both on and off serpentine soils throughout much of its range while two closely related species, *M. nudatus* and *M. pardalis*, have restricted ranges and are found exclusively on serpentine soils (Gardner & Macnair, 2000; Hughes et al., 2001).

Reciprocal transplant and common garden studies show that *M. guttatus* is locally adapted to serpentine soils (Palm et al., 2012; Selby, 2014). When planted at serpentine field sites (Selby, 2014) or on serpentine soil in the lab (Palm et al., 2012; Selby, 2014), plants from non-serpentine populations died in the juvenile stage while serpentine populations had high survival. In contrast, a study by Meindl et al. (2013) found no survival differences between *M. guttatus* plants from serpentine and non-serpentine populations when planted on a mixture of native serpentine and potting soils. These contrasting results could be due to the different soil matrices that were used: full serpentine soil versus a mix of serpentine and potting soils. It is also possible that the non-serpentine populations investigated by Meindl et al. (2013) had a higher frequency of tolerance alleles segregating due to ongoing gene flow with nearby serpentine populations.

Hydroponic experiments are often conducted to determine the specific soil chemical variables that are important selective agents in serpentine habitats. Hydroponic studies using *M. guttatus* have revealed differential tolerance of serpentine and non-serpentine populations to low Ca:Mg ratio (Palm et al., 2012; Selby, 2014) and high Ni (A. Jeong, unpublished) growth environments. These results suggest that the low Ca and high Mg and Ni levels that characterize serpentine soils are likely driving local adaptation of *M. guttatus* populations to these habitats (but see Gardner & Macnair, 2000; Murren et al., 2006).

Adaptation to serpentine soils has also led to the evolution of new species within the *M. guttatus* complex, resulting in two serpentine endemic species: the outcrossing *M. nudatus* restricted to Napa and Lake Counties and the obligately selfing *M. pardalis* found in Calaveras and Tuolumne Counties. The serpentine endemics often grow sympatrically with *M. guttatus*, but inhabit drier microsites (Gardner & Macnair, 2000; Hughes et al., 2001). Accelerated development and flowering time are often selected for in rapidly drying sites as a means of drought escape (Franks, 2011; McKay et al., 2003). Differences in flowering time (Figure 1) likely contribute to RI between the serpentine endemics and *M. guttatus*. Self-fertilization in *M. pardalis* further contributes to RI with *M. guttatus*. In contrast, *M. nudatus* is outcrossing. However, pollinator constancy causes strong prezygotic isolation (RI = 0.947) between *M. guttatus* and *M. nudatus*: *Dialictus* species preferentially visit *M. nudatus* flowers while honeybees preferentially visit *M. guttatus* (Gardner & Macnair, 2000; Lowry et al., 2008a). In addition to ecological causes of RI, postzygotic isolation in the form of hybrid seed lethality (RI = 0.958) is a strong barrier to gene flow between *M. nudatus* and *M. guttatus* (Gardner & Macnair, 2000; Macnair & Gardner, 1998; Lowry et al., 2008a).

Several other *Mimulus* species have also adapted to serpentine soils in western North America. *Mimulus glaucescens*, a member of the *M. guttatus* complex, and *M. primuloides*, a sister species to the complex, can both be found growing on and off of serpentine soils. Additionally, more distantly related species (*M. floribundus*, *M. layneae*, *M. douglasii*, *M. congdonii*, *M. kelloggii*, and *M. mephiticus*) have populations occurring both on and off serpentine soils. The repeated evolution of serpentine tolerance within the *Mimulus* genus provides a rich opportunity to explore whether shared or unique physiological and genetic mechanisms underlie serpentine adaptation in these different species.

Cu Mine Adaptations

Copper ore mining has resulted in high concentrations of heavy metals in surface soils and water (Chapters 14, 15) which exert strong selection on local plant populations (Bradshaw, 1991; Wu et al., 1975). Similar to serpentine soils, plant adaptations to mine tailings have occurred independently multiple times within species (Christie & Macnair, 1984; Macnair et al., 1989; Schat et al., 1996). However, in contrast to most serpentine habitats, mine tailings are often quite young and have only recently been colonized. *M. guttatus* has adapted to copper contaminated sites in western North America within the last 150 years. Populations of *M. guttatus* grow on copper mine tailings at multiple sites near Copperopolis, CA in the foothills of the Sierra Nevada (Allen & Sheppard, 1971), on the Bingham mine near Salt Lake City, UT (Christie & Macnair, 1984), and at mine sites in Shasta and El Dorado counties in northern CA (R. O'Dell & K. Wright, unpublished data). These mine populations of *M. guttatus* are located in close geographic proximity to populations living on uncontaminated soils, creating the potential for migration and hybridization (Allen & Sheppard, 1971; Macnair et al., 1993). Lab-based, hydroponic studies have demonstrated that populations of *M. guttatus* from Cu-contaminated soils are more tolerant of elevated Cu levels than plants from uncontaminated sites (Macnair & Christie, 1983). A survey of populations near Copperopolis found that Cu tolerance is nearly fixed in four mine populations (99.77%, N=2796), at intermediate frequency (12-45%; N=197) in three uncontaminated but adjacent sites, and at low frequency (0-2%; N=1118; 12 populations) in the majority of uncontaminated sites in the region (Macnair et al., 1993). These results suggest strong selection for tolerance in Cu contaminated habitats and little or no selection against tolerant plants in uncontaminated soils. Reciprocal transplant experiments show that genotypes from mine populations have greater fitness than off-mine genotypes in the Cu-contaminated habitat (K. Wright, unpublished data), providing further evidence that *M. guttatus* is locally adapted to Cu mine tailings.

Adaptation to the Cu mine environment has potentially resulted in a speciation event within the *M. guttatus* complex. *Mimulus cupriphilus* is a recently derived, morphologically distinct, and highly selfing species found only on two small Cu mines near Copperopolis (Macnair & Gardner, 1998). Recent morphological-based taxonomic research has hypothesized that *M. cupriphilus* may be derived from the serpentine endemic *M. pardalis* (Nesom, 2012). Reproductive isolation between *M. guttatus* and *M. cupriphilus* has not been investigated in the field, but greenhouse experiments reveal that *M. cupriphilus* flowers under shorter day-lengths (Friedman & Willis, 2013), which may contribute to RI with *M. guttatus* (Macnair & Gardner, 1998; K. Wright, unpublished results).

Coastal Habitat Adaptations

Perennial populations of *M. guttatus* grow along the Pacific coast of North America (from southern California to the far western islands of Alaska) where they must cope with both salt spray and saline soils. Coastal perennial *M. guttatus* is morphologically the largest member of the species complex and has previously been classified as a distinct ecotype (Lowry, 2012), variety (Pennell, 1947), and species (Heller, 1904; Nesom, 2012). A series of laboratory experiments have confirmed that coastal perennial plants have evolved a high level of salt tolerance compared to other *M. guttatus* populations (Lowry et al., 2008b, 2009) and are even able to live in sites directly splashed by ocean waves. In the field, a reciprocal transplant study revealed that plants from inland *M. guttatus* populations sustain a high level of leaf necrosis and subsequent mortality when transplanted to coastal habitats (Lowry et al., 2008b).

There are three major mechanisms by which plants evolve salt tolerance: 1) Plant exclusion of Na^+ ions; 2) osmotic stress tolerance; and 3) tissue tolerance to Na^+ ions (Chapter 4; Munns & Tester, 2008). While some plants have evolved mechanisms that exclude toxic Na^+ ions from entering their stem tissue (Boyce, 1954; Munns & Tester, 2008), such exclusion often results in a major osmotic gradient between the environment and plant cells which can cause osmotic stress. Many plants actually uptake Na^+ ions to come into osmotic balance with their environment. However, high levels of Na^+ can be toxic to leaf tissues (Munns et al., 2006; Rus et al., 2006). Therefore, some plants have evolved mechanisms of ion stress tolerance that either allow cells to tolerate higher concentrations of Na^+ ions or to exclude these ions. Such tissue tolerance is often mediated by the sequestration of Na^+ ions in the vacuole of leaf cells (Munns & Tester, 2008; Zhu, 2001).

Lowry et al. (2009) conducted a series of physiological experiments to determine which salt tolerance mechanism was involved in adaptation to coastal habitats in *M. guttatus*. Both coastal and inland inbred lines accumulate similar concentrations of Na^+ ions in their leaves when grown under saline hydroponic conditions, suggesting that salt tolerance is not mediated by whole plant exclusion of Na^+. However, coastal plants are far more tolerant of elevated leaf Na^+ levels, implicating tissue tolerance as the likely mechanism of salt tolerance in coastal plants. The exact mechanism of leaf tissue tolerance in *M. guttatus* is currently unknown.

Flowering Time Escape from Harsh Environmental Conditions

Reciprocal transplant experiments have demonstrated that natural selection often favors different flowering times in different environments (Ågren & Schemske, 2012; Anderson et al., 2011; Hall & Willis, 2006; Leinonen et al., 2013; Verhoeven et al., 2008; Weinig et al., 2002).

Figure 1. A) Under natural field conditions, a higher proportion of *M. nudatus* plants flower earlier in the season than *M. guttatus*. Data collected by J. Selby (unpublished) from multiple transects established in a single mixed population at the University of California McLaughlin Reserve in 2010. B) Two sympatric populations of *M. nudatus* and *M. guttatus* were grown under eight hour days. Only the *M. nudatus* plants flowered.

In the *M. guttatus* complex, differences in flowering time are often associated with differences in water availability between habitats. Adaptive differences in flowering time have been shown to occur between several early flowering species in the complex (*M. nudatus, M. pardalis, M. cupriphilus, M. laciniatus*, and *M. nasutus*) and sympatric *M. guttatus* populations (Friedman & Willis, 2013), as well as between annual and perennial populations of *M. guttatus* (Kiang & Hamrick, 1978; Lowry et al., 2008a; Martin & Willis 2007; Wu et al., 2010).

The developmental stage at which a plant will flower and the day-length required for flowering (critical photoperiod), differ between annual and perennial populations of *M. guttatus*. Annual populations of *M. guttatus* grow in habitats with very low soil moisture availability in the summer months, while perennial populations grow in streams and seeps with relatively high year-round soil moisture (Hall & Willis, 2006; Lowry, 2012; Lowry et al., 2008a). This difference in soil water availability results in divergent selection on both flowering time and critical photoperiod between annual and perennial populations (Friedman & Willis, 2013; Hall & Willis, 2006; Lowry & Willis, 2010; Lowry et al., 2008a). Populations of *M. guttatus* and closely related edaphic endemics—*M. cupriphilus, M. laciniatus, M. nudatus,* and *M. pardalis*—often occur sympatrically, but differ in microhabitat and critical photoperiod. While many populations of *M. guttatus* live in moist seeps and streams, the edaphic endemics often inhabit soils with low water-holding capacity that dry out earlier in the summer dry season. For example, *M. laciniatus* grows on moss patches in granitic outcrops (Sexton et al., 2011) that dry out rapidly after annual snow melts in the Sierra Nevada Mountains. *Mimulus nudatus* inhabits upland serpentine outcrops that dry out following the end of the California wet season. Most populations of *M. guttatus* require long days to flower; but sympatric populations of edaphic endemics flower under shorter day lengths possibly to avoid drought conditions in their rapidly drying microhabitats (Friedman & Willis, 2013). Both the serpentine endemic *M. nudatus* and the Cu mine endemic *M.*

cupriphilus can flower under day lengths as short as 8 hours, while sympatric populations of *M. guttatus* require over 11-13 hours (Figure 1; Friedman & Willis, 2013). These flowering time differences likely contribute to RI between the edaphic endemics and *M. guttatus*. Furthermore, *M. cupriphilus, M. laciniatus*, and *M. pardalis* are all self-fertilizing and these shifts of mating system may have evolved as forms of reproductive assurance (Franks, 2011; Ivey & Carr, 2012; Macnair & Gardner, 1998; Martin & Willis, 2007; Wu et al., 2010).

The Genetic Architecture of Adaptation to Harsh Environments

Much of the research aimed at understanding the genetic basis of adaptations to harsh environments in *Mimulus* has utilized quantitative trait locus (QTL) mapping approaches. However, recent technological advances provide new opportunities for using population genomics to determine the genetic basis of adaptive traits. These two approaches have different strengths and when used together they have the potential to provide a more complete picture of the genetics of adaptation than either approach in isolation.

Quantitative Trait Locus Mapping

Often, the first step to investigating the genetic basis of adaptation is to conduct QTL mapping in hybrids between divergent populations or ecotypes. Quantitative trait locus mapping has recently been criticized as an approach for understanding the genetic basis of adaptive traits based on its inability to detect small effect functional polymorphisms (Rockman, 2012) and disinterest by some evolutionary biologists in the "molecular details" of adaptation (Travisano & Shaw, 2013). Despite its detractors, QTL mapping has been crucial in identifying important loci that have advanced the understanding of fundamental evolutionary questions with regard to adaptation and speciation (reviewed in Bomblies, 2013; Coyne & Orr, 2004; Lee et al., 2014).

The genetic and genomic toolkit available for *Mimulus* has enabled numerous QTL mapping projects aimed at characterizing the genetic basis of adaptive phenotypic traits (Fishman et al., 2002; Hall et al., 2006; Lowry et al., 2009; 2013; Sweigart et al., 2006; Wright et al., 2013). These studies use traditional genotyping approaches (e.g., PCR-based markers) to investigate phenotypic variation segregating in a hybrid mapping population (F2, RIL, etc.). The availability of next-generation sequencing technologies has accelerated QTL mapping projects using a bulk segregant analysis (BSA) approach (Magwene et al., 2011; Michelmore et al., 1991). Bulk segregant analysis works through the selection of hybrid individuals from both tails of the phenotypic distribution for a particular trait. The DNA from hybrid individuals from each tail is pooled in equimolar concentrations and each pools is then sequenced. Allele frequencies at polymorphic sites across the genome are calculated for each pool. Allele frequencies at sites not associated with the phenotype should not differ between the pools, while allele frequencies at sites associated with the phenotype (QTLs) will diverge. Quantitative trait loci for photoperiod differences between *M. nasutus* and *M. guttatus*, as well as between annual and perennial populations of *M. guttatus*, have been efficiently and rapidly identified using BSA (Fishman et al., 2013; Friedman & Willis, 2013). These techniques have also been used to map QTLs for salt tolerance, serpentine tolerance, flowering time differences, and leaf shape, not only in *M. guttatus* but also in other species such as *M. laciniatus* and *M. nudatus* (J. Selby and K. Ferris, unpublished). To confirm the presence of a QTL, individual F2s are genotyped at markers in regions of the genome that showed allele frequency differences in the BSA. Genotyping F2s individually also enables estimation of the effect size of QTLs. To ultimately identify the causal functional variant underlying these QTLs, additional fine mapping studies in larger mapping populations or association mapping/genome scan approaches are needed (e.g., Yuan et al., 2013).

To investigate the genetic basis of adaption to the Cu mine habitat, Wright et al. (2013) conducted a QTL mapping experiment for Cu tolerance, measured using a lab-based hydroponic assay, and identified a single, large effect locus, *Tol1* (Wright et al., 2013). This experiment revealed strong genetic differentiation at markers in tight linkage with *Tol1*, consistent with the hypothesis that this locus was strongly selected during Cu mine colonization. Interestingly, adaptation to the mine environment has resulted in the development of a post-zygotic intrinsic reproductive isolating barrier (Macnair & Christie, 1983). The Copperopolis population of *M. guttatus* is fixed for an allele that results in F1 hybrid necrosis in crosses to plants from multiple off-mine populations (Christie & Macnair, 1984; Macnair & Christie, 1983). This incompatibility factor, *Nec1*, was fine mapped to a region in tight linkage (<1cM) with the major Cu tolerance locus, *Tol1* (Wright et al., 2013). The distribution of genetic variation between the Copperopolis population and adjacent off-mine populations suggests that strong selection on *Tol1* caused the hybrid incompatibility allele at *Nec1* to hitchhike to fixation at Copperopolis (Wright et al., 2013). This study demonstrates that natural selection on a locally adaptive trait can indirectly drive a hybrid incompatibility allele to high frequency due to tight genetic linkage.

Recent studies have also made major progress in mapping QTLs for variation in critical photoperiod between different populations and species of *Mimulus*. Differences in critical photoperiod for flowering between annual and perennial populations of *M. guttatus* are caused by two large effect QTLs (Friedman & Willis, 2013; Hall et al., 2006; Hall et al., 2010). Annual and perennial populations also differ in their vernalization requirements, and mapping studies have identified a mixture of large and small effect QTLs which contribute to these differences. The selfing species *M. nasutus* and the outcrossing species *M. guttatus* differ in the critical photoperiods at which they transition from vegetative to reproductive growth. These differences are controlled by two major effect QTLs, one of which co-localizes with one of the photoperiod QTLs identified between annual and perennial populations of *M. guttatus* (Fishman et al., 2013). One of the QTLs controlling critical photoperiod differences between *M. nasutus* and *M. guttatus* mapped near an ortholog of *Arabidopsis thaliana* FLOWERING LOCUS T/TERMINAL FLOWER 1, while the other mapped near an ortholog of the DELLA gene GIBBERELLIC ACID INSENSITIVE (Fishman et al., 2013). In all cases, these flowering time differences likely reflect adaptive divergence between habitats due to differences in the timing of low soil moisture availability.

Combining QTL and Field Experiments to Understand Adaptation

In order to identify QTLs that actually contribute to adaptive differences, mapping studies should be conducted in the field or QTLs that have been identified in laboratory-based studies should be tested for their fitness effects in native habitats. Reciprocal transplant studies that incorporate hybrid mapping populations (F2s, backcrosses, etc.) can identify adaptive traits which are strongly correlated with survival and fecundity. Hybrids can then be genotyped to identify the loci that contribute to fitness in the field. This approach has been used to map QTLs for the ability to survive on serpentine soils in the field in *M. guttatus* (Selby, 2014). Near isogenic lines (NILs), genetic lines that are identical with the exception of introgressed genetic regions of interest, can also be used to test whether specific traits and genetic loci are adaptive. For instance, a NIL may consist of a line that is identical to a local genotype except for a single foreign QTL. Near isogenic lines can be planted across environments and the fitness of each NIL can be measured to determine whether individual genetic loci are adaptive.

Lowry et al. (2009) identified three major QTLs that contribute to differences in salt spray tolerance in the laboratory between inland annual and coastal perennial *M. guttatus* populations. These salt tolerance QTLs were then evaluated for their effects on fitness in a field reciprocal transplant study using recombinant inbred lines (RILs). Interestingly, all three salt tolerance QTLs discovered in the lab had a significant effect on fitness in coastal habitats but no detectable effect on fitness in inland habitats. The fitness effect of the major Cu tolerance QTL, *Tol1*, was similarly investigated via reciprocal transplant of NILs possessing mine and off-mine alleles at *Tol1* (K. Wright, unpublished). The mine allele at *Tol1* significantly increased the probability a plant would survive to flower in the mine environment, but had no detectable effect on survival in the off-mine environment. This genotype by environment interaction in which a locus has significant fitness effects in one habitat but little or no fitness effects in a different habitat is called conditional neutrality (reviewed in Anderson et al., 2011; Kawecki & Ebert, 2004). Alleles with conditionally neutral effects are likely to reach high frequency in the population in which they are under strong selection, but have the opportunity to diffuse to other populations through gene flow.

Lowry & Willis (2010) conducted a reciprocal transplant experiment using NILs, in which a chromosomal inversion polymorphism was introgressed reciprocally into coastal perennial and inland annual *M. guttatus* genetic backgrounds. In contrast to the three salt tolerance loci that only had fitness effects in the coastal habitat, the inversion locus had contrasting fitness effects across habitats. In the inland habitat, the inland orientation of the inversion contributed to higher fitness by facilitating escape from seasonal drought via earlier flowering time. In the coastal habitat, the coastal perennial orientation of the inversion increased fitness by shifting the allocation of plant resources from flowering to growth and multi-season survival, which is advantageous in the coastal habitat because there is year-round soil moisture availability. This genotype by environment pattern of opposite fitness effects of a locus across habitats is called "antagonistic pleiotropy" (reviewed in Anderson et al., 2011; Kawecki & Ebert, 2004). Antagonistic pleiotropy will reduce gene flow between habitats at a particular locus because local alleles are advantageous over foreign ones in each environment.

Population Genomics

Recent technological advances have made whole genome population sequencing feasible for many systems, including *Mimulus* (Brandvain et al., 2014; Flagel et al., 2014; Hellsten et al., 2013). Such sequence data offer excellent opportunities to conduct population genomic analyses to identify loci involved in adaptation to harsh environments. Compared to QTL mapping, population genomic approaches can leverage numerous natural recombination events and therefore offer the potential to more precisely identify causative functional alleles. Large-scale population genomic studies have detected associations between nucleotide variation and climate across broad geographic spaces (Coop et al., 2009; Hancock et al., 2011; Lasky et al., 2014).

However, there are potential problems associated with population genomic studies. For example, the demographic history and population structure of sampled genomes can lead to false positive associations of alleles with environmental variables (Coop et al., 2010). Many methods have been developed to control for demography and population structure, but these can lead to false negatives if adaptive alleles are correlated spatially with population structure. Furthermore, it is very difficult to establish demographic history with confidence and virtually impossible to control for phenomena like "allelic surfing" (Excoffier & Ray, 2008). In addition, most population genomic studies have failed to account for the structure of the genome itself which can also lead to the discovery of false positive "outlier" loci. For example, intrinsic differences in recombination rate across the genome can skew the fixation index (F_{ST}) and other summary statistics by affecting local levels of nucleotide diversity (Cruickshank & Hahn, 2014; Lowry et al., 2013; Renaut et al., 2013). Finally, population genomics studies cannot distinguish between loci of major effect versus those that may be subtle modifiers.

Given the potential limitations of population genomic studies, careful consideration is advised in developing the sampling strategy of focal populations. For instance, sampling adjacent populations in habitats that differ sharply in environmental conditions will minimize the effects of population structure.

Additionally, by sampling multiple independent pairs of adjacent populations in different habitats, researchers can more reliably identify alleles that are selected in each environment as well as test whether the same or different genes have been used by different populations in adapting to similar habitats. Investigating the genomic basis of adaptation to edaphic conditions is ideally suited to this experimental design because the environmental gradients between soil types are often discontinuous and multiple pairs of divergently adapted populations are often located within a single region. We further advocate that comparing results from genome resequencing studies with other lines of evidence (QTL mapping, functional molecular biology, and reciprocal transplant experiments) is the most thorough way to confidently identify loci involved in adaptation to environmental variation across space.

We are currently conducting such experiments to investigate the genomic basis of adaptation to Cu and serpentine soils in *M. guttatus*. Finally, it should be noted that many analytical methods and technologies are still being developed which will likely improve the utility of population genomics.

High-Throughput Phenotyping

High-throughput, laboratory-based assays can assist in elucidating the function of QTLs that affect fitness in the field, in identifying specific traits that contribute to adaptive divergence, and in assisting fine scale genetic mapping to identify causal loci. However, development of robust laboratory-based assays that mimic the selective environment experienced by plants in native habitats is challenging. In this section, we discuss experimental designs we have used to study edaphic adaption and salt tolerance as well as highlight some of the challenges we encountered during the development of these assays.

Laboratory Edaphic Assays

The handful of QTL mapping studies that have evaluated the genetic architecture of serpentine adaptation (Bratteler et al., 2006; Burrell et al., 2012) have exclusively mapped QTLs that confer tolerance to a single soil chemical variable isolated in altered liquid nutrient feeds. We recognize the powerful insight that such hydroponic methods can provide; however, none of the QTLs that confer tolerance to an isolated soil chemical variable have been tested for their effects on plant fitness in native soils. The interactions between different ions (Brooks, 1987; Gabbrielli & Pandolfini, 1984) as well as the physical properties of the soils are likely to contribute to adaptation to serpentine habitats. To account for the full suite of selective factors associated with serpentine soils or mine tailings, mapping experiments should be conducted in native soils. For example, the genetic basis of serpentine tolerance in *M. guttatus* was originally investigated by planting F2s in the field and conducting a BSA on the survivors from serpentine and non-serpentine field sites. We have also grown F2 mapping populations on native serpentine soil in the lab. By planting seeds either in plug trays or on tissue-culture plates filled with serpentine soil, we are easily able to screen 1000s of F2s for juvenile survival which has enabled rapid fine-mapping of a major serpentine tolerance QTL in *M. guttatus* (Selby, 2014).

Mapping QTLs for hydroponic tolerance to isolated soil chemical variables and testing for co-localization of these QTLs with field fitness QTLs will enable researchers to begin to identify the mechanisms that contribute to fitness differences between populations. We have developed a high-throughput hydroponic platform to assay plant tolerance to individual soil chemical variables. Tolerance manifests as a differential response to a treatment medium. This differential response is typically observed by measuring a plant growth parameter (height, biomass, etc.) in both a treatment and a control solution. The ratio of plant size in treatment versus control solution is used as an index of tolerance to control for inherent size differences. This design requires that genetically identical individuals be grown in both treatment and control solutions. However, taking clones from large, robust plants that have already acclimated to a benign growth environment fails to mimic how plants would experience soils with altered nutrient profiles in the field. Therefore, we have developed a seedling assay based on the sequential testing method of Schat and ten Bookum (1992). Single genotypes are grown in increasingly severe treatment solutions and root growth rate is scored in each treatment level. For each individual, the treatment concentration that stops root growth, referred to as the "Effective Concentration 100%" (EC100), is scored. This strategy controls for inherent differences in root growth rate without the requirement that individuals also be grown in a control solution.

We designed a growth platform that has enabled high-throughput hydroponic studies. Watertight boxes were constructed out of PVC foam board (11.5" x 5" x 7.5") with removable lids with holes (4 rows of 17) through which drinking straws are suspended into the solution (Figure 2). Seeds are sown on an inert rockwool medium stuffed into 200uL PCR tubes with the tips clipped off. The tubes are then placed into the holes in the lid of each box and seeds are allowed to germinate and grow in a nutrient solution (¼ strength Hoagland's, prepared as described by Epstein (1972) until most seedlings have roots protruding through the bottom of the rockwool (~7 days).

The position of root tips is then tracked for each plant by sliding a small dental rubber band around the straws. Every two days, the position of the root tip is marked and the treatment solution is changed. At the end ofa series of increasingly severe treatments, the distance between the rubber bands is measured, providing root growth rates in each treatment level from which EC100 can be calculated. We have successfully used this design to assay populations of *M. guttatus* for differences in tolerance to low Ca:Mg ratio, high Ni, and high Cu, and have also used this method to map QTLs for these tolerance differences (A. Jeong, J. Selby, & K. Wright, unpublished).

Figure 2. Photos of high-throughput hydroponic growth platform for root growth assays. A) Roots growing in straws with rubber bands marking positions of root tips in each treatment level. B) Front view of box at end of experiment with tubes connected to an air pump and rosettes of plants on top. C) Top of box showing *M. guttatus* seedlings at the start of the experiment.

Laboratory Salt Tolerance Assays

We have developed multiple methods for assessing the salt tolerance of individual plants for the purpose of genetic mapping in *M. guttatus*. Lowry et al. (2009) initially used a salt spray assay to map QTLs in RILs from a cross between inland annual and coastal perennial lines of *M. guttatus*. Plants were sprayed with 5mL of 500mM NaCl solution every other day. Plants were then scored for the day which they no longer had any green tissue (see Lowry et al., 2008b, 2009 for complete methods). This experiment identified three major salt tolerance QTLs, which were subsequently tested for their fitness effects in the field.

While the salt spray assay yielded the localization of three QTLs, it was difficult to conduct without introducing considerable random heterogeneity into the experiment due to some plants receiving more spray than others. A larger subsequent QTL mapping experiment was ultimately abandoned because the date of death for the salt spray assay was highly variable within inbred lines.

Following this setback, we sought a better methodology for assessing salt tolerance of individual plants. We first developed two types of assays with agar plates that contained various levels of NaCl in the media. One assay involved growing seedlings on plates and the other involved transferring hole punches from adult leaves to plates following the methods of Prasad et al. (2000). Both assays proved useful for detecting differences in tolerance between coastal perennial and inland annual populations. However, neither method was well suited for QTL mapping with large populations of hybrids. Assays involving plates with NaCl were generally plagued by large block effects between plates. Further, the transfer of seedlings to plates is problematic in general as it introduces a major shock effect since plants have no time to acclimate to the stress as they would in nature (Juenger et al., 2010; Munns & Tester, 2008)

We recently developed a hydroponic assay that is much more promising for future genetic studies of salt tolerance. This method involves growing plants in perlite with ½ strength Hoagland's solution as a nutrient media. The plants are initially grown for 2-4 weeks at a 0mM concentration of NaCl. The salinity treatment is then increased by 25mM increments each day to allow the plants to acclimate. Once the treatment solution reaches a concentration of 150mM NaCl it is no longer increased, but simply replaced every three days. Appropriate levels of calcium must be added to the solution once the NaCl treatments are initiated because sodium can interfere with the assimilation of calcium (Al-Harbi et al., 1995; Wakeel et al., 2009). Using this set-up, there are clear differences in date of death between coastal perennial and inland annual *M. guttatus* plants at the final treatment concentration of 150mM NaCl (D. Lowry, unpublished).

Conclusion

Mimulus has been established as a model system for investigating the genetic basis of adaptation to harsh environmental conditions. Significant progress has been made in understanding the genetic architecture of adaptive flowering time escape from seasonal drought, adaptation to toxic soils, and salt tolerance. The combination of classic reciprocal transplant experiments with modern molecular genetics has led to a deeper understanding of how individual loci contribute to adaptations across habitats (Hall et al., 2010; Lowry & Willis, 2010). The detailed genetic dissection of tolerance to edaphic conditions has revealed new insights into the mechanisms by which natural selection can drive the formation of reproductive isolation (Wright et al., 2013). The importance of critical photoperiod in timing the initiation of flowering to avoid seasonally harsh conditions (Fishman et al., 2013;

Friedman & Willis, 2013) has also been demonstrated and genetic dissection of these critical photoperiod differences is now underway in multiple *Mimulus* laboratories. While QTLs have been localized for many traits involved in adaptation to harsh environments, the actual genes that underlie these adaptations have thus far remained elusive. Fine genetic mapping of these QTLs has brought us closer to identifying the causal genes. However, these efforts can sometimes take over a decade to accomplish, even in model systems such as *Arabidopsis thaliana* (Des Marais et al., 2014). Further difficulties in identifying the causal locus can arise from genome assembly issues, as has been the case for identifying the major Cu tolerance locus in *M. guttatus* (Wright et al., 2013). Despite these challenges, we are optimistic that the combination of genetic mapping with new phenotyping methods and population genomic approaches will yield new insights into the evolution of adaptation to harsh environmental conditions. Beyond improving our understanding of evolutionary mechanisms, studies of the genetic basis of adaptation to harsh environmental conditions have many potential applied benefits. Many of the selective pressures present in these extreme habitats—water-limitation; toxic, nutrient poor soils; thermal extremes—are likely to increase as a result of climate change (see Chapters 7, 13), pollution, and other human-induced environmental impacts. Elucidating the genetic basis of adaptation to these stresses will have important applications for food security, restoration of polluted sites, and conservation of critical habitats that are the drivers of biological diversity.

References

Ågren, J. & Schemske, D. W. (2012) Reciprocal transplants demonstrate strong adaptive differentiation of the model organism *Arabidopsis thaliana* in its native range. *New Phytologist 194*, 1112-1122.

Ågren, J., Oakley, C. G., McKay, J. K., Lovell, J. T. & Schemske, D. W. (2013) Genetic mapping of adaptation reveals fitness tradeoffs in *Arabidopsis thaliana*. *Proceedings of the National Academy of Sciences (USA) 110*, 21077-82.

Al-Harbi, A. R. (1995) Growth and nutrient composition of tomato and cucumber seedlings as affected by sodium chloride salinity and supplemental calcium. *Journal of Plant Nutrition 18*, 1403-1416.

Alexander, E. B., Coleman, R. G., Keeler-Wolf, T. & Harrison, S. P. (2007) *Serpentine geoecology of Western North America: Geology, soils, and vegetation*. New York, NY: Oxford University Press.

Allen, W. R. & Sheppard, P. M. (1971) Copper tolerance in some Californian populations of the Monkey Flower, *Mimulus guttatus*. *Philosophical Transactions of the Royal Society Series B 177*, 177–96.

Anderson, J. T., Willis, J. H. & Mitchell-Olds, T. (2011) Evolutionary genetics of plant adaptation. *Trends in Genetics 27*, 258-266.

Antonovics, J. & Bradshaw, A. D. (1970) Evolution in closely adjacent plant populations VIII. Clinal patterns at a mine boundary. *Heredity 25*, 349-362.

Beardsley, P. M. & Olmstead, R. G. (2002) Redefining Phrymaceae: The placement of *Mimulus*, tribe Mimuleae, and *Phryma*. *American Journal of Botany 89*, 1093-1102.

Bomblies, K. (2013) Genes causing postzygotic hybrid incompatibility in plants: A window into co-evolution. In: Z. J. Chen & J. A. Birchler (Eds.). *Polyploid and hybrid genomics* (pp. 223-239). Oxford: John Wiley & Sons, Inc.

Boyce, S. G. (1954) The salt spray community. *Ecological Monographs, 24*, 29-67.

Baxter, I., Brazelton, J. N., Yu, D., Huang, Y. S., Lahner, B., Yakubova, E., Li, Y., Bergelson, J., Borevitz, J. O., Nordborg, M., Vitek, O. & Salt, D. E. (2010) A coastal cline in sodium accumulation in *Arabidopsis thaliana* is driven by natural variation of the sodium transporter AtHKT1; 1. *PLoS Genetics 6*, e1001193.

Bradshaw, A. D. (1991) The Croonian lecture, 1991: Genostasis and the limits to evolution. *Philosophical Transactions of the Royal Society Series B 333*, 289-305.

Brady, K. U., Kruckeberg, A. R. & Bradshaw, H. D. Jr. (2005) Evolutionary ecology of plant adaptation to serpentine soils. *Annual Review of Ecology, Evolution, and Systematics 36*, 243–66.

Brandvain, Y., Kenney, A. M., Flagel, L., Coop, G. & Sweigart, A. L. (2014) Speciation and introgression between *Mimulus nasutus* and *Mimulus guttatus*. *PLOS Genetics 10*, e1004410.

Bratteler, M., Lexer, C. & Widmer, A. (2006) Genetic architecture of traits associated with serpentine adaptation of *Silene vulgaris. Journal of Evolutionary Biology 19*, 1149-1156.

Brooks, R. R. (1987) *Serpentine and its vegetation: A multidisciplinary approach*. Portland, OR: Dioscorides Press.

Bunn, R. A. & Zabinski, C. A. (2003) Arbuscular mycorrhizae in thermal-influenced soils in Yellowstone National Park. *Western North American Naturalist 63*, 409-415.

Burrell, A. M., Hawkins, A. K. & Pepper, A. E. (2012) Genetic analyses of nickel tolerance in a North American serpentine endemic plant, *Caulanthus amplexicaulis var. barbarae* (Brassicaceae). *American Journal of Botany 99*, 1875-1883.

Christie, P. & Macnair, M. R. (1984) Complementary lethal factors in two North American populations of the yellow monkey flower. *Journal of Heredity 75*, 510-511.

Clausen, J. (1951) *Stages in the evolution of plant species*. Ithaca, NY: Cornell University Press.

Clausen, J. & Hiesey, W. M. (1958) *Experimental studies on the nature of species. IV. Genetic structure of ecological races*. Washington, DC: Carnegie Institution of Washington Publication 615.

Clausen, J., Keck, D. D. & Heisey, W. M. (1940) *Experimental studies on the nature of species. I. Effect of varied environments on western North American plants*. Washington, DC: Carnegie Institute Publication no. 520.

Clemens, S. (2001) Molecular mechanisms of plant metal tolerance and homeostasis. *Planta 212*, 475-486.

Colangelo, E. P. & Guerinot, M. L. (2006) Put the metal to the petal: Metal uptake and transport throughout plants. *Current Opinion in Plant Biology 9*, 322-330.

Consortium of California Herbaria (2014). Online: http://ucjeps.berkeley.edu/consortium/

Cooley, A. M., Modliszewski, J. L., Rommel, M. L. & Willis, J. H. (2011) Gene duplication in *Mimulus* underlies parallel floral evolution via independent *trans*-regulatory changes. *Current Biology 21*, 700-704.

Coop, G., Pickrell, J. K., Novembre, J., Kudaravalli, S., Li, J., Absher, D., Myers, R. M., Cavalli-Sforza, L. L., Feldman, M. W. & Pritchard, J. K. (2009) The role of geography in human adaptation. *PLoS Genetics 5*, e1000500.

Coop, G., Witonsky, D., Di Rienzo, A. & Pritchard, J. K. (2010) Using environmental correlations to identify loci underlying local adaptation. *Genetics*, *185*, 1411-1423.

Coyne, J. A. & Orr, H. A. (2004) *Speciation*. Sunderland, MA: Sinauer Associates.

Cruickshank, T. E. & Hahn, M. W. (2014) Reanalysis suggests that genomic islands of speciation are due to reduced diversity, not reduced gene flow. *Molecular Ecology 23*, 3133-3157.

Delmer, D. P. (1974) Studies on the nature of the adaptations of the monkey flower, *Mimulus guttatus*, to a thermophilic environment. *Canadian Journal of Botany 52,* 1509-1514.

Des Marais, D. L., Auchincloss, L. C., Sukamtoh, E., McKay, J. K., Logan, T., James, H., Richards, J. H. & Juenger, T. E. (2014) Variation in MPK12 affects water use efficiency in *Arabidopsis* and reveals a pleiotropic link between guard cell size and ABA response. *Proceedings of the National Academy of Sciences (USA) 111,* 2836–2841.

Douglas, D. A. (1981) The balance between vegetative and sexual reproduction of *Mimulus primuloides* (Scrophulariaceae) at different altitudes in California. *Journal of Ecology 69*, 295-310.

Epstein, E. (1972) *Mineral nutrition of plants: Principles and perspectives*. New York, NY: John Wiley and Sons.

Excoffier, L. & Ray, N. (2008) Surfing during population expansions promotes genetic revolutions and structuration. *Trends in Ecology & Evolution 23*, 347-351.

Ferris, K., Sexton, J. P. & Willis, J. H. (*In press)* Speciation of a rare and cryptic rock outcrop specialist in the yellow monkey flowers. *Philosophical Transactions of the Royal Society Series B*.

Fishman, L., Kelly, A. J. & Willis, J. H. (2002) Minor quantitative trait loci underlie floral traits associated with mating system divergence in *Mimulus. Evolution 56*, 2138-2155.

Fishman, L., Sweigart, A. L., Kenney, A. M. & Campbell, S. (2013) Major quantitative trait loci control divergence in critical photoperiod for flowering between selfing and outcrossing species of monkeyflower (*Mimulus*). *New Phytologist 201*, 1498-1507.

Flagel, L. E., Willis, J. H & Vision, J. T. (2014) The standing pool of genomic structural variation in a natural population of *Mimulus guttatus*. *Genome Biology and Evolution 6*, 53-64.

Franks, S. J. (2011) Plasticity and evolution in drought avoidance and escape in the annual plant *Brassica rapa*. *New Phytologist 190*, 249-257.

Friedman, J. & Willis, J. H. (2013) Major QTLs for critical photoperiod and vernalization underlie extensive variation in flowering in the Mimulus guttatus species complex. *New Phytologist 199*, 571-583.

Gabbrielli, R. & Pandolfini, T. (1984) Effect of Mg2+ and Ca2+ on the response to nickel toxicity in a serpentine endemic and nickel-accumulating species. *Physiologia Plantarum 62*, 540-544.

Gardner, M. & Macnair, M. (2000) Factors affecting the co-existence of the serpentine endemics *Mimulus nudatus* Curran and its presumed progenitor, *Mimulus guttatus* Fisher ex DC. *Biological Journal of the Linnean Society 69*, 443-459.

Gillespie, J. H. & Turelli, M. (1989) Genotype-environment interactions and the maintenance of polygenic variation. *Genetics 121*, 129-38.

Goodstein, D. M., Shu, S., Howson, R., Neupane, R., Hayes, R. D., Fazo, J., Mitros, T., Dirks, W., Hellsten, U., Putnam, N. Ä. & Rokhsar, D. S. (2012) Phytozome: A comparative platform for green plant genomics. *Nucleic Acids Research 40*, D1178-D1186.

Griffin, N. (2010) Inbreeding depression and competition in the square-stemmed monkey-flower (*Mimulus ringens*). PhD Dissertation. St. Louis, MO: Washington University in St. Louis.

Hall, M. C. & Willis, J. H. (2006) Divergent selection on flowering time contributes to local adaptation in *Mimulus guttatus* populations. *Evolution 60*, 2466-2477.

Hall, M. C., Basten, C. & Willis, J.H. (2006) Pleiotropic quantitative trait loci contribute to population divergence in traits associated with life-history variation in *Mimulus guttatus. Genetics 172*, 1829-1844.

Hall, M. C., Lowry, D. B. & Willis, J. H. (2010) Is local adaptation in *Mimulus guttatus* caused by trade-offs at individual loci? *Molecular Ecology 19, 2739*-2753.

Hancock, A. M., Brachi, B., Faure, N., Horton, M. W., Jarymowycz, L. B., Sperone, F. G., Toomajian, C., Roux, F. & Bergelson, J. (2011) Adaptation to climate across the *Arabidopsis thaliana* genome. *Science 334*, 83-86.

Hanikenne, M., Talke, I. N., Haydon, M. J., Lanz, C., Nolte, A., Motte, P., Kroymann, J., Weigel, D. & Krämer, U. (2008) Evolution of metal hyperaccumulation required cis-regulatory changes and triplication of HMA4. *Nature 453*, 391-395.

Hedrick, P.W. (1986) Genetic polymorphism in heterogeneous environments: A decade later. *Annual Review of Ecology and Systematics 17*, 535-66.

Heller, A. A. (1904) Western species, new and old. II. *Muhlenbergia* 1, 47-62.

Hellsten, U., Wright, K. M., Jenkins, J., Shu, S., Yuan, Y., Wessler, S. R. & Rokhsar, D. S. (2013) Fine-scale variation in meiotic recombination in *Mimulus* inferred from population shotgun sequencing. *Proceedings of the National Academy of Sciences (USA) 110*, 1947819482.

Hereford, J. (2009) A quantitative survey of local adaptation and fitness trade-offs. *The American Naturalist 173*, 579-588.

Hughes, R., Bachmann, K., Smirnoff, N. & Macnair, M. R. (2001) The role of drought tolerance in serpentine tolerance in the *Mimulus guttatus* Fischer ex DC. complex. *South African Journal of Science 97*, 581-586

Ivey, C. T. & Carr, D. E. (2012) Tests for the joint evolution of mating system and drought escape in *Mimulus*. *Annals of Botany 109*, 583-598.

Juenger, T. E., Sen, S., Bray, E., Stahl, E., Wayne, T., McKay, J. K. & Richards, J. H. (2010) Exploring genetic and expression differences between physiologically extreme ecotypes: comparative genomic hybridization and gene expression studies of Kas-1 and Tsu-1 accessions of *Arabidopsis thaliana. Plant, Cell and Environment 33*, 1268-84.

Kawecki, T. J. & Ebert, D. (2004) Conceptual issues in local adaption. *Ecology Letters 7*, 1225-1241.

Kiang, Y. T. & Hamrick, J. L. (1978) Reproductive isolation in the *Mimulus guttatus – M. nasutus* complex. *American Midland Naturalist 100*, 269–276.

Kobayashi, Y. & Weigel, D. (2007) Move on up, it's time for change—mobile signals controlling photoperiod-dependent flowering. *Genes & Development,* 21, 2371-2384.

Kruckeberg, A. R. (1951) Intraspecific variability in the response of certain native plant species to serpentine soil. *American Journal of Botany 38*, 408-419.

Lasky, J. R., Des Marais, D. L., Lowry, D. B., Povolotskaya, I., McKay, J. K., Richards, J. H., Keitt, T. H. & Juenger, T. E. (2014) Natural variation in abiotic stress responsive gene expression and local adaptation to climate in Arabidopsis thaliana. Molecular Biology and Evolution, DOI: 10.1093/molbev/msu170.

Lee, Y. W., Gould, B. A. & Stinchcombe, J. R. (2014) Identifying the genes underlying quantitative traits: A rationale for the QTN programme. *AoB Plants 6*, plu004.

Leimu, R. & Fischer, M. (2008) A meta-analysis of local adaptation in plants. *PLoS One 3*, e4010.

Leinonen, P. H., Remington, D. L., LeppÄLÄ, J. & Savolainen, O. (2013) Genetic basis of local adaptation and flowering time variation in *Arabidopsis lyrata. Molecular Ecology 22*, 709-723.

Lekberg, Y., Roskilly, B., Hendrick, M., Zabinski, C., Barr, C. & Fishman, L. (2012) Phenotypic and genetic differentiation among yellow monkeyflower populations from thermal and non-thermal soils in Yellowstone National Park. *Oecologia, 170*, 111-122.

Levene, H. (1953) Genetic equilibrium when more than one ecological niche is available. *The American Naturalist 87*, 331-333.

Lexer, C., Welch, M. E., Durphy, J. L. & Rieseberg, L. H. (2003) Natural selection for salt tolerance quantitative trait loci (QTLs) in wild sunflower hybrids: Implications for the origin of *Helianthus paradoxus*, a diploid hybrid species. *Molecular Ecology 12*, 1225-1235.

Lowry, D. B. (2012) Ecotypes and the controversy over stages in the formation of new species. *Biological Journal of the Linnean Society 106*, 241-257.

Lowry, D. B. & Willis, J. H. (2010) A widespread chromosomal inversion polymorphism contributes to a major life-history transition, local adaptation, and reproductive isolation. *PLoS Biology 8*, e1000500.

Lowry, D. B., Modliszewski, J. L., Wright, K. M., Wu, C. A. & Willis, J. H. (2008a) The strength and genetic basis of reproductive isolating barriers in flowering plants. *Philosophical Transactions of the Royal Society Series B 363*, 3009-3021.

Lowry, D. B., Rockwood, R. C. & Willis, J. H. (2008b) Ecological reproductive isolation of coast and inland races of *Mimulus guttatus*. *Evolution 62*, 2196-2214.

Lowry, D. B., Hall, M. C., Salt, D. E. & Willis, J. H. (2009) Genetic and physiological basis of adaptive salt tolerance divergence between coastal and inland *Mimulus guttatus. New Phytologist 183*, 776–88.

Lowry, D. B., Logan, T. L., Santuari, L., Hardtke, C. S., Richards, J. H., DeRose-Wilson, L. J., McKay, J. K., Sen, S. & Juenger, T. E. (2013) Expression QTL mapping across water availability environments reveals contrasting associations with genomic features in *Arabidopsis thaliana. The Plant Cell 25*, 3266–3279.

Ludlow, M. M. (1989) Strategies of response to water stress. In: K. Kreeb, C. Richter & T. Hinckley (Eds.). *Structural and functional responses to environmental stresses: Water shortage* (pp. 269–281). The Hague: Academic Publishing.

Macnair, M. R. (1981) The uptake of copper by families of *Mimulus guttatus* differing in genotype primarily at a single copper tolerance locus. *New Phytologist 88*, 723-730.

Macnair, M. R. (1987) Heavy metal tolerance in plants: A model evolutionary system. *Trends in Ecology and Evolution 2*, 354-359.

Macnair, M. R. (1989) The potential for rapid speciation in plants. *Genome 31*, 203–10.

Macnair, M. R. & Christie, P. (1983) Reproductive isolation as a pleiotropic effect of copper tolerance in *Mimulus guttatus*. *Heredity 50*, 295–302.

Macnair, M. R. & Gardner, M. (1998) The evolution of edaphic endemics. In: D. J. Howard & S. H. Berlocher (Eds.). Endless forms: Species and speciation (pp. 157-171). Oxford: Oxford University Press.

Macnair, M. R., Macnair, V. E., & Martin, B. E. (1989) Adaptive speciation in *Mimulus*: An ecological comparison of *M. cupriphilus* with its presumed progenitor, *M. guttatus. New Phytologist 112*, 269-279.

Macnair, M. R., Smith, S. E. & Cumbes, Q. J. (1993) Heritability and distribution of variation in degree of copper tolerance in *Mimulus guttatus* at Copperopolis, California. *Heredity 71*, 445-445.

Magwene, P. M., Willis, J. H. & Kelly, J. K. (2011) The statistics of bulk segregant analysis using next generation sequencing. *PLoS Computational Biology 7*, e1002255.

Martin, N. H. & Willis, J. H. (2007) Ecological divergence associated with mating system causes nearly complete reproductive isolation between sympatric *Mimulus* species. *Evolution 61*, 68-82.

McKay, J. K., Richards, J. H. & Mitchell-Olds, T. (2003) Genetics of drought adaptation in *Arabidopsis thaliana*: I. Pleiotropy contributes to genetic correlations among ecological traits. *Molecular Ecology 12*, 1137-1151.

Meindl, G., Bain, D. & Ashman, T.-L. (2013) Edaphic factors and plant–insect interactions: direct and indirect effects of serpentine soil on florivores and pollinators. *Oecologia 173*, 1355-1366.

Mendez-Vigo, B., Pico, F. X., Ramiro, M., Martinez-Zapater, J. M. & Alonso-Blanco, C. (2011) Altitudinal and climatic adaptation is mediated by flowering traits and FRI, FLC, andPHYC genes in *Arabidopsis*. *Plant Physiology 157*, 1942–1955.

Michelmore, R. W., Paran, I. & Kesseli, R. V. (1991) Identification of markers linked to disease-resistance genes by bulked segregant analysis: a rapid method to detect markers in specific genomic regions by using segregating populations. *Proceedings of the National Academy of Sciences (USA) 88*, 9828-9832.

Munns, R. & Tester, M. (2008) Mechanisms of salinity tolerance. *Annual Review of Plant Biology 59*, 651-681.

Munns, R., James, R. A. & Läuchli, A. (2006) Approaches to increasing the salt tolerance of wheat and other cereals. *Journal of Experimental Botany 57*, 1025-1043.

Murren, C. J., Douglass, L., Gibson, A. & Dudash, M. (2006) Individual and combined effects of Ca/Mg ratio and water availability on trait expression in *Mimulus guttatus*. *Ecology 87*, 2591-2602.

Nesom, G. (2012) Taxonomy of Erythranthe sect. Simiolus (Phrymaceae) in the USA and Mexico. *Phytoneuron 40*, 1–123.

O'Dell, R. E. & Rajakaruna, N. (2011) Intraspecific variation, adaptation, and evolution. In: S. H. Harrison & N. Rajakaruna (Eds.). *Serpentine: The evolution and ecology of a model system* (pp. 97-137). Berkeley, CA: University of California Press.

Palm, E., Brady, K. & Van Volkenburgh, E. (2012) Serpentine tolerance in Mimulus guttatus does not rely on exclusion of magnesium. *Functional Plant Biology 39*, *679*-688.

Pennell, F. W. (1947) Some hitherto undescribed Scrophulariaceae of the Pacific states. *Proceedings of the Academy of Natural Sciences of Philadelphia 99*, 151-171.

Peterson, M. L., Rice, K. J. & Sexton, J. P. (2013) Niche partitioning between close relatives suggests trade-offs between adaptation to local environments and competition. *Ecology and Evolution 3*, 512-522.

Prasad, K. V. S. K., Sharmila, P., Kumar, P. A. & Saradhi, P. P. (2000) Transformation of *Brassica juncea* (L.) Czern with bacterial codA gene enhances its tolerance to salt stress. *Molecular Breeding 6*, 489-499.

Prasad, K. V. S. K, Song, B-H., Olson-Manning, C., Anderson, J. T., Lee, C.-R., Schranz, M. E., Windsor, A. J., Clauss, M. J., Manzaneda, A. J., Naqvi, I., Reichelt, M., Gershenzon, J., Schuler, M. A. & Mitchell-Olds, T. (2012) A gain-of-function polymorphism controlling complex traits and fitness in nature. *Science 337*, 1081-1084.

Renaut, S., Grassa, C. J., Yeaman, S., Moyers, B. T., Lai, Z., Kane, N. C., Bowers, E., Burke, J. M. & Rieseberg, L. H. (2013) Genomic islands of divergence are not affected by geography of speciation in sunflowers. *Nature Communications 4*, 1827. DOI: 10.1038/ncomms2833.

Rockman, M. V. (2012) The qtn program and the alleles that matter for evolution: All that's gold does not glitter. *Evolution 66*, 1-17.

Rundle, H. D. & Nosil, P. (2005) Ecological speciation. *Ecology Letters* 8, 336-352.

Rus, A., Baxter, I., Muthukumar, B., Gustin, J., Lahner, B., Yakubova, E. & Salt, D. E. (2006) Natural variants of At*HKT1* enhance Na+ accumulation in two wild populations of *Arabidopsis*. *PLoS Genetics 2*, e210.

Schluter, D. & Conte, G. L. (2009) Genetics and ecological speciation. *Proceedings of the National Academy of Sciences (USA) 106(Supplement 1)*, 9955-9962.

Selby, J. P. (2014) The genetic basis of local adaptation to serpentine soils in *Mimulus guttatus*. PhD Dissertation. Durham, SC: Duke University.

Sexton, J. P., Strauss, S. Y. & Rice, K. J. (2011) Gene flow increases fitness at the warm edge of a species' range. *Proceedings of the National Academy of Sciences (USA) 108*, 11704-11709.

Schat, H. & ten Bookum, W. M. (1992) Genetic control of copper tolerance in *Silene vulgaris*. *Heredity 68*, 219-229.

Schat, H., Vooijs, R. & Kuiper, E. (1996) Identical major gene loci for heavy metal tolerances that have independently evolved in different local populations and subspecies of *Silene vulgaris*. *Evolution 50*, 1888-1895.

Streisfeld, M. A., Young, W. N. & Sobel, J. M. (2013) Divergent selection drives genetic differentiation in an R2R3-Myb transcription factor that contributes to incipient speciation in *Mimulus aurantiacus*. *PLoS Genetics 9*, e1003385.

Susič, N., Bohanec, B. & Murovec, J. (2014) Agrobacterium tumefaciens-mediated transformation of bush monkey-flower (*Mimulus aurantiacus* Curtis) with a new reporter gene ZsGreen. *Plant Cell, Tissue and Organ Culture 116*, 243-251.

Sweigart, A. L., Fishman, L. & Willis, J. H. (2006) A simple genetic incompatibility causes hybrid male sterility in *Mimulus*. *Genetics 172*, 2465-2479.

Tilstone, G. H. & Macnair, M. R. (1997) Nickel tolerance and copper-nickel co-tolerance in *Mimulus guttatus* from copper mine and serpentine habitats. *Plant and Soil 191*,173–80.

Travisano, M. & Shaw, R. G. (2013) Lost in the map. *Evolution 67*, 305-314.

Verhoeven, K. J. F., Poorter, H., Nevo, E. & Biere, A. (2008) Habitat-specific natural selection at a flowering-time QTL is a main driver of local adaptation in two wild barley populations. *Molecular Ecology 17*, 3416–3424.

Vickery, R. K. (1952) A study of the genetic relationships in a sample of the *Mimulus guttatus* complex. PhD Dissertation. Menlo Park, CA: Stanford University.

Vickery, R. K. (1964) Barriers to gene exchange between members of the *Mimulus guttatus* complex. *Evolution 18*, 52–69.

Vickery, R. (1978) Case studies in the evolution of species complexes in *Mimulus*. *Evolutionary Biology 11*, 405–507.

Wakeel, A., Abd-El-Motagally, F., Steffens, D. & Schubert, S. (2009) Sodium-induced calcium deficiency in sugar beet during substitution of potassium by sodium. *Journal of Plant Nutrition and Soil Science 172*, 254-260.

Weinig, C., Ungerer, M. C., Dorn, L. A., Kane, N. C., Halldorsdottir, S. S., Toyonaga, Y., Mackay, T. F. C., Purugganan, M. D. & Schmitt, J. (2002) Novel loci control reproductive timing in *Arabidopsis thaliana* in natural environments. *Genetics 162*, 1875-1884.

Wright, K. M., Lloyd, D., Lowry, D. B., Macnair, M. R. & Willis, J. H. (2013) Indirect evolution of hybrid lethality due to linkage with selected locus in *Mimulus guttatus*. *PLoS Biology 11*, e1001497.

Wu, L., Bradshaw, A. D., Thurman, D.A. (1975) The potential for evolution of heavy metal tolerance in plants: III. The rapid evolution of copper tolerance in *Agrostis stolonifera*. *Heredity 34*, 165-187.

Wu, C. A., Lowry, D. B., Cooley, A. M., Wright, K. M., Lee, Y. W. & Willis, J. H. (2008) *Mimulus* is an emerging model system for the integration of ecological and genomic studies. *Heredity 100*, 220-230.

Wu, C. A., Lowry, D. B., Nutter, L. I. & Willis, J. H. (2010) Natural variation for drought response traits in the *Mimulus guttatus* species complex. *Oecologia 162*, 23-33.

Yuan, Y.-W., Sagawa, J. M., Young, R. C., Christensen, B. J. & Bradshaw, H. D. (2013) Genetic dissection of a major anthocyanin QTL contributing to pollinator-mediated reproductive isolation between sister species of *Mimulus*. *Genetics* 194, 255-263.

Zhu, J.-K. (2001) Plant salt tolerance. *Trends in Plant Science 6*, 66-71.

In: Plant Ecology and Evolution in Harsh Environments ISBN: 978-1-63321-955-7
Editors: N. Rajakaruna, R. S. Boyd and T. B. Harris

Chapter 12

BRYOPHYTES: SURVIVAL IN A DRY WORLD THROUGH TOLERANCE AND AVOIDANCE

Dale H. Vitt*, Barbara Crandall-Stotler and Andrew Wood
Department of Plant Biology, Southern Illinois University,
Carbondale, IL, USA

ABSTRACT

Drought provides a major stress in the lives of many plants, especially those with limited abilities to retain water for long periods of time for physiological processes. Mosses, liverworts, and hornworts are among plants that are poikilohydric, yet many species within these three lineages occupy drought-prone habitats and in fact can occur in some of the harshest conditions on the planet. Their success in establishing and occupying these habitats is largely due to their physiological tolerance to desiccation, whereby individuals survive complete loss of free water. Additionally, some species have a number of morphological adaptations or life history features that enhance their abilities to withstand dehydration. We argue here that many members of both moss and liverwort lineages are desiccation tolerant, allowing them to survive in periodically dry habitats, and that in many of the moss lineages this physiological desiccation tolerance is remarkably high. However, in other lineages with less tolerance to desiccation, morphological or life history adaptations are present that also allow survival under periodic drought. These avoidance strategies are somewhat different in mosses than in liverworts and hornworts.

INTRODUCTION

Drought is an abiotic environmental situation that occurs when potential evapotranspiration is greater than incoming precipitation and is often associated with the loss of water from tissues and cells causing dehydration stress. Plants can avoid water deficit stress either through the evolution of a complex set of morphological (succulent growth

* Corresponding author Email: dvitt@siu.edu.

forms) and physiological features (*e.g.,* C_4 and CAM photosynthetic pathways and anatomy) or by temporally escaping drought through shortened life cycles. Another strategy for avoiding water deficit stress is physiological tolerance wherein plants have evolved the physiological abilities to survive cellular desiccation.

Mosses, with over 12,700 species (Magill, 2010), and liverworts, with about 7,500 species (von Konrat et al., 2010) form the second and fourth largest groups of land plants, respectively (ferns are third), and in the past have been considered (along with hornworts) to comprise a Division of plants called Bryophyta. Although many textbooks still incorrectly define bryophytes as non-vascular green plants, these three groups of plants are at best distantly related. These three groups form a grade of organization best defined by similarities in life cycle and ecological roles - wherein the sporophyte generation remains attached and is partially or wholly dependent on the photosynthetic gametophyte generation. These groups differ fundamentally in both gametophyte and sporophyte development and have few, if any, shared derived features (Crandall-Stotler, 1974; 1984). In mosses and most liverworts, the gametophyte has stems and leaves, and most species of mosses have at least some form of internal water conducting tissues, but these are usually poorly developed. In contrast, the gametophytes of all hornworts are fairly simple thalli.

Mosses and leafy liverworts have other features in common that apparently have evolved independently. Among these features are: 1) leaves only a single cell in thickness; 2) leaves that never have a petiole and are attached to the stem along a wide insertion; and 3) a lack of ability to retain water (poikilohydric) and cell turgor for an extended length of time (but a remarkable physiological ability to tolerate desiccation). This tolerance to desiccation can be illustrated by comparing an African violet to a moss, each growing on a window sill and left without water. The African violet remains photosynthetically active for an extended period of time – perhaps a week or two, avoiding the lack of water by retarding water loss by a thick impervious cuticle and by active water movement through well-developed water-conducting tissues from underground roots to photosynthetically active, aboveground stems and complex leaves that lose water only through transpiration from stomates. After some period of weeks the cells lose turgor and the plant shuts down physiologically, but when rewetted does not revive and subsequently dies. The moss, in contrast, also begins this trial photosynthetically active, and after a few hours dries out and becomes quiescent, but when rewetted becomes active once again. These two different strategies for survival, one of avoiding desiccation through various morphological traits and the second of tolerating desiccation by physiologically shutting down cell functions, allow vascular plants and bryophytes to occupy and function in very different habitats and ecological situations. Although most mosses and liverworts are desiccation tolerant to some degree, they also avoid drought through a number of morphological avoidance mechanisms.

DESICCATION TOLERANCE—A PRIMER

Strategies of Water Economy

As exemplified by our African violet and moss examples, land plants have evolved two major strategies for water economy: homoiohydry and poikilohydry. Homoiohydry is defined as "striving to maintain a high water potential under water limiting conditions" and poikilohydry as "the inability to control water loss to the environment" (Wood, 2005). Homoiohydry is the dominant land plant strategy, and homoiohydric plants (such as the African violet) have evolved a variety of physiological and anatomical features that maintain a high internal water potential (Bewley, 1979; Nobel, 1983). Poikilohydry, on the other hand, is a defining feature of mosses, liverworts, and hornworts, and poikilohydric plants are unable to control water loss to the environment with the result that cellular water content fluctuates in concert with external water availability. Land plants are constantly losing water to the surrounding environment because the surrounding air is extremely "dry" relative to the plant. Some of the adaptations required for invasion of the land by aquatic plants would have required several adaptations including: 1) tolerance of vegetative tissues to desiccation; 2) reproductive and fertilization strategies in non-aqueous environments; 3) gas exchange across a liquid-air interface; and 4) enhanced ion and metabolite transport.

The vast majority of plants cannot survive equilibrium with dry air and will die upon complete drying; however, a number of plant species can survive complete drying (i.e., desiccation). Desiccation is the complete loss of "free" water from an organism that corresponds to <0.1 g H_2O g^{-1} dry mass (10% water content or less) (Alpert, 2005; Gaff, 1971). These "desiccation-tolerant" plants equilibrate with dry air (50% RH and 20° C) and resume normal metabolic function upon rehydration (Bewley, 1979; Proctor et al., 2007). Wood (2005) proposed that the successful land-invading plant would have been compact, poikilohydric, and desiccation tolerant. In addition, the first successful land-invading plant would have been able to tolerate high irradiation and temperature fluctuations.

Desiccation tolerance, or "drying without dying," is one of the most amazing phenomena in biology and has been the topic of a number of reviews (see Alpert, 2005; Bewley, 1979; Oliver et al., 2005; Proctor et al., 2007; Wood, 2007) and books (Black & Pritchard, 2002; Jenks & Wood, 2007). Over the past 30 years of research, several general concepts have emerged: 1) desiccation tolerance is observed in each of the three domains of life (Archaea, Bacteria, and Eukarya); 2) desiccation tolerance is often associated with tolerance to other abiotic stresses; 3) in many plants and animals, desiccation tolerance is limited to specialized structures such as cysts, tubers, spores, or seeds; 4) desiccation tolerant reproductive structures are found within the mosses, liverworts, pteridophytes, gymnosperms, and angiosperms; 5) in many plants and animals, vegetative tissues (*i.e.*, non-specialized structures) are desiccation tolerant; and 6) vegetative desiccation tolerance is a common feature of mosses and liverworts.

Distribution of Desiccation Tolerance

Desiccation tolerance within vegetative tissues is widely distributed in land plants. Approximately 320 species of vascular plants (less than 0.15% of the total) possess vegetative desiccation tolerance (Porembski & Barthlott, 2000). They reside within nine pteridophyte families (Adiantaceae, Aspleniaceae, Davalliaceae, Grammitidaceae, Hymenophyllaceae, Isoëtaceae, Polypodiaceae, Schizaeaceae, and Selaginellaceae) and ten angiosperm families (Acanthaceae, Cactaceae, Cyperaceae, Gesneriaceae, Labiatae, Liliaceae, Myrothamnaceae, Poaceae, Scrophulariaceae, and Velloziaceae). Interestingly, vegetative desiccation tolerance is undocumented in the gymnosperms. Of the approximately 21,000 species of mosses, liverworts, and hornworts, the majority are postulated to be desiccation tolerant and able to survive brief desiccation of modest intensity (Proctor & Pence, 2002; Proctor et al., 2007; Wood, 2007). Two hundred and ten bryophyte species (approximately 1% of the total) have been experimentally determined to be desiccation tolerant (Wood, 2007) and have been identified within four of the eight classes of mosses (Andreaeopsida, Bryopsida, Polytrichopsida, and Tetraphidopsida), two of the three classes of liverworts (Marchantiopsida and Jungermanniopsida), and one of the two classes of hornworts (Anthocerotopsida).

Experimental Determination of Desiccation Tolerance

The ability of vegetative tissues to revive from the air-dried state is a common characteristic of mosses (Proctor et al., 2007); however, the speed and intensity of the drying event is critical. Oliver & Bewley (1997) proposed that "fully desiccation tolerant" species can survive extremely rapid water loss, and that "modified desiccation tolerant" species can only survive slow water loss. The vast majority of desiccation tolerant bryophytes are "fully" tolerant while the vast majority of desiccation tolerant vascular plants are "modified" tolerant. Stable drying environments over a wide range of relative humidity (RH) can be created within an enclosed space using known solutions. Wood (2007) proposed five categories of desiccation tolerant bryophytes. Category 1 is the majority of documented desiccation tolerant species and they can survive equilibrium with extremely dry air (*i.e.*, 0–30% RH or less than -162 MPa). Category 2 species can survive equilibrium with moderately dry air (i.e., 70–80% RH or -30 to -48 MPa). Category 3 species can survive equilibrium with 70–80% RH, die at 0–30% RH, but will survive severe water deficit stress if the gametophores are ''hardened'' by either slow drying or exposure to a mild desiccation pre-treatment (*i.e.*, 96% RH for 24 h). Category 4 can survive 0-30% RH and Category 5 can survive 70-80% RH only if the gametophores are hardened.

The 'Austin Protocol' has been proposed as a simple and standardized procedure for determining vegetative desiccation tolerance (Wood, 2007). This protocol evaluates the modulated chlorophyll fluorescence in plant tissues equilibrated at two relative humidity set points (*i.e.*, 67% RH and 23% RH). Mosses and liverworts that recover from equilibration at 67% RH are classified as "modified" desiccation tolerant and those that do not recover are classified as desiccation sensitive.

Species able to recover from equilibration at 23% RH are classified as "fully" desiccation tolerant. Based upon this system of classification, essentially every desiccation tolerant species in the Andreaeales, Bryales, Dicranales, Grimmiales, Hookeriales, Hypnales, Orthotrichales, Polytrichales, and Pottiales are either Category 1, 2, or 3. This means that (with few notable exceptions) all desiccation tolerant mosses can survive equilibration with 23% RH (either with or without pretreatment) and are fully desiccation tolerant. Similar results are seen in the liverworts. Every documented desiccation tolerant species in the Marchantiopsida and most of the species in the Jungermanniopsida can survive equilibration with 23% RH (either with or without pretreatment).

Mosses – A Primer

Mosses, Division Bryophyta, seemingly appear in the fossil record relatively late, with the earliest confirmed fossils from the Upper Carboniferous (Krassilov & Schuster, 1984). By Cretaceous time, many of the modern families and genera were present, and by early Tertiary, even some modern species seem to have fossil occurrences.

Moss Habitats

Mosses occur in all of the major biomes of the world. They dominate the ground layer in the boreal forest and are foundational species of northern peatlands. They are among the most abundant epiphytes in higher elevation tropical rainforests and with liverworts and algae form soil crusts in prairies and deserts. In alpine and montane areas they may form complete cover on boulders and dominate many cliff faces, ledges, and rock falls. In temperate and oceanic regions, they form conspicuous carpets on the forest floor and are epiphytic on tree trunks. Their canopies have a rich internal structure and are the homes for numerous insects, rotifers, tardigrades, and other invertebrates. Only a few species are aquatic and few have any ability to tolerate salt. If one were to place moss species onto Grime's triangle (Grime, 1988) of ecological strategies, they would largely fall either as ruderals or as stress tolerators, and almost none would be classified as competitors (Grime et al., 1990). Mosses occupy many relatively harsh habitats that periodically are subject to drought, including habitats that are present in early succession, where desiccation is the key limiting factor for most species.

Moss Phylogeny

Currently there are a number of robust molecular-based reconstructions of moss phylogeny at the family rank and above (Cox et al., 2010). Morphologically, the major lineages of mosses are defined by features of the peristome (Vitt, 1984). The phylogeny as presently constructed shows eight major lineages (Figure 1) along with several more isolated ones with only a few taxa. Lineage 1 (Sphagnales) - contains the genus *Sphagnum*, with many unique sporophytic characters and gametophores having complex leaves of dead hyaline cells and living green cells. There is some evidence that although *Sphagnum* as a genus may have evolved quite early compared to other moss lineages, its diversification occurred only in the Miocene (Shaw et al., 2010). Much of the diversity of *Sphagnum* is in tropical and temperate regions where it occurs overhanging on cliff faces and forming mats on moist soil, but a small number of species have evolved decomposition-resistant gametophores and are a major component of peat and foundation species of northern peatlands. Lineage 2 (Andreaeales) - contains the rock mosses with only two genera that are highly desiccation tolerant and occur only on periodically dry rock surfaces in arctic and alpine areas of the world. It is in this early lineage that desiccation tolerance is clearly evident. Lineage 3 (Polytrichales) - contains the structurally complex Polytrichaceous mosses, with well-developed water and food conducting systems, complex multi-layered leaves with cuticle and photosynthetic lamellae (resembling the palisade layer of vascular plant leaves) and underground rhizoidal mats.

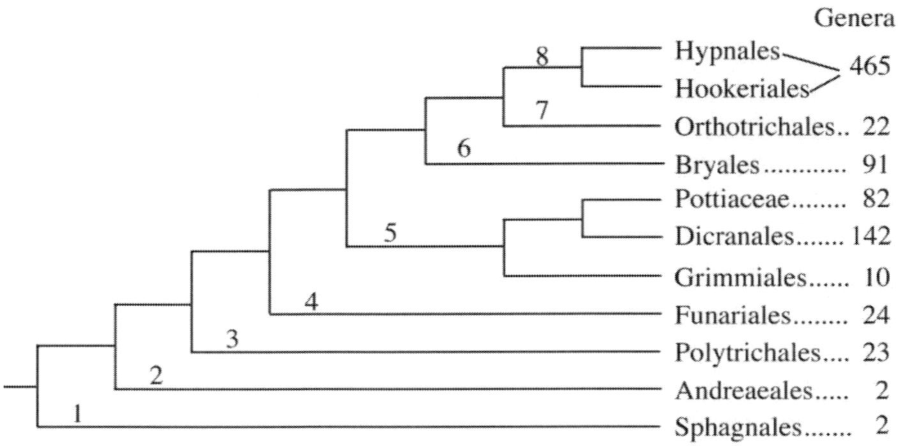

Figure 1. The major lineages (as Orders) of the mosses (Division Bryophyta) summarized from Cox et al. (2010). Numbers above the branches refer to the lineage numbers discussed in the text; 1 = mosses with nematodontous peristomes; 4 = mosses with diplolepideous opposite peristomes; 5 = mosses with haplolepideous peristomes; 6-8 = mosses with diplolepideous alternate peristomes (after Vitt, 1984). Numbers to right are approximate number of genera based largely on Cox et al. (2010).

The species occur on soil, occasionally on organic matter over rocks, and include the largest of the mosses (*Dawsonia*), reaching 1m in height.

Lineage 4 (Funariales) -contains genera related to *Funaria*, a lineage morphologically defined by peristome features (arthrodontous, diplolepideous with opposite endostome and exostome) (Edwards, 1984). The species occur on moist soil and organic matter, and have large, thin-walled cells, a strong costa, and are acrocarpous (producing perichaetia terminal on stems). Epiphytic and rock substrates are not colonized by species in this lineage and tolerance to desiccation is limited.

Lineage 5 (Grimmiales, Dicranales, Pottiaceae) - contains the Dicranoids, a large assemblage of genera all with a unique peristome (arthrodontous, haplolepideous), acrocarpous placement of archegonia, and mostly sympodial growth (erect and with few branches). This is the first lineage that contains species able to occupy an epiphytic habitat, although most species occur on mineral soil or on organic soil over rocks.

Lineage 6 (The Bryales) – This lineage has acrocarpous plants with a diplolepideous, ciliate peristome with endostome segments positioned alternate to exostome teeth (Vitt et al., 1998). The species are specialized to habitats of moist shaded soils, stream sides, and disturbed soils. Although a few species are epiphytes, these are rare.

Lineage 7 (Orthotrichales) – This is a lineage with peristome features similar to the Hypnales and Bryales, but with no endostomial cilia. These genera have species highly specialized for epiphytic habitats, especially in tropical forest canopies.

Lineage 8 (Hypnales, Hookerales) – Defined by arthrodontous, diplolepideous, ciliate, and alternate peristomes, these are pleurocarpous (with lateral placement of archegonia), have complex branched gametophytes, many with differentiated stems and branches, and occupy a wide range of habitats. In particular they seemingly are highly adapted for epiphytism and competitive forest floor habitats. One specialized lineage within the group, the Hookeriales, has apparently lost much of their desiccation tolerance and occupy shaded, mesic habitats that rarely if ever experience drought.

MOSSES – AVOIDANCE ENHANCEMENTS TO DESICCATION TOLERANCE

The evolution of increased physiological tolerance to desiccation is paramount to mosses being able to occupy drier and drier habitats.

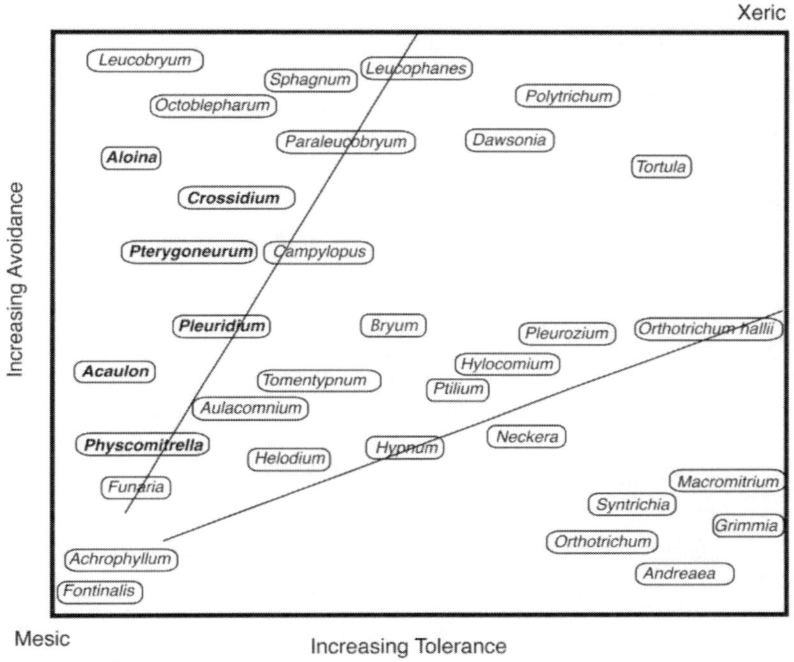

Figure 2. Conceptual model showing the distribution of select moss genera along a dryness gradient (mesic to xeric). Oblique lines delineate major strategies from largely avoidance (upper left) to largely tolerance (lower right). Boldface names are genera with escape strategies.

Additionally, mosses have enhanced their physiological tolerance with a suite of morphological or life history features. These morphological adaptations increase the ability of some species of mosses to occupy even more desiccation prone habitats. In most cases, specific avoidance features have evolved convergently in two or more lineages. In agreement with Oliver et al. (2000) we argue here that physiological tolerance to desiccation was to some extent present in the ancient moss lineage from the beginning. Adaptation to drought advanced either through increasing physiological desiccation tolerance or through morphological and life history changes leading to the ability to avoid desiccation. Conceptually, many moss lineages have increased their physiological tolerance to drought, but also have developed morphological avoidance mechanisms that enhance habitat availability (Figure 2). Surprisingly, few mosses are truly aquatic and, as expected, those that are have little tolerance to drought, but even the aquatic *Fontinalis* (of streams and pools) can be induced if slowly dehydrated (Cruz de Carvahlo et al., 2011; 2012; 2014). A few species occur in deeply shaded, wet habitats such as seeps, beneath overhanging vegetation, and on exposed tree roots. These species also have little or no drought tolerance and occur sporadically through the moss lineages. Especially noteworthy are species in the Hookeriales that are characterized by flattened leaf arrangements and large thin-walled leaf cells (*Achrophyllum, Distichophyllum,* and *Hookeria*).

Protection of Photosynthetically Active Leaf Cells

At least three distinct lineages have leaves in which enlarged, dead, porose hyaline cells (often termed leucocysts) partially or completely encase smaller, photosynthetically active (green) cells (often termed chlorocysts). In *Sphagnum*, the leaves are one cell layer thick with large hyaline cells that alternate with, and partially or completely surround, the smaller green cells (Figure 3).

The green cells are a variety of shapes (in transverse section), ranging from circular, trapezoidal, to triangular. Most species have green cells exposed on one side of the leaf, which may facilitate gas exchange as the surface is exposed to air rather than water of the adjacent hyaline cells.

In *Leucobryum*, small green cells are sandwiched between 2-3 (and up to 7) layers of dead porose hyaline cells (Figure 4). Cardot (1900) demonstrated that the structure of this multi-layered leaf is derived from modification of costal cells. In *Leucobryum* and related genera (*Cladopodanthus*, *Ochrobryum*, *Schistomitrion*, and *Terrestria*) the hyaline cells lie opposite to one another, with the green cells diamond-shaped. Like *Leucobryum* and related genera, *Paraleucobryum* has peristomial affinities with genera having a 'Dicranoid' peristome (long, bifid peristome teeth with vertical lines of ornamentation); however, the most recent molecular phylogeny of mosses (Cox et al., 2010) places *Paraleucobryum*, along with *Dicranum*, in a lineage separate from that of *Leucobryum* (which is sister to *Campylopus*, a genus with some species having a costa with some hyaline/green cell development as well as a distinct leaf lamina). *Paraleucobryum* has a very similar leaf structure to that of *Leucobryum* with green cells sandwiched between a dorsal and ventral layer of hyaline cells; however, in *Paraleucobryum*, the green cells are large, of similar size to the hyaline cells, and there are some green cells in the outer dorsal layer of cells; like *Leucobryum* the hyaline cells are opposite to one another and alternate to the diamond-shaped green cells (Figure 5).

In *Leucobryum* and *Paraleucobryum*, the multi-layered part of the leaf, consisting of a layer of green cells enveloped by two or more layers of hyaline cells, is always associated with a narrow, unistratose leaf lamina and appears to have been derived from costal elements. Compared to these genera is the situation in *Leucophanes* (Leucophanaceae, but closely related to the Calymperaceae) where small diamond-shaped green cells are sandwiched between a dorsal and ventral layer of large hyaline cells; however, in this case there is a well-developed costa with stereids (fibers) present as well (Figure 6) - thus here the multi-layered leaf is seemingly derived from leaf laminal origins. A number of genera, apparently related to the Calymperaceae (with short, single, stubby, papillose peristome teeth), have a superficially comparable leaf structure; among these are *Exodictyon*, *Exostratum*, *Arthrocormus*, and *Octoblepharum*.

These genera have 3-7 sided green cells sandwiched between multiple (up to 8-9) layers of hyaline cells and no costa or unistratose lamina (Figure 7). *Exodictyon* also has green cells on both outer leaf surfaces in addition to a central layer. *Octoblepharum albidum* is a very common tropical moss that occurs at the bases of palm trunks in coconut plantations, an extremely dry and exposed habitat.

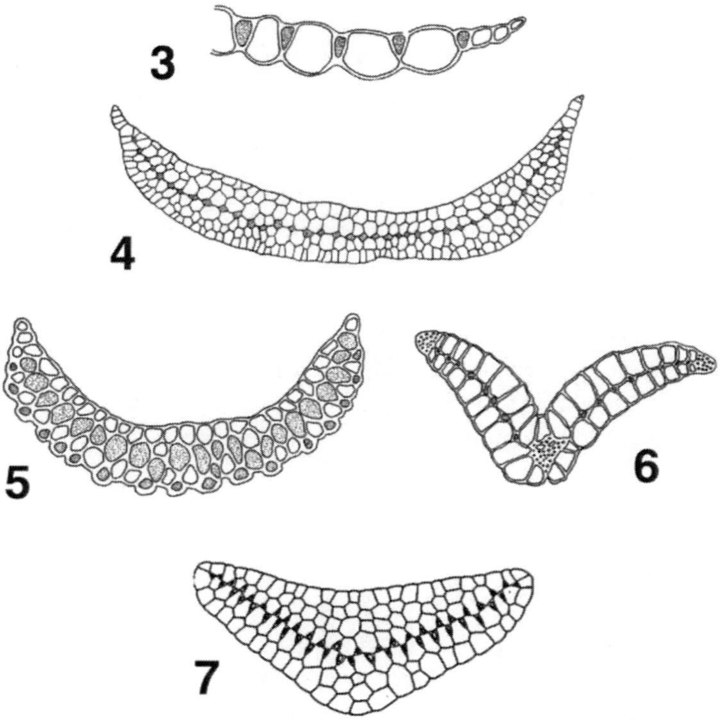

Figures 3-7. Transverse sections through mid-portions of leaves from genera with dead hyaline cells protecting living green cells (stippled). 3. *Sphagnum*; 4. *Leucobryum*; 5. *Paraleucobryum*; 6. *Leucophanes*; 7. *Octoblepharum*.

In addition to these genera with multistratose leaves, a large number of mostly tropical moss genera have single-layered leaves with photosynthetically active green cells in the distal portion and dead porose hyaline cells (in one layer) in the proximal portion (*e.g.*, *Syrrhopodon*, *Leucophanella*, and *Mitthyridium*). It appears that porose hyaline cells, in many cases surrounding active living green cells, have appeared in a number of distinct lineages, serving to extend the activity of a reduced number of photosynthetically active cells for a longer period of time under drought conditions.

Complex Canopies

Mosses, like vascular plants, have variable canopies of leaves, branches, and stems, all of these important in controlling water balance and gas exchange; especially important are surface roughness and stem density and position (Rice & Schneider, 2004; Rice et al., 2001). Mosses found in the earlier moss lineages are acrocarpic (perichaetia terminal on main stems) and have sympodial branching (thus after a fertilization event growth of the gametophore stops with a new stem arising from beneath the terminal perichaetium). This situation severely limits branching frequency, and hence complex canopy development is often not well-developed in these species (but see LaFarge-England, 1996 for a detailed discussion). Pleurocarpy (perichaetia formed from lateral buds) is a foundational feature of the crown lineage of mosses (the Hypnales) and it is in pleurocarpous mosses that monopodial branching (branching from lateral buds) along with lateral perichaetia that complex branching patterns can provide protection from evaporative stress. Additionally, complex canopy development allows for the expansion of colonies in both horizontal and vertical directions providing a competitive advantage. If one envisions a boreal forest floor, dominated by *Hylocomium splendens* (pleurocarpic sympodial), *Pleurozium schreberi* (pleurocarpic monopodial), and *Ptilium crista-castrensis* (pleurocarpic monopodial), the daily routine is for the moss canopy to be moist and functioning in early morning owing to dew formation; however, by mid-day on a sunny day the moss canopy is evaporating moisture and the mosses rapidly become non-functional. The amount of time on a daily basis that the moss is hydrated and photosynthetically active is strongly correlated to the growth rate and to carbon fixation. Thus there is a strong advantage to the presence of three-dimensional complex canopies composed of richly branched plants where growth cannot only be vertical from the apical bud of the primary stem, but also continue for two or more years from lateral branches (Benscoter & Vitt, 2007).

Endohydric Water Conduction

Genera within the Polytrichales extend their tolerance to drought by having highly complex gametophores. More so than any other mosses these species have a well-developed cuticle, a complex leaf laminal structure and dense vertical lamellae on the adaxial surface of the leaf that increases photosynthetic ability, a differentiated underground rhizoidal mass, and an internal conducting system of dead water-conducting cells (hydroids) and living nutrient/food conducting cells (leptoids) that connect leaves to stems through leaf traces (unlike other mosses that may have a costa and stem conducting elements, but these do not connect through the stem cortex). When active, these species have leaves that are straight and bent 90 degrees to the stem, but as they desiccate, they twist and curl against the stems and on rewetting unfurl and are alive and well - thus they are clearly desiccation tolerant, but they extend their active period by a series of mechanisms that effectively move water from below ground to the above ground structures. Species in this lineage are among the largest of mosses, attaining nearly 3-10 decimeters in height. Smaller species (1-10 mm in height) are often not only reduced in stature, but have less well-developed conducting systems, fewer lamellae, and are more desiccation tolerant - these genera (*Atrichum*, *Psilopilum*) are among the most derived in the family lineage (Cox et al., 2010; Hyvonen et al., 2004).

Ectohydric Water Conduction

Many mosses occur emergent from wetland substrates. They are largely dependent on wicking water upward from a permanent supply, yet they occur well above the water level (Busby et al., 1978). Vitt & Glime (1984) called these mosses 'emergents.' Few if any of these species have efficient internal water conducting systems, and they rely on a set of external stem features that conduct water upward owing to evaporative pressure. For example, *Aulacomnium palustre*, some species of *Philonotis*, *Paludella squarrosa*, and *Tomentypnum* species have a dense mat of rhizoids that completely covers the stems. Some mosses have branched or filamentous, green, stem outgrowths called paraphyllia that in some species form a tomentum over the stems (*e.g.*, *Helodium* spp., *Cratoneuron commutatum*, and some *Thuidium* species). Species of *Sphagnum* have complex branching structure of hanging and spreading branches that provide numerous, small spaces for capillary movement.

Shortened Functional Life Cycle and Enhanced Photosynthetic Apparatus

One life history strategy widely used by mosses is to avoid drought by shortening the functional life cycle, thereby temporally escaping periods of drought. Annuals and species with even shorter, ephemeral life styles largely avoid periods of drought by completing the gametophytic life cycle phase (or by having short-lived gametophores) during a brief, moist growing season and weathering the dry season as a diaspore (or in some cases as protonema or in a desiccated state). The problem with this strategy is that photosynthetic activity is severely limited by length of the wet period. One strategy to remedy this problem is to increase the structural components in which photosynthesis takes place. Several mechanisms are seemingly present in mosses, including leaves with the adaxial costal surface having: 1) photosynthetic pads; 2) filaments; 3) lamellae; or 4) adaxial leaf laminal surface having convex cell walls. Shortened life cycles not only include modification to the gametophyte generation (Vitt, 1981), but also to the sporophyte generation. Such modifications are reduction in seta length resulting in capsules immersed in the vegetative leaves, capsule axis compressed to a globose shape, loss of peristome function, and lack of an operculum (cleistocarpy) that results in spores being deposited in the same place as the parent plant. Escape also occurs by shortening the complete life cycle to an ephemeral status (life cycle completed in a few weeks only) and is present in a number of moss families. Examples are species of *Physcomitrium*, *Physcomitrella patens*, *Aphanorregma* and some species of *Entosthodon* in the Funariaceae, and *Discelium nudum* and *Gigaspermum* in closely related families of the lineage. The Dicranoid lineage, species with tiny, bulbiform plants and cleistocarpous, immersed capsules, include *Pleuridium*, *Eccremidium*, *Pseudephemerum*, and *Cleistocarpidium* in the Ditrichaceae and *Bruchia* in the Bruchiaceae. The classic genus, *Ephemerum*, with tiny (less than one mm) high gametophores arising from a persistent protonema has in the past been placed near Funariaceae, but recent molecular data place these species in the Pottiaceae, in the Dicranoid lineage. This last family contains well over 15 genera with shortened life cycles and these genera are found in at least 2-3 separate lineages (Zander, 1993). Examples include *Acaulon* and the distantly related species of *Astomum* (currently placed with species in *Weissia*) and *Aschisma*.

Although these species with shortened life histories are evident in at least three ordinal lineages, only species in one has structural photosynthetic enhancements. Within the Pottiaceae, *Crossidium* has a costal pad of photosynthetic filaments (Figures 8-9), *Aloina* has not only filaments but the leaf laminae are inrolled over the filaments providing a pseudo multi-layered leaf (Figure 10), *Pterygoneurum* species have a costal pad with wavy lamellae (Figure 11), *Pseudocrossidium* in one lineage and *Stegonia*, some *Tortula* species (Figure 12) and *Microbryum* in another have costal pads with enlarged thin-walled photosynthetic cells. Finally *Plaubelia*, *Hyophila*, and *Timmiella* (Figure 13) have laminal cells with highly convex ventral walls - these apparently acting as lenses concentrating sunlight to the chloroplasts. Most of these genera with shortened life cycles differ from their close relatives in greatly reduced or lack of papillae on leaf cells and sporophytes that have lost any functional spore dispersal abilities. In most of these species, peristomes are lacking and have no organized capsule dehiscence (Vitt, 1981).

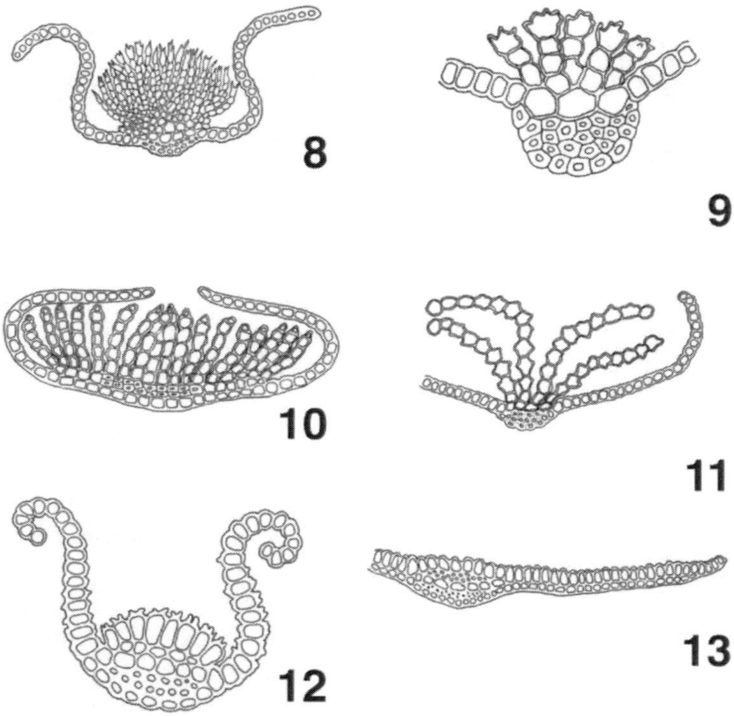

Figures 8-13. Transverse sections through the upper portion of leaves from species with an enhanced photosynthetic apparatus. 8. *Crossidium squamiferum*; 9. *Crossidium aberrans*; 10. *Aloina pilifera*; 11. *Pterygoneurum ovatum*; 12. *Tortula atrovirens (= Desmatodon convolutus)*; 13. *Timmiella anomala.*

Multi-Stratose Leaves

Scattered among genera inhabiting a variety of xeric habitats are a number of species that have photosynthetic leaf cells in 2-3 layers (or strata). These species occur in a number of independent lineages and usually as isolated species. For example, *Orthotrichum hallii* and *O. bolanderi*, both species growing on dry rocks, have bistratose leaves and are in separate lineages within the family (Vitt, 1971). *Syntrichia caninervis*, a desert soil species, has bistratose leaves, as does *Didymodon bistratosa*, a species of rock ledges. *Macromitrium*, a genus wherein many species occur within tropical forest canopies and on tree branches, has a number of species with bistratose leaves (*e.g.*, *M. tongense*). Species of *Grimmia* that occur exclusively on xeric, exposed rock faces have many species with bistratose leaves. Although there are no data on actual tolerances, the correlation between habitat and bistratose leaves suggests that this could be a method of extending the length of time that cells can be active by reducing the ratio of evaporative surface to cell volume.

Falcate-Secund Leaves

Occurring throughout the lineages of mosses are species that have all leaves curved to one side of the stem forming semi-circles. Most of these species have the leaves pointed downward towards their substrate – such as vertical tree trunks, horizontal log surfaces, or the forest floor (*e.g.*, *Hypnum*). As mosses desiccate, often on a daily basis, the retention of the last remaining water in an area of photosynthetic activity is of utmost importance as this allows a longer period of positive carbon gain. Falcate-secund leaves characteristically are also curved laterally and form a hollow cavity in the upper inside portion of the leaf, and it is here that water is retained longest.

Leaf Cell Structural Modifications

Almost all moss species are poikilohydric (lack of ability to retain water). They possess only a thin cuticle and although many have conducting elements in the stem and leaves these do not connect through the cortex of the stem; thus efficient internal water movement is lacking in most species. As a result of these inadequacies in being able to move or retain water for any length of time, mosses rely on being able to quickly absorb water from atmospheric sources, enable photosynthesis to rapidly initiate, and continue at a high rate (equal to that of vascular plants: Martin & Adamson, 2001) while the moss plants are hydrated. However, if the unistratose leaf cells are immersed in a water film, the exchange of gases is severely inhibited and if species are growing in sunny habitats leaf temperatures can become high. As a result, not only do moss leaves absorb and move water quickly, they must also provide cell surfaces exposed to the atmosphere and therefore dry out quickly in order to reduce cell temperatures after a wetting event (Proctor, 2009). Additionally, mosses are active at any temperature above 0°C when moist and carry out effective photosynthesis throughout the year, even under the melting snow pack in northern areas.

Mosses that occur in harsh environments such as rock faces, dry land soil crusts, and tree trunks and branches all appear to have a high level of desiccation tolerance, but also have a set of morphological attributes that are associated with these severe habitats. These species have thick cell walls; thus a higher amount of carbon is allocated to structural components than species of mesic habitats. They mostly have papillae - cell wall protuberances in a variety of forms, from mere finger-like bumps to branched, forked, or c- to u-shaped structures. The papillae may serve a variety of purposes. They may create capillary channels that effectively move water along the leaf surface allowing rewetting of the leaf to occur more quickly (Proctor, 1979). They are usually hollow and create an increased cell surface to volume ratio allowing water to be absorbed and lost more quickly, and they create an elevated surface for enhanced gas exchange. Species having papillae often have cells in the upper portion of leaves differentiated from those below - the upper ones photosynthetic with abundant chloroplasts and papillae; while the lower ones lack chloroplasts and are non-papillose, larger, thinner-walled, and clear — perhaps with greater water retention abilities. In some cases these lower cells have openings (pores) in their walls (Proctor, 1979). Many species occurring in xeric habitats have leaves that are wide-spreading (exposing the adaxial surface to the atmosphere) when moist, but upon drying become crisped and twisted around the stems. This leaf movement is rapid, often occurring within minutes of wetting or drying (*e.g.*, *Syntrichia ruralis*). Although perhaps related more to temperature than moisture, many xeric species have leaves ending in clear, long or short hair points, these apparently providing a protective layer from harsh, bright sunlight and/or reducing surface temperatures. Since any conducting tissues present in the costa (midrib) of the leaves do not connect to those in the stems, each leaf appears to act as an independent entity, photosynthesizing in its upper portion, transferring the photosynthate to the lower clear cells and storing it there, with no interactions with nearby leaves.

This is fundamentally different from the highly interactive situation in vascular plant leaves wherein turgor is maintained by a strong interconnected water connecting system and carbon is readily translocated. Whereas in vascular plants water maintenance is from the bottom-up throughout the entire plant, in these xeric mosses, water is obtained and transferred only locally from the top-down. It may be that the costa in these xeric species, which is almost always well-developed, serves as a strengthening organ rather than one that provides for water and/or photosynthate movement. When desiccation tolerant moss species dry out, the cell membrane of the individual cells shrinks and contracts away from the rigid, thick cell wall. The cell sap becomes concentrated in small spherical vesicles within each cell surrounded by an intact cell membrane, which is separated from the cell wall. This separation of cell membrane from cell wall can be seen not only in the xeric species of mosses (*e.g.*, *Orthotrichum*, *Ulota*, and *Grimmia*), but also in more mesic species (*e.g.*, *Mnium*, *Bryum*, *Fabronia*, and *Dicranum*).

Resource Allocation

Turetsky et al. (2008) have shown that, in *Sphagnum*, species living on hummocks allocate more of their carbon to structural components compared to species living in wetter lawns and carpets - these latter species allocating more carbon to soluble cell sap components. Thus lawn species have higher primary production in habitats that do not require highly structured canopies, but decompose more rapidly, while drier hummock habitats require more structural organization that leads to increased drought avoidance, but less production and less decomposition. This appears to be true for mosses in general.

The ability of species to tolerate desiccation not only is dependent on a species' physiological tolerance, but also on morphological enhancements. These enhancements require a trade-off in where carbon is allocated. Increased cell wall thickness, cell ornamentation, external structure for water movement, and water-storage cells all require allocation of carbon to non-photosynthetic processes; however, these increase the length of time a cell can be active.

Living in drought-prone habitats is not without costs. Sexual allocation, sporophyte abortion, and vegetative propagation are all influenced by dehydration (Stark et al., 1998; 2000). In many cases, these reproduction traits and ecological strategies are strongly influenced by the rate of drying as well as variation in the number of drying events (Stark et al., 2013). Likewise periodic climatic changes can influence the survival abilities of species occurring in desert habitats (Reed et al., 2012).

LIVERWORTS – A PRIMER

Division Marchantiophyta appear much earlier in the fossil record than mosses, with *Metzgeriothallus sharonae* and *Pallaviciniites devonicus*, simple thalloid fossils that are morphologically referable to the Pelliidae, known from the Devonian. Molecular clock estimates of divergence times, which are constrained by fossil placeholders, suggest that the backbone lineages of the division were all established by the end of the Carboniferous and that most of the modern family lineages had diverged by the mid-Cretaceous (Cooper et al., 2012). Despite many lineages having a long evolutionary history, much of the extant species level diversity of liverworts appears to have arisen in the Cenozoic, particularly in the last 80 Myr (Heinrichs et al., 2007; Wilson et al., 2007).

Liverwort Habitats

Liverworts occur on all continents, and inhabit a wide variety of micro-niches from the Arctic to the Antarctic. While abundant in the temperate forests of the northern hemisphere, they are most diversified in the southern hemisphere, especially in the temperate rain forests of the Austral region and the montane, tropical cloud forests of the Andean chain. They often grow intermixed with mosses, along stream banks, on rock walls and bluff tops, on shaded soil on the forest floor or over decomposing logs, as epiphytes on forest trees, and even on or intermixed with *Sphagnum* in peatlands. In the tropical rain and cloud forests of the southern hemisphere, species of the Lejeuneaceae and Metzgeriidae form dense communities on the exposed, upper surface of leaves in the forest canopy, and other taxa from the speciose Lepidoziaceae, Plagiochilaceae, and Lophocoleaceae form thick mats on branches and tree trunks.

Despite abundant moisture in these forest communities, such epiphyllous and epiphytic liverworts undergo repeated cycles of rapid drying and re-wetting (Proctor et al., 2007). Although many liverworts grow in moist, protected habitats, many other genera (*e.g.*, *Riccia*, *Fossombronia*, and *Exormotheca*) have species that tolerate harsh, seasonally dry habitats, including the deserts of southwestern Africa; northern, central, and western Australia; and the Mediterranean climates of Europe and North America. In fact, in *Riccia* the range of habitat tolerances is extremely broad, ranging from the aquatic habitat of floating *R. fluitans* to the extremely dry, exposed bluff top habitat of *R. dictyospora*.

Field bryologists have long been aware of the vast diversity of moisture, light, and temperature regimes occupied by liverworts and have postulated that many liverworts possess some level of drought tolerance; *i.e.*, they can tolerate some degree of drying and metabolic slow-down without dying (Campbell, 1904; Clausen, 1952; 1964). Like mosses, they are poikilohydric plants, in which cellular water potential is basically the same as their surroundings. Their ability to tolerate intermittent, as well as extended dry periods, involves two components: 1) reducing water loss to maintain metabolically suitable cell water content through physiological or structural adaptations; and 2) being physiologically desiccation tolerant, *i.e.*, being able to recover cell function after nearly complete loss of free cell water (Alpert & Oliver, 2002; Proctor et al., 2007). As in mosses, how well a species can adapt to desiccation limits to a great extent the habitat in which it grows.

Liverwort Phylogeny

Liverwort gametophores are of three different body types, referred to as simple thalloid, complex thalloid, and leafy morphologies. Although traditional classification schemes suggested that these body forms represented natural groups, phylogenetic reconstructions based on nucleotide sequence data have consistently resolved five backbone lineages, with the most speciose lineage, the leafy liverworts, further divided into two major sub-groups (Forrest et al., 2006; Shaw et al., 2011). Thus, six major lineages are currently recognized (Figure 14).

Lineage 1 (Haplomitriopsida) - This lineage is the first divergence within liverworts, and as such, is also the first lineage of extant embryophytes. This is a small lineage, consisting of three genera in two subclasses, *Haplomitrium* (Haplomitriidae), *Treubia*, and *Apotreubia* (Treubiidae). All species of these genera grow on soil in constantly moist environments and are not tolerant to drought (Wood, 2007).

The gametophytes of *Haplomitrium* consist of erect, radially symmetric leafy shoots that arise from a branched, hyaline, subterranean stolon system, while *Treubia* and *Apotreubia* consist of little-branched, prostrate, lobed thalli. Both stolons and leafy shoots of *Haplomitrium* contain a small, central strand of thin-walled, hydrolyzed cells that are presumed to serve in water conduction and/or storage. In contrast to the hydroids of mosses, these cells bear numerous small wall perforations on both the terminal and lateral walls. All three taxa produce copious secretions of slime that coat the stolons in *Haplomitrium* and ventral midrib surface in *Treubia* and *Apotreubia*. These secretions, which are released through the epidermis, contain abundant insoluble carbohydrates and substantial amounts of water that likely protect the plants during brief periods of drying (Duckett et al., 2006). Since the secretions are more abundant at the growing tips of the stolons and thalli, when plants are exposed to drying, these tips may remain hydrated. Tests of tolerance in *Treubia lacunosa*, using the Austin Protocol (Wood, 2007) show, in fact, about 15% cell survival in shoot apices at 67% humidity level, but no survival or recovery at 23% RH (Crandall-Stotler & Wood, unpublished data).

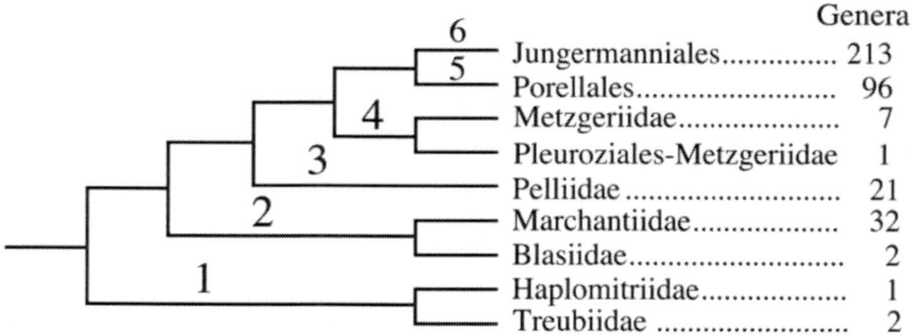

Figure 14. Phylogenetic reconstruction of the backbone lineages of liverworts (Division Marchantiophyta), based on the analyses of Forrest et al. (2006). Numbers above the branches refer to the lineage numbers discussed in the text; lineage 1 = Haplomitriopsida; 2 = Marchantiopsida; 3–6 = Jungermanniopsida. Estimates of generic diversity in each lineage are based on Crandall-Stotler et al. (2009).

Lineage 2 (Marchantiopsida) - This lineage consists of all complex thalloid genera (Marchantiidae) and two simple thalloids, *Blasia* and *Cavicularia* (Blasiidae). The gametophytes of most members of the complex thalloid clade possess multistratose thalli, with air chambers and air pores, two or more rows of large ventral scales and both pegged and smooth rhizoids. Thalli in *Blasia* and *Cavicularia*, in contrast, are differentiated into a midrib and unistratose wing, are without air chambers or pegged rhizoids, and are unique among liverworts in bearing two rows of *Nostoc*-containing auricles along the ventral side of the midrib. To date, there is no evidence of drought tolerance in the Blasiidae, but several derived genera of complex thalloid liverworts (the Marchantiidae) have been confirmed to be drought and/or desiccation tolerant (Bischler, 1998; Wood, 2007). Among these are *Riccia*, *Corsinia*, *Cronisia*, *Exormotheca*, *Targionia*, and *Monocarpus*; all taxa with fairly small, thick thalli and short life cycles, as well as the much larger plants of the Aytoniaceae, such as *Mannia*, *Asterella*, and *Reboulia*. When one of these plants dries, the sides and apex of the thallus enroll over the dorsal surface, so that the thicker central midrib of the thallus is completely covered by the darkly pigmented ventral scales and rhizoids. This process, of course, greatly decreases the surface area exposed for evaporation and provides protection to the meristematic cells of the apex. When seasonal rains return, the thallus quickly unrolls and growth is renewed, especially at the thallus apex. This scenario, perhaps coupled with intracellular mechanisms of desiccation tolerance, can allow for very long-term dormancy as exemplified by the report that a few specimens of *Riccia macrocarpa* from Tunisia were able to renew apical growth after being stored more than 23 years in dry conditions (Breuil-Sée, 1993). It is interesting that there are no drought tolerant taxa in the Marchantiaceae, which is the first diverging lineage of the true complex thalloids (Bischler, 1998).

Lineage 3 (Pelliidae) - Most of the diversity of simple thalloid hepatics is found in this lineage, including such familiar taxa as *Pellia*, *Fossombronia*, *Moerckia*, *Pallavicinia*, and *Symphyogyna*. Most genera in the lineage grow in constantly moist habitats and are drought intolerant, although they are able to withstand repeated cycles of brief drying and re-wetting. In contrast, *Fossombronia*, the most speciose genus in the lineage, typically inhabits harsh, seasonally dry habitats that require some mechanism for drought tolerance. Notable among these is a short, annual life cycle, production of large, desiccation resistant spores, and formation of swollen, tuberous apices or pendent, subterranean tubers that allow for perennation during the dry season. *Pallavicinia* and *Symphyogyna* as well as other genera of the Pallaviciniales are the only group of liverworts outside of *Haplomitrium* to have water-holding conducting strands in the thallus midrib. As in *Haplomitrium*, these cells bear numerous pits on their lateral walls, but unlike *Haplomitrium*, their walls are thick and fibrous. It is probable that these cells serve as an internal water reservoir that releases water to the photosynthetic wing cells during short periods of drying, as occurs even in rain forest habitats, where many of them grow. When ample water again returns, these dead, hollow cells rapidly re-fill as water diffuses in from the surrounding thallus cells. In genera in which the winged thalli arise from subterranean rhizomes, *e.g.*, *Jensenia*, *Hymenophyton*, and dendroid species of *Pallavicinia* and *Symphyogyna*, there is a fairly continuous influx of water from the central strand of the rhizome system into the midrib strand cells, so water is always available, even when ambient air humidity levels fall.

Lineage 4 (Metzgeriidae) - This lineage comprises the two most derived families of simple thalloid liverworts, the Metzgeriaceae and Aneuraceae, as well as the leafy liverwort *Pleurozia*. Most species of *Pleurozia* and *Metzgeria* are epiphytes, with some species of

Metzgeria occurring as epiphylls, while the Aneuraceae are found on moist soil, rocks, and rotting logs in both temperate and tropical forests. Several species of *Riccardia* also occur in the peatlands of the Andean parámos. The phylogenetic link between *Pleurozia* and the Metzgeriales is puzzling, but well supported by numerous studies. This strange genus bears two rows of leaves, each of which bears a complex dorsal water-sac, postulated to serve as an insect trap.

Although morphologically very different from the small, simple thalli of *Metzgeria*, *Pleurozia* grows via a 2-sided, or lens-shaped, apical cell like all other members of the lineage. Clausen (1964) has confirmed that *Pleurozia purpurea* and three species of *Metzgeria* are drought tolerant, although only a few cells of *Metzgeria* survived treatment at 15% RH levels. The Aneuraceae appears to be drought intolerant although only a few species and none of those from the páramo have been tested (Clausen, 1952; Crandall-Stotler, unpublished data).

Lineage 5 (Jungermanniidae, Porellales) - The "true" leafy liverworts are the last backbone divergence of the division. Their dichotomy into two morphologically well-defined sister groups, the Porellales (lineage 5) and the Jungermanniales (lineage 6) is estimated to have occurred in the early Carboniferous.

The greatest diversity of the Porellales is found in the rain and cloud forests of the tropics, where they occur as epiphytes. All members of the order have incubous leaf insertions, unequally bilobed leaves, with the small ventral lobe often forming an inflated water sac, and precocious, endosporic spore germination. Despite growing in a moist to wet macrohabitat, most members of the Porellales are very desiccation tolerant; *Porella* and *Frullania*, in fact, rival the tolerance of extremely tolerant mosses like *Syntrichia* and *Grimmia* (Crandall-Stotler, unpublished data; Marschall & Proctor, 1999).

Lineage 6 (Jungermanniidae, Jungermanniales) - The Jungermanniales are the largest and most diverse lineage of the leafy liverworts, both morphologically and ecologically. Although there are some epiphytic taxa in the lineage, soil, rock, and fallen logs are more common substrates. Complicate unequally bilobed leaves occur in a few taxa (*e.g.*, *Nowellia*, *Scapania*, and *Schistochila*), but the smaller lobe is dorsal, rather than ventral as in the Porellales. Leaf insertions may be transverse, incubous, or succubous, with the latter being the most common expression in the order. Spore sizes vary and germination is almost always exosporic. Drought tolerance is of scattered occurrence, and is only moderately expressed in many of the taxa in which it has been confirmed (Clausen, 1952; 1964). Genera able to tolerate 12 hours at 15% RH, *i.e.*, highly tolerant taxa, include *Ptilidium*, *Herbertus*, *Mastigophora*, *Jamesoniella*, *Cephaloziella*, *Isopaches*, *Southbya*, and *Gymnomitrion*.

These genera are distributed across the phylogeny of the Jungermanniales, *Ptilidium* being a very early divergence and *Gymnomitrion* in one of the crown groups; however, fewer than 8% of the genera in this lineage have been studied, generally with a single species tested per genus. Increased sampling that includes genera, like *Stephaniella* and *Gongylanthus*, from exposed rocks and dry soils of high elevations, and tropical elements of the most speciose families, including the Lepidoziaceae, Lophocoleaceae, and Plagiochilaceae, are needed to evaluate the distribution and significance of desiccation tolerance in this large, polymorphic lineage.

LIVERWORTS - AVOIDANCE ENHANCEMENTS TO DESICCATION TOLERANCE

Tolerance to extreme drying has been less well studied in liverworts than in mosses, and almost nothing is known about physiological mechanisms at the cell level that allow desiccation tolerant taxa to recover cell function after prolonged drying in controlled conditions. It has been shown that abscisic acid (ABA) enhances recovery of desiccated thalli in normally tolerant *Exormotheca holstii* (Hellewege et al., 1994) and induces desiccation tolerance in the otherwise intolerant *Riccia fluitans* (Pence et al., 2005). Pressel et al. (2009) have shown that the cytological changes that occur during desiccation and recovery in *Southbya nigrella* are comparable to changes also seen in mosses during these processes, including an increase in soluble carbohydrates in the cytosol. Taken together, these results suggest that liverworts likely possess a cell mechanism similar to that of mosses, but this has never been confirmed. Since liverworts are poikilohydric, like mosses, it is not surprising that they possess many of the same structural and life history adaptations to enhance the efficiency of water retention and thereby increase the chances of surviving drought by avoiding desiccation (Figure 15).

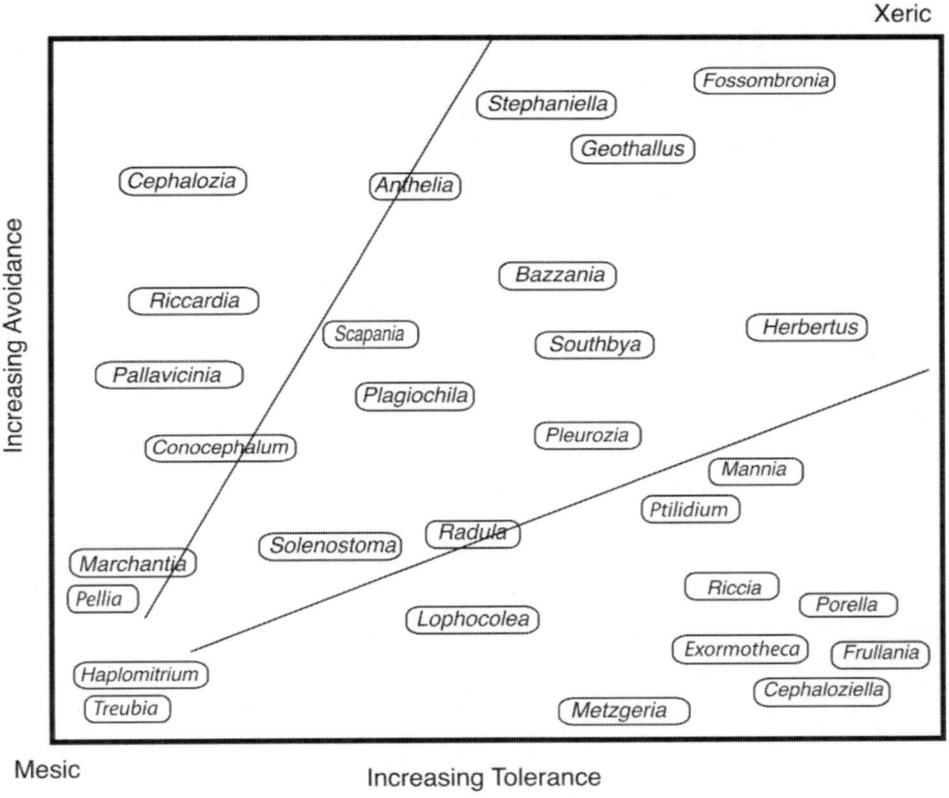

Figure 15. Conceptional model, showing the distribution of select liverwort genera along a dryness gradient. Oblique lines delineate major strategies from largely avoidance (upper left) to largely tolerance (lower right).

These include, but are not limited to, possessing thickened cell walls and special wall ornamentation, increasing the capacity for external water uptake with dense rhizoidal mats, forming large external capillary spaces through which water can move and be stored without compromising gas exchange and photosynthetic potential, reducing surface to volume ratios, increasing the degree of overlap between leaves, and escaping through shortening the life cycle. Considering the structural features of taxa that typically inhabit sites with extended dry periods, several additional characters that promote desiccation tolerance through avoidance have also been postulated. These include slime secretions, especially over meristematic areas; highly divided, toothed, ciliated or highly rucked leaves; ventral water sacs; rhizomes or stolons; and perennating tubers or tuberous apices.

Leaf Insertions, Spacing, and Form

The leaves of liverworts are generally unistratose and in most cases are inserted laterally along an oblique line, in either a succubous (basal or basiscopic margin dorsal) or an incubous (apical or acroscopic margin dorsal) pattern. Typically, they overlap along their margins, resulting in a shingling arrangement. Experiments by Clee (1937) suggested that capillary water movement occurs from base to the apex in plants with succubous leaves and from apex to base in those with incubous leaves. Although this would seem to explain the predominance of incubous insertions in epiphytic taxa, this hypothesis does not seem to hold true (Mägdefrau, 1937; Schuster, 1966). It does appear, however, that water evaporates more quickly from plants with succubous leaves than plants with incubous leaves, a phenomenon that Schuster (1966) equates with the observation that succubous insertions allow more of the dorsal surface of the stem to be exposed. Incubous leaves, in contrast, are more tightly appressed and cover most of the exposed stem surface; *i.e.*, a tighter shingling is possible in incubous leaf arrangements. With a tighter, more closed leaf arrangement, the rate of water loss from internal tissues is decreased, and brief intermittent periods of drying, such as those found in epiphytic habitats, are better tolerated. According to Schuster (1966), plants with succubous leaves have little drought tolerance and are, therefore, most commonly found in moist, shaded habitats. Transverse leaf insertions occur primarily in plants of continuously moist habitats, but they are also found in *Herbertus* and *Cephaloziella*, two highly desiccation tolerant genera. In *Cephaloziella*, plants are extremely small, leaves are often papillose, distantly spaced, and tightly clasp the stem, which is the major photosynthetic structure of the plant. The plants are often darkly pigmented and typically form small, tightly interwoven mats on crusty soils, although they may also be interspersed throughout mats or cushions of other bryophytes. In contrast, *Herbertus* plants are robust, erect, radially symmetric cushion-forming plants in which the leafy shoots arise from a branched rhizome system that is densely covered with rhizoids. Water from below can be held in capillary spaces between the individual leafy shoots and atmospheric water can be held in the spaces between the erect, appressed, deeply bifid leaves and the stem. Rates of water uptake from the surroundings and short-term storage are also enhanced by elongate, thick-walled cells with large trigones and intermediate thickenings that form branched vittae in the lamina of each leaf and underleaf, and by the thick-walled cells of the stem.

Liverwort leaves vary greatly in form, but are usually divided into two or more equal or unequal lobes. Leaves that are highly divided and/or ciliated, such as those of *Ptilidium* and

Trichocolea, show slower rates of evaporation than undivided leaves of similar surface area, presumably because they create capillary spaces for water uptake and retention (Clausen, 1952). This may explain why perichaetial leaves and other structures, which surround the archegonia, are usually divided and toothed to ciliate even in taxa with undivided vegetative leaves. A few other modifications of leaf form that enhance survival in exposed harsh environments include the following: formation of broad channels, or pleats, on the adaxial surface of the highly crispate leaves of *Fossombronia*; polystratose leaves in the *Patagonian* taxa, *Herzogiara* (Pseudolepicoleaceae), and *Pachyglossa* (Lophocoleaceae); hyaline marginal and submarginal cells, with the larger photosynthetic cells restricted to lower parts of the leaf as in some species of *Gymnomitrion* (Gymnomitriaceae); and small completely hyaline, leaves covering branched stem paraphyllia, the only photosynthetic tissues in *Stephaniella* (Stephaniellaceae), a plant of dry, high elevation sites in the Andes.

Water Sacs and Ventral Appendages

The universal occurrence of inflated ventral lobules or water sacs in the Porellales was viewed by Goebel (1905) as evidence of their role as water-storing structures. Such a function is, in fact, inherent in them being called water sacs. Although this notion of water sac function persists, experiments on *Frullania* demonstrated that when wetted stems are air-dried at ambient temperatures, water quickly evaporates from the ventral water sacs (Blomquist, 1929); water will also rapidly re-fill the sacs when dried stems are wetted. Clausen (1952) postulated instead that ventral lobules, whether highly inflated or not, create capillary spaces that allow for rapid uptake of water when it is available, but her experiments also show that they are not water storage structures. It is more likely, as postulated by Herzog (1926), that they serve instead to capture and concentrate nutrients from throughfall.

Like water sacs, large ventral underleaves and foliar scales also increase capillarity along the ventral side of the stem in prostrate taxa. In erect, isophyllous genera they function in photosynthesis just like the lateral leaves, but in the prostrate taxa, they likely are not effective in photosynthesis although they still contain plastids. Underleaf insertion is always transverse, and many underleaves are divided or ornamented with slime papillae. Their presence creates capillary spaces along the ventral side of the stem and the papillae help protect the stem apex. In general, increasing the volume of capillary spaces promotes the maintenance of turgidity through brief periods of atmospheric drying, but is likely not a mechanism to avoid long-term drought (Clausen, 1952).

Rhizomes, Stolons, and Other Subterranean Structures

Rhizomes and stolons are branched subterranean stems. In hepaticology, the term stolon is reserved for the leafless, hyaline subterranean system of the Haplomitriales and the term rhizome is applied to underground stems bearing reduced leaves, thick-walled epidermal cells, and dense coverings of rhizoids. Rhizome systems can be extensive horizontal and vertical systems, with positively geotropic branches sometimes penetrating several centimeters into the soil (*e.g.*, *Lembidium* and *Wettsteinia*). Whether horizontal or vertical, these systems can absorb water from the upper layers of the substrate and move it into the leafy shoots by a combination of capillarity and diffusion even when atmospheric moisture is limited. Rhizomes also function as perennating structures during periods of extended dryness, renewing production of leafy shoots from quiescent branch initials when moisture conditions return.

The modification of leafy stem apices into tubers is another means of persisting by perennation during drought or other unfavorable growing conditions. This avoidance mechanism is particularly well developed in *Fossombronia* and *Petalophylllum*. This developmental modification of normal stem growth is induced by drying. Tubers, like rhizomes, are densely covered with rhizoids and bear thick-walled epidermal cells, but they are fleshy, determinant in growth, and filled with starch-filled parenchyma cells.

Cell Wall Modifications

Liverworts of xerophytic habitats typically possess leaf and stem cells with especially large trigones, or corner thickenings, and nodular intermediate thickenings on their interior walls and/or protruding papillae on their dorsally exposed, superficial walls. Consisting primarily of hemicelluloses, pectins, and cellulose, such thickenings increase the potential for apoplastic water uptake, movement, and retention (Proctor, 1979). In addition, the depressions between papillae form a network of small capillaries that transport water over the surface of the leaf. When aqueous dye is used to wet papillose leaves, the dye solution quickly fills the depressions, but the tips of the papillae remain dry, allowing gas exchange to occur through them under otherwise saturated conditions. In contrast, taxa of moist, but intermittently dry habitats often have exposed, outer cell walls that are quite water-repellent. For example, in taxa with concave, overlapping leaves (*e.g.*, *Nowellia*, *Marsupella*, and *Solenostoma*) this is likely due to the physical effects of surface tension (Proctor, 2009), but in other taxa, these walls are coated with thin layers of epicuticular waxes (Duckett & Soni, 1972; Heinrichs et al., 2000). In *Anthelia*, these waxes are visible with a hand-lens as a layer of white crystalline threads or needles, but in other taxa waxy deposits are only visible in SEM preparations (Heinrichs et al., 2000). According to Heinrichs et al. (2000), the main component of the wax in *Anthelia* is a diterpene, while waxes in *Plagiochila* contain a mix of steryl esters, triacylglycerals, and free fatty acids. Neither these waxes nor various wall ornamentations are part of a cuticle, despite references to the contrary in taxonomic descriptions. In contrast to cuticle, which limits both water uptake and evaporation in vascular plant cells, the thin waxy deposits of liverworts do not prevent the plant from drying out. Rather, they help maintain an interface for gas exchange and also prevent wetting of fully dry plants when water is insufficient for recovery.

Pegged Rhizoids

One of the diagnostic characters of the complex thalloid lineage (Marchantiopsida, subclass Marchantiidae) is the occurrence of two types of unicellular rhizoids in most taxa: living, but highly vacuolated smooth rhizoids, and non-living rhizoids bearing internal wall thickenings termed pegs (Duckett et al., 2014). In general, smooth rhizoids emerge perpendicular to the thallus and grow downward into the substrate, while pegged rhizoids extend in bundles parallel to the ventral thallus surface, where they are covered by the ventral scales. Bundles of pegged rhizoids also occur in internal furrows of the upright stalks of the gametangiophores and carpocephala. According to Duckett et al. (2014) the smooth rhizoids of complex thalloid liverworts function, as in all other liverworts, in anchorage and as fungal conduits and are not at all desiccation tolerant. Whereas smooth rhizoids will collapse when dried, pegged rhizoids, with their elastic walls and internal thickenings, do not. Dye studies show, conclusively, that pegged rhizoids form an extensive conducting system that is structurally unaffected by desiccation, which is especially important in the carpocephala. In addition, pegged rhizoids allow the vegetative thalli of taxa growing in periodically dry habitats (*e.g.*, *Mannia*, *Plagiochasma*, and *Targionia*) to resist desiccation (Duckett et al., 2014).

DESICCATION TOLERANCE IN HORNWORTS

Estimated to include approximately 250 species in 14 genera, hornworts are currently hypothesized to be sister to the tracheophytes, based on molecular phylogenetic studies (Qiu et al., 2006; Shaw et al., 2011). Their thalloid gametophores always possess ventral mucilage-filled cavities that harbor symbiotic cyanobacteria and their large plastids are unique among embryophytes in having channel thylakoids, with central pyrenoids present in many, but not all, taxa (Renzaglia et al., 2009). Hornworts typically occur on moist soil in fairly shaded habitats, with *Dendroceros* being the only epiphytic member of the group; however, some taxa, like *Phymatoceros*, are found in areas with persistent drought or long periods of seasonal dryness.

Whether any hornworts are actually desiccation tolerant, or are desiccation avoiders, is equivocal. In Oliver et al. (2000) hornworts are scored as desiccation tolerant on a phylogenetic tree, but no data are cited to support this conclusion. Alpert & Oliver (2002) cited this reference for the statement: 'There are also desiccation-tolerant hornworts.' However, Oliver et al. (2005) reconstruct hornworts as lacking vegetative desiccation tolerance, and further suggest that the earliest expression of desiccation tolerance in some of these early lineages was in spores, not in vegetative tissues. According to Wood (2007), *Dendroceros granulosus* was found to be desiccation tolerant by Proctor (personal communication), but the experimental proof of this has never been published, not even in subsequent publications written by Proctor (*e.g.*, Proctor et al., 2007; Proctor, 2009; 2010).

Some hornworts found in seasonally dry localities produce abundant swollen, marginal and/or apical tubers on their thalli (*e.g.*, several species of *Phaeoceros* from Washington, Oregon, and California), while the genus *Phymatoceros* produces long-stalked, subterranean ventral tubers. Tuber production of both types is initiated as sporophytes mature and the tubers persist in the soil crust or subterranean soil bank after the vegetative thallus deteriorates. When rains return, new plants germinate from these quiescent structures (Crandall-Stotler et al., 2006). At least in *Phymatoceros*, such tubers contain high levels of abscisic acid (Hartung et al., 1994), which has been shown to induce desiccation tolerance in other bryophytes (Pence et al., 2005), and personal observations confirm that both marginal and subterranean tubers can germinate after nine months of dryness. In addition to tuber production as a means of surviving periods of drought, hornworts also can have very short vegetative cycles, followed by rapid spore release, as in the case of *Notothylas*. Hornwort spores are able to germinate even after many years of storage in dried herbarium packets

(Renzaglia et al., 2009); these observations suggest that spores are likely desiccation tolerant, but this hypothesis has never been tested.

Conclusion

Species of mosses, liverworts, and hornworts occur in remarkably dry habitats. Most species of mosses and some species of liverworts tolerate drought by being tolerant to desiccation. At least in mosses it appears that this physiological tolerance to desiccation was present in the earliest lineages and in some of these lineages is a key feature for inhabiting extreme habitats. Many liverworts have less tolerance to desiccation, but there are notable exceptions. In addition to having a physiological tolerance to desiccation, some mosses and liverworts also have a suite of morphological features that enhance water retention and hence avoidance of drought, or have life history changes that allow the species to effectively escape drought. It is this combination of desiccation tolerance and drought avoidance (Figures 2, 15) that determines drought tolerance and is a key factor in determining habitat limitations for both moss and liverwort species.

Acknowledgments

We dedicate this paper to Dr. Raymond Stotler (1940-2013) who would have loved to be involved in writing this chapter. Ray was a friend, colleague, and husband whose enthusiasm for liverworts was infectious.

References

Alpert, P. (2005) The limits and frontiers of desiccation-tolerant life. *Integrative and Comparative Biology 45*, 685-695.

Alpert, P. & Oliver, M. J. (2002) Drying without dying. In: M. Black & H. W. Pritchard (Eds.). Desiccation and survival in plants: Drying without dying (pp. 1-43). Wallingford, NY: CAB International.

Benscoter, B. W. & Vitt, D. H. (2007) Evaluating feathermoss growth: a challenge to traditional methods and implications for the boreal carbon budget. *Journal of Ecology 95*, 151-158.

Bewley, J. D. (1979) Physiological aspects of desiccation tolerance. *Annual Review of Plant Physiology 30*, 195-238.

Bischler, H. (1998) Systematics and evolution of the genera of the Marchantiales. *Bryophytorum Bibliotheca 51*, 1-201.

Black, M. & Pritchard, H. W. (Eds.). (2002) *Desiccation and survival in plants: Drying without dying*. New York, NY: CABI Publishing.

Blomquist, H. L. (1929) The relation of capillary cavities in the Jungermanniaceae to water absorption and storage. *Ecology 10*, 556-557.

Breuil-Sée, A. (1993) Recorded desiccation-survival times in bryophytes. *Journal of Bryology 17*, 679-680.

Busby, J. R, Bliss, L. C. & Hamilton, C. D. (1978) Microclimate control of growth rates and habitats of the boreal forest mosses, *Tomenthypnum nitens* and *Hylocomium splendens*. *Ecological Monographs 48*, 95-110.

Campbell, D. H. (1904) Resistance of drought by liverworts. *Torreya 4*, 81-86.

Cardot, J. (1900) Recherches anatomiques sur les Leucobryacees. *Mémoires de la Société Nationale des sciences naturelles de Cherbourg 32*, 1-84.

Clausen, E. (1952) Hepatics and humidity: a study on the occurrence of hepatics in a Danish tract and the influence of relative humidity on their distribution. *Dansk Botanisk Arkiv 15*, 5-80.

Clausen, E. (1964) The tolerance of hepatics to desiccation and temperature. *Bryologist 67*, 411-417.

Clee, D. A. (1937) Leaf arrangement in relation to water conduction in the foliose Hepaticae. *Annals of Botany (London), n.s. 1*, 325-328.

Cooper, E. D., Henwood, M. J. & Brown, E. A. (2012) Are liverworts really that old? Cretaceous origins and Cenozoic diversifications in Lepidoziaceae reflect a recurrent theme in liverwort evolution. *Biological Journal of the Linnean Society 107*, 425-441.

Cox, C. J., Goffinet, B., Wickett, N. J., Boles, S. B. & Shaw, A. J. (2010) Moss diversity: A molecular phylogenetic analysis of genera. *Phytotaxa 9*, 175-195.

Crandall-Stotler, B. (1984) Musci, hepatics and anthocerotes – An essay on analogues. In: R. M. Schuster (Ed.*)*. New manual of bryology. Volume 2 (pp. 1093-1129). Nichinan: Hattori Botanical Laboratory.

Crandall-Stotler, B. (1974) Morphogenetic designs and a theory of bryophyte origins and divergence. *Bioscience 30*, 580-585.

Crandall-Stotler, B. J., Stotler, R. E. & Doyle, W. T. (2006) On Anthoceros phymatodes M. Howe and the hornwort genus Phymatoceros Stotler, W.T. Doyle & Crand.-Stotl. (Anthocerotophyta). *Cryptogamie, Bryologie 27*, 59–73.

Crandall-Stotler, B. J., Stotler, R. E. & Long, D. G. (2009) Phylogeny and classification of the Marchantiophyta. *Edinburgh Journal of Botany 66*, 155-198.

Cruz de Carvalho, R., Branquinho, C. & Marques da Silva, J. (2011) Physiological consequences of desiccation in the aquatic bryophyte *Fontinalis antipyretica*. *Planta 234*, 195–205.

Cruz de Carvalho, R., Catalá, M., Marques da Silva, J., Branquinho, C. & Barreno, E. (2012) The impact of dehydration rate on the production and cellular location of reactive oxygen species in an aquatic moss. *Annals of Botany 110*, 1007–1016.

Cruz de Carvalho, R., Silva, A. B., Soares, R., Almeida, A. M., Coelho, A. V., Marques da Silva, J. & Branquinho, C. (2014) Differential proteomics of dehydration and rehydration in bryophytes: *Evidence towards a common desiccation tolerance mechanism. Plant, Cell & Environment 37*, DOI: 10.1111/pce.12266.

Duckett, J. G. & Soni, S. L. (1972) Scanning electron microscope studies on the leaves of Hepaticae. I. Ptilidiaceae, Lepidoziaceae, Calypogeiaceae, Jungermanniaceae, and Marsupellaceae. *Bryologist 75*, 536–549.

Duckett, J. G., Carafa, A. & Ligrone, R. (2006) A highly differentiated glomeromycotean association with the mucilage-secreting, primitive Antipodean liverwort *Treubia*

(Treubiaceae): clues to the origins of mycorrhizas. *American Journal of Botany 93*, 797-813.

Duckett, J. G., Ligrone, R., Renzaglia, K. S. & Pressel, S. (2014) Pegged and smooth rhizoids in complex thalloid liverworts (Marchantiopsida): structure, function and evolution. *Botanical Journal of the Linnaean Society 174*, 68–92.

Edwards, S. (1984) Homologies and inter-relations of moss peristomes. In: R. M. Schuster (Ed.). New manual of bryology. Volume 2 (pp. 658-695). Nichinan: Hattori Botanical Laboratory.

Forrest, L. L., Davis, E. C., Crandall-Stotler, B. J., Clark, B. A. & Hollingsworth, M. L. (2006) Unraveling the evolutionary history of the liverworts (Marchantiophyta): multiple taxa, genomes and analysis. *Bryologist 109*, 303-334.

Gaff, D. F. (1971) Desiccation tolerant plants in southern Africa. *Science,* 174, 1033–1034.

Goebel, K. (1905) Organography of plants. Part II. Special organography. [translated by IB Balfour]. Oxford: Oxford University Press.

Grime, J. P. (1988) The C-S-R model of primary plant strategies-origins, implications, and tests. In: L. D. Gottlieb & S. K. Jain (Eds.). Plant evolutionary biology (pp. 371-393). London: Chapman and Hall.

Grime, J. P., Rincon, E. R. & Wickerson, B. E. (1990) Bryophytes and plant strategy theory. *Botanical Journal of the Linnean Society 104*, 175-186.

Hartung, W., Hellwege, E. M. & Volk, O. H. (1994) The function of abscisic acid in bryophytes. *Journal of the Hattori Botanical Laboratory 76*, 59–65.

Heinrichs, J., Anton, H., Gradstein, S. R., Mues, R. & Holz, I. (2000) Surface wax, a new taxonomic feature in Plagiochilaceae. *Plant Systematics and Evolution 225*, 225-233.

Heinrichs, J., Hentschel, J., Wilson, R., Feldberg, K. & Schneider, H. (2007) Evolution of leafy liverworts (Jungermanniidae, Marchantiophyta): estimating divergence times from chloroplast DNA sequences using penalized likelihood with integrated fossil evidence. *Taxon 56*, 31-44.

Hellewege, E. M., Dietz, K.-J., Heinrich, O. H. & Hartung, W. (1994) Abscisic acid and the induction of desiccation tolerance in the extremely xerophilic liverwort *Exormotheca holstii. Planta 194*, 525-531.

Herzog, T. (1926) Geographie der Moose. Jena: Verlag von Gustav Fischer.

Hyvonen, J., Kosinen, S., Smith-Merrill, G. L., Hedderson, T. A. & Stenroos, S. (2004) Phylogeny of the Polytrichales (Bryophyta) based on simultaneous analysis of molecular and morphological data. *Molecular Phylogenetics and Evolution 31*, 915-928.

Jenks, M. A. & Wood, A. J. (Eds.). (2007) Plant desiccation tolerance. Oxford: Blackwell Press.

Krassilov, V. A. & Schuster, R. M. (1984) Paleozoic and Mesozoic fossils. In: R. M. Schuster (Ed.). New manual of bryology. Volume 2 (pp. 1172-1193). Nichinan: Hattori Botanical Laboratory.

LaFarge-England, C. (1996) Growth form, branching pattern, and perichaetial position in mosses: Cladocarpy and pleurocarpy redefined. *Bryologist 99*, 170-186.

Mägdefrau, K. (1937) Der Wasserhaushalt der Moose. *Annales Bryologici 10*, 141-150.

Magill, R. E. (2010) Moss diversity: new look at old numbers. *Phytotaxa 9*, 167-174.

Marschall, M. & Proctor, M. C. F. (1999) Desiccation tolerance and recovery of the leafy liverwort Porella platyphylla (L.) Pfeiff.: Chlorophyll-fluorescence measurements. *Journal of Bryology 21*, 257-262.

Martin, C. E. & Adamson, V. J. (2001) Photosynthetic capacity of mosses relative to vascular plants. *Journal of Bryology 23*, 319-324.

Nobel, P. S. (1983) Biophysical plant physiology and ecology. New York, NY: W.H. Freeman.

Oliver, M. J. & Bewley, J. D. (1997) Desiccation-tolerance of plant tissues: a mechanistic overview. *Horticultural Reviews 18*, 171–212.

Oliver, M. J., Tuba, Z. & Mishler, B. D. (2000) The evolution of desiccation tolerance in land plants. *Journal of Plant Ecology 151*, 85-110.

Oliver, M. J., Velten, J. & Mishler, B. D. (2005) Desiccation tolerance in bryophytes: a reflection of the primitive strategy for plant survival in dehydrating habitats? *Integrative and Comparative Biology 45*, 788–799.

Pence, V. C., Dunford, S. S. & Redella, S. (2005) Differential effects of abscisic acid on desiccation tolerance and carbohydrates in three species of liverworts. *Journal of Plant Physiology 162*, 1331-1337.

Porembski, S. & Barthlott, W. (2000) Granitic and gneissic outcrops (inselbergs) as centers of diversity for desiccation tolerant vascular species. *Plant Ecology 151*, 19-28.

Pressel, S., Duckett, J. G., Ligrone, R. & Proctor, M. C. F. (2009) Effects of de- and rehydration in desiccation-tolerant liverworts: a cytological and physiological study. *International Journal of Plant Sciences 170*, 182-199.

Proctor, M. C. F. (1979) Structure and eco-physiological adaptation in bryophytes. In: G. C. S. Clarke & J. G. Duckett (Eds.). Bryophyte systematics (pp. 479-509). London, UK: Academic Press.

Proctor, M. C. F. (2009) Physiological ecology. In: A. J. Shaw & B. Goffinet (Eds.). Bryophyte biology. Second edition (pp. 237-297). Cambridge, UK: Cambridge University Press.

Proctor, M. C. F. (2010) Recovery rates of chlorophyll-fluorescence parameters in desiccation-tolerant plants: fitted logistic curves as a versatile and robust source of comparative data. *Plant Growth Regulation 62*, 233–240.

Proctor, M. C. F. & Pence, V. C. (2002) Vegetative tissues: bryophytes, vascular plants, resurrection plants and vegetative propagules. In: M. Black & H. W. Pritchard (Eds.). Desiccation and survival in plants: Drying without dying (pp. 207-237). Wallingford, NY: CAB International.

Proctor, M. C. F., Oliver, M. J., Wood, A. J., Albert, P., Stark, L. R., Cleavitt, N. L. & Mishler, B. D. (2007) Desiccation-tolerance in bryophytes: A review. *Bryologist 110*, 595-621.

Qiu, Y.-L., Li, L. B., Wang, B., Chen, Z., Knoop V., Groth-Malonek, M., Dombrovska, O., Lee, J., Kent, L., Rest, J., Estabrook, G. F., Hendry, T. A., Taylor, D. W., Testa, C. M., Ambros, M., Crandall-Stotler, B., Duff, R. J., Stech, M., Frey, W., Quandt, D. & Davis, C. C. (2006) The deepest divergences in land plants inferred from phylogenomic evidence. *Proceedings of the National Academy of Sciences (USA) 103*, 15511-15516.

Reed, S. C., Coe, K. K., Sparks, J. P., Housman, D. C., Zelikova, T. J. & Belnap, J. (2012) Changes to dryland rainfall result in rapid moss mortality and altered soil fertility. *Nature Climate Change 2*, 752–755.

Renzaglia, K. S., Villarreal, J. C. & Duff, R. J. (2009) New insights into morphology, anatomy, and systematics of hornworts. In: A. J. Shaw & B. Goffinet (Eds.). Bryophyte biology. Second edition (pp. 139-171). Cambridge: Cambridge University Press.

Rice, S. K. & Schneider, N. (2004) Cushion size, surface roughness and the control of water balance and carbon flux in the cushion moss Leucobryum glaucum (Leucobryaceae). *American Journal of Botany 91*, 1164-1172.

Rice, S. K., Collins, D. & Anderson, A. M. (2001) Functional significance of variation in bryophyte canopy structure. *American Journal of Botany 88*, 1568-1576.

Schuster, R. M. (1966) The Hepaticae and Anthocerotae of North America. Vol. 1. Columbia, NY: Columbia University Press.

Shaw, A. J., Devos, N., Cox, C. J., Boles, S. B., Shaw, B., Buchanan, A. M., Cave, L. & Seppelt, R. (2010) Peatmoss (*Sphagnum*) diversification associated with Miocene Northern Hemisphere climatic cooling? *Molecular Phylogenetics and Evolution 55*, 1139-1145.

Shaw, A. J., Szövényi, P. & Shaw, B. (2011) Bryophyte diversity and evolution: Windows into the early evolution of land plants. *American Journal of Botany 98*, 352-369.

Stark, L. R., Mishler, B. D. & McLetchie, D. N. (1998) Sex expression and growth rates in natural populations of the desert soil crustal moss *Syntrichia caninervis*. *Journal of Arid Environments 40*, 401-416.

Stark, L. R., Mishler, B. D. & McLetchie, D. N. (2000) The cost of realized sexual reproduction: Assessing patterns of reproductive allocation and sporophyte abortion in a desert moss. *American Journal of Botany 87*, 1599-1608.

Stark, L. R., Greenwood, J. L., Brinda, J. C. & Oliver, M. J. (2013) The desert moss Pterygoneurum lamellatum (Pottiaceae) exhibits an inducible ecological strategy of desiccation tolerance: Effects of rate of drying on shoot damage and regeneration. *American Journal of Botany 100*, 1522-1531.

Turetsky, M. R., Crow, S., Evans, R. J., Vitt, D. H. & Wieder, R. K. (2008) Tradeoffs in resource allocation among moss species control decomposition in boreal peatlands. *Journal of Ecology 96*, 1297-1305.

Vitt, D. H. (1971) The infrageneric evolution, phylogeny, and taxonomy of the genus *Orthotrichum* (Musci) in North America. *Nova Hedwigia,* 21, 683-711.

Vitt, D. H. (1981) Adaptive modes of the moss sporophyte. *Bryologist 83*, 162-182.

Vitt, D. H. (1984) Classification of the Bryopsida. In: R. M. Schuster (Ed.). *New manual of bryology.* Volume 2 (pp. 696-758). Nichinan: Hattori Botanical Laboratory.

Vitt, D. H. & Glime, J. M. (1984). The structural adaptations of aquatic Musci. *Lindbergia 10*, 95-110.

Vitt, D. H., Goffinet, B. & Hedderson, T. (1998) The ordinal classification of the mosses: Questions and answers for the 1990s. In: J. W. Bates, N. W. Ashton & J. G. Duckett (Eds.). Bryology for the 21st Century (pp. 113-123). London: British Bryological Society.

Von Konrat, M., Söderström, L., Renner, M. A. M., Hagborg, A., Briscoe, L. & Engel, J. J. (2010) Early land plants today (ELPT): How many liverwort species are there? *Phytotaxa 9*, 22-40.

Wilson, R., Gradstein, S. R., Schneider, H. & Heinrichs, J. (2007) Unraveling the phylogeny of Lejeuneaceae (Jungermanniopsida). Evidence of four main lineages. *Molecular Phylogeny and Evolution 43*, 270-282.

Wood, A. (2005) Eco-physiological adaptations to limited water environments. In: P. M. Hasegawa & M. A. Jenks (Eds.). Plant abiotic stress (pp. 1-13). Oxford: Blackwell Press.

Wood, A. J. (2007) The nature and distribution of vegetative desiccation-tolerance in hornworts, liverworts and mosses. *Bryologist 110*, 163-177.

Zander, R. (1993) Genera of the Pottiaceae: Mosses of harsh environments. *Bulletin of the Buffalo Society of Natural Sciences 32*, 1-378.

In: Plant Ecology and Evolution in Harsh Environments
Editors: N. Rajakaruna, R. S. Boyd and T. B. Harris
ISBN: 978-1-63321-955-7

Chapter 13

CLIMATE CHANGE AND THE FUTURE OF EDAPHIC FLORAS

Barbara Fernandez-Going*

Department of Ecology, Evolution, and Marine Biology, University of California, Santa Barbara, Santa Barbara, CA, USA
Wildland Resources Department, Utah State University, Logan, UT, USA

ABSTRACT

Climate change is one of the most important threats to biodiversity and has motivated a tremendous body of research in plant ecology and biogeography. However, much of this research has focused on ecological communities on relatively fertile soils, despite the importance of edaphic floras to global and regional biodiversity. Edaphic floras share certain characteristics that may cause them to respond differently to climate change than other communities. These floras are supported by poor, infertile soil types, such as limestone, gypsum, or serpentine, which are often deficient in limiting resources or have high concentrations of chemicals inimical to plant growth. Further, they are often patchy and spatially isolated which may limit their ability to track climate through migration. In this chapter, I discuss the possible implications of these characteristics to the fate of edaphic floras and review the associated literature in the hopes of generalizing predictions for the persistence of these special floras.

INTRODUCTION

Anthropogenic climate change is one of the most pervasive threats to the current distribution and abundance of species and probably the most impacting driver of global change. The global climate system has experienced an unprecedented increase the mean and inter-annual variability in temperature since the 1950s (Stocker et al., 2013). Precipitation patterns have also been measurably altered over the last decades. In addition to these

* Corresponding author: bar.fern34@gmail.com.

directional patterns, there has been a rise in the number and intensity of extreme events, such as heat waves or heavy precipitation events (Stocker et al., 2013). The effects of current climate change on species, communities, and ecosystems have been widely documented and can be quite substantial (Walther, 2010).

The response to altered temperature and precipitation regimes varies among plant species in ways that cascade through the community and ecosystem as a whole. In some cases, species have responded to climate change with phenotypic plasticity (*i.e.*, alterations of phenology or growth patterns) or through rapid evolution, which can lead to changes in range distributions and may feed back to influence community composition via species interactions (Walther, 2010). For example, in a study of altered precipitation regimes in grasslands, species responses and interactions drove changes in community composition over time (Suttle et al., 2007). In the first year of the study nitrogen-fixing species responded strongly with increased productivity to an extension of the rainy season (late spring precipitation). Annual grass production rose dramatically the following year, presumably facilitated by fertilization by the nitrogen-fixing species, and forb diversity declined through the remainder of the experiment as annual grass litter accumulated. These effects cascaded to the invertebrate community, which also declined in diversity over time (Suttle et al., 2007). However, most of the work on climate and community structure and dynamics has been conducted in relatively fertile soil types, which may respond differently from habitats in which temperature and precipitation are not primary limiting factors.

Distribution and Characteristics of Edaphic Floras

The last decade has seen a small but steady increase in research on the effects of climate change on edaphic floras. Those floras occurring on soils with unusual or extreme characteristics (*e.g.,* high calcium carbonate or heavy metal concentrations, low macronutrient concentrations, soils derived from ultramafic parent materials, or other azonal soils) occur in nearly all the major climatic areas of the world. Notable examples include serpentine floras in Mediterranean California and tropical Cuba, gypsum floras in semi-arid and arid climates, and limestone floras in temperate Europe and subtropical China. These floras are easily recognized: they often stand out as patchy, isolated outcrops in stark contrast to the more productive surrounding vegetation on 'normal' soils, particularly in productive climates. Soils supporting edaphic floras are often inimical to plant growth, resulting in patches of stunted, sparse vegetation. In addition, harsh, infertile soils exclude most species in the local species pool and host specialist species on those particular soils, which often are narrow endemics. For example, nearly 80% of the plant species in California avoid serpentine soil, while the state boasts nearly 250 specialist endemics (Safford et al., 2005). Species adapted to these unique soils are often slow-growing, with suites of traits associated with resource-use efficiency and slow resource acquisition (see Chapter 6). Thus, special soil floras are often compositionally, functionally, and structurally distinct from surrounding communities.

The importance of patchiness is probably most well-studied in serpentine plant communities, but is also known for some calcareous substrates. Harrison (1999) observed that plant species diversity was greater on small patches (0.5-3 ha) of serpentine compared to large patches (>5 km^2) and there was greater beta diversity for endemics on small patches.

Figure 1. Map of California showing the patchiness of serpentine soil at a regional scale. Modified from Fernandez-Going et al. (2013).

However, small patches were also more likely to be invaded by non-native species (Harrison, 1999). In contrast, the richness of both habitat specialists and generalists increased with patch size calcareous grasslands (Krauss et al., 2004). Patches of bedrock for these unique soils can sometimes be highly spatially isolated, surrounded by kilometers of more productive vegetation where (presumably) competition from generalist species may be strong (Figure 1).

These two main characteristics of harsh substrates: nutrient or chemical stress and patchiness, are likely to have a strong influence on the response of edaphic species and communities to climate change. Soil infertility may be strong enough to limit the responsiveness of these communities, thereby providing some resistance to climate change, or climate change may exacerbate the effects of soil infertility, potentially putting a major proportion of these floras at risk of extinction. An indirect effect of abiotic stress is the selection of stress-tolerant species in edaphic floras, which may be less responsive to climate and thus be less extinction prone than species with more competitive life histories. However, the patchy nature of the bedrock may limit the ability of species to migrate and track their climate niche if the patch distance is greater than the ability of species to disperse or if these species are unable to persist in the matrix of non-serpentine soil. This chapter explores the influence of abiotic stress and habitat patchiness on the response of edaphic communities to climate change and reviews the current literature with the goal of identifying trends for these communities. I focus on plant communities on serpentine, limestone, and gypsum soils because these are the most widely studied special soils.

NUTRIENT LIMITATION

Nutrient limitation is the most widely cited reason for the optimistic view that edaphic floras will persist in the face of fairly large changes in climate (and in the absence of other disturbances). This hypothesis assumes that plants in these special edaphic habitats are proximally limited by major nutrients, either because the soils are deficient in nutrients such as nitrogen or phosphorus or because excess ions inhibit nutrient uptake, and thus are limited by temperature and precipitation secondarily. Some experimental evidence from grasslands on serpentine-derived soil seems to support this hypothesis. Additions of nitrogen and phosphorus improved biomass production of native and exotic species in California grasslands, while water addition alone had no effect on richness or biomass production (Going et al., 2009; Huenneke et al., 1990).

Further evidence for nutrients as the primary limiting factor comes from a study in serpentine garrigue in which plant community composition in response to phosphorus addition was monitored for 12 years (Chiarucci & Maccherini, 2007). Phosphorus addition in two consecutive years at the start of the study altered community composition patterns, with most species generally increasing in cover. While there was no effect of inter-annual variability in climate on community composition in fertilized or control plots, effects on species richness were strongest when phosphorus limitation was removed. Serpentine grasslands also experienced less turnover in species composition over a ten-year period than communities on adjacent, more fertile soils (Fernandez-Going et al., 2012). There was mixed evidence for a direct effect of soil. The frequency and abundance of 41 soil generalists (Figure 2a), species occurring on both soils, was greater on serpentine than on non-serpentine soil, but a composite index of soil fertility was unrelated to compositional turnover. These species also had lower specific leaf areas on serpentine soils (Figure 2b), suggesting harsh soils select for more stress-tolerant traits (discussed below). Recently, evidence has emerged that heterogeneity in soil depth and resources, rather than absolute amounts of resources, may be the driving mechanism stabilizing limestone communities (Fridley et al., 2011). Measurements of soil moisture, nutrient availability, and pH across soil depth gradients within experimental plots exposed to different climate manipulations showed that individual species had contrasting responses to treatments in shallow versus deep microsites. Biomass increased or decreased at the microsite scale, but not at the plot scale (Fridley et al., 2011). Serpentine, as well as other unique substrates, are often reported to be heterogeneous in several variables, including depth, texture, and resources and more research is needed to determine if heterogeneity contributes to the stability of these communities as well.

The refugium hypothesis, most often described for relic species occurring on serpentine soil but also noted for ironstone ranges and gypsum habitats, holds that outcrops of harsh, low nutrient soils have served as important refugia for some species during historical shifts in climate (Gibson et al., 2012; Kruckeberg, 1984; Palacio et al., 2007). That is, species were able to persist through large climatic fluctuation only by surviving on azonal soils. However, there is only limited paleontological evidence to suggest that soil fertility and texture can influence the response of plant communities to centennial or millennial climate change (Brubaker, 1975; Bush & Flenley, 1987). In Michigan, pollen records indicate that forest development and changes in community composition were slower on sandy soils compared to other soil textures (Bernabo, 1981).

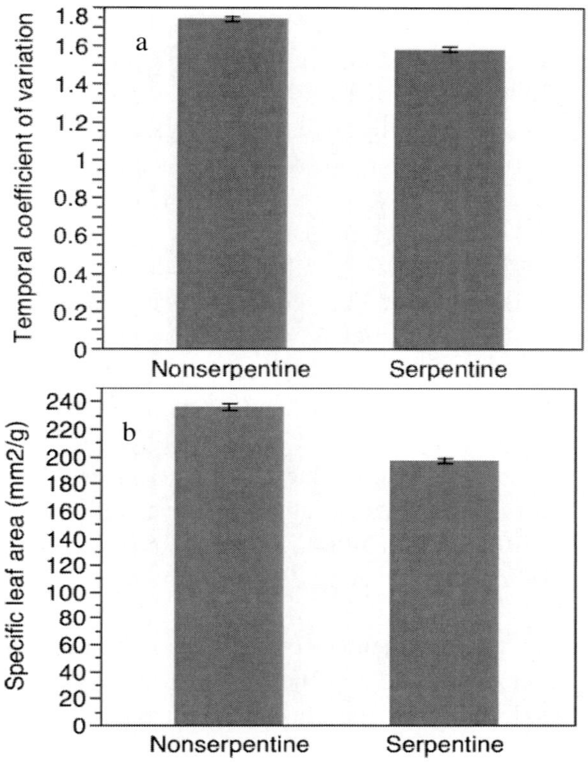

Figure 2. a) Temporal (over a ten year survey) coefficient of variation in cover and b) mean specific leaf area of 41 grassland species commonly found on serpentine and non-serpentine soils. Modified from Fernandez-Going et al. (2013). Bars in both figures are statistically significant at the $P<0.01$ level.

The only paleontological study known from ultramafic soils, located in the Klamath Mountains of California, shows that communities on serpentine and peridotite substrates continued to support herbaceous species indicative of parkland vegetation over millennia, while mixed conifer forest eventually developed on non-ultramafic substrates in response to an increase in climatic productivity (Briles et al., 2011). Although this work suggests that communities on harsh soils can resist changes in climate over long timescales, researchers have only recently explored the issue of rapid anthropogenic climate change.

Recent experiments in serpentine and limestone grasslands suggest that these floras may be relatively resistant to rapid changes in temperature and precipitation. In one of the foundational studies on the subject, Grime et al. (2000) reported that after five years of temperature and precipitation manipulations (winter warming, summer drought, and supplemented summer rainfall) the biomass and composition of grassland communities on ancient, relatively infertile limestone substrate (Buxton) varied little compared to grasslands on arable limestone soil (Wytham). The Buxton site continued to show resistance to climate manipulations after 12 years (Grime et al., 2008). Similar results were observed in serpentine grasslands in response to three years of experimental rainfall reduction (Fernandez-Going & Harrison, 2013) and to ten years of natural variation in precipitation (Fernandez-Going et al., 2012).

In addition to the direct effect of climate on edaphic floras, changes in climate may interact with soil fertility to influence species responses via nutrient cycling and organic matter accumulation. Accumulation of soil organic matter is primarily controlled by climate, vegetation, and parent material, so that decomposition and organic matter quantity is low in nutrient-poor habitats (Coûteaux et al., 1995; Trumbore, 1997). In rainy biomes soils can become impoverished through leaching of macronutrients, but in arid biomes increases in precipitation often lead to increases in soil fertility via increased plant productivity and organic matter accumulation (Singer & Munns, 2006). However, in nutrient-poor habitats accumulation of organic matter may be primarily limited by the low quality of recalcitrant litter (Coûteaux et al., 1995), potentially resulting in a slower response to climate change. Although no studies have specifically addressed this question, one recent study suggests that soil-based feedbacks may influence community response to precipitation. In a comparison of diversity patterns in serpentine and non-serpentine communities along California's precipitation gradient (spanning 20-185 cm annual precipitation), Fernandez-Going et al. (2013) found that plant productivity increased more strongly with precipitation on the more fertile non-serpentine soils compared to productivity on serpentine. This result may partially reflect an increase in soil organic matter in non-serpentine soils, but not in serpentine soils (Fernandez-Going et al., 2013). Although other factors are likely to vary spatially and thus limit inference to climate change, the study does suggest a potential role for soil fertility-climate feedbacks.

Notably, there are very few other studies of the responses of communities on soils other than serpentine or limestone to natural climatic variation or manipulated climate. However, there is mixed indirect evidence for the response of gypsum plant communities (see Chapter 5) to precipitation. Several studies have shown that early drought can negatively impact the emergence and establishment of gypsophiles, but that the response to drought is species specific (Escudero et al., 2000; Matesanz et al., 2008). Yet other work in gypsum drylands has shown that biotic interactions can shift from facilitation to competition, driven largely by water availability and temperature (Maestre et al., 2010). These studies suggest that the effect of climate change on gypsum communities may depend on which species occur in the community and their sensitivity to climate. However, other work shows that gypsum soils hold more soil moisture than adjacent calcareous soil so that aridity plays a diminished role in structuring plant community types and cover (Pueyo & Alados, 2007). Although high gypsum content can interfere with ion exchange and inhibit nutrient uptake, it remains unclear how important this interference is relative to aridity and topographic features for structuring gypsum community composition (Castillejo et al., 2011; Pueyo & Alados, 2007). In arid regions water holding capacity is limited in calcareous soils compared to valley bottoms with relatively salty soils.

Plant Traits and Community Structure

It is generally accepted that species growing on chemically harsh or nutrient poor soils (general characteristics of special soils) tend to have traits associated with stress tolerance (Chapin, 1991; Grime, 1977). Broadly, species adapted to these habitats have slow growth, low capacity for nutrient absorption (particularly phosphate) and slow photosynthesis, and they tend to allocate more resources to survivorship compared to fast-growing, competitive species (Chapter 9; Chapin, 1991; Cornelissen et al., 2003). Indeed, a trade-off between tolerance and competitive ability has long been hypothesized to be the proximal cause of edaphic endemism in many systems (Escudero et al., 1999; Kruckeberg, 1984; Palacio et al., 2007). Many of the morphological features of plants in low nutrient environments, such as narrow leaves with thick waxy layers, are also found in plants of water-limited environments (Cunningham et al., 1999) and thus adaptation to nutrient stress may confer resistance to changes in precipitation. This has been observed or inferred for many edaphic endemics, including the serpentine endemic, *Mimulus nudatus* (Phrymaceae). Researchers have observed that this uncommon, locally restricted endemic occupies drier habitats and is more drought tolerant than its more widespread assumed progenitor, *Mimulus guttatus* (Hughes et al., 2001; also see Chapter 11). In another study in California serpentine grasslands, three endemics with traits strongly suggestive of stress tolerance were found to be unresponsive to decreases in spring precipitation (Fernandez-Going, unpublished data). However, some evidence suggests that endemics may be more susceptible to climate change. To assess the effect of the last 50 years of climate warming on endemic-rich communities, Damschen et al. (2010) re-sampled Robert Whittaker's plots on diorite and serpentine in the Klamath-Siskiyou region of southern Oregon. The composition of both communities had shifted with nearly the same magnitude to be more similar to xeric communities found on south-facing slopes. Further, they observed large declines in the cover of serpentine endemic species, but not widespread species, suggesting that endemics were more sensitive to warming (Damschen et al., 2010). The study site was relatively well protected from nitrogen deposition and the authors avoided logged sites, suggesting that other global change factors likely played only a small role in structuring these communities.

Observations of edaphic endemic response to temperature and precipitation can only suggest a role for functional traits: they do not rule out other factors that may drive the responses of these species to climate change. In addition, although endemics are an important component of edaphic communities, they often represent a minor proportion of the total cover and diversity. To date only one study has specifically examined the role of functional traits in edaphic community response to natural climatic variation. In a ten-year study of community composition, Fernandez-Going et al. (2012) found that species richness and composition serpentine grassland communities was much less variable over time than communities on adjacent non-serpentine soils, although this variability was influenced by exotic annual grasses. Serpentine communities were shorter-statured, had lower specific leaf areas and leaf water content, and had lower foliar nitrogen than non-serpentine communities: these traits conferred resistance to change (Fernandez-Going et al., 2012).

SPATIAL ISOLATION

Modeling approaches to the issues of climate change have suggested that species with narrow, specialized ranges may be especially vulnerable to climate change (Dirnböck et al., 2011; Malcolm et al., 2006), a result that may have profound implications for edaphic floras and endemic species. Broadly, the models calculate multivariate correlations between occurrence or abundance and current climate data, which then define a species' realized climate niche. That climate niche is then projected into future geographic space under a variety of climate change scenarios and dispersal abilities. For example, Loarie et al. (2008)

projected large extinction risks and shifts in diversity towards more mesic local climates for species endemic to the California Floristic Province. Much of the risk hinged on the assumption that species were unable to disperse and diversity loss declined considerably when that assumption was relaxed. Dispersal ability is a concern for species occurring in habitat islands, such as species that occupy patchy outcrops with special soils. Indeed, a recent study of 12 serpentine endemic plant species in California suggests that minimum dispersal jumps of 596-1891 m, depending on the climate change scenario, would be required for those species to track their climate niche within their newly suitable range (Damschen et al., 2012). These distances reflect the minimum shortest edge-to-edge distance between patches of serpentine soil.

Long-distance seed dispersal is uncommon for most species and can be especially rare in floras adapted to spatially isolated habitats (Cain et al., 2000; Riba et al., 2009), a pattern that is strikingly similar to evolution of short-distance dispersal in species on oceanic islands (Schenk, 2013). The few studies of seed dispersal in edaphic floras suggest chemical and nutrient stresses are more influential on dispersal mode than habitat patchiness and that species adapted to special soils are generally not good dispersers. Schenk (2013) observed a reduction in wing area on seeds from gypsophilic species compared to non-gypsum species in *Mentzelia* (section *Bartonia*; Loasaceae). Interestingly, he did not observe a concurrent increase in seed size as is often noted in island endemics, which may reflect the stressful conditions of gypsum habitats (Schenk, 2013; Stebbins, 1971; also see Chapter 5). Spasojevic et al. (2014) compared dispersal syndromes for more than 450 species on serpentine and non-serpentine soils from three vegetation types: forest, chaparral, and grassland. Serpentine communities had a lower proportion of directed dispersal mechanisms (dispersal by vertebrates) and a significantly higher proportion of wind-dispersed species than non-serpentine communities. The results were consistent among vegetation types and were not influenced by patch size, suggestion that habitat quality is the proximal driver of dispersal modes in these species (Spasojevic et al., 2014). Interestingly, there was no difference between soil types in the proportion of species with long-distance dispersal mechanisms.

Aside from the large influence of dispersal ability on prediction of species distributions in the future, a major limitation of using bioclimatic models to forecast extinction risk for edaphic floras is that these models do not account for non-climatic factors that may influence species' distributions (Schwartz et al., 2006). Further, as discussed above, the response to harsh, nutrient-limited soils may alter species' climatic niches. Early evidence for edaphic influence on species' climate niches can be seen in a comparative vegetation study of limestone and acid soil communities in the Santa Catalina Mountains of Arizona. For example, pine-oak woodland occurred in the same elevational range that would support a more productive pine forest on diorite, and open oak woodland on diorite was replaced by less productive grassland on limestone (Whittaker & Niering, 1968). Using herbarium data for species with populations on and off of serpentine soil in California ($n = 814$ species), Burge & Salk (2013) found that serpentine populations had lower elevational limits, lower maximum temperature thresholds, and a more narrow range of minimum temperature limits than non-serpentine populations. These results suggest that serpentine populations not only have different climate niches than their non-serpentine counterparts, but the results are also evidence for reduced extreme temperature tolerance on serpentine (Burge & Salk, 2013). They also strongly suggest that simple bioclimatic models may be insufficient to predict future distributions of edaphic endemic species and communities.

Climate Change in the Context of Other Global Change Factors

Many edaphic floras are currently threatened by a number of other factors, including habitat fragmentation, the spread of invasive species, and nitrogen deposition, that have the potential to interact with climate change, yet we know very little about how these floras will respond to these interactions. A number of studies have examined the effects of these factors singly but considerably fewer have examined the effects of interactions among some of the factors. Although we are in critical need of more studies that focus on interactions between climate change and other global change drivers, we can try to glean some patterns from the current literature.

As suggested in the section on nutrient limitation, the addition of limited resources, such as through nitrogen deposition, can dramatically alter the composition and structure of edaphic floras. Nitrogen deposition can reduce community diversity by promoting the increased dominance of one or a few species and through losses of nitrogen-fixing species (Going et al., 2009; Weiss, 1999). In some cases, the effects of nitrogen deposition seem to be driven by the increased abundance of non-native and invasive species, at least on serpentine substrates. In serpentine grasslands, non-native annual grasses are primarily nitrogen-limited (Going et al., 2009; Huenneke et al., 1990) and nitrogen deposition in some grasslands has led to the spread of an invasive grass, a decline in native species richness, and a significant threat to an endangered butterfly (Weiss, 1999). In California, increased invasibility of edaphic floras under nitrogen deposition may be inhibited by reduced soil moisture from a drying climate (through both decreased precipitation and increased evapotranspiration from warmer temperatures).

One recent study (Eskelinen & Harrison, 2014) of the effects of increased precipitation and fertilization across a gradient of productivity found that adding an NPK fertilizer and supplementing spring precipitation promoted the success of two noxious grassland invaders [*Centaurea solstitialis* (Asteraceae) and *Aegilops triuncialis* (Poaceae)] along the entire gradient. However, these interactive effects were greatest in the most productive non-serpentine soil, intermediate in the clay-rich 'lush' serpentine soil, and weakest in the least productive rocky serpentine soil. In their study comparing temporal stability of communities on serpentine and non-serpentine soil, Fernandez-Going et al. (2012) showed that non-native annual grasses are highly dynamic on serpentine and may increase compositional turnover in response to climatic variability. Increased variability in species composition could potentially leave communities more susceptible to stochastic factors.

Although many edaphic floras occur in patchy landscapes, habitat fragmentation may influence the persistence of some species. Several studies have shown that the effect of patch size and isolation can influence the reproductive output of endemic species, but these effects may be highly variable among years (Rabasa et al., 2009; Wolf & Harrison, 2001). For example, one serpentine endemic (*Helianthis exilis*; Asteraceae) was found to be restricted to large serpentine outcrops with seeps while another (*Calystegia collina*; Convolvulaceae) showed reduced reproductive success on small outcrops (Wolf, 2001). Reduced fitness in *Calystegia* was largely driven by pollen limitation with the species apparently maintaining populations on small outcrops through vegetative reproduction.

As with communities on more fertile substrates, anthropogenic habitat fragmentation can reduce species diversity and may be exacerbated by other factors. In a study of gypsum floras, Pueyo et al. (2008) observed that communities on fragmented gypsum soils had lower diversity than communities on continuous gypsum soils, but that livestock grazing ameliorated the effects of fragmentation through increased seed dispersal. Gypsum specialists were also less abundant in fragmented sites. The only known study to examine the effect of climate change and fragmentation in an edaphic community (Matesanz et al., 2009) suggests that the effects of drought and fragmentation may be additive. To examine the effects of three global change factors, researchers in that study compared the response of *Centaurea hyssopifolia* to water addition on small versus large fragments and in good versus poor quality patches of gypsum during two consecutive drought years. There was a significant decline in reproductive output of as a result of fragmentation, which is striking considering that most of the smaller fragments in the region (Belinchón, Spain) were created 25 years prior to the study as a result of agricultural policy. Drought and habitat quality had no effect on reproductive success, but did drive an advance in phenology (Matesanz et al., 2009). These results suggest that the resistance to climate change observed in several edaphic floras could be limited under disturbance by other global change drivers.

CONCLUSION

Accurately predicting the effect of climate change on species and communities is one of our greatest challenges and yet is of utmost importance if we are to conserve biodiversity and ecosystem functioning. Recent reviews have shown that species are already responding to current (and relatively mild) changes in climate (*e.g.*, Walther, 2010). One meta-analysis found that species have shifted their range poleward by an average of 6.1 km/year and that there was an average advance of phenological events of 2.3 days/decade (Parmesan & Yohe, 2003). However, there is a large amount of variability in these responses, indicating that other factors may have influenced how species have responded to climate change. The work reviewed in this chapter suggests communities also vary in their sensitivity to climate and that the nature and distribution of soils are two factors that may explain some of this variation.

Several lines of evidence point to greater resistance to climate change for floras on harsh, unique substrates. Observational studies show that plant communities on ultramafic soils, such as serpentine, change little in response to natural climatic variation (*e.g.*, Fernandez-Going et al., 2012) or to historical climate change (*e.g.*, Briles et al., 2011). This finding is supported by experimental work showing that productivity and community composition change less over time in response to precipitation manipulations on serpentine or infertile limestone soils (Fernandez-Going et al., 2012; Grime et al., 2000). This apparent resistance seems to be driven, at least in part, by an indirect effect of soil on species composition. Special edaphic communities are often composed of long-lived, slow growing evergreen species or species with stress tolerant traits that respond slowly to environmental change. Soil-climate feedbacks may be another indirect effect of harsh soils that contribute to climatic resistance (*e.g.*, Fernandez-Going et al., 2013).

Although it is tempting to express optimism about the persistence of edaphic communities into the future based on the above evidence, there is still insufficient research from which to draw generalizations and there are a handful of studies that suggest we still do not fully understand how soils and climate interact to influence community response to directed climate change. For example, much of the research suggesting that these harsh, infertile communities are resistant to climatic variation has come from serpentine soils in semi-arid climates. The one study conducted on serpentine in a relatively mesic climate found that serpentine communities responded to 50 years of climate warming with the same magnitude as non-serpentine communities (Damschen et al., 2010). This is also in contrast to the finding that infertile limestone soils were less responsive to perturbations in precipitation than more fertile limestone soils (Grime et al., 2000). Further, species growing on gypsum soils were shown to be sensitive to drought (Matesanz et al., 2009), in striking contrast to the finding that some serpentine endemics are insensitive to reductions in precipitation (Figure 3).

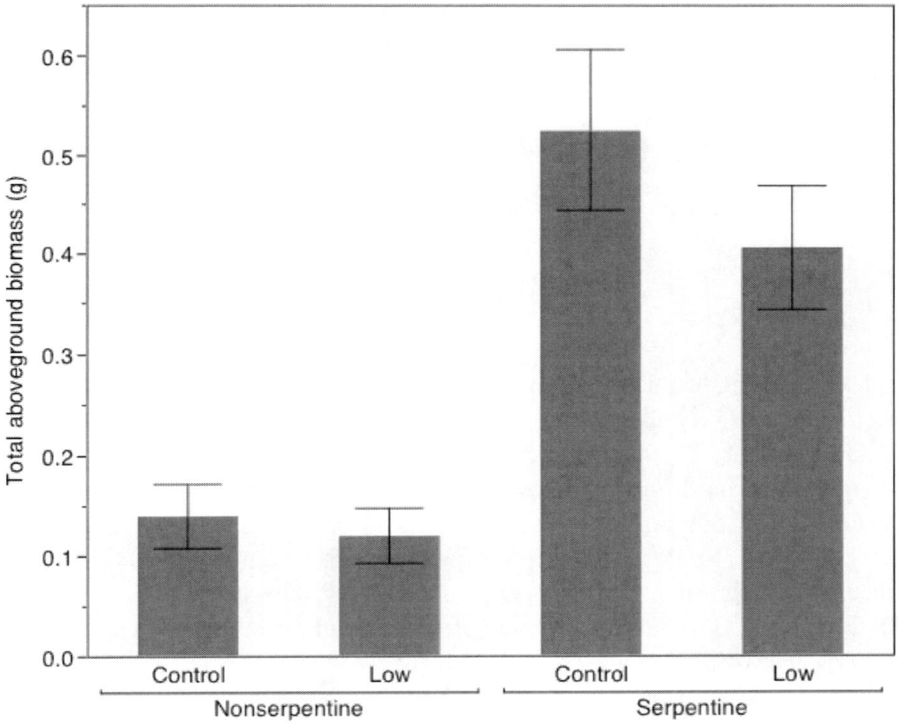

Figure 3. Biomass of three serpentine endemics grown under ambient (control) and decreased (low) precipitation in non-serpentine and serpentine grasslands from 2009 to 2011. Bars represent mean ± SE (n=10 per treatment). There were no significant differences in endemic biomass on either soil type.

The discrepancy among these results highlights the need for more long-term experiments, including experiments that focus on the basic ecology of edaphic communities. In particular, it is unclear to what degree edaphic communities are climate-limited. For example, research suggests that nutrient limitation is primary for serpentine communities, while work with gypsum species suggests that topography and aridity are important for these communities. We also do not fully understand the relative importance of limiting factors other than nutrient and water availability such as the presence of heavy metals in serpentine, for the response of edaphic communities to climate.

Finally, while there is a growing awareness of the importance of plant-soil feedbacks in the context of global change, there are nearly no studies that have been conducted *in situ* in special soils (van der Putten et al., 2013, but see Casper & Castelli, 2007). If we are to accurately predict the effects of climate change on edaphic communities as a group, we need more research on the responses of communities on different substrates. Ideally, climate change experiments should be conducted across a variety of both climates and soils, including soils that do not fall under the category of 'special soils,' to tease out the effects of climate and soil. More research is also needed on the indirect effects of climate change. In addition to soil-climate feedbacks, which are far from understood for most systems, it remains unclear how individual species will be affected indirectly by climate change through changes in the competitive environment. Although competition is presumed (and often is) to be important for the persistence of soil endemics on less stressful soils, competition can reduce growth even on harsh infertile soils (Fernandez-Going & Harrison, 2013). Thus, despite a minimal direct response to climate, endemic success may depend more strongly on the response of the community to climate. In conclusion, the increased focus on the fate of special soil communities to climate change is encouraging, but we have far to go before we can have confidence in our ability to develop conservation strategies for these important and unique communities.

REFERENCES

Bernabo, J. C. (1981) Quantitative estimates of temperature changes over the last 2700 years in Michigan based on pollen data. *Quaternary Research 15*, 143-159.

Briles, C. E., Whitlock, C., Skinner, C. N. & Mohr, J. (2011) Holocene forest development and maintenance on different substrates in the Klamath Mountains, northern California, USA. *Ecology 92*, 590-601.

Brubaker, L. B. (1975) Postglacial forest patterns associated with till and outwash in Northcentral Upper Michigan. *Quaternary Research 5*, 499-527.

Burge, D. O. & Salk, C. F. (2013) Climatic niche shifts in the serpentine flora of California. *Journal of Vegetation Science,* DOI: 10.1111/jvs.12144.

Bush, M. B. & Flenley, J. R. (1987) The age of the British Chalk grassland. *Nature 329*, 434-436.

Cain, M. L., Milligan, B. G. & Strand, A. E. (2000) Long-distance seed dispersal in plant populations. *American Journal of Botany 87*, 1217-1227.

Casper, B. B. & Castelli, J. P. (2007) Evaluating plant-soil feedbacks together with competition in a serpentine grassland. *Ecology Letters 10*, 394-400.

Castillejo, J. M., Castéllo, R., San Cristobal, A. G. & Abad, S. (2011) Soil-plant relationships along a semiarid gypsum gradient (Rio de Aguas, SE Spain). *Plant Ecology 212*, 1287-1297.

Chapin, F. S. (1991) Integrated responses of plants to stress. *BioScience 41*, 29-36.

Chiarucci, A. & Maccherini, S. (2007) Long-term effects of climate and phosphorus fertilisation on serpentine vegetation. *Plant and Soil 293*, 133-144.

Cornelissen, J. H. C., Lavorel, S., Garnier, E., Díaz, S., Buchmann, N., Gurvich, D. E., Reich, P. B., ter Steege, H., Morgan, H. D., van der Heijden, M. G. A., Pausus, J. G. & Poorter,

H. (2003) A handbook of protocols for standardised and easy measurement of plant functional traits worldwide. *Australian Journal of Botany 51*, 335-380.

Coûteaux, M., Bottner, P. & Berg, B. (1995) Litter decomposition, climate and litter quality. *Trends in Ecology and Evolution 10*, 63-66.

Cunningham, S. A., Summerhayes, B. & Westoby, M. (1999) Evolutionary divergences in leaf structure and chemistry, comparing rainfall and soil nutrient nutrients. *Ecology 69*, 569-588.

Damschen, E. I., Harrison, S. & Grace, J. B. (2010) Climate change effects on an endemic-rich edaphic flora: Resurveying Robert H. Whittaker's Siskiyou sites (Oregon, USA). *Ecology 91*, 3609-3619.

Damschen, E. I., Harrison, S., Ackerly, D. D., Fernandez-Going, B. M. & Anacker, B. L. (2012) Endemic plant communities on special soils: early victims or hardy survivors of climate change? *Journal of Ecology 100*, 1122-1130.

Dirnböck, T., Essl, F. & Rabitsch, W. (2011) Disproportional risk for habitat loss of high-altitude endemic species under climate change. *Global Change Biology 17*, 990-996.

Escudero, A., Somolinos, R. C., Olano, J. M. & Rubio, A. (1999) Factors controlling the establishment of Helianthemum squamatum, an endemic gypsophile of semi-arid Spain. *Journal of Ecology 87*, 290-302.

Escudero, A., Iriondo, J. M., Olano, J. M., Rubio, A. & Somolinos, R. C. (2000) Factors affecting the establishment of a gypsophyte: The case of *Lepidium subulatum* (Brassicaceae). *American Journal of Botany 87*, 861-871.

Eskelinen, A. & Harrison, S. (2014) Exotic plant invasions under enhanced rainfall are constrained by soil nutrients and competition. *Ecology 95*, 682-692.

Fernandez-Going, B. M. & Harrison, S. (2013) Effects of experimental water addition depend on grassland community characteristics. *Plant Ecology 214*, 777-786.

Fernandez-Going, B. M., Anacker, B. L. & Harrison, S. P. (2012) Temporal variability in California grasslands: Soil type and species functional traits mediate response to precipitation. *Ecology 93*, 2104-2114.

Fernandez-Going, B. M., Harrison, S. P., Anacker, B. L. & Safford, H. D. (2013) Climate interacts with soil to produce beta diversity in Californian plant communities. *Ecology 94*, 2007-2018.

Fridley, J. D., Grime, J. P., Askew, A. P., Moser, B. & Stevens, C. J. (2011) Soil heterogeneity buffers community response to climate change in species-rich grassland. *Global Change Biology 17*, 2002-2011.

Gibson, N., Meissner, R., Markey, A. S. & Thompson, W. A. (2012) Patterns of plant diversity in ironstone ranges in arid south western Australia. *Journal of Arid Environments 77*, 25-31.

Going, B. M., Hillerislambers, J. & Levine, J. M. (2009) Abiotic and biotic resistance to grass invasion in serpentine annual plant communities. *Oecologia 159*, 839-847.

Grime, J. P. (1977) Evidence for the existence of three primary strategies in plants and its relevance to ecological and evolutionary theory. *The American Naturalist 111*, 1169-1174.

Grime, J. P., Brown, V. K., Thompson, K., Masters, G. J., Miller, S. H., Hillier, S. H., Clarke, I. P., Askew, A. P., Corker, D. & Kielty, J. P. (2000) The response of two contrasting limestone grasslands to simulated climate change. *Science 489*, 762-765.

Grime, J. P., Fridley, J. D., Askew, A. P., Thompson, K., Hodgson, J. G. & Bennett, C. R. (2008) Long-term resistance to simulated climate change in an infertile grassland. *Proceedings of the National Academy of Sciences (USA) 105*, 10028-10032.

Harrison, S. (1999) Local and regional diversity in a patchy landscape: Native, alien, and endemic herbs on serpentine. *Ecology 80*, 70-80.

Huenneke, L. F., Hamburg, S. P., Koide, R., Mooney, H. A. & Vitousek, P. M. (1990) Effects of soil resources on plant invasion and community structure in Californian serpentine grassland. *Ecology 71*, 478-491.

Hughes, R., Bachmann, K., Smirnoff, N. & Macnair, M. R. (2001) The role of drought tolerance in serpentine tolerance in the Mimulus guttatus Fischer ex DC. complex. *South African Journal of Science 97*,581-586.

Krauss, J., Klein, A., Steffan-Dewenter, I. & Tscharntke, T. (2004) Effects of habitat area, isolation, and landscape diversity on plant species richness of calcareous grasslands. *Biodiversity and Conservation 13*, 1427-1439.

Kruckeberg, A. R. (1984) California serpentines: Flora, vegetation, geology, soils and management problems. Berkeley, CA: University of California Press.

Loarie, S. R., Carter, B. E., Hayhoe, K., McMahon, S., Moe, R., Knight, C. A. & Ackerly, D. D. (2008) Climate change and the future of California's endemic flora. *PLoS ONE,* DOI: 10.1371/journal.pone.0002502.

Maestre, F. T., Bowker, M. A., Escolar, C., Puche, M. D., Soliveres, S., Maltez-Mouro, S., García-Palacios, P., Castillo-Monroy, A. P., Martínez, I. & Escudero, A. (2010) Do biotic interactions modulate ecosystem functioning along stress gradients? Insights from semi-arid plant and biological soil crust communities. *Philosophical Transactions of the Royal Society 365*, 2057-2070.

Malcolm, J. R., Liu, C., Neilson, R. P., Hansen, L. & Hannah, L. (2006) Global warming and extinctions of endemic species from biodiversity hotspots. *Conservation Biology 20*, 538-548.

Matesanz, S., Escudero, A. & Valladares, F. (2008) Additive effects of a potentially invasive grass and water stress on the performance of seedlings of gypsum specialists. *Applied Vegetation Science 11*, 287-296.

Matesanz, S., Escudero, A. & Vallardares, F. (2009) Impact of three global change drivers on a Mediterranean shrub. *Ecology 90*, 2609-2621.

Palacio, S., Escudero, A., Montserrat-Martí, G., Maestro, M., Milla, R. & Albert, M. J. (2007) Plants living on gypsum: Beyond the specialist model. *Annals of Botany 99* 333-343.

Parmesan, C. & Yohe, G. (2003) A globally coherent fingerprint of climate change impacts across natural systems. *Nature 421*, 37-42.

Pueyo, Y. & Alados, C. L. (2007) Abiotic factors determining vegetation patterns in a semi-arid Mediterranean landscape: Different responses on gypsum and non-gypsum substrates. *Journal of Arid Environments 69*, 490-505.

Pueyo, Y., Alados, C. L., Barrantes, O., Komac, B. & Rietkerk, M. (2008) Differences in plant communities associated with habitat fragmentation and livestock grazing. *Ecological Applications 18*, 954-964.

Rabasa, S. G., Gutiérrez, D. & Escudero, A. (2009) Temporal variation in the effects of habitat fragmentation on reproduction of the Mediterranean shrub *Colutea hispanica. Plant Ecology 200*, 241-254.

Riba, M., Mayol, M., Giles, B. E., Ronce, O., Imbert, E., van der Velde, M., Chauvet, S., Ericson, L., Bijlsma, R., Vosman, B., Smulders, M. J. M. & Olivieri, I. (2009) Darwin's wind hypothesis: Does it work for plant dispersal in fragmented habitats? *New Phytologist 183*, 667-677.

Safford, H. D., Viers, J. H. & Harrison, S. P. (2005) Serpentine endemism in the California flora: A database of serpentine affinity. *Madroño 52*, 222-257.

Schenk, J. J. (2013) Evolution of limited seed dispersal ability on gypsum islands. *American Journal of Botany 100*, 1811-1822.

Schwartz, M. W., Iverson, L. R., Prasad, A. M., Matthews, S. N. & O'Connor, R. J. (2006) Predicting extinctions as a result of climate change. *Ecology 87*, 1611-1615.

Singer, M. J. & Munns, D. N. (2006) Soils: An introduction. Sixth edition. Upper Saddle River, NJ: Prentice Hall.

Spasojevic, M. J., Damschen, E. I. & Harrison, S. (2014) Patterns of seed dispersal syndromes on serpentine soils: Examining the roles of habitat patchiness, soil infertility and correlated functional traits. *Plant Ecology & Diversity 7*, 401-410.

Stocker, T. F., Qin, D., Plattner, G. K., Tignor, M., Allen, S. K., Boschung, J., Nauels, A., Xia, Y., Bex, V. & Midgley, P. M. (2013) Climate change 2013: The physical basis. Contribution of the Working Group I to the Fifth Assessment Report of the Intergovernmental Panel on Climate Change. New York, NY: Cambridge University Press.

Stebbins, G. L. (1971) Adaptive radiation of reproductive characteristics in angiosperms, II: Seeds and seedlings. *Annual Review of Ecology, Evolution and Systematics,* 2, 237-260.

Suttle, K. B., Thomsen, M. A. & Power, M. E. (2007) Species interactions reverse grassland responses to changing climate. *Science 315*, 640-642.

Trumbore, S. E. (1997) Potential responses of soil organic carbon to global environmental change. *Proceedings of the National Academy of Sciences (USA) 94*, 8284-8291.

van der Putten, W. H., Bardgett, R. D., Bever, J. D., Bezemer, T. M., Casper, B. B., Fukami, T., Kardol, P., Klironomos, J. N., Kulmatiski, A., Schweitzer, J. A., Suding, K. N., Van de Voorde, T. F. J. & Wardle, D. A. (2013) Plant-soil feedbacks: the past, the present and future challenges. *Journal of Ecology 101*, 265-276.

Walther, G. (2010) Community and ecosystem responses to recent climate change. *Philosophical Transactions of the Royal Society 365*, 2019-2024.

Weiss, S. B. (1999) Cars, cows, and checkerspot butterflies: Nitrogen deposition and management of nutrient-poor grasslands for a threatened species. *Conservation Biology 13*, 1476-1486.

Whittaker, R. H. & Niering, W. A. (1968) Vegetation of the Santa Catalina Mountains, Arizona: IV. Limestone and acid soils. *Journal of Ecology 56*, 523-544.

Wolf, A. T. (2001) Conservation of endemic plants in serpentine landscapes. *Biological Conservation 100*, 35-44.

Wolf, A. T. & Harrison, S. P. (2001) Effects of habitat size and patch isolation on reproductive success of the serpentine morning glory. *Conservation Biology 15*, 111-121.

In: Plant Ecology and Evolution in Harsh Environments ISBN: 978-1-63321-955-7
Editors: N. Rajakaruna, R. S. Boyd and T. B. Harris

Chapter 14

CONSERVATION AND RESTORATION OF CHEMICALLY EXTREME EDAPHIC ENDEMIC FLORA IN THE WESTERN US

Ryan E. O'Dell*

Bureau of Land Management, Hollister, California, USA

ABSTRACT

Chemically extreme natural ecosystems in the western US include serpentine, gabbro, carbonate, saline, guano, and hydrothermally-altered substrates. Each one of these chemically extreme substrates has a unique combination of adverse conditions that is physiologically stressful for plant establishment and productivity. Edaphic endemic plant species have evolved in response to the strong physiological adaptation selection pressure and isolation of the chemically extreme edaphic substrates. Due to the limited extent of chemically extreme substrates, many edaphic endemic plant species have small populations and highly localized distributions and consequently are vulnerable to extinction from human impacts including urban, road, energy and water development; mining; logging; livestock grazing; off-highway vehicle recreation; alteration of fire regime; invasive plant species; and climate change. The diversity of rare, local edaphic endemic plant species in chemically extreme ecosystems is clearly recognized and valued by private nonprofit, city, state, county, and federal land managers with numerous conservation areas established to protect them. Despite the network of established conservation areas throughout the western US to protect chemically extreme ecosystems and their rare edaphic endemic plant species, state laws to protect rare plants on private or state lands are virtually nonexistent in several western US states including Washington, Idaho, Montana, Wyoming, Colorado, and Utah. Laws to protect rare plant species on private lands in all western states are limited to nonexistent. The lack of coordination for species included on the federal and state endangered species lists, as well as the lack of a single, standardized ranking system used by all of the nonprofit conservation organizations to consistently rank species for conservation priority, is yet another substantial challenge in the protection of rare edaphic endemic plant species. With ever increasing levels of resource extraction and a burgeoning human population, only

* Corresponding author: rodell@blm.gov.

continued diligent environmental activism will ensure the protection of chemically extreme ecosystems under current laws and promote the establishment of future laws to protect these unique ecosystems in the other western US states.

INTRODUCTION

The interaction of extreme climate, topography, and geology has resulted in an exceptionally high diversity of endemic plant species in the western US, reaching its pinnacle in California (Harrison, 2013; Stein et al., 2000). Chemically extreme substrates such as serpentine, gabbro, carbonate, saline, guano, and hydrothermally-altered substrates often occur as isolated edaphic islands within broad landscapes of substantially less stressful substrates (Harrison & Rajakaruna, 2011; Kruckeberg, 1986; 1991; 2002; Rajakaruna, 2004). The strong physiological selection pressure and isolation of chemically extreme substrates has resulted in a diversity of edaphic endemic species (Harrison, 2013; Harrison & Rajakaruna, 2011; Kruckeberg, 2002; 2006; Rajakaruna, 2004). The distribution of edaphic endemic species may vary from highly localized (with only one small population at one locality) to numerous large disjunct populations associated with the scattered distribution of the extreme substrate. Chemically extreme ecosystems are at threat of numerous human impacts including urban, road, energy, and water development; mining; logging; livestock grazing (this can have positive or negative effects); off-highway vehicle recreation (authorized or not); alteration of fire regime; invasive plant species; and climate change (Alexander et al., 2007; Harrison, 2013; Kruckeberg, 2002; 2006; Stein et al., 2000). The high botanical diversity of chemically extreme ecosystems is widely recognized by private nonprofit, city, county, state, and federal land managers with numerous conservation areas established throughout the western US to protect them.

In addition to being important to regional biodiversity, the conservation of edaphic endemic plant species is recognized as being an important biological resource for land management practices such as restoration, revegetation, or phytoremediation of chemically extreme substrates to reduce air, water, and soil environmental hazards. Examples include the use of the serpentine endemic *Quercus durata* var. *durata* (Fagaceae) and other serpentine tolerant plant species to revegetate the Atlas Asbestos Mine in California to reduce air and water contamination with chrysotile asbestos (USEPA, 2011) and the use of saline endemic *Distichlis spicata* (Poaceae) to revegetate large areas of the Owens Dry Lake playa in California to reduce airborne dust pollution in the Owens Valley (Dickey et al., 2005a; b; LADWP, 2013). Other environmental applications of chemically extreme edaphic endemic plant species include phytoremediation and heavy metal phytomining with serpentine endemic Ni hyperaccumulators such as *Streptanthus polygaloides* (Brassicaceae; Brooks, 1998; Chapter 15).

The purpose of this chapter is to highlight achievements that have been made and the continued challenges associated with the conservation and restoration of chemically extreme ecosystems and the rare edaphic endemic plant species that they harbor in the western US. This chapter: 1) discusses the history and laws governing the conservation of chemically extreme edaphic endemic plant species and ecosystems; 2) discusses the various types of conservation areas established by nonprofit, state, and federal organizations to protect edaphic endemic plant species; 3) discusses basic restoration and revegetation methods for chemically extreme substrates; and 4) reviews conservation areas, edaphic endemic plant species, and the restoration and revegetation of the different chemically extreme ecosystems including serpentine, gabbro, carbonate, saline, guano, and hydrothermally-altered substrates.

ENVIRONMENTAL LAWS AND THE HISTORY OF EXTREME EDAPHIC CONSERVATION

Recognition and protection of chemically extreme ecosystems in the western US began with the designation of Yellowstone National Park along with its prominent hydrothermal features in 1872 (Stein et al., 2000). Conservation interest in chemically extreme ecosystems and their associated floras picked up steam in the following century with the rise of the environmental and conservation movement and the passage of the Endangered Species Act (ESA) in 1973 that afforded protection to listed threatened and endangered plants on federal lands (Kline, 2011; Martinez, 2013; see Table 1). The ESA forbids "take" of federally listed species which is defined as to "harass, harm, pursue, hunt, shoot, wound, kill, trap, capture, or collect, or to attempt to engage in any such conduct."

Unfortunately, the federal ESA does not afford protection to federally listed plant species on state or private lands. Many western states recognized this deficiency of the federal ESA and eventually established their own state ESAs. Passage of state ESAs (regulation; respective state management division) included California in 1984 (Title 14, Division 1, Subdivision 3, Chapter 6; CA Department of Fish and Wildlife; Preceded by the Native Plant Protection Act of 1977); Oregon in 1987 (Division 73, Chapter 564; OR Department of Agriculture); Nevada in 1969 (before the federal ESA; Title 47, Chapter 527; NV Division of Forestry); Arizona in 1993 (Title 3, Chapter 7; AZ Department of Agriculture); and New Mexico in 1985 (Title 19, Chapter 21; NM Forestry Division; see Table 1). Although Washington, Idaho, Montana, Wyoming, Colorado, and Utah have state natural resource divisions to coordinate management and recovery of federally listed animal species on state lands, none have state laws to protect rare plants on state or private lands. It should be noted that the federal ESA and state ESA lists are independent of each other. Species that are included on the federal ESA list are often not included on the state ESA lists and *vice versa*.

Land ownership of chemically extreme ecosystems includes private individual, private nonprofit conservation organization (land trust), and government (city, county, state and federal) (Stein et al., 2000). The largest nonprofit conservation organization in the western US is The Nature Conservancy (TNC; Stein et al., 2000). City and county public land includes city and county rights-of-ways, parks, open space, reserves, and preserves. State public land includes state highway rights-of-ways, state parks, state recreation areas, state universities, and other undesignated state lands.

Federal public land management agencies include the National Park Service (NPS, DOI); United States Fish and Wildlife Service (USFWS, DOI); United States Forest Service (USFS, Department of Food and Agriculture); and the Bureau of Land Management (BLM, Department of the Interior). The federal ESA has the authority to protect federally listed plant species only on federal lands except in the case of a federal nexus where a project on private or state owned lands has been undertaken with federal funds or requires a federal permit. Federal land managers formally consult with the USFWS on management actions which may impact federally listed species. State ESAs have the authority to protect state listed plants on city, county, and state lands. State land managers formally consult with the respective state management divisions that enforce the state ESA on management actions that may impact state listed species. State ESAs differ in their authority to protect state listed plants on private lands.

The Oregon ESA clearly states that it has no authority over state listed plant species on private lands. Other state ESAs including California, Nevada, Arizona, and New Mexico assert some authority to protect state listed plants on private lands; however, enforcement of the state ESA to protect state listed plant species on private lands relies primarily on the construction project permitting process (California, Nevada) and regulation of the plant trade (Arizona, New Mexico). ESA take permits are issued to private landowners by the state for removal or destruction of state listed plant species in California and Nevada. California often negotiates mitigation for the plants and habitat that will be destroyed before issuing construction and take permits.

In Arizona and New Mexico, state listed plants on private lands can legally be destroyed without a take permit, but a permit is required if the plants are collected and sold. The definition of "take" in Arizona and New Mexico is different than the Federal, California, and Nevada ESAs in that it only includes the acts of removal from private land with the intent to possess, transport, export, sell, or offer for sale, and does not include "kill" (population and habitat destruction).

Exceptions in the California and Nevada ESAs allow private landowners to take state listed plant species through routine vegetation removal and land maintenance. As a consequence of their narrow, edaphically restricted distribution, high degree of habitat impacts relative to habitat size, and poor protection on private lands in the western states, many chemically extreme edaphic endemic plant species are highly vulnerable to extinction. Conservation easements and acquisition by conservation organizations are some of the limited options for ensuring protection of chemically extreme ecosystems on private lands (Harrison, 2013; Stein et al., 2000).

Environmental activism is at the core of conservation on public and private lands. Nonprofit organizations such as the various state Native Plant Societies, Sierra Club, Center for Biological Diversity, and The Nature Conservancy monitor activities on public and private lands and work to protect endangered plant species and sensitive ecosystems. The various state Native Plant Societies, other species ranking organizations, and NatureServe (spin off of The Nature Conservancy) have each established their own independent ranking systems to list and prioritize monitoring and protection of rare plant species (Table 1). Although the resulting rankings for each rare plant species are roughly equivalent between the organizations, there is still considerable ranking discontinuity as is evident in the serpentine endemic species status rankings in Table 2 between Del Norte county in California (California Native Plant Society; CNPS) and adjacent Josephine and Curry counties in Oregon (Oregon Biodiversity Information Center; ORBIC). Serpentine ecosystems and their rare endemic plant species are continuous across the state border and impacts to those ecosystems are similar across the border, but there are many species that are included on the California Rare Plant Rank list (CRPR) of CNPS that are not included on the Rare Species Ranking list of ORBIC.

Table 1. Plant species conservation status for federal, state, and nonprofit conservation organizations and whether the conservation status conveys protection of the species on federal, state, or private land

Authority	Conservation Status	Conservation Status Code	Conservation Status Code Definition	Status Conveying Protection on Land Owned by:		
				Federal	State	Private
Federal	Endangered	FE	In danger of extinction throughout all or any significant portion of its range.	Yes	No	No
State: CA, OR, NV, NM	Endangered	SE	In danger of extinction throughout all or any significant portion of its range.	No	Yes	Yes, except OR
Federal	Threatened	FT	Likely to become endangered within the foreseeable future throughout all or any significant portion of its range	Yes	No	No
State: CA, OR, NV	Threatened	ST	Likely to become endangered within the foreseeable future throughout all or any significant portion of its range	No	Yes	Yes, except OR
California	Rare	SR	Not presently threatened with extinction, but occurs in small numbers throughout its range	No	Yes	Yes
Arizona	Highly safeguarded	SHS	No collection allowed - Federally listed species and plants highly targeted for the plant trade	No	Yes	Yes
	Salvage restricted	SSR	Collection with permit only - Succulents and other rare plants collected for the plant trade	No	Yes	Yes
	Export restricted	SER	Transport out of state prohibited - Succulents and other rare plants targeted for the plant trade	No	Yes	Yes
	Salvage assessed	SA	Permits required to remove live trees	No	Yes	Yes
	Harvest restricted	SHR	Permits required to remove plant by-products	No	Yes	Yes

Table 1. (Continued)

Authority	Conservation Status	Conservation Status Code	Conservation Status Code Definition	Status Conveying Protection on Land Owned by:		
				Federal	State	Private
Federal	Candidate	FC	Proposed for addition to the Federal threatened and endangered species list	No	No	No
CA Native Plant Society California Rare Plant Rank	Extinct in CA	CRPR 1A	Presumed extinct in CA	No	No	No
	Rare and endangered in CA and elsewhere	CRPR 1B	Rare, threatened, or endangered in CA and elsewhere	No	No	No
	Extirpated in CA, but occurs elsewhere	CRPR 2A	Presumed extirpated in CA, but more common elsewhere	No	No	No
	Rare and endangered in CA only	CRPR 2B	Rare, threatened, or endangered in CA, but more common elsewhere	No	No	No
	Need more info	CRPR 3	More information is needed - Review list	No	No	No
	Rare	CRPR 4	Limited distribution - Watch List	No	No	No
		+ 0.1	Seriously threatened in CA	No	No	No
		+ 0.2	Moderately threatened in CA	No	No	No
		+ 0.3	Not very threatened in CA	No	No	No
OR Native Plant Society defers to: OR Biodiversity Info. Center		ORBIC	Uses the NatureServe state rank system and ranking method (see below). ORBIC ranks species on its list independently from NatureServe.	No	No	No

Authority	Conservation Status	Conservation Status Code	Conservation Status Code Definition	Status Conveying Protection on Land Owned by:		
				Federal	State	Private
WA Native Plant Society defers to: WA Natural Heritage Program	Presumed extinct	WNHP EX	Presumed extinct in WA	No	No	No
	Endangered	WNHP E	In danger of becoming extinct in WA	No	No	No
	Threatened	WNHP T	Likely to become endangered in the near future in WA	No	No	No
	Sensitive	WNHP S	Vulnerable and declining in WA (= rare)	No	No	No
	Watch	WNHP W	Watch list (= limited distribution)	No	No	No
NV Native Plant Society	Possibly extirpated	NNPS PE	Possibly extirpated from NV	No	No	No
	Endangered	NNPS E	In danger of becoming extinct in NV	No	No	No
	Threatened	NNPS T	Likely to become endangered in the near future in NV	No	No	No
	Marginal or rare	NNPS M	Marginal, disjunct, rare and/or distinct in NV	No	No	No
	Watch	NNPS W	Watch list (= limited distribution)	No	No	No
	Research needed	NNPS R	Research - More information is needed	No	No	No
UT Native Plant Society	Extremely high	UNPS EXH	Extremely high priority (= endangered) in UT	No	No	No
	High	UNPS H	High priority (= threatened) in UT	No	No	No
	Medium	UNPS M	Medium priority (= rare) in UT	No	No	No
	Watch	UNPS W	Watch and evaluate (= limited distribution)	No	No	No
	Need data	UNPS ND	Need more information	No	No	No

Table 1. (Continued)

Authority	Conservation Status	Conservation Status Code	Conservation Status Code Definition	Status Conveying Protection on Land Owned by:		
				Federal	State	Private
	Global rank	G	Global rank	No	No	No
	National rank	N	National rank	No	No	No
The AZ, NM, ID, MT, WY, and CO Native Plant Societies defer entirely to NatureServe	State rank	S	State rank	No	No	No
	Subspecies and variety	T	Subspecies and varieties	No	No	No
	Critically imperiled	1	Critically imperiled (= endangered)	No	No	No
		2	Imperiled (= threatened) Imperiled	No	No	No
	Vulnerable	3	Vulnerable (= rare)	No	No	No
	Apparently secure	4	Apparently secure (= limited distribution)	No	No	No
	Secure	5	Secure (= common and widespread)	No	No	No
	Extinct or extirpated	X	Extinct or extirpated	No	No	No
	Possibly extinct	H	Possibly extinct or extirpated	No	No	No
	Indeterminate	R or ?	Known, but status indeterminate	No	No	No
	Not yet ranked	NR	Not yet ranked	No	No	No

Table 2. Selected serpentine ecosystem conservation areas and the serpentine endemic plant species they contain, including their conservation status and anthropogenic threats. * = historic impact in conservation areas, but current threat in surrounding or nearby serpentine ecosystems. See Table 1 for conservation status codes. Serpentine endemic species and threats for each locality determined from Alexander et al. (2007); Barker (1984); Borgias (1994); Brian (2004); Calflora (2014); Callizo & Clifton (1984); Callizo & Ruygt (1984); Carter (2004); Coleman & Kruckeberg (1999); Consortium of California Herbaria (2014); Goforth (1984); Google Earth (2014); Griffin (1984); Kruckeberg (1969; 1984a; b; 2006); Latimer (1984); McLeod (1984); McCormick (1984); Medeiros (1984); Mosseaux (2004); NatureServe (2014); O'Dell (personal field observations 2000 – 2014); Oregon Flora Atlas (2014); Safford (2011); Safford et al. (2005); Sawyer (1984); Sommers (1984); USFWS (1984; 1998; 2006a; b); Whipple (1984); and Wilson & Norris (1989)

County State	Locality: Conservation Area	Manager(s)	Serpentine Endemic Species	Threats
Fresno CA	Tivy Mountain: Tivy Mountain Preserve	Sierra Foothill Conservancy	*Streptanthus polygaloides* (NNR, G5)	Livestock grazing*, invasive plant species
Mariposa CA	Bagby: Bagby Serpentine ACEC	BLM Motherlode FO	*Garrya congdonii* (N2N4, G2G4); *Streptanthus polygaloides* (NNR, G5)	Road development/maintenance, mining/ quarrying*, off-highway vehicles, invasive plant species
Tuolumne CA	Red Hills: Red Hills ACEC	BLM Motherlode FO	*Allium tuolumnense* (N2, G2); *A. jepsonii* (CRPR 1B.2, N1, G1); *Brodiaea pallida* (CRPR 1B.1, N1, G1, FE), *Chlorogalum grandiflorum* (CRPR 1B.2, N3, G3); *Cryptantha mariposae* (CRPR 1B.3, N2, G2); *C. hispidula* (NNR, G4?); *Eriogonum tripodum* (CRPR 4.2, N3, G3); *Githopsis pulchella* subsp. *serpentinicola* (CRPR 4.3, N3, G4T3); *Lomatium congdonii* (CRPR 1B.2, N2, G2); *L. marginatum* (NNR, G5); *Lupinus spectabilis* (CRPR 1B.2, N2, G2); *Packera clevelandii* (N4?, G4?); *P. layneae* (CRPR 1B.2, N2, G2, FT); *Perideridia bacigalupii* (CRPR 4.2, N3, G3); *Streptanthus polygaloides* (NNR, G5); *Trichostema rubisepalum* (CRPR 4.3, N3, G3); *Verbena californica* (CRPR 1B.1, N2, G2, FT)	Urban development*, road development/maintenance, water development (reservoir)*, mining/quarrying*, livestock grazing, off-highway vehicles, invasive plant species

Table 2. (Continued)

County State	Locality: Conservation Area	Manager(s)	Serpentine Endemic Species	Threats
El Dorado CA	Traverse Creek: Traverse Creek Botanical SIA	USFS Eldorado NF	*Eriogonum tripodum* (CRPR 4.2, N3, G3); *Lomatium marginatum* (NNR, G5); *Packera layneae* (CRPR 1B.2, N2, G2, SR, FT); *Quercus durata* var. *durata* (NNR, G4T4?); *Streptanthus polygaloides* (NNR, G5)	Urban development*, road development/maintenance, mining/quarrying*, off-highway vehicles, invasive plant species
Santa Barbara CA	Figueroa Mountain: Sedgewick Ranch Reserve; Figueroa Mountain Recreation Area	UC Natural Reserve System; USFS Los Padres NF	*Allium diabolense*; *Calystegia collina* subsp. *venusta* (CRPR 4.3, N3, G4T3); *Caulanthus amplexicaulis* var. *barbarae* (CRPR 1B.1, N1, G4T1); *Quercus durata* var. *durata* (NNR, G4T4?)	Road development/maintenance, livestock grazing, invasive plant species
San Luis Obispo CA	San Luis Obispo: Cuesta Ridge Botanical SIA; Cal Poly Technical Institute; Irish Hills Natural Reserve (City); Johnson Ranch Open Space (City); Reservoir Canyon Natural Reserve (City); South Hills Open Space (City); Stenner Springs Natural Reserve (City)	USFS Los Padres NF; Cal Poly State University; City of San Luis Obispo	*Arctostaphylos obispoensis* (CRPR 4.3, N3?, G3?); *Aspidotis carlotta-halliae* (CRPR 4.2, N3, G3); *Calochortus obispoensis* (CRPR 1B.2, N2, G2); *Carex obispoensis* (CRPR 1B.2, N2, G2); *Chlorogalum pomeridianum* var. *minus* (CRPR 1B.2, N2, G5T2); *Chorizanthe breweri* (CRPR 1B.3, N2, G2); *Dudleya abramsii* subsp. *murina* (CRPR 1B.3, N2, G3T2); *Hesperocyparis sargentii* (N4, G4); *Monardella palmeri* (CRPR 1B.2, N2, G2); *Quercus durata* var. *durata* (NNR, G4T4?); *Salix breweri* (NNR, G3?); *Sidalcea hickmanii* subsp. *anomala* (CRPR 1B.2, N1, G3T1)	Urban development*, road development/maintenance, mining/quarrying*, livestock grazing, invasive plant species
San Luis Obispo CA	Cypress Mountain: Cypress Mountain ACEC	BLM Bakersfield FO	*Arctostaphylos obispoensis* (CRPR 4.3, N3?, G3?); *Galium hardhamiae* (CRPR 1B.3, N2, G2); *Hesperocyparis macnabiana* (N4, G4)	Road development/maintenance

County State	Locality: Conservation Area	Manager(s)	Serpentine Endemic Species	Threats
San Benito CA	San Benito Mountain: San Benito Mountain RNA; Serpentine ACEC	BLM Hollister FO	*Allium diabolense*; *Aspidotis carlotta-halliae* (CRPR 4.2, N3, G3); *Astragalus clevelandii* (CRPR 4.3, N3?, G3?); *A. rattanii* var. *jepsonianus* (CRPR 1B.2, N3, G4T3); *Camissonia benitensis* (CRPR 1B.1, N1, G1, FT); *Cryptantha mariposae* (CRPR 1B.3, N2, G2); *Fritillaria falcata* (CRPR 1B.2, N2, G2); *F. viridea* (CRPR 1B.2, N2, G2); *Garrya congdonii* (N2N4, G2G4); *Hesperolinon disjunctum* (NNR, G3?); *Layia discoidea* (CRPR 1B.1, N2, G2); *Leptosiphon ambiguus* (CRPR 4.2, N3, G3); *Solidago guiradonis* (CRPR 4.2, N3, G3); *Quercus durata* var. *durata* (NNR, G4T4?); *Salix breweri* (NNR, G3?); *Trichostema rubisepalum* (CRPR 4.3, N3, G3); *Streptanthus breweri* (NNR, G5)	Road development/maintenance, mining/quarrying*, logging*, off-highway vehicles, invasive plant species
Santa Clara CA	Coyote Ridge - Santa Teresa Hills - New Almaden: Coyote Ridge Butterfly Trust Area (Nonprofit); Tulare Hill ER (Nonprofit); CR Open Space Preserve (County) Santa Teresa County Park; New Almaden Quicksilver County Park	Silicon Valley Land Conservancy (Nonprofit); Santa Clara Valley Open Space Authority (County); Santa Clara County Parks	*Calystegia collina* subsp. *collina* (NNR, G4T3T4); *Castilleja affinis* subsp. *neglecta* (CRPR 1B.2, N1, G4G5T1, ST, FE); *Ceanothus ferrisiae* (CRPR 1B.1, N1, G1, FE); *Cirsium fontinale* var. *campylon* (CRPR 1B.2, N2, G2T2); *Dudleya abramsii* subsp. *setchellii* (CRPR 1B.1, N2, G2, FE); *Lessingia micradenia* var. *glabrata* (CRPR 1B.2, N2, G2T2); *Quercus durata* var. *durata* (NNR, G4T4?); *Streptanthus glandulosus* subsp. *albidus* (CRPR 1B.1, N1, G2T1, FE)	Urban development*, road development/maintenance, water development (reservoir)*, mining/quarrying*, livestock grazing, invasive plant species
San Mateo CA	Edgewood - Crystal Springs: Edgewood County Park and Natural Reserve; Crystal Springs County Park	San Mateo County Parks	*Acanthomintha duttonii* (CRPR 1B.1, N1, G1, SE, FE); *Cirsium fontinale* var. *fontinale* (CRPR 1B.1, N1, G2T1, SE, FE); *Hesperolinon congestum* (CRPR 1B.1, N2, G2, ST, FT); *Leptosiphon ambiguus* (CRPR 4.2, N3, G3); *Lessingia arachnoidea* (CRPR 1B.2, N3, G3); *Quercus durata* var. *durata* (NNR, G4T4?)	Urban development*, road development/maintenance, water development (reservoir)*, invasive plant species

Table 2. (Continued)

County State	Locality: Conservation Area	Manager(s)	Serpentine Endemic Species	Threats
San Francisco CA	San Francisco Presidio: Golden Gate National Recreation Area; Presidio Trust	NPS	*Arctostaphylos franciscana* (CRPR 1B.1, N1, G3T1, FE); *A. montana* subsp. *ravenii* (CRPR 1B.1, N1, G3T1, SE, FE); *Clarkia franciscana* (CRPR 1B.1, N1, G1, SE, FE); *Hesperolinon congestum* (CRPR 1B.1, N2, G2, ST, FT)	Urban development*, road development/maintenance, invasive plant species
Alameda CA	Oakland Hills: Redwood Regional Park	East Bay Regional Parks	*Clarkia franciscana* (CRPR 1B.1, N1, G1, SE, FE); *Eriogonum luteolum* var. *caninum* (CRPR 1B.2, N2, G5T2)	Urban development*, road development/maintenance, invasive plant species
Contra Costa CA	Mount Diablo: Mount Diablo State Park	State Parks	*Cordylanthus nidularius* (CRPR 1B.1, N1, G1); *Quercus durata* var. *durata* (NNR, G4T4?)	Road development/maintenance, livestock grazing*, invasive plant species
Marin CA	Tiburon: Old Saint Hilary's Open Space Preserve (City); Ring Mountain Open Space Preserve (County)	City of Tiburon; Marin County Parks	*Aspidotis carlotta-halliae* (CRPR 4.2, N3, G3), *Calochortus tiburonensis* (CRPR 1B.1, N1, G1, ST, FT); *Castilleja affinis* subsp. *neglecta* (CRPR 1B.2, N1, G4G5T1, ST, FE); *Calystegia collina* subsp. *collina* (NNR, G4T3T4); *Eriogonum luteolum* var. *caninum* (CRPR 1B.2, N2, G5T2); *Hesperolinon congestum* (CRPR 1B.1, N2, G2, ST, FT); *Quercus durata* var. *durata* (NNR, G4T4?); *Streptanthus glandulosus* subsp. *niger* (CRPR 1B.1, N1, G1, SE, FE)	Urban development*, road development/maintenance, mining/quarrying*, invasive plant species

County State	Locality: Conservation Area	Manager(s)	Serpentine Endemic Species	Threats
Marin CA	Mount Tamalpais-Carson Ridge: Mount Tamalpais State Park; Marin Municipal Water District	State Parks; Marin County Municipal Water District	*Arctostaphylos montana* subsp. *montana* (CRPR 1B.3, N2, G3T2); *Aspidotis carlotta-halliae* (CRPR 4.2, N2, G3); *Calamagrostis ophitidis* (CRPR 4.3, N3, G3); *Calystegia collina* subsp. *collina* (NNR, G4T3T4); *C. c.* subsp. *oxyphlla* (CRPR 4.2, N3, G4T3); *Campanula griffinii* (N3?, G3?); *Ceanothus jepsonii* (NNR, G3T1T3); *Cirsium hydrophilum* var. *vaseyi* (CRPR 1B.2, N1, G1T1); *Collomia diversifolia* (CRPR 4.3, N3, G3); *Eriogonum luteolum* var. *caninum* (CRPR 1B.2, N2, G5T2); *Hesperocyparis sargentii* (N4, G4); *Hesperolinon congestum* (CRPR 1B.1, N2, G2, ST, FT); *Lessingia micradenia* var. *micradenia* (CRPR 1B.2, N1, G2T1); *Navarretia rosulata* (CRPR 1B.2, N2?, G2?); *Quercus durata* var. *durata* (NNR, G4T4?); *Sidalcea hickmanii* subsp. *viridis* (CRPR 1B.3, N1N2, G3T1T2); *Streptanthus batrachopus* (CRPR 1B.3, N1, G1); *S. glandulosus* subsp. *pulchellus* (CRPR 1B.2, N1, G4T1); *Viola cuneata* (NNR, G4)	Urban development*, road development/maintenance, water development (reservoir)*, mining/quarrying*, invasive plant species
Sonoma CA	Mount Hood-Sugarloaf Ridge: Hood Mountain Regional Park and Open Space Pres. (County Parks); Sugarloaf Ridge SP	Sonoma County Parks; State Parks	*Calystegia collina* (N4, G4); *Ceanothus jepsonii* (NNR, G3); *Cordylanthus tenuis* subsp. *brunneus* (CRPR 4.3, N3, G4G5T3); *Hesperocyparis sargentii* (N4, G4); *Packera greenei* (NNR, G3?); *Quercus durata* var. *durata* (NNR, G4T4?)	Urban development*, road development/maintenance, invasive plant species
Sonoma CA	The Cedars: Austin Creek State Recreation Area; Cedars ACEC	State Parks; BLM Ukiah FO	*Arctostaphylos bakeri* subsp. *bakeri* (CRPR 1B.1, N2, G2T2); *A. b.* subsp. *sublaevis* (CRPR 1B.2 N2, G2T2); *A. viscida* subsp. *pulchella* (N3N4, G5T3T4); *Asclepias solanoana* (CRPR 4.2, N3, G3); *Calochortus raichei* (CRPR 1B.2, N1, G1); *Calystegia collina* subsp. *oxyphlla* (CRPR 4.2, N3, G4T3); *Ceanothus jepsonii* (NNR, G3T1T3); *Collomia diversifolia* (CRPR 4.3, N3, G3); *Cordylanthus pringlei* (NNR, G2); *Cypripedium californicum* (CRPR 4.2, N3, G3); *Erigeron serpentinus* (CRPR 1B.3, N1, G1); *Eriogonum cedrorum* (CRPR 1B.3, N1, G1); *E. ternatum* (CRPR 4.3, N4, G4); *Garrya congdonii* (N2N4, G2G4); *Hesperocyparis macnabiana* (N4, G4); *H. sargentii* (N4, G4); *Hesperolinon disjunctum* (NNR, G3?); *H. spergulinum* (NNR, G3?); *Phacelia*	Road development/maintenance, mining/quarrying*, invasive plant species

Table 2. (Continued)

County State	Locality: Conservation Area	Manager(s)	Serpentine Endemic Species	Threats
			corymbosa (N4?, G4G5); *Quercus durata* var. *durata* (NNR, G4T4?); *Salix breweri* (NNR, G3?); *Streptanthus barbiger* (CRPR 4.2, N3, G3); *S. glandulosus* subsp. *hoffmanii* (CRPR 1B.3, NH, G4TH); *S. g.* subsp. *sonomensis* (N3?, G4T2T3Q); *S. morrisonii* (CRPR 1B.2, N2, G2)	
Napa CA	Lake Hennesey: Lake Hennesey Recreation Area	State Parks	*Astragalus claranus* (CRPR 1B.1, N1, G1, ST, FE); *Ceanothus jepsonii* (NNR, G3T1T3); *Garrya congdonii* (N2N4, G2G4); *Leptosiphon jepsonii* (CRPR 1B.2, N2, G2); *Lomatium marginatum* (NNR, G5T4?); *Quercus durata* var. *durata* (NNR, G4T4?); *Salix breweri* (NNR, G3?); *Streptanthus barbiger* (CRPR 4.2, N3, G3); *S. hesperidis* (CRPR 1B.2, N2, G2)	Water development (reservoir), invasive plant species
Napa CA	Cedar Roughs: Knoxville Wildlife Area (State); Cedar Roughs RNA	CDFW (State); BLM Ukiah FO	*Arctostaphylos viscida* subsp. *pulchella* (N3N4, G5T3T4); *Astragalus claranus* (CRPR 1B.1, ST, FE); *A. clevelandii* (CRPR 4.3, N3?, G3?); *Calamagrostis ophitidis* (CRPR 4.3, N3, G3); *Calycadenia pauciflora* (NNR, G3?); *Calystegia collina* subsp. *collina* (NNR, G4T3T4); *C. c.* subsp. *oxyphlla* (CRPR 4.3, N3, G4T3); *Campanula griffinii* (N3?, G3?); *Ceanothus jepsonii* (NNR, G3T1T3); *Clarkia gracilis* subsp. *tracyi* (CRPR 4.2, N3, G5T3); *Collomia diversifolia* (CRPR 4.3, N3, G3); *Cordylanthus tenuis* subsp. *brunneus* (CRPR 4.3, N3, G4G5T3); *Cryptantha hispidula* (NNR, G4?); *Erythronium helenae* (CRPR 4.2, N3, G3); *Frangula californica* subsp. *crassifolia* (N3?, G5T3?); *Garrya congdonii* (N2N4, G2G4); *Helianthus exilis* (CRPR 4.2, NNR, G4); *Hesperocyparis macnabiana* (N4, G4); *H. sargentii* (N4, G4); *Hesperolinon bicarpellatum* (CRPR 1B.2, N2, G2); *H. sharsmithae* (CRPR 1B.2); *H. tehamense* (CRPR 1B.3, N1, G1); *Leptosiphon jepsonii* (CRPR 1B.2, N2, G2); *Lessingia ramulosa* (NNR, G3?); *Navarretia jepsonii* (CRPR 4.3, N3, G3); *N. rosulata* (CRPR 1B.2, N2?, G2?); *Packera clevelandii* (N4?, G4?); *Quercus durata* var. *durata* (NNR, G4T4?); *Salix breweri* (NNR, G3?); *Streptanthus breweri* (NNR, G5); *S. hesperidis* (CRPR 1B.2, N2, G2)	Road development/maintenance, Invasive plant species

County State	Locality: Conservation Area	Manager(s)	Serpentine Endemic Species	Threats
Napa, Lake CA	Knoxville: McLaughlin Reserve; Knoxville ACEC	UC Natural Reserve System; BLM Ukiah FO	*Allium fimbriatum* var. *purdyi* (CRPR 4.3, N3, G4G5T3); *Arctostaphylos viscida* subsp. *pulchella* (N3N4, G5T3T4); *Asclepias solanoana* (CRPR 4.2, N3, G3); *Astragalus clevelandii* (CRPR 4.3, N3?, G3?); *A. rattanii* var. *jepsonianus* (CRPR 1B.2, N3, G4T3); *Balsamorhiza sericea* (CRPR 1B.3, N4, G4Q); *Calamagrostis ophitidis* (CRPR 4.3, N3, G3); *Calycadenia pauciflora* (NNR, G3?); *Calyptridium quadripetalum* (CRPR 4.3, N3, G3); *Calystegia collina* subsp. *collina* (NNR, G4T3T4); *C. c.* subsp. *oxyphylla* (CRPR 4.2, N3, G4T3); *Ceanothus jepsonii* (NNR, G3T1T3); *Clarkia gracilis* subsp. *tracyi* (CRPR 4.2, N3, G5T3); *Collinsia greenei* (NNR, G3G4); *Collomia diversifolia* (CRPR 4.3, N3, G3); *Cordylanthus tenuis* subsp. *brunneus* (CRPR 4.3, N3, G4G5T3); *Cryptantha dissita* (CRPR 1B.2, N1, G5T1); *C. hispidula* (NNR, G4?); *Delphinium uliginosum* (CRPR 4.2, N3, G3); *Eriogonum nervulosum* (CRPR 1B.2, N2, G2); *E. tripodum* (CRPR 4.2, N3, G3); *Frangula californica* subsp. *crassifolia* (N3?, G5T3?); *Fritillaria purdyi* (CRPR 4.3, N3, G3); *Garrya congdonii* (N2N4, G2G4); *Harmonia hallii* (CRPR 1B.2, N2?, G2?); *Helianthus exilis* (CRPR 4.2, NNR, G4); *Hesperocyparis macnabiana* (N4, G4); *H. sargentii* (N4, G4); *Hesperolinon bicarpellatum* (CRPR 1B.2, N2, G2); *H. disjunctum* (NNR, G3?); *H. drymarioides* (CRPR 1B.2, N2, G2); *H. sharsmithiae* (CRPR 1B.2, N2, G2Q); *H. spergulinum* (NNR, G3?); *Leptosiphon jepsonii* (CRPR 1B.2, N2, G2); *Lessingia micradenia* var. *glabrata* (CRPR 1B.2, N2, G2T2); *L. ramulosa* (NNR, G3?); *Lomatium hooveri* (CRPR 4.3, N3, G3); *L. marginatum* (NNR, G5T4?); *Mimulus nudatus* (N3, G3); *Navarretia jepsonii* (CRPR 4.3, N3, G3); *Nemacladus montanus* (NNR, G2G3); *Packera clevelandii* (N4?, G4?); *P. greenei* (NNR, G3?); *Phacelia corymbosa* (N4?, G4G5); *Quercus durata* var. *durata* (NNR, G4T4?); *Salix breweri* (NNR, G3?); *Streptanthus barbiger* (CRPR 4.2, N3, G3); *S. brachiatus* (CRPR 1B.2, NNR, G2); *S. breweri* (NNR, G5); *S. hesperidis* (CRPR 1B.2, N2, G2); *S. morrisonii* (CRPR 1B.2, N2, G2)	Road development/maintenance, water development (reservoir)*, mining/quarrying*, livestock grazing, off-highway vehicles, invasive plant species

Table 2. (Continued)

County State	Locality: Conservation Area	Manager(s)	Serpentine Endemic Species	Threats
Colusa, Lake CA	Walker Ridge - Frenzel Creek: Bear Creek BMA (State); Indian Valley Wildlife Area (State); Indian Valley Brodiea ACEC; Frenzel Creek RNA	CalTrans (State); CDFW (State); USFS Mendocino NF; BLM Ukiah FO	*Allium fimbriatum* var. *purdyi* (CRPR 4.3, N3, G4G5T3); *Arctostaphylos viscida* subsp. *pulchella* (N3N4, G5T3T4); *Asclepias solanoana* (CRPR 4.2, N3, G3); *Astragalus clevelandii* (CRPR 4.3, N3?, G3?); *A. rattanii* var. *jepsonianus* (CRPR 1B.2, N3, G4T3); *Brodiaea rosea* (CRPR 1B.1, N1, G1, SE); *Calyptridium quadripetalum* (CRPR 4.3, N3, G3); *Calystegia collina* subsp. *oxyphylla* (CRPR 4.3, N3, G4T3); *C. c.* subsp. *tridactylosa* (CRPR 1B.2, N1, G4T1); *Campanula griffinii* (N3?, G3?); *Ceanothus jepsonii* (NNR, G3T1T3); *Collinsia greenei* (NNR, G3G4); *Collomia diversifolia* (CRPR 4.3, N3, G3); *Cordylanthus tenuis* subsp. *brunneus* (CRPR 4.3, N3, G4G5T3); *Cryptantha hispidula* (NNR, G4?); *Eriogonum nervulosum* (CRPR 1B.2, N2, G2); *E. tripodum* (CRPR 4.2, N3, G3); *Frangula californica* subsp. *crassifolia* (N3?, G5T3?); *Fritillaria purdyi* (CRPR 4.3, N3, G3); *Garrya congdonii* (N2N4, G2G4); *Harmonia hallii* (CRPR 1B.2, N2?, G2?); *Helianthus exilis* (NNR, G4); *Hesperocyparis macnabiana* (N4, G4); *H. sargentii* (N4, G4); *Lessingia ramulosa* (NNR, G3?); *Lomatium hooveri* (CRPR 4.3, N3, G3); *Mimulus nudatus* (CRPR 4.3, N3, G3); *Navarretia jepsonii* (CRPR 4.3, N3, G3); *Nemacladus montanus* (NNR, G2G3); *Packera clevelandii* (N4?, G4?); *P. greenei* (NNR, G3?); *Phacelia corymbosa* (N4?, G4G5); *Quercus durata* var. *durata* (NNR, G4T4?); *Salix breweri* (NNR, G3?); *Streptanthus brachiatus* (CRPR 1B.2, NNR, G2); *S. breweri* (NNR, G5); *S. morrisonii* (CRPR 1B.2, N2, G2)	Road development/maintenance, energy development (wind), water development (reservoir)*, mining/quarrying*, livestock grazing, off-highway vehicles, invasive plant species
Lake, Colusa, Glenn CA	Snow Mountain: Snow Mountain Wilderness Area	USFS Mendocino NF	*Asclepias solanoana* (CRPR 4.2, N3, G3); *Eriogonum nervulosum* (CRPR 1B.2, N2, G2); *Lomatium ciliolatum* (NNR, G1G3); *Quercus durata* var. *durata* (NNR, G4T4?); *Streptanthus breweri* (NNR, G5)	Road development/maintenance*

County State	Locality: Conservation Area	Manager(s)	Serpentine Endemic Species	Threats
Mendocino CA	Red Mountain: South Fork Eel River Wilderness (Red Mountain Unit); Red Mountain ACEC; Little Red Mountain ER (State)	BLM Arcata FO; CDFW (State)	*Arabis mcdonaldiana* (CNPS 1B.1, N2, G2, SE, FE); *Ceanothus pumilus* (NNR, G3?); *Cypripedium californicum* (CRPR 4.2, N3, G3); *Eriogonum kelloggii* (CNPS 1B.2, N1, G1, SE, FC); *Galium ambiguum* subsp. *siskiyouense* (NNR, G4TNR); *Garrya congdonii* (N2N4, G2G4); *Gentiana setigera* (CRPR 1B.2, N2, G2); *Hesperocyparis sargentii* (N4, G4); *Lomatium congdonii* (CRPR 1B.2, N2, G2); *L. engelmannii* (CRPR 4.3, N3, G3); *L. tracyi* (CRPR 4.3, N3, G3); *Pseudoleskeella serpentinense* (N3, G3); *Quercus durata* var. *durata* (NNR, G4T4?); *Sedum laxum* subsp. eastwoodiae (CRPR 1B.2, N1, G1, FC); *Silene campanulata* subsp. *campanulata* (CNPS 4.2, N3, G5T3Q, SE); *Viola cuneata* (NNR, G4)	Road development/maintenance, mining/quarrying*, logging*
Tehama, Trinity, Mendocino CA	Yolla Bolly: Yolla Bolly-Middle Eel Wilderness	USFS Mendocino NF	*Allium hoffmanii* (CRPR 4.3, N3, G3); *Asclepias solanoana* (CRPR 4.2, N3, G3); *Astragalus whitneyi* var. *siskiyouensis* (N3N4, G5T3T4); *Collinsia greenei* (NNR, G3G4); *Ericameria ophitidis* (CRPR 4.3, N3, G3); *Eriogonum libertini* (CRPR 4.2, N3, G3); *E. strictum* var. *greenei* (CRPR 4.3, NNR, G5); *E. umbellatum* var. *humistratum* (CRPR 4.3, N3, G5T3); *Frangula californica* subsp. *crassifolia* (N3?, G5T3?); *Fritillaria purdyi* (CRPR 4.3, N3, G3); *Galium ambiguum* subsp. *siskiyouense* (NNR, G4TNR); *Garrya buxifolia* (NNR, G4); *Harmonia stebbinsii* (CRPR 1B.2, N2, G2); *Leptosiphon nuttallii* subsp. *howellii* (CRPR 1B.3, N2, G5T2); *Lewisia stebbinsii* (CRPR 1B.2, N1, G1); *Lomatium ciliolatum* (NNR, G1G3); *Minuartia rosei* (CRPR 4.2, N3, G3); *Phacelia corymbosa* (N4?, G4G5); *Quercus durata* var. *durata* (NNR, G4T4?)	Road development/maintenance, invasive plant species

Table 2. (Continued)

County State	Locality: Conservation Area	Manager(s)	Serpentine Endemic Species	Threats
Siskiyou, Trinity CA	Mount Eddy - Scott Mountain - Gibson Peak: Kangaroo Lake Botanical SIA; Mount Eddy RNA; Crater Creek RNA; Preacher Meadows RNA; Cedar Basin RNA; Trinity Alps Wilderness	USFS Klamath NF; USFS Shasta-Trinity NF	*Allium hoffmanii* (CRPR 4.3, N3, G3); *Arctostaphylos viscida* subsp. *pulchella* (N3N4, G5T3T4); *Arnica spathulata* (CRPR 4.3, N3?, G3?); *Astragalus whitneyi* var. *siskiyouensis* (N3N4, G5T3T4); *Balsamorhiza sericea* (CRPR 1B.3, N4, G4Q); *Boechera koehleri* (CRPR 1B.3, N3, G3); *Calochortus greenei* (CRPR 1B.2, N3, G3); *Campanula rotundifolia* (NNR, G5); *Carex scabriuscula* (CRPR 4.3, N3N4, G3G4); *C. serratodens* (NNR, G5); *Ceanothus pumilus* (NNR, G3?); *Chaenactis suffrutescens* (CRPR 1B.3, N3, G3); *Collinsia greenei* (NNR, G3G4); *Cordylanthus tenuis* subsp. *viscidus* (N4, G4G5T4); *Cypripedium californicum* (CRPR 4.2, N3, G3); *Darlingtonia californica* (CRPR 4.2, N3N4, G3G4); *Draba carnosula* (CRPR 1B.3, N2, G2); *Epilobium siskiyouense* (CRPR 1B.3, N3, G3); *Eriogonum alpinum* (CRPR 1B.2, N2, G2, SE); *E. congdonii* (CRPR 4.3, N3, G3); *E. siskiyouense* (CRPR 4.3, N3, G3); *E. strictum* var. *greenei* (CRPR 4.3, NNR, G5); *E. ternatum* (CRPR 4.3, N4, G4); *E. umbellatum* var. *humistratum* (CRPR 4.3, N3, G5T3); *Erythronium citrinum* var. *roderickii* (CRPR 1B.3, N3, G4T3); *Fritillaria purdyi* (CRPR 4.3, N3, G3); *Galium ambiguum* subsp. *siskiyouense* (NNR, G4TNR); *G. serpenticum* subsp. *scotticum* (CRPR 1B.2, N2, G4G5T2); *Garrya buxifolia* (NNR, G4); *G. congdonii* (N2N4, G2G4); *Gentiana setigera* (CRPR 1B.2, N2, G2); *Hastingsia serpentinicola* (NNR, G3?); *Helianthus exilis* (CRPR 4.2, NNR, G4); *Howellanthus dalesianus* (CRPR 4.3, N3, G3); *Iris bracteata* (CRPR 3.3, N4N5, G4G5); *I. innominata* (CRPR 4.3, N4N5, G4G5); *Ivesia pickeringii* (CRPR 1B.2, N2, G2); *Lomatium engelmannii* (CRPR 4.3, N3, G3); *L. tracyi* (CRPR 4.3, N3, G3); *Minuartia stolonifera* (CRPR 1B.3, N1, G1); *Penstemon filiformis* (CRPR 1B.3, N3, G3); *Phacelia corymbosa* (N4?, G4G5); *P. greenei* (CRPR 1B.2, N2, G2); *Pinguicula macroceras* (CRPR 2.2, NNR, G5); *Polemonium eddyense* (CRPR 1B.2); *Polystichum lemmonii* (N4, G4); *Raillardella pringlei* (CRPR 1B.2, N3, G3); *Rudbeckia glaucescens* (N3, G3); *Streptanthus barbatus* (NNR, G3G4); *Trifolium longipes* subsp. *elmeri* (NNR, G5T4?); *Veronica copelandii* (CRPR 4.3, N3, G3); *Viola cuneata* (NNR, G4)	Road development/maintenance, water development (reservoir)*, mining/quarrying*, logging*, livestock grazing*, off-highway vehicles*

County State	Locality: Conservation Area	Manager(s)	Serpentine Endemic Species	Threats
Siskiyou CA	Yreka: China Hill	City of Yreka	*Phlox hirsuta* (CRPR 1B.2, N1, G1, SE, FE); *Phacelia corymbosa* (N4?, G4G5)	Urban development*, road development/maintenance*, off-highway vehicles, invasive plant species
Humboldt, Trinity CA	Mount Lassic: Lassics Botanical SIA; Mount Lassic Wilderness	USFS Six Rivers NF	*Allium hoffmanii* (CRPR 4.3, N3, G3); *Arnica spathulata* (CRPR 4.3, N3?, G3?); *Ceanothus pumilus* (NNR, G3?); *Collinsia greenei* (NNR, G3G4); *Eriogonum strictum* var. *greenei* (CRPR 4.3, NNR, G5); *Frangula californica* subsp. *crassifolia* (N3?, G5T4?); *Fritillaria purdyi* (CRPR 4.3, N3, G3); *Galium ambiguum* subsp. *siskiyouense* (NNR, G4TNR); *Lomatium ciliolatum* (NNR, G1G3); *Lupinus constancei* (CRPR 1B.2, N1, G1); *Phacelia corymbosa* (N4?, G4G5)	Road development/maintenance, logging*, off-highway vehicles
Humboldt CA	Horse Mountain: Horse Mountain Botanical SIA	USFS Six Rivers NF	*Arnica cernua* (CRPR 4.3, N5, G5); *Cypripedium californicum* (CRPR 4.2, N3, G3); *Fritillaria purdyi* (CRPR 4.3, N3, G3); *Garrya buxifolia* (NNR, G4); *Hastingsia serpentinicola* (NNR, G3?); *Lomatium tracyi* (CRPR 4.3, N3, G3); *Phacelia corymbosa* (N4?, G4G5); *Viola cuneata* (NNR, G4)	Road development/maintenance, mining/quarrying*, logging*, off-highway vehicles

Table 2. (Continued)

County State	Locality: Conservation Area	Manager(s)	Serpentine Endemic Species	Threats
Del Norte CA	Gasquet: Myrtle Creek Botanical SIA; LE Horton RNA; North Fork Smith River Botanical Area; Craig's Creek RNA; Bear Basin Butte Botanical SIA	USFS Six Rivers NF	*Antennaria suffrutescens* (CRPR 4.3, N4, G4); *Arabis aculeolata* (CRPR 2.2, NNR, G4); *A. mcdonaldiana* (CRPR 1B.1, N2, G2, SE, FE); *Arctostaphylos hispidula* (N3, G3); *A. viscida* subsp. *pulchella* (N3N4, G5T3T4); *Arnica cernua* (CRPR 4.3, N5, G5); *A. spathulata* (CRPR 4.3, N3?, G3?); *Boechera koehleri* (CRPR 1B.3, N3, G3); *Cardamine nuttallii* var. *gemmata* (CRPR 3.3, N3, G5T3); *Carex serpenticola* (CRPR 2.3, NNR, G4); *Castilleja miniata* subsp. *elata* (CRPR 2.2, N3, G5T3); *Ceanothus pumilus* (NNR, G3?); *Cordylanthus tenuis* subsp. *viscidus* (N4, G4G5T4); *Cypripedium californicum* (CRPR 4.2, N3, G3); *Darlingtonia californica* (CRPR 4.2, N3N4, G3G4); *Dicentra formosa* subsp. *oregana* (CRPR 4.2, N4, G5T4); *Epilobium rigidum* (CRPR 4.3, N3N4, G3G4); *Eriogonum pendulum* (CRPR 2.2, N4, G4); *E. ternatum* (CRPR 4.3, N4, G4); *Galium ambiguum* subsp. *siskiyouense* (NNR, G4TNR); *Garrya buxifolia* (NNR, G4); *G. congdonii* (N2N4, G2G4); *Gentiana setigera* (CRPR 1B.2, N2, G2); *Hastingsia serpentinicola* (NNR, G3?); *Horkelia sericata* (CRPR 4.3, NNR, G3G4); *Iris bracteata* (CRPR 3.3, N4N5, G4G5); *I. innominata* (CRPR 4.3, N4N5, G4G5); *Lathyrus delnorticus* (CRPR 4.3, NNR, G4); *Lewisia oppositifolia* (CRPR 2.2, N4, G4); *Lilium bolanderi* (CRPR 4.2, N4, G4); *Lomatium howellii* (CRPR 4.3, N4N5, G4G5); *L. tracyi* (CRPR 4.3, N3, G3); *Micranthes howellii* (CRPR 4.3, NNR, G4); *Minuartia howellii* (CRPR 1B.3, N4, G4); *Packera hesperia* (CRPR 2.2, N3, G3); *P. macounii* (CRPR 4.3, NNR, G5); *Phacelia corymbosa* (N4?, G4G5); *Pinguicula macroceras* (CRPR 2.2, N3N4, G5); *Poa piperi* (CRPR 4.3, NNR, G4); *Pseudoleskeella serpentinense* (N3, G3); *Pyrrocoma racemosa* var. *congesta* (CRPR 2.3, NNR, G5T4); *Rosa gymnocarpa* var. *serpentina* (CRPR 1B.3, N2, G5T2); *Rudbeckia glaucescens* (N3?, G4T3?); *Salix delnortensis* (CRPR 4.3, NNR, G4); *Sanicula peckiana* (CRPR 4.3, N4, G4); *Silene campanulata* subsp. *campanulata* (CRPR 4.2, N3, G5T3Q, SE); *S. serpentinicola* (CRPR 1B.2, N2, G2); *Streptanthus howellii* (CRPR 1B.2, N2N3, G2G3); *Trifolium longipes* subsp. *elmeri* (NNR, G5T4?); *Vancouveria chrysantha* (CRPR 4.3, N4, G4); *Viola cuneata* (NNR, G4); *V. primulifolia* subsp. *occidentalis* (CRPR 1B.2, T2, G5T2)	Road development/maintenance, mining/quarrying*, logging*, off-highway vehicles, invasive plant species

County State	Locality: Conservation Area	Manager(s)	Serpentine Endemic Species	Threats
Josephine, Curry OR	Rough and Ready Creek - Eight Dollar Mountain: E.D.M. ACEC; E.D.M. Preserve; E.D.M. Botanical Wayside; Illinois Valley Botanical Emphasis Area; Cedar Log Flat RNA; Woodcock Bog RNA; North Fork Silver Creek RNA; Lemmingsworth Gulch RNA; Limpy Botanical Trail; Kalmiopsis Wilderness; R. and R. Cr. Preserve; R. and R. Cr. Botanical Wayside; R. and R. Cr. ACEC; R. and R. Flat Botanical Area	BLM Medford FO; USFS Rogue River-Siskiyou NF; State Parks; TNC	*Antennaria suffrutescens* (N4, G4); *Arabis aculeolata* (NNR, G4); *A. macdonaldiana* (ORBIC 1, N2, G2, SE, FE); *Arctostaphylos hispidula* (N3, G3); *Arnica cernua* (N5, G5); *Balsamorhiza sericea* (ORBIC 4, N4, G4Q); *Boechera koehleri* (ORBIC 1, N3, G3); *Calochortus howellii* (ORBIC 1, N3, G3, ST); *C. umpquaensis* (ORBIC 1, N3, G3, SE); *Cardamine nuttallii* var. *gemmata* (ORBIC 4, N3, G5T3); *Ceanothus pumilus* (NNR, G3?); *Cypripedium californicum* (ORBIC 4, N3, G3); *Darlingtonia californica* (ORBIC 4, N3N4, G3G4); *Dicentra formosa* subsp. *oregana* (ORBIC 4, N4, G5T4); *Epilobium rigidum* (ORBIC 4, N3N4, G3G4); *Eriogonum pendulum* (ORBIC 4, N4, G4); *E. ternatum* (N4, G4); *Galium ambiguum* var. *siskiyouense* (NNR, G4TNR); *Gentiana setigera* (ORBIC 1, N2, G2); *Hastingsia bracteosa* (ORBIC 1, N2, G2, ST); *H. serpentinicola* (NNR, G3?); *Iris bracteata* (N4N5, G4G5); *I. innominata* (N4N5, G4G5); *Lewisia howellii* (N4, G4T4Q); *L. oppositifolia* (ORBIC 4, N4, G4); *Lilium bolanderi* (N4, G4); *Microseris howellii* (ORBIC 4, N3, G3, ST); *Minuartia howellii* (N4, G4); *Packera hesperia* (ORBIC 4, N3, G3); *Phacelia corymbosa* (N4?, G4G5); *Poa piperi* (NNR, G4); *Polystichum lemmonii* (N4, G4); *Pseudoleskeella serpentinense* (N3, G3); *Pyrrocoma racemosa* var. *congesta* (NNR, G5T4); *Rudbeckia glaucescens* (N3?, G4T3?); *Salix delnortensis* (ORBIC 4, NNR, G4); *Sanicula peckiana* (ORBIC 4, N4, G4); *Silene bolanderi* (N4?, G4T4?); *Streptanthus howellii* (ORBIC 1, N2N3, G2G3); *Vancouveria chrysantha* (ORBIC 4, N4, G4); *Viola cuneata* (NNR, G4); *V. lobata* (NNR, G4); *V. primulifolia* var. *occidentalis* (N2, G5T2)	Road development/maintenance, mining/quarrying*, logging*, livestock grazing*, off-highway vehicles, invasive plant species

Table 2. (Continued)

County State	Locality: Conservation Area	Manager(s)	Serpentine Endemic Species	Threats
Josephine OR	Waldo: Waldo-Takilma ACEC; French Flat ACEC	BLM Medford FO	*Arabis aculeolata* (NNR, G4); *Arctostaphylos hispidula* (N3, G3); *Arnica cernua* (N5, G5); *Balsamorhiza sericea* (ORBIC 4, N4, G4Q); *Calochortus howellii* (ORBIC 1, N3, G3, ST); *Cardamine nuttallii* var. *gemmata* (ORBIC 4, N3, G5T3); *Ceanothus pumilus* (NNR, G3?); *Cypripedium californicum* (ORBIC 4, N3, G3); *Darlingtonia californica* (ORBIC 4, N3N4, G3G4); *Dicentra formosa* subsp. *oregana* (ORBIC 4, N4, G5T4); *Epilobium rigidum* (ORBIC 4, N3N4, G3G4); *Eriogonum pendulum* (ORBIC 4, N4, G4); *E. ternatum* (N4, G4); *Galium ambiguum* var. *siskiyouense* (NNR, G4TNR); *Gentiana setigera* (ORBIC 1, N2, G2); *Iris bracteata* (N4N5, G4G5); *Iris innominata* (N4N5, G4G5); *Lathyrus delnorticus* (NNR, G4); *Lewisia howellii* (N4, G4T4Q); *L. oppositifolia* (ORBIC 4, N4, G4); *Lilium bolanderi* (N4, G4); *Microseris howellii* (N3, G3); *Minuartia howellii* (N4, G4); *Packera hesperia* (ORBIC 4, N3, G3); *Phacelia corymbosa* (N4?, G4G5); *Pseudotrillium rivale* (N4, G4); *Pyrrocoma racemosa* var. *congesta* (NNR, G5T4); *Rudbeckia glaucescens* (N3?, G4T3?); *Sanicula peckiana* (ORBIC 4, N4, G4); *Viola cuneata* (NNR, G4); *V. lobata* (NNR, G4)	Road development/maintenance, mining/quarrying*, logging*, off-highway vehicles, invasive plant species
Curry OR	Snow Camp: Hunter Creek Bog ACEC; Red Flat/Hunter Creek Bog SIA; Big Craggies SIA; Sourgame SIA	BLM Medford FO; USFS Rogue River-Siskiyou NF	*Antennaria suffrutescens* (N4, G4); *Arctostaphylos hispidula* (N3, G3); *Arnica cernua* (N5, G5); *Cardamine nuttallii* var. *gemmata* (ORBIC 4, N3, G5T3); *Carex serpenticola* (NNR, G4); *Cypripedium californicum* (ORBIC 4, N3, G3); *Darlingtonia californica* (ORBIC 4, N3N4, G3G4); *Galium ambiguum* subsp. *siskiyouense* (NNR, G4TNR); *Gentiana setigera* (ORBIC 1, N2, G2); *Horkelia sericata* (ORBIC 3, NNR, G3G4); *Lathyrus delnorticus* (NNR, G4); *Lomatium howellii* (N4N5, G4G5); *Pseudotrillium rivale* (N4, G4); *Phacelia corymbosa* (N4?, G4G5); *Salix delnortensis* (ORBIC 4, NNR, G4); *Viola cuneata* (NNR, G4)	Road development/maintenance, mining/quarrying*, logging*

County State	Locality: Conservation Area	Manager(s)	Serpentine Endemic Species	Threats
Douglas OR	Riddle: Beatty Creek RNA; *Calchortus coxii* conservation easement	BLM Roseburg FO	*Calochortus coxii* (ORBIC 1, N1, G1, SE); *Pseudoleskeella serpentinense* (ORBIC 4, N3, G3)	Road development/maintenance, logging, livestock grazing, invasive plant species
Grant OR	Baldy Mountain -Strawberry Mountain: Baldy Mountain PRNA; Canyon Creek RNA; Strawberry Mountain Wilderness	USFS Malheur NF	*Polystichum lemmonii* (N4, G4)	Road development/maintenance, mining/quarrying*, logging*, off-highway vehicles, invasive plant species
Kittitas WA	Wenatchee Mountains: El Dorado Creek RNA	USFS Okanogan-Wenatchee NF	*Chaenactis thompsonii* (WNHP S, N3, G3); *Claytonia megarhiza* var. *nivalis* (N3, G4G5T3); *Lomatium cuspidatum* (N2N3, G2G3); *Poa curtifolia* (N3, G3); *Polystichum lemmonii* (N4, G4)	Road development/maintenance, logging*
Skagit WA	Twin Sisters: Olivine Bridge Natural Area Preserve	Washington State Department of Natural Resources	*Polystichum lemmonii* (N4, G4)	Mining/quarrying*, logging*

One of the earliest examples of public environmental activism in the western US to preserve a chemically extreme ecosystem for its rare endemic flora is the serpentine of Ring Mountain in Marin County, California. Under the threat of urban development and at the behest of concerned local citizens, The Nature Conservancy purchased much of the private lands on undeveloped Ring Mountain and established it as a preserve in 1982 to protect the unique serpentine ecosystem and its rare, local endemic plant species including state and federally listed threatened *Calochortus tiburonensis* (Liliaceae), *Hesperolinon congestum* (Linaceae), and *Castilleja affinis* subsp. *neglecta* (Orobanchaceae; Ellman, 1982; Kenwood Press, 2009; Stein et al., 2000). The Ring Mountain Preserve harbors the only known population of *C. tiburonensis*.

Another local community campaign began in 1979 to protect the serpentine grassland at Edgewood in San Mateo County, California from development and eventually resulted in the establishment of the Edgewood Natural Preserve in 1993, protecting the state and federally listed local serpentine endemic plant species *Acanthomintha duttonii* (Lamiaceae) and *H. congestum* (Curtis, 2008). The Edgewood Natural Preserve harbors the last known population of *A. duttonii*. The Nature Conservancy negotiated acquisition of Ash Meadows in Nye County, Nevada from a private developer in 1983. The USFWS purchased the land and established the Ash Meadows National Wildlife Refuge in 1984 which protects several state listed endangered and federally listed threatened, local saline endemic plant species including *Astragalus phoenix* (Fabaceae), *Enceliopsis nudicaulis* var. *corrugata* (Asteraceae), *Grindelia fraxinipratensis* (Asteraceae), *Ivesia kingii* var. *eremica* (Rosaceae), *Mentzelia leucophylla* (Loasaceae), *Nitrophila mohavensis* (Amaranthaceae), and *Zeltnera namophila* (Gentianaceae; Stein et al., 2000; USFWS, 1990a).

Conservation Area Designations

Designated conservation areas under nonprofit conservation organization, city, or county ownership are usually open space, reserves, or preserves, but also include parks under county ownership. Designated conservation areas under state ownership are usually reserves or parks, but also include recreation areas, botanical management areas, and botanical waysides. Conservation areas of private nonprofit conservation organizations, city, county, and state usually allow light recreation on established trails (hiking, mountain biking, horseback riding) and often use livestock grazing as a vegetation management tool for invasive annual grass species in California. There are substantial differences between the land management mandates of the different federal agencies. Both NPS (National Park, Grassland, Seashore, Monument, or Recreation Area) and USFWS (National Wildlife Refuge; NWR) lands are managed primarily for light recreation on established trails (hiking, horseback riding) and are highly conservation-oriented. A few National Parks and NWRs utilize livestock grazing to manage grassland habitat. USFS and BLM manage the largest areas of federal public land in the western US and are under more intensive multiple-use mandates that include energy development, mining, logging, livestock grazing, and motorized and nonmotorized recreation, as well as conservation. The USFS and BLM designate specific areas within their large land holdings as having unique natural values for conservation.

Both USFS and BLM designate conservation areas as Wilderness Study Areas (WSAs) and Research Natural Areas (RNAs). A WSA is land of primeval character and influence, without permanent improvements such as roads or structures that is managed to preserve its natural conditions. RNAs are designated for research to preserve significant ecosystems in their natural condition to conduct ecological studies. USFS designates conservation areas as Special Interest Areas (SIAs). SIAs are designated to include a broader range of values than a RNA including botanical, geological, and scenic resources. Botanical SIAs are typically established to protect rare or endangered plant species and to encourage the public to learn about those species and their ecosystems. The BLM designates Areas of Critical Environmental Concern (ACEC). ACECs are the equivalent of the USFS's SIAs and are designated for special management to protect natural values including rare, endemic, or endangered plant species; rare geological features; and unique landscapes. The President designates National Monuments (NM) for the NPS, USFWS, USFS, and BLM. NMs are designated for their significant historical and scientific interest. Congress designates Wilderness Areas based on WSAs. NMs and Wilderness Areas are managed primarily for light recreation (hiking, horseback riding) and are highly conservation oriented. Some Wilderness areas permit livestock grazing since it was grandfathered into the Wilderness Act of 1964.

Restoration and Revegetation Methods

There are multiple gradations in the human-mediated recovery of disturbed or degraded chemically extreme substrates including restoration, reclamation, rehabilitation, and revegetation. Restoration is the process of returning an ecosystem to a condition very similar to the ecological structure and function that existed prior to disturbance (SER, 2004). Restoration is undertaken on mildly or moderately degraded ecosystems where some to most ecological function is still intact and typically involves manipulation of vegetation cover, structure, composition, and/or diversity and may also involve additional minor manipulation of disturbance regime, soil conditions, and/or hydrology to restore original conditions. In contrast, reclamation, rehabilitation, and revegetation are typically undertaken on severely degraded ecosystems such as construction sites and mined lands where most to all ecological function has been destroyed (Bradshaw & Chadwick, 1980; Tordoff et al., 2000; Van Kekerix & Kay, 1986; Williamson et al., 1982). These ecosystem recovery methods often involve intensive soil structure and function building, followed by replanting. Reclamation, rehabilitation, and revegetation have the common primary goal of landscape stability and land use, rather than a high degree of original ecological function, therefore the term revegetation is used to encompass all three processes in this chapter.

Restoration of chemically extreme substrates may be focused on recovery of a single edaphic endemic plant species (especially a federally listed one) or oriented towards improvement of the entire ecosystem. Federally listed plant species restoration is guided by species specific USFWS Recovery Plans. Restoration and recovery of a single plant species is typically small-scale (local population level) and usually focuses on research, propagation, introduction, and/or improvement of habitat conditions that benefit that particular species. Ecosystem approaches in restoration and revegetation, in contrast, generally occur at a larger scale than single species, are more resource intensive, and may benefit multiple edaphic endemic plant species and the ecosystem as a whole. Restoration practices on chemically extreme ecosystems may include manipulation of vegetative cover and composition through livestock grazing, invasive species control, prescribed fire and native species planting; modification of soil properties through ripping and tillage for decompaction; and alteration of hydrology by reestablishing surface water and/or tidal flow.

Revegetation practices for drastically disturbed substrates may include physical site stabilization including slope engineering, landscape recontouring and erosion control; substrate physical and chemical condition modification including tillage or ripping, topsoiling, and/or amendment with chemical fertilizers or organic matter; and plant and microbial materials selection with extensive planting (Bradshaw, 1997; Bradshaw & Chadwick, 1980; Tordoff et al., 2000; Van Kekerix & Kay, 1986; Williamson et al., 1982). At sites where no soil remains, topsoil application is the most ecologically appropriate method. Topsoil application reintroduces original soil physical and chemical function of the ecosystem along with the microbiological community and seed bank of locally adapted species. The topsoil source should be soil stripped from the site prior to construction or mining disturbance. Use of topsoil from a differing geological source is not advised as it will permanently alter the substrate chemistry of the site and has high potential to introduce invasive or other ecologically inappropriate species (O'Dell & Claassen, 2011). Substrate amendment with chemical fertilizers and organic matter should be approached with great caution and rigorously tested in the greenhouse and at small field scales prior to broad application at the landscape scale. Chemical modification of the substrate can fundamentally and permanently alter the physically and chemically stressful conditions that support the uniquely adapted plant community, resulting in unexpected and undesirable affects such as nonnative plant species invasion (O'Dell & Claassen, 2011).

Careful plant materials selection is of particular importance for both restoration and revegetation of chemically extreme ecosystems. Local, native plant species and ecotypes growing on undisturbed, geologically identical substrate adjacent to the target restoration or revegetation area have unique morphological and physiological adaptations to the local climate and edaphic conditions (O'Dell & Claassen, 2011; Zedler, 2000). Plant species selection for the revegetation community should focus on primary succession (pioneer) species observed on drastically disturbed substrates growing on identical geology in the same microclimate in the local area. Primary succession species have adaptations to the high level of climate exposure and elevated disturbance at revegetation sites (Dahlgren et al., 1997; Khater et al., 2003; Mota et al., 2004; O'Dell & Claassen, 2011; Rufo & de la Fuente, 2010). A mix of grasses and other fibrous rooted species and deep rooted woody species is ideal for the revegetation of slopes, with the former providing surface erosion control and the latter providing deep substrate anchoring to reduce the risk of landslide as well as the development of appropriate understory microclimate (Bradshaw & Chadwick, 1980; Van Kekerix & Kay, 1986; Williamson et al., 1982). Once the primary revegetation community is well established, more sensitive plant species requiring the microclimate conditions provided by the primary succession species can be planted to increase diversity.

Conservation areas of extreme edaphic substrates are widely scattered throughout the western US and each contains a unique suite of endemic species adapted to the local climatic and edaphic conditions. Restoration and revegetation of serpentine, gabbro, carbonate, saline, guano, and hydrothermally-altered substrates each require a unique approach based upon substrate chemistry, climate, and other local site conditions. Due to the highly variable methods used for the restoration and recovery of individual federally listed plant species, restoration and revegetation discussion in this chapter will primarily focus on the ecosystem level.

SERPENTINE FLORA CONSERVATION AND RESTORATION

Serpentine is ultramafic rock composed of ferromagnesian silicates and includes peridotite and its hydrothermally-altered derivative, serpentinite (Alexander et al., 2007; Brooks, 1987; Kruckeberg, 1984a). Serpentine is rich in Fe and Mg but has a low Si content. Due to its origins from oceanic mantle emplaced at subduction zones along continental plate margins, most serpentine rock exposures are located in the far western states of California, Oregon and Washington. Significant serpentine exposures are located in the Coast Ranges, Sierra Nevada, and Klamath Mountains in California; Klamath Mountains and Blue Mountains in Oregon; and the Wenatchee Mountains and between Mount Baker and the San Juan Islands in Washington (Alexander et al., 2007). Serpentine weathers to generate soils that are deficient in plant essential macronutrients (N, P, K and Ca) and contain an excess of the plant essential macronutrient Mg (very low Ca:Mg molar ratio), as well as an excess of phytotoxic heavy metals including Co, Cr and Ni (Chapter 6; Alexander et al., 2007; Brooks, 1987; Kruckeberg, 1984a).

Significant adverse impacts to serpentine ecosystems in the western US occur from urban, road, energy (wind), and water (reservoir) development; mining and quarrying; logging; livestock grazing; off-highway vehicle recreation; and invasive plant species (Alexander et al., 2007; O'Dell & Claassen, 2011). A short documentary film, The Invisible Peak, chronicles the history of military development on serpentine on the West Peak of Mount Tamalpais in Marin County, California, the resulting environmental impacts, and the current public campaign to restore it (Invisible Peak, 2014).

Serpentine endemic plant species and conservation areas are presented in Table 2. USFWS Recovery Plans for federally listed serpentine endemic plant species include those for species of the San Francisco Bay Area (Marin, San Francisco, San Mateo, Alameda, Santa Clara counties; USFWS, 1998), *Camissonia benitensis* (Onagraceae; USFWS, 2006a), *Arabis macdonaldiana* (Brassicaceae; USFWS, 1984), and *Phlox hirsuta* (Polemoniaceae; USFWS, 2006b). There are currently no USFWS recovery plans for *Astragalus claranus*, *Brodiaea pallida* (Themidaceae), or *Verbena californica* (Verbenaceae).

Serpentine restoration and revegetation in the western US has previously been reviewed in detail by O'Dell & Claassen (2009, 2011). Most serpentine ecosystem restoration in the western US has focused on grasslands adversely impacted by fire regime alteration, livestock grazing, pollution from atmospheric N and invasive plant species. Restoration methods on serpentine grasslands have included tilling or ripping to relieve soil compaction, as well as manipulation of vegetation structure and species composition through carefully controlled livestock grazing regime, mechanical vegetation removal, and prescribed fire. Vegetation manipulation has primarily focused on reducing invasive annual grass competition effects on native herbaceous serpentine plant species and is an important management tool for the recovery of federally listed serpentine endemic plant species of the San Francisco Bay Area

(Curtis, 2008; Mayall, 2008; Naumovich et al., 2008; 2009; 2010; 2011; 2012; USFWS, 1998).

Serpentine revegetation has focused on stabilization and revegetation of drastically disturbed sites including road banks, building pads, landfills, quarries, and mines. Revegetation methods include slope engineering and erosion control; substrate decompaction through tillage and deep ripping; serpentine topsoil application or subgrade substrate amendment with chemical fertilizer, organic matter and mychorrhizal inoculum to improve fertility; and these steps are followed by seeding and planting with local, native, serpentine tolerant plant species (O'Dell & Claassen, 2011).

Gabbro Flora Conservation and Restoration

Gabbro is similar to serpentine in its mineralogy and origins. Gabbro is emplaced at subduction zones and, like serpentine, it is rich in ferromagnesian silicates, but has a higher Si content (Alexander, 1993; 2011; Alexander et al., 2007; Wilson et al., 2009). Most significant gabbro exposures in the western US are located in the Sierra Nevada foothills and in the Peninsular Range of California.

Due to the mineralogical similarity between serpentine and gabbro, gabbro soils are chemically similar to serpentine soils, but are less deficient in Ca and have lower concentrations of heavy metals such as Ni (Alexander, 1993; 2011; Hunter & Horenstein, 1992; Wilson et al., 2009). Significant adverse impacts to gabbro ecosystems in the western US occur from urban and road development, off-highway vehicle recreation, fire suppression, and invasive plant species (USFWS, 2002).

Gabbro endemic plant species and conservation areas are presented in Table 3. US Fish and Wildlife Service Recovery Plans for federally listed plant species include one for gabbro soil endemic plants of the Central Sierra Nevada Foothills (USFWS, 2002). This Recovery Plan includes the serpentine-gabbro endemic *Packera layneae* (Asteraceae). There is currently no USFWS recovery plan for *Ceanothus ophiochilus* (Rhamnaceae).

The chaparral and herbaceous understory vegetation of gabbro soils is fire adapted and dependent upon fire for regeneration (Boyd, 1987; 2007; Oberhauer, 1993; Wilson et al., 2009). Prescribed fire has been used as a restoration tool for federally listed gabbro endemic plant species at the Pine Hill Preserve, El Dorado County, California (Hinshaw, 2008; USFWS, 2002). No revegetation research is known to have been conducted for gabbro plant communities, but given the chemical similarity between gabbro and serpentine revegetation, the methods would theoretically be similar for both.

Carbonate Flora Conservation and Restoration

Carbonate substrates are marine in origin and include limestone (calcium carbonate), its metamorphosed derivative marble, and the precipitate minerals dolomite (calcium magnesium carbonate) and gypsum (calcium sulfate - not a carbonate but chemically similar; Chapter 5). Limestone and marble are derived from the skeletal remains of coral or foraminifera.

Dolomite and gypsum are minerals that precipitate out in anaerobic marine environments. Carbonate substrates are most abundant in the desert southwest as a result of their marine origins during a time when much of the region was a shallow sea (Kiver & Harris, 2011). Carbonate substrates have the opposite extreme chemical condition of serpentine with respect to plant-available Ca and Mg: there is an excess of Ca relative to Mg, resulting in an exceptionally high Ca:Mg molar ratio (Mg deficiency: Kruckeberg, 2002; 2006). Significant adverse impacts to carbonate ecosystems in the western US occur from urban, agriculture, road, energy (fossil fuel), and water development; mining and quarrying; livestock grazing; off-highway vehicle recreation; and invasive plant species.

Carbonate endemic plants and conservation areas are presented in Table 4. US Fish and Wildlife Service Recovery Plans for federally listed plant species includes that for San Bernardino Mountains carbonate endemic plants (USFWS, 1997), *Astragalus montii* (USFWS, 1995a), *Primula maguirei* (Primulaceae; USFWS, 1990b), *Astragalus cremnophylax* var. *cremnophylax* (USFWS, 2006c), *Pediocactus bradyi* (Cactaceae; USFWS, 1985a), and *Purshia subintegra* (Rosaceae; USFWS, 1995b). There is currently no USFWS Recovery Plan for *Eriogonum gypsophilum* (Polygonaceae).

Despite the abundance of carbonate rock ecosystems in the western US, there is a scarcity of information regarding their restoration and revegetation. Globally, there appear to be two strategies for the revegetation of carbonate substrates: 1) a passive strategy which allows natural recruitment and succession to run its own course (Gentili et al., 2011; Khater et al., 2003; Mota et al., 2003; Robson et al., 2009); and 2) an active strategy involving human mediated revegetation to rapidly advance vegetation establishment and succession (Ballesteros et al., 2012; Clemente et al., 2004; Cohen-Fernández & Naeth, 2013a; b; Cohen-Fernández et al., 2013; Hall et al., 2003; Kyriazopoulos et al., 2012; Ortiz et al., 2012). Carbonate substrate revegetation involves topsoil, organic matter and fertilizer application, followed by planting of local, native carbonate tolerant plant species. Passive and active revegetation strategies result in different plant communities. Human mediated revegetation results in substantially greater vegetative cover within a shorter period of time, but generally results in lower species richness as compared to natural revegetation (Moreno-Penaranda et al., 2004; Tropek et al., 2010).

Considering that the majority of carbonate substrates in the desert southwest are subject to arid climate regimes and support only spare vegetative cover even in undisturbed condition, the optimal revegetation strategy of carbonate substrates may be to recontour the disturbance, apply carbonate topsoil (if available), and broadcast seeds of local, native carbonate tolerant species or allow natural recruitment to progress on its own (Gonella & Neel, 1993; Mistretta & White, 2000; Robson et al., 2009). More research on the restoration and revegetation methods of carbonate substrates in the southwestern US is needed.

Table 3. Selected gabbro ecosystem conservation areas and the gabbro endemic plant species they contain, including their conservation status and anthropogenic threats. * = historic impact in conservation areas, but current threat in surrounding or nearby gabbro ecosystems. † = species is also endemic to serpentine. See Table 1 for conservation status codes. Gabbro species and threats for each locality determined from Alexander (1993; 2011); Alexander et al. (2007); Burge (2013); Calflora (2014); Consortium of California Herbaria (2014); Google Earth (2014); Hinshaw (2008); NatureServe (2014); Oberhauer (1993); USFWS (2002); and Wilson et al. (2007)

County State	Locality: Conservation Area	Manager(s)	Gabbro Endemic Species	Threats
El Dorado CA	Pine Hill: Pine Hill ER	CDFW (State); BLM Motherlode FO	*Calystegia stebbinsii* (CRPR 1B.1, N1, G1, SE, FE); *Ceanothus roderickii* (CRPR 1B.2, N2, G2, SR, FE); *Chlorogalum grandiflorum* (CRPR 1B.2, N3, G3); *Fremontodendron decumbens* (CRPR 1B.1, N1, G1, SR, FE); *Galium californicum* subsp. *sierrae* (CRPR 1B.2, N1, G5T1, SR, FE); *Packera layneae*† (CRPR 1B.2, N2, G2, SR, FT); *Wyethia reticulata* (CRPR 1B.2, N2, G2)	Urban development*, road development/maintenance*, off-highway vehicles*, fire suppression, invasive plant species
Kern CA	Bodfish: Piute Cypress ACEC/WSA; Bodfish Piute Cypress Botanical SIA	BLM Bakersfield FO; USFS Sequoia NF	*Hesperocyparis nevadensis* (CRPR 1B.2, N2, G4T2); *Streptanthus cordatus* var. *piutensis* (CRPR 1B.2, N1, G5T1)	Road development/maintenance, fire suppression, invasive plant species
Riverside CA	Agua Tibia Mountain: Agua Tibia Wilderness	USFS Cleveland NF	*Ceanothus ophiochilus* (CRPR 1B.1, N1, G1, SE, FT)	Urban development*, road development/maintenance, off-highway vehicles, fire suppression, invasive plant species
San Diego CA	Guatay Mountain: Guatay Mountain RNA; Tecate Cypress Botanical SIA	USFS Cleveland NF	*Calochortus dunnii* (CRPR 1B.2, N2, G2, SR); *Hesperocyparis forbesii* (CRPR 1B.1, N1, G2)	Off-highway vehicles, fire suppression, invasive plant species
San Diego CA	King Creek: King Creek RNA	USFS Cleveland NF	*Calochortus dunnii* (CRPR 1B.2, N2, G2, SR); *Hesperocyparis stephensonii* (CRPR 1B.1, N1, G4T1)	Fire suppression, invasive plant

County State	Locality: Conservation Area	Manager(s)	Gabbro Endemic Species	Threats
San Diego CA	McGinity Mountain: McGinty Mountain ER (State); San Diego NWR - McGinity Mountain	CDFW (State); USFWS	*Carex obispoensis*† (CRPR 1B.2, N2, G2); *Chamaebatia australis* (CRPR 4.2, N4, G4); *Nolina interrata* (CRPR 1B.1, N1, G1); *Packera ganderi* (CRPR 1B.2, N2, G2)	Urban development*, road development/maintenance, off-highway vehicles, fire suppression, invasive plant species
San Diego CA	Viejas Mountain: Viejas Mountain RNA	USFS Cleveland NF	*Calochortus dunnii* (CRPR 1B.2, N2, G2, SR); *Sibaropsis hammittii* (CRPR 1B.2, N2, G2)	Fire suppression, invasive plant species, road development/ maintenance, off-highway vehicles, fire suppression, invasive plant species
San Diego CA	Otay Mountain: Otay Mountain ER (State); Otay Mountain Wilderness; Cedar Canyon ACEC; San Diego NWR	CDFW (State); BLM Palm Springs FO; USFWS	*Calochortus dunnii* (CRPR 1B.2, N2, G2, SR); *Chamaebatia australis* (CRPR 4.2, N4, G4); *Hesperocyparis forbesii* (CRPR 1B.1, N1, G2)	Urban development*, road development/maintenance, off-highway vehicles, fire suppression, invasive plant species
San Diego CA	Black Mountain: Black Mountain Open Space Park	City of San Diego	*Packera ganderi* (CRPR 1B.2, N2, G2)	Urban development*, road development/maintenance, off-highway vehicles, fire suppression, invasive plant species
San Diego CA	Tecate Peak: Kuchamaa ACEC	BLM Palm Springs FO	*Calochortus dunnii* (CRPR 1B.2, N2, G2, SR); *Chamaebatia australis* (CRPR 4.2, N4, G4); *Hesperocyparis forbesii* (CRPR 1B.1, N1, G2); *Packera ganderi* (CRPR 1B.2, N2, G2)	Road development/maintenance, off-highway vehicles, fire suppression, invasive plant species

Table 3. (Continued)

County State	Locality: Conservation Area	Manager(s)	Gabbro Endemic Species	Threats
San Diego CA	Cuyamaca Peak: Cuyamaca Rancho State Park	State Parks	*Calochortus dunnii* (CRPR 1B.2, N2, G2, SR); *Chamaebatia australis* (CRPR 4.2, N4, G4); *Hesperocyparis stephensonii* (CRPR 1B.1, N1, G4T1); *Packera ganderi* (CRPR 1B.2, N2, G2)	Road development/maintenance, fire suppression, invasive plant species
San Diego CA	Sycuan Peak: Sycuan Peak ER	CDFW (State)	*Carex obispoensis*† (CRPR 1B.2, N2, G2); *Nolina interrata* (CRPR 1B.1, N1, G1); *Packera ganderi* (CRPR 1B.2, N2, G2)	Urban development, road development/maintenance, off-highway vehicles, fire suppression, invasive plant species
San Diego CA	San Miguel Mountain: San Diego NWR	USFWS	*Carex obispoensis* (CRPR 1B.2, N2, G2)†; *Calochortus dunnii* (CRPR 1B.2, N2, G2, SR); *Nolina interrata* (CRPR 1B.1, N1, G1)	Urban development*, road development/maintenance, fire suppression, invasive plant species

Table 4. Selected carbonate ecosystem conservation areas and the carbonate endemic plant species they contain, including their conservation status and anthropogenic threats. * = historic impact in conservation areas, but current impact in surrounding or nearby carbonate ecosystems. See Table 1 for conservation status codes. Carbonate endemic species and threats for each locality determined from Arizona Rare Plant Field Guide (2014); Calflora (2014); Consortium of California Herbaria (2014); Google Earth (2014); Kruckeberg (2002, 2006); Lloyd & Mitchell (1973); NatureServe (2014); Nevada Rare Plant Atlas (2014); New Mexico Rare Plants (2014); Utah Rare Plant Guide (2014); Utah Vascular Plant Atlas (2014); and York (2001)

County State	Locality: Conservation Area	Manager(s)	Substrate	Carbonate Endemic Species	Threats
Shasta CA	Devils Rock: Devils Rock - Hosselkus RNA	USFS Shasta-Trinity NF	Limestone	*Ageratina shastensis* (CRPR 1B.2, N2, G2); *Erythranthe taylori*; *Neviusia cliftonii* (CRPR 1B.2, N2, G2)	Invasive plant species
Fresno CA	Boyden Cavern: Sequoia National Monument	NPS	Limestone, marble	*Eriogonum ovalifolium* var. *monarchense* (CRPR 1B.3, N1, G5T1); *Gilia yorkii* (CRPR 1B.2, N1, G1); *Heterotheca monarchensis* (CRPR 1B.3, N1, G1); *Streptanthus fenestratus* (CRPR 1B.3, N2, G2)	Invasive plant species
San Bernardino CA	San Bernardino Mountains: Carbonate Endemic Plants RNA	BLM Barstow FO	Dolomite, limestone	*Acanthoscyphus parishii* var. *goodmaniana* (CRPR 1B.1, N1, G4?T1, FE); *Astragalus albens* (CRPR 1B.1, N1, G1, FE); *Erigeron parishii* (CRPR 1B.1, N2, G2, FE); *Eriogonum ovalifolium* var. *vineum* (CRPR 1B.1, N1, G5T1, FE); *Physaria kingii* subsp. *bernardina* (CRPR 1B.1, N1, G5T1, FE)	Road development/maintenance, mining/quarrying*, off-highway vehicles, invasive plant species
Mono CA	White Mountains: White Mountains Wilderness	USFS Inyo NF; BLM Bishop FO	Dolomite	*Eriogonum gracilipes* (NRR, G3G4); *Penstemon barnebyi* (CRPR 2.1, N3N4, G3G4); *P. calcareus* (CRPR 1B.3, N2N3, G2G3)	Road development/maintenance, livestock grazing, off-highway vehicles

Table 4. (Continued)

County State	Locality: Conservation Area	Manager(s)	Substrate	Carbonate Endemic Species	Threats
Clark, Nye NV	Mount Charleston: Mount Charleston Wilderness Area	USFS Humboldt-Toiyabe NF; BLM Las Vegas FO	Limestone	*Antennaria soliceps* (NNPS W, G1G2, N1N2); *Astragalus oophorus* var. *clokeyanus* (NNPS W, G4T2, N2); *Draba jaegeri* (NNPS W, G2, N2); *Draba paucifructa* (NNPS W, G1G2, N1N2); *Glossopetalon clokeyi* (NNPS W, G2, N2); *Ivesia jaegeri* (NNPS W, G2G3, N2N3)	Urban development*, road development/maintenance, water development*, invasive plant species
Clark NV	Sheep Range: Desert National Wildlife Range	USFWS	Limestone	*Astragalus ackermanii* (NNPS W, N2, G2); *A. amphioxys* var. *musimonum* (NNPS W, N2, G5T2); *Chrysothamnus eremobius* (NNPS W, N1, G1); *Eremogone stenomeres* (NNPS W, N2, G2); *Erigeron ovinus* (NNPS W, N2, G2); *Gilia ripleyi* (NNPS W, N3, G3); *Sphaeromeria compacta* (NNPS W, N2, G2)	Road development*/maintenance, off-highway vehicles, invasive plant species
White Pine NV	Mount Washington: Great Basin National Park	NPS	Limestone	*Jamesia tetrapetala* (NNPS W, N2, G2); *Penstemon rhizomatosus* (NNPS W, N1, G1); *Primula nevadensis* (NNPS W, N2, G2); *Silene nachlingerae* (NNPS W, N2, G2)	Road development/maintenance, mining/quarrying*, off-highway vehicles, invasive plant species
Cache UT	City of Logan region: Logan Canyon; Mount Naomi Wilderness Area; Wellsville Mountains Wilderness Area	USFS Wasatch-Cache NF	Limestone	*Erigeron cronquistii* (UNPS W, N2, G2); *Musineon lineare* (N2, G2); *Penstemon compactus* (N2, G2); *Primula maguirei* (UNPS H, N1, G1, FT); *Viola frank-smithii* (UNPS W, N1, G1)	Road development/ maintenance, water development

County State	Locality: Conservation Area	Manager(s)	Substrate	Carbonate Endemic Species	Threats
Salt Lake, Utah UT	Lone Peak: Lone Peak Wilderness Area	USFS Uinta NF; USFS Wasatch-Cache NF	Limestone	*Erigeron garrettii* (N2, G2); *Jamesia americana* var. *macrocalyx* (UNPS H, NNR, G5T2); *Physaria garrettii* (UNPS W, N2, G2)	Urban development*, road development/ maintenance
Beaver UT	Wah Wah Mountains: Wah Wah Mountains ACEC	BLM Fillmore FO	Limestone	*Eriogonum soredium* (UNPS H, N1, G1, FC); *Lepidium ostleri* (UNPS H, N1, G1, FC); *Trifolium friscanum* (UNPS H, N1, G1, FC)	Off-highway vehicles
Sanpete UT	Heliotrope Mountain: Heliotrope Mountain	USFS Manti- La Sal NF	Limestone	*Astragalus montii* (UNPS W, N1, G2T1, FT); *Packera musiniensis* (UNPS H, NNR, G1); *Silene petersonii* (UNPS M, N3, G3)	Road development/ maintenance, energy development (fossil fuel), mining/quarrying*
Garfield,Iron, Kane UT	Cedar City region: Table Cliff RNA; Box-Death Hollow Wilderness Area; Cedar Breaks NM; Bryce Canyon NP; Escalante Petrified Forest State Park	USFS Dixie NF; NPS; State Parks	Limestone	*Castilleja revealii* (UNPS H, N2, G2); *Cymopterus minimus* (UNPS W, N1N2, G1G2Q); *Erigeron proselyticus* (N3, G3); *Eriogonum aretioides* (UNPS W, N2, G2); *Happlopappus zionis* (N3, G3); *Pediomelum pariense* (N2N3, G2G3); *Penstemon bracteatus* (UNPS W, N2, G2); *Silene petersonii* (UNPS M, N2N3, G2G3); *Sphaeromeria capitata* (UNPS M, N3, G3)	Road development/ maintenance, invasive plant species

Table 4. (Continued)

County State	Locality: Conservation Area	Manager(s)	Substrate	Carbonate Endemic Species	Threats
Mohave, Grand Canyon AZ	Grand Canyon region: Grand Canyon National Park; Grand Canyon-Parashant National Monument; Vermillion Cliffs National Monument; Lake Mead National Recreation Area; Vermillion Cliffs Wilderness Area; Piute Wilderness; Mount Logan Wilderness; Mount Trumball Wilderness Kanab Creek Wilderness	NPS; BLM Arizona Strip FO	Limestone	*Astragalus cremnophylax* var. *cremnophylax* (N1, G1G2T1,S HS, FE); *A. c.* var. *hevronii* (N1, G1G2T1); *A. c.* var. *myriorrhaphis* (N1, G1G2T1, S SR); *Pediocactus bradyi* (N1, G1, S HS, FE); *P. paradinei* (N1N2, G1G2, S HS); *P. peeblesianus* var. *fickeiseniae* (N1N2, G1G2T1T2, S HS); *Penstemon distans* (N2, G2, S SR); *P. petiolatus* (N2N3, G1G3); *Rosa stellata* subsp. *abyssa* (NNR, G4T2, S SR)	Road development/ maintenance, water development*, livestock grazing, off-highway vehicles, invasive plant species
Mohave AZ	Burro Creek: Upper Burro Creek Wilderness	BLM Kingman FO	Limestone	*Astragalus newberryi* var. *aquarii* (NNR, G5T1); *Purshia subintegra* (N1, GNA, S HS, FE)	Road development/ maintenance, mining/quarrying*, invasive plant species
Yavapai, Coconino AZ	Clarkdale-Camp Verde region: Tuzigoot National Monument; Montezuma Castle National Monument; Red Rock-Secret Mountain Wilderness; Sycamore Canyon Wilderness; Red Rock State Park	NPS; USFS Coconino NF; USFS Kaibab NF; USFS Prescott NF; State Parks	Limestone	*Eriogonum ericifolium* var. *ericifolium* (N2, G3T2); *Hedeoma diffusum* (N3, G3, S SR); *Purshia subintegra* (N1, GNA, S HS, FE); *Salvia dorrii* subsp. *mearnsii* (N3, G5T3, S SR)	Urban development*, agriculture development*, road development/ maintenance, water development*, mining/quarrying*, off-highway vehicles, invasive plant species

County State	Locality: Conservation Area	Manager(s)	Substrate	Carbonate Endemic Species	Threats
Harding NM & Dallas TX	Texas Panhandle region: Kiowa National Grasslands	USFS Cibola NF	Limestone	*Astragalus wittmannii* (N3, G3); *Packera spellenbergii* (N2, G2)	Agriculture development*, road development/maintenance, off-highway vehicles, invasive plant species
Dona Ana NM	San Andres Mountains: San Andres Mountains NWR	USFWS	Limestone	*Penstemon alamosensis* (N3, G3); *Perityle staurophylla* var. *homoflora* (N2, G4T2); *P. s.* var. *staurophylla* (N2, G4T2); *Salvia summa* (N3?, G3?)	Road development/ maintenance, invasive plant species
Eddy NM & Culberson TX	Guadalupe Mountains: Carlsbad Caverns National Park; Guadalupe Mountains National Park	NPS	Limestone, gypsum	*Anulocaulis leiosolenus* var. *howardii* (NNR, G4T2); *Chaetopappa hersheyi* (N3, G3); *Hedeoma apiculata* (N3, G3); *Justicia wrightii* (N2, G2); *Lepidospartum burgessii* (NR, G2, SE); *Mentzelia humilis* var. *guadalupensis* (NNR, G4T2); *Perityle quinqueflora* (N4, G4); *Polygala rimulicola* var. *rimulicola* (N3, G3T3); *Salvia summa* (N3?, G3?); *Solidago wrightii* var. *guadalupensis* (N3, G4T3); *Streptanthus sparsiflorus* (N2, G2)	Urban development*, road development/maintenance, energy development (fossil fuel)*, invasive plant species
Eddy NM	Yeso Hills: Yeso Hills RNA	BLM Carlsbad FO	Gypsum	*Anulocaulis leiosolenus* var. *gypsogenus* (NNR, G4); *Astragalus gypsodes* (N2, G2); *Eriogonum gypsophilum* (N1, G1, SE, FT); *Linum allredii* (NNR, G1G2)	Energy development (fossil fuel)*, mining/quarrying*, invasive plant species

Table 5. Selected saline ecosystem conservation areas and the saline endemic plant species they contain, including their conservation status and anthropogenic threats. * = historic impact in conservation areas, but current impact in surrounding or nearby saline ecosystems. See Table 1 for conservation status codes. Saline endemic species and threats for each locality determined from Calflora (2014); Consortium of California Herbaria (2014); Cooper & Wolf (2007); Google Earth (2014); NatureServe (2014); Nevada Rare Plant Atlas (2014); O'Dell (personal field observations 2000 – 2014); Palaima (2012); USFWS (1985b; 1990b; 2014); and Zedler (2000)

County State	Locality: Conservation Area	Manager(s)	Saline Endemic Species	Threats
Shasta CA	Crystal Creek: Whiskeytown National Recreation Area	NPS	*Distichlis spicata* (N5, G5); *Puccinellia howellii* (CRPR 1B.1, N1, G1)	Road development/maintenance, invasive plant species
Nye NV	Ash Meadows: Ash Meadows NWR; Ash Meadows ACEC	USFWS; BLM Las Vegas FO	Numerous species in Chenopodiaceae; *Astragalus phoenix* (NNPS T, N2, G2, SE, FT); *Calochortus striatus* (CRPR 1B.2, N2, G2); *Chloropyron tecopense* (CRPR 1B.2, NNPS T, N2, G2); *Distichlis spicata* (N5, G5); *Enceliopsis nudicaulis* var. *corrugata* (NNPS T, N2, G5T2, SE, FT); *Grindelia fraxinipratensis* (CRPR 1B.2, NNPS T, N2, G2, FT); *Ivesia kingii* var. *eremica* (NNPS T, N1N2, G4T1T2Q, SE, FT); *Mentzelia leucophylla* (NNPS T, N1, G1Q, SE, FT); *Nitrophila mohavensis* (CRPR 1B.1, NNPS E, N1, G1, SE, FE); *Spartina gracilis* (NNR, G5); *Spiranthes infernalis* (N1, G1); *Zeltnera namophila* (NNPS T, N2, G2Q, SE, FT)	Urban development*, road development/maintenance*, water diversion (hydrology alteration)*, mining/quarrying*, livestock grazing*, invasive plant species
Western US deserts, multiple counties and states	Inland aridland saline soils - numerous conservation areas	USFWS, NPS, USFS, BLM, State, County, Private Nonprofit	*Atriplex coronata* var. *notatior* - CA (CRPR 1B.1, N1, G4T1, FE); *Atriplex tularensis* – CA (CRPR 1A, NX, GX, SE); *Distichlis spicata* (N5, G5); *Puccinellia parishii* (N2N3, G2G3) - CA (CRPR 1B.1), AZ (SHS), and NM (SE)	Urban development*, road development/maintenance, energy development (solar), water development (reservoir), water diversion (hydrology alteration), mining/quarrying, livestock grazing, off-highway vehicles, invasive plant species

County State	Locality: Conservation Area	Manager(s)	Saline Endemic Species	Threats
Western US coastline, multiple counties and states	Coastal tidal wetlands - numerous conservation areas	USFWS, NPS, USFS, BLM, State, County, Private Nonprofit	Numerous species in Chenopodiaceae; *Chloropyron maritimum* subsp. *maritimum* - CA (CRPR 1B.2, N2, G4?T2, SE, FE); *C. m.* subsp. *palustre* (N2, G4?T2) - CA (CRPR 1B.2), OR (ORBIC 1, SE); *C. molle* subsp. *hispidum* - CA (CRPR 1B.1, N2, G2T2); *C. m.* subsp. *molle* – CA (CRPR 1B.2, N1, G2T1, SR, FE); *Distichlis spicata* (N5, G5); *Puccinellia* sp.; *Spartina* sp.; *Suaeda californica* - CA (CRPR 1B.1, N1, G1, FE)	Urban development*, road development/maintenance, water diversion (hydrology alteration), sedimentation or pollution; invasive plant species

Table 6. Selected guano ecosystem conservation areas and the guano endemic plant species they contain, including their conservation status. See Table 1 for conservation status codes. Guano endemic species at each locality determined from Calflora (2014); Consortium of California Herbaria (2014); Google Earth (2014); NatureServe (2014); Oregon Flora Atlas (2014); and Vasey (1985)

County State	Locality	Conservation Area	Manager	Guano Endemic Species
San Mateo, San Francisco, Marin, Mendocino, Humboldt CA	California coast	San Pedro Rock, Farallon Islands NWR, Chimney Rock (Point Reyes National Seashore), Anchor Rock, Sugarloaf Rock	USFWS, NPS	*Lasthenia maritima* (N4, G4)
Curry, Coos, Lincoln, Tillamook OR	Oregon coast	Oregon Islands NWR - Hunters Rock, Table Rock, Yaquina Rocks, Pyramid Rock	USFWS	*Lasthenia maritima* (N4, G4)
Clallam WA	Washington coast	Washington Islands NWR - Carroll Island, Seal Rock	USFWS	*Lasthenia maritima* (N4, G4)

Table 7. Selected hydrothermally-altered ecosystem conservation areas and the hydrothermal endemic plant species they contain, including their conservation status and anthropogenic threats. * = historic impact in current conservation areas, but current impact in surrounding or nearby hydrothermally-altered ecosystems. See Table 1 for conservation status codes. Hydrothermal endemic species and threats for each locality determined from Billings (1950; 1992); Calflora (2014); Consortium of California Herbaria (2014); Google Earth (2014); Kruckeberg (2006); NatureServe (2014); Nevada Rare Plant Atlas (2014); Pavlik & Enberg (2001a; b); Stout et al. (1997); Stout & Al-Niemi (2002); and USFWS (1995c)

County State	Locality: Conservation Area	Manager	Hydrothermal endemic species	Threats
Active hydrothermal vents				
Sonoma CA	Little Geyser: The Geysers	Calpine	*Panicum acuminatum* var. *thermale* (CRPR 1B.2, N5, G5, SE)	Road development/maintenance, energy development (geothermal)
Park WY	Firehole Valley: Yellowstone NP	NPS	*Agrostis rossiae* (N1, G1); *Panicum acuminatum* var. *sericeum* (N5, G5)	Road development/maintenance
Hydrothermal spring precipitates and hydrothermal sinter				
Eureka NV	Hot Spring Hill: Hot Spring Hill – Conservation Agreement	BLM Mount Lewis FO	*Castilleja salsuginosa* (NNPS E, N1, G1Q, SE)	Energy development (geothermal)*
Washoe NV	Steamboat Hot Springs: Steamboat Hot Springs ACEC	BLM Sierra Front FO	*Eriogonum ovalifolium* var. *williamsiae* (NNPS E, N1, G5T1, SE, FE); *E. robustum* (NNPS W, N2N3, G2); *Mimulus ovatus* (NNPS T, N1N2, G1G2Q)	Road development/maintenance, energy development (geothermal)*, off-highway vehicles
Hydrothermally-altered rocks				
Washoe NV	City of Reno region: Mount Rose Wilderness; Hidden Valley County Park	USFS Humboldt-Toiyabe NF; County Parks	*Eriogonum robustum* (NNPS W, N2N3, G2); *Plagiobothrys glomeratus* (NNPS W, N2N3, G2G3)	Urban development*, road development/maintenance, mining/ quarrying*, off-highway vehicles
Storey NV	Virginia City region: Virginia City Highlands County Park	County Parks	*Eriogonum robustum* (NNPS W, N2N3, G2); *Plagiobothrys glomeratus* (NNPS W, N2N3, G2G3)	Urban development*, road development/maintenance, mining/quarrying*, off-highway vehicles

SALINE FLORA CONSERVATION AND RESTORATION

Saline substrates contain high concentrations of salt and typically have alkaline pH (Chapter 4; Dahlgren et al., 1997; Zedler, 2000). The salt originates from seawater in coastal tidal wetlands of California, Oregon, and Washington and from evaporative concentration in arid inland basins (sinks, playas) of the San Joaquin Valley, Great Basin, Mojave, Sonoran, and Chihuahuan deserts. Dissolved salt lowers the osmotic potential of the substrate, decreasing the ability of plants to obtain water and maintain adequate tissue hydration (Marschner, 2002). Members of Chenopodiaceae are highly adapted to saline soils and are particularly well-represented in saline ecosystems. Significant adverse impacts to coastal tidal wetland ecosystems in the western US include urban development (wetland infilling), alteration of tidal flow, sedimentation, pollution, and invasive plant species (Palaima, 2012; Zedler, 2000). Significant adverse impacts to desert playas and saline wetland ecosystems include urban, road, energy (solar), and water development; mining; livestock grazing; off-highway vehicle recreation; and invasive plant species.

Saline endemic plants and conservation areas are presented in Table 5. US Fish and Wildlife Service Recovery Plans for federally listed plant species include that for the endangered and threatened species of Ash Meadows, Nevada (USFWS, 1990a), salt marsh bird's-beak [*Chloropyron maritimum* subsp. *maritimum* (Orobanchaceae); USFWS, 1985b], and tidal marsh ecosystems of northern and central California [*Chloropyron molle* subsp. *molle* and *Suaeda californica* (Chenopodiaceae); USFWS, 2014]. There is currently no Recovery Plan for *Atriplex coronata* var. *notatior* (Chenopodiaceae).

Coastal tidal wetland restoration and revegetation in the western US has been previously reviewed in detail by Boyer & Thornton (2012), Josselyn & Bucholz (1984), and Zedler (1996; 2000). Physical restoration of coastal tidal wetlands has focused on reestablishment of channels and tidal flow, and floodplain elevation balance with sediment influx and efflux. Following restoration of tidal hydrologic balance, the wetlands are revegetated by natural recruitment and/or planting of native, local saline tolerant plant species. Invasive plant species control with herbicide is conducted for *Spartina* sp. (Poaceae; Roberts & Pullin, 2008; SFEISP, 2014) and *Lepidium latifolium* (Brassicaceae; Boyer & Burdick, 2010; Whitcraft & Grewell, 2012).

Large-scale restoration and revegetation of arid inland basin saline vegetation has focused on Owens Dry Lake, Inyo County, California, where salt tolerant *Distichlis spicata* was planted from container stock over large areas of the playa to reduce airborne dust pollution (Dickey et al., 2005a; b; LADWP, 2013). In addition to extremely saline substrate conditions, revegetation of playas is challenging due to B toxicity, high substrate pH (>10), and intense soil surface aridity coupled with a shallow, anoxic water table that severely restricts rooting depth (Breen, 2005; Breen & Richards, 2008; Dahlgren et al., 1997; James et al., 2005). Playa revegetation success is improved by substrate tillage to roughen the surface (seed capture); N, P and K fertilization; drip irrigation to leach salt and B and to increase surface water availability; and planting local, saline tolerant plant species (Breen, 2005; Breen & Richards, 2008; Dickey et al., 2005a; b; James et al., 2005). Due to high rates of N volatilization from high substrate pH, polymer-coated, slow-release, nitrate-based fertilizer is used rather than uncoated ammonium-based fertilizer. Additional inland saline substrate restoration and revegetation techniques in the western US are described by Beauchamp et al. (2009), Blank & Young (2004), Cooper & Wolf (2007), and Weber & Hanks (2006).

Guano Flora Conservation and Restoration

Guano substrates are created from the accumulation of bird excrement on oceanic coastal rocks and pinnacles. The substrate is saline, acidic, and N-rich (Rajakaruna, 2004; Rajakaruna et al., 2009). The only known strict guano substrate endemic in the western US is *Lasthenia maritima* (Asteraceae; Ornduff, 1965; Vasey, 1985). The species only grows on guano substrate on widely scattered coastal rocks and small islands in close proximity to the coasts of California, Oregon, and Washington, as far north as Vancouver Island in British Columbia, Canada. The largest population of the species is on the Farrallon Islands, San Francisco County, California (Coulter, 1971; Vasey, 1985). Greatest threats to *L. maritima* are from invasive plant species and possibly sea level rise due to climate change.

Guano substrate endemic plants and conservation areas are presented in Table 6. Restoration of *Lasthenia maritima* habitat on the Farallon Islands has included manual removal of the invasive plant species *Tetragonia tetragonioides* (Aizoaceae), *Malva* sp. (Malvaceae), and annual grasses (USFWS, 2005). No revegetation research is known to have been conducted for guano soils in the western US.

Hydrothermally-Altered Flora Conservation and Restoration

Hydrothermally-altered substrates occur at active or extinct hydrothermal steam vents and hot springs. Active hydrothermal activity subjects both the substrate and air locally around the vent to elevated temperatures (thermal stress; Pavlik & Enberg, 2001a; b; Stout & Al-Niemi, 2002; Stout et al., 1997). Hydrothermal fluids alter the surrounding parent rock with precipitates of elemental S, metal sulfide minerals, and Si sinter (Billings, 1950; Blecker et al., 2012; Salisbury, 1954; 1964). Upon exposure to the atmosphere, microbially-mediated oxidation of S and sulfide minerals results in the generation of sulfuric acid and substrate acidification (pH <5) which exacerbates plant macronutrient deficiency and toxicity of heavy metals (Billings, 1950; Salisbury, 1954; 1964; Tordoff et al., 2000; Williamson et al., 1982). This is the same process that results in the chemically extreme condition of metal sulfide ore mine spoils. Significant adverse impacts to active hydrothermal areas are from geothermal energy development. Major impacts to hydrothermally-altered rocks occur as the result of elemental S and metal sulfide ore mining. Other significant impacts to hydrothermally-altered ecosystems include urban and road development, mining, and off-highway vehicle recreation.

Hydrothermally-altered substrate endemic plants and conservation areas are presented in Table 7. USFWS Recovery Plans for federally listed hydrothermal endemic plant species include that for *Eriogonum ovalifolium* var. *williamsae* (USFWS, 1995c). Although substantial substrate chemistry and ecological research has been conducted on hydrothermally-altered rock plant communities in Nevada (Billings, 1950; 1992; DeLucia et al., 1989; Schlesinger et al., 1989) and Utah (Salisbury, 1954; 1964), and extensive revegetation research and literature review has been conducted for metal sulfide ore mine spoils (Bradshaw, 1997; Hudson, 1998; Schaller, 2000; Tordoff et al., 2000; Williamson et al., 1982; Williams & Schuman, 1987), no restoration research is known to have been conducted for active hydrothermal vent or hydrothermally-altered rock ecosystems in the western US. In the US, abandoned mine lands that cause air, water, and soil pollution are treated as environmental hazards with many designated as United States Environmental Protection Agency superfund sites. Major S and metal sulfide mine revegetation projects have included the Sulphur Bank Mercury Mine (Lake County; Heeraman et al., 2001; USEPA, 2014a) and the Leviathan Mine superfund sites (Alpine County; LRWQCB, 2014; USEPA, 2014b) in California. Both projects used physical stabilization, substrate amendment with lime to increase pH, and organic matter to increase fertility, and then seeding with plant species tolerant of the hydrothermally-altered substrate.

Abandoned metal ore mine lands represent perhaps one of the least appreciated ecosystems in the western US, generally being viewed as an unsightly environmental hazard rather than unique, chemically extreme ecosystems in their own right. Although some abandoned mine lands are deserving of their environmental hazard designation as a major source of air and water pollution, they are also grand studies of biological adaptation and evolution. For example, the annual plant *Mimulus cupriphilus* (Phrymaceae) evolved from its presumed progenitor *M. guttatus* within a period of 150 years in response to the challenge of growing on acidic, Cu-Zn mine spoil at the McNulty Mine in Calaveras County, California (Macnair, 1989).

The Río Tinto district in southern Spain has been mined for the past 5000 years and is one of the largest metal sulfide ore mining areas in the world, with 20 km^2 of mine pits and spoil piles (Asensi et al., 2011). *Erica andevalensis* (Ericaceae) is a local endemic, primary succession shrub adapted to the acidic and heavy metal rich mine spoils. Despite its appearance of a barren, chemical wasteland and the river (Río) of acid mine drainage ($pH < 3$) the color of red wine (Tinto) that originates from it and flows 70 km before reaching the Atlantic Ocean, the Río Tinto mining district is designated a Protected Natural Landscape and has been proposed as an International Geopark to UNESCO (Asensi et al., 2011). Perhaps in 5000 years, with the continued evolution of mine spoil endemic plant species and a change in public perception, some of the abandoned mine lands of the western US will also gain the status of unique, chemically extreme ecosystems worthy of conservation.

Conclusion

Chemically extreme ecosystems are limited in their extent, support rare, local edaphic endemic plant species, and are highly vulnerable to human impacts. Although numerous conservation areas are established for chemically extreme ecosystems and their edaphic endemic plant species on federal and state lands throughout the western US, there are few legal protections for these unique ecosystems and their plant species on state lands in Washington, Idaho, Montana, Wyoming, Colorado, and Utah or on private lands anywhere in the western US. The lack of coordination for which species are included on the federal and state endangered species lists, as well as the lack of a single, standardized ranking system used by all of the nonprofit conservation organizations to consistently rank species for conservation priority, are substantial challenges in the protection of rare edaphic endemic plant species.

Much progress has been made towards ecologically appropriate methods for the restoration and revegetation of serpentine, gabbro, and saline ecosystems in the western US, but there has been almost no study of methods for the restoration or revegetation of carbonate substrates of the arid southwestern US. Given the extensive area and distribution of carbonate substrates in the southwestern US, their high proportion of associated strict, local carbonate endemic plant species, and their vulnerability to mining, further research is needed for ecologically appropriate restoration and revegetation methods for carbonate abandoned mine lands and other disturbed carbonate landscapes.

With ever increasing levels of resource extraction and a burgeoning human population, diligent environmental activism is crucial for the protection of chemically extreme ecosystems under current laws and to promote the establishment of future laws to protect them in the other western states. Continued research, development, and application of ecologically appropriate methods for the restoration and revegetation of chemically extreme ecosystems will continue to ensure the sustainability of chemically extreme ecosystems and persistence of the rare, endemic plant species that they support.

REFERENCES

Alexander, E. B. (1993) Gabbro and its soils. *Fremontia 21*, 8-10.

Alexander, E. B. (2011) Gabbro soils and plant distributions on them. *Madroño 58*, 113–122.

Alexander, E. B., Coleman, R. G., Keeler-Wolf, T. & Harrison, S. P. (2007) *Serpentine geoecology of western North America.* New York, NY: Oxford University Press.

Arizona Rare Plant Field Guide (2014) *Arizona rare plant field guide.* Online: http://aznps.com/rareplants.php. Accessed April 13, 2014.

Asensi, A., Aguiar, C., Sánchez-Mata, D. & Monteiro-Henriques, T. (2011) *Flora and vegetation of Iberian ultramafics. Excursion guide*. Coimbra, Portugal: Universidade de Coimbra.

Ballesteros, M., Canadas, E. M., Foronda, A., Fernandez-Ondono, E., Penas, J. & Lorite, J. (2012) Vegetation recovery of gypsum quarries: short-term sowing response to different soil treatments. *Applied Vegetation Science 15*, 187-197.

Barker, L. (1984) Scott Mountain-China Mountain crest zone. *Fremontia 11*, 13-14.

Beauchamp, V. P., Walz, C. & Shafroth, P. B. (2009) Salinity tolerance and mycorrhizal responsiveness of native xeroriparian plants in semi-arid western USA. *Applied Soil Ecology 43*, 175-184.

Billings, W. D. (1950) Vegetation and plant growth as affected by chemically altered rocks in the western Great Basin. *Ecology 31*, 62-74.

Billings, W. D. (1992) Islands of Sierran plants on the arid slopes of Peavine Mountain. *Mentzelia 6*, 32-39.

Blank, R. R. & Young, J. A. (2004) Revegetation of saline playa margins. In: A. L. Hild, N. L. Shaw, S. E. Meyer, T. D. Booth & E. D. McArthur (Eds.). *Seed and soil dynamics in shrubland ecosystems: Proceedings.* Volume 31 (pp. 37-39). Fort Collins, CO: USDA Forest Service Rocky Mountain Research Station.

Blecker, S. W., Stillings, L. L., Amacher, M. C., Ippolito, J. A. & DeCrappeo, N. M. (2012) Development of vegetation based soil quality indices for mineralized terrane in arid and semi-arid regions. *Ecological Indicators 20*, 65–74.

Borgias, D. (1994) Oregon plants, Oregon places: Rough and Ready Creek. *Kalmiopsis 4*, 1-6.

Boyd, R. (1987) The effects of controlled burning on three rare plants. In: T. S. Elias (Ed.). *Conservation and management of rare and endangered plants* (pp. 513-517). Sacramento, CA: California Native Plant Society Press.

Boyd, R. S. (2007) Response of fire of *Ceanothus roderickii* (Rhamnaceae), a federally endangered California endemic shrub. *Madroño 54*, 13-21.

Boyer, K. E. & Burdick, A. P. (2010) Control of *Lepidium latifolium* (perennial pepperweed) and recovery of native plants in tidal marshes of the San Francisco Estuary. *Wetlands Ecology and Management 18*, 731-743.

Boyer, K. E. & Thornton, W. J. (2012) Natural and restored tidal marsh communities. In: A. Palaima (Ed.). *Ecology, conservation, and restoration of tidal marshes: The San Francisco Estuary* (pp. 233-252). Berkeley, CA: University of California Press.

Bradshaw, A. D. (1997) Restoration after mining for minerals: An ecological review. In: T. Jaffré, R. D. Reeves & T. Becquer (Eds.). *Écologie des milieu sur roches ultramafiques et sur sols métallifères* (pp. 239-248). Nouméa: Documents Scientifiques et Techniques Volume Special III2.

Bradshaw, A. D. & Chadwick, M. J. (1980) *The restoration of land: The ecology and reclamation of derelict and degraded land*. Berkeley, CA: University of California Press.

Breen, A. N. (2005) *Improving seedling establishment on the saline playa at Owens Lake, California through soil amendment, improved seed quality, and increased seed entrapment.* PhD Dissertation. Davis, CA: University of California.

Breen, A. N. & Richards, J. H. (2008) Seedling growth and nutrient content of two desert shrubs growing in amended soil. *Arid Land Research and Management 22*, 46-61.

Brian, N. J. (2004) Hunter Creek. *Kalmiopsis 11*, 36-45.

Brooks, R. R. (1987) *Serpentine and its vegetation: A multidisciplinary approach*. Portland, OR: Dioscorides Press.

Brooks, R. R. (Ed.) (1998) *Plants that hyperaccumulate heavy metals: Their role in phytoremediation, microbiology, archaeology, mineral exploration and phytomining*. Wallingford, Oxon: CAB International.

Burge, D. O. (2013) The Pine Hill region: An island of plant diversity in the Sierra Nevada. *Manzanita 17*, 1-7.

Calflora (2014) *Calflora*. Online: http://www.calflora.org. Accessed April 12, 2014.

Callizo, J. & Clifton, G. (1984) Yolo County. *Fremontia 11*, 16-17.

Callizo, J. & Ruygt, J. (1984) Cedar Roughs. *Fremontia 11*, 17-18.

Carter, S. (2004) Beatty Creek Research Natural Area. *Kalmiopsis 11*, 54-62.

Clemente, A. S., Werner, C., Maguas, C., Cabral, M. S., Martins-Loucao, M. A. & Correia, O. (2004) Restoration of a limestone quarry: Effect of soil amendments on the establishment of native Mediterranean sclerophyllous shrubs. *Restoration Ecology 12*, 20-28.

Cohen-Fernández, A. C. & Naeth, M. A. (2013a) Erosion control blankets, organic amendments and site variability influenced the initial plant community at a limestone quarry in the Canadian Rocky Mountains. *Biogeosciences 10*, 5243-5253.

Cohen-Fernández, A. C. & Naeth, M. A. (2013b) Increasing woody species diversity for sustainable limestone quarry reclamation in Canada. *Sustainability 5*, 1340-1355.

Cohen-Fernández, A. C., Naeth, M. A. & Wilkinson, S. R. (2013) Anthroposol development from limestone quarry substrates. *Canadian Journal of Soil Science 93*, 555-566.

Coleman, R. G. & Kruckeberg, A. R. (1999) Geology and plant life of the Klamath-Siskiyou Mountain region. *Natural Areas Journal 19*, 320-340.

Consortium of California Herbaria (2014) *Consortium of California Herbaria*. Online: http://ucjeps.berkeley.edu/consortium/. Accessed April 12, 2014.

Cooper, D. J. & Wolf, E. C. (2007) *A restoration and monitoring plan for* Puccinellia howellii *Whiskeytown National Recreation Area, California*. Fort Collins, CO: Colorado State University, Department of Forest, Rangeland and Watershed Stewardship.

Coulter, M. (1971) A flora of the Farallon Islands, California. *Madroño 21*, 131-137.

Curtis, C. (2008) Edgewood County Park and Natural Preserve: How it happened. *Fremontia 36*, 3-11.

Dahlgren, R. A., Richards, J. H. & Yu, Z. (1997) Soil and groundwater chemistry and vegetation distribution in a desert playa, Owens Lake, California. *Arid Soil Research and Rehabilitation 11*, 221–244.

DeLucia, E. H., Schlesinger, W. H. & Billings, W. D. (1989) Edaphic limitations to growth and photosynthesis in Sierran and Great Basin vegetation. *Oecologia 78*, 184-190.

Dickey, J., Hall, M., Madison, M., Smesrud, J., Griswold, M., Cotton, Q., Heilmann, M., Roland, G., Jordahl, J., Harasick, R., Bamossy, W., Coles, R., Wheeler, L., Brown, P., Burton, K., Fornelli, R., Anderson, I., Riedel-Lehrke, M., Tiller, R., & Richards, J. (2005a) Stabilizing Owens Dry Lake surface with irrigated saltgrass. Part 1: Existing conditions and the challenges of establishing saltgrass on the playa & Part II: The managed vegetation project. *Ecesis 15 (1),* pp. 1, 6, 7, 9 and *15 (2),* pp. 1, 5, 6, 7, 9.

Dickey, J., Hall, M., Madison, M., Smesrud, J., Griswold, M., Cotton, Q., Heilmann, M., Roland, G., Jordahl, J., Harasick, R., Bamossy, W., Coles, R., Wheeler, L., Brown, P., Burton, K., Fornelli, R., Anderson, I., Riedel-Lehrke, M., Tiller, R. & Richards, J. (2005b) *Managing salt to stabilize the Owens Playa with saltgrass*. In: Proceedings of the International Salinity Forum: Managing Saline Soils and Water: Science, Technology, and Social Issues (pp. 147-150). Riverside, CA.

Ellman, P. (1982) Report from Ring Mountain. *Fremontia 10*, 9-10.

Gentili, R., Sgorbati, S. & Baroni, C. (2011) Plant species patterns and restoration perspectives in the highly disturbed environment of the Carrara marble quarries (Apuan Alps, Italy). *Restoration Ecology 19*, 32-42.

Goforth, D. (1984) Gasquet Mountain. *Fremontia 11*, 11-12.

Gonella, M. P. & Neel, M. C. (1993) Characterizing rare plant habitat for restoration in the San Bernardino National Forest. *Wildland shrub and arid land restoration symposium proceedings*. Gen. Tech. Rep. INT-GTR-315. Ogden, UT: Department of Agriculture, Forest Service, Intermountain Research Station.

Google Earth (2014) *Google Earth version 7.0.2.8415*. Online: http://www.google.com/earth/. Accessed April 12, 2014.

Griffin, J. (1984) San Benito Mountain. *Fremontia 11*, 22-23.

Hall, A. R., du Preez, D. R. & Campbell, E. E. (2003) Recovery of thicket in a revegetated limestone mine. *South African Journal of Botany 69*, 434-445.

Harrison, S. P. (2013) *Plant and animal endemism in California*. Berkeley, CA: University of California Press.

Harrison, S. P. & Rajakaruna, N. (Eds.) (2011) *Serpentine: The evolution and ecology of a model system*. Berkeley, CA: University of California Press.

Heeraman, D. A., Claassen, V. P. & Zasoski, R. J. (2001) Interaction of lime, organic matter and fertilizer on growth and uptake of arsenic and mercury by Zorro fescue (*Vulpia myuros* L.). *Plant and Soil 234*, 215-231.

Hinshaw, G. (2008) *Pine Hill Preserve management plan*. Bureau of Land Management. El Dorado Hills, CA: Bureau of Land Management, Motherlode Field Office.

Hudson, T. (1998) *Remediation of historical mine sites: Technical summaries and bibliography*. Littleton, CO: Society for Mining, Metallurgy, and Exploration.

Hunter, J. C. & Horenstein, J. E. (1992) The vegetation of the Pine Hill area (California) and its relation to substratum. In: A. J. M. Baker, J. Proctor & R. D. Reeves (Eds.). *The vegetation of ultramafic (serpentine) soils* (pp. 197-206). Andover, Hampshire: Intercept.

Invisible Peak. 2014. *The Invisible Peak. The project to restore the West Peak of Mount Tamalpais to its natural state*. Online: http://invisiblepeak.com. Accessed April 1, 2014.

James, J. J., Tiller, R. L. & Richards, J. H. (2005) Multiple resources limit plant growth and function in a saline-alkaline desert community. *Journal of Ecology 93*, 113-126.

Josselyn, M. N. & Bucholz, J. W. (1984) *Marsh restoration in San Francisco Bay: A guide to design and planning*. Technical Report #3. Tiburon, CA: Tiburon Center for Environmental Studies, San Francisco State University.

Kenwood Press (2009) *Obituaries. Phyllis Ellman, 1923-2009*. Kenwood, CA: Kenwood Press.

Khater, C., Martin, A. & Maillet, J. (2003) Spontaneous vegetation dynamics and restoration prospects for limestone quarries in Lebanon. *Applied Vegetation Science 6*, 199-204.

Kiver, E. P. & Harris D. V. (2011) *Geology of US Parklands*. New York, NY: John Wiley and Sons.

Kline, B. (2011) *First along the river: A brief history of the US environmental movement*. Lanham, MD: Rowman and Littlefield Publishers.

Kruckeberg, A. R. (1969) Plant life on serpentinite and other ferromagnesian rocks in northwestern North America. *Syesis 2*, 15-114.

Kruckeberg, A. R. (1984a) *California serpentines: Flora, vegetation, geology, soils, and management problems*. Berkeley, CA: University of California Press.

Kruckeberg, A. R. (1984b) The flora on California's serpentine. *Fremontia 11*, 3-10.

Kruckeberg, A. R. (1986) An essay: The stimulus of unusual geologies for plant speciation. *Systematic Botany 11*, 455-463.

Kruckeberg, A. R. (1991) An essay: Geoedaphics and island biogeography for vascular plants. *Aliso 13*, 225-238.

Kruckeberg, A. R. (2002) *Geology and plant life: The effects of landforms and rock types on plants*. Seattle, WA: University of Washington Press.

Kruckeberg, A. R. (2006) *Introduction to California soils and plants: Serpentine, vernal pools, and other geobotanical wonders*. Berkeley, CA: University of California Press.

Kyriazopoulos, A. P., Abraham, E. M., Parissi, Z. M. & Tsiouvaras, C. N. (2012) Evaluation of different herbaceous species used for revegetation of a limestone quarry with a thin topsoil layer in the Mediterranean region. *Journal of Food, Agriculture and Environment 10*, 1152-1155.

LADWP (2013) *Owens Lake Master Project*. Online: http://www.ladwp.com/docs/OPLADWP050832. Accessed March 25, 2014.

Latimer, H. (1984) Eastern Fresno County. *Fremontia 11*, 29-30.

LRWQCB. 2014. *Lahontan Regional Water Quality Control Board. Leviathan Mine Project.* Online:http://www.waterboards.ca.gov/lahontan//water_issues/projects/leviathan_project/. Accessed March 25, 2014.

Lloyd, R. M. & Mitchell, R. S. (1973) *A flora of the White Mountains, California and Nevada.* Berkeley, CA: University of California Press, Berkeley.

Macnair, M. R. (1989) A new species of *Mimulus* endemic to copper mines in California. *Botanical Journal of the Linnean Society 100*, 1-14.

Marschner, H. (2002) *Mineral nutrition of higher plants*. San Diego, CA: Academic Press.

Martinez, J. M. (2013) *American environmentalism: Philosophy, history, and public policy*. Boca Raton, FL: CRC Press.

Mayall, D. (2008) Protecting Coyote Ridge. *Fremontia 36*, 12-19.

McCormick, S. (1984) Ring Mountain. *Fremontia 11*, 18-19.

McLeod, M. (1984) Cuesta Ridge West. *Fremontia 11*, 24.

Medeiros, J. (1984) The Red Hills. *Fremontia 11*, 28-29.

Mistretta, O. & White, S. D. (2000) Introducing two federally listed carbonate-endemic plants onto a disturbed site in the San Bernardino Mountains, California. *Southwestern rare and endangered plants: Proceedings of the third conference, Flagstaff, AZ.* Proceedings RMRS-P-23 (pp. 20-26). Fort Collins, CO: U.S. Department of Agriculture, Forest Service, Rocky Mountain Research Station.

Moreno-Penaranda, R., Lloret, F. & Alcainz, J. M. (2004) Effects of sewage sludge on plant community composition in restored limestone quarries. *Restoration Ecology 12*, 290-296.

Mosseaux, M. (2004) French Flat. *Kalmiopsis 11*, 46-53.

Mota, J. F., Sola, A. J., Dana, D. E. & Jiménez-Sánchez, M. L. (2003) Plant succession in abandoned gypsum quarries. *Phytocoenologia 33*, 13-28.

Mota, J. F., Sola, A. J., Jiménez-Sánchez, M. L., Pérez-García, F. J. & Merlo, M. E. (2004) Gypsicolous flora, conservation and restoration of quarries in the southeast of the Iberian Peninsula. *Biodiversity and Conservation 13*, 1797-1808.

NatureServe (2014) *NatureServe*. Online: http://www.natureserve.org/. Accessed April 12, 2014.

Naumovich, L., Niederer, C. & Weiss, S. B. (2008) *Serpentine prairie restoration project, Redwood Regional Park*. Menlo Park, CA: Creekside Center for Earth Observation.

Naumovich, L., Niederer, C. & Weiss, S.B. (2009) *Serpentine prairie restoration project, Redwood Regional Park Year 1*. Menlo Park, CA: Creekside Center for Earth Observation.

Naumovich, L., Niederer, C. & Weiss, S.B. (2010) *Serpentine prairie restoration project, Redwood Regional Park Year 2*. Menlo Park, CA: Creekside Center for Earth Observation.

Naumovich, L., Niederer, C. & Weiss, S. B. (2011) *Serpentine prairie restoration project, Redwood Regional Park. Year 3*. Menlo Park, CA: Creekside Center for Earth Observation.

Naumovich, L., Niederer, C. & Weiss, S.B. (2012) *Serpentine prairie restoration project, Redwood Regional Park. Year 4*. Menlo Park, CA: Creekside Center for Earth Observation.

Nevada Rare Plant Atlas (2014) *Nevada rare plant atlas*. Online: http://heritage.nv.gov/atlas/. Accessed April 12, 2014.

New Mexico Rare Plants (2014) *New Mexico rare plants*. Online: http://nmrareplants.unm.edu/rarelist.php. Accessed April 12, 2014.

Oberhauer, T. A. (1993) Soils and plants of limited distribution in the Peninsular Ranges. *Fremontia 21*, 3-8.

O'Dell, R. E. & Claassen, V. P. (2009) Serpentine revegetation: A review. *Northeastern Naturalist 16*, 253-271.

O'Dell, R. E. & Claassen, V. P. (2011) Restoration and revegetation of harsh soils. In: S. Harrison & N. Rajakaruna (Eds.). *Serpentine: The evolution and ecology of a model system* (pp. 383-413). Berkeley, CA: University of California Press.

Oregon Flora Atlas (2014) *Oregon flora atlas*. Online: http://www.oregonflora.org/atlas.php. Accessed April 12, 2014.

Ornduff, R. (1965) Ornithocoprophilous endemism in Pacific Basin angiosperms. *Ecology 46*, 864-867.

Ortiz, O., Ojeda, G., Espelta, J. M. & Alcaniz, J. M. (2012) Improving substrate fertility to enhance growth and reproductive ability of a *Pinus halepensis* Mill. afforestation in a restored limestone quarry. *New Forests 43*, 365-381.

Palaima, A. (2012) *Ecology, conservation, and restoration of tidal marshes: The San Francisco Estuary*. Berkeley, CA: University of California Press.

Pavlik, B. M. & Enberg, A. (2001a) Developing an ecosystem perspective from experimental monitoring programs: I. Demographic responses of a rare geothermal grass to soil temperature. *Environmental Management 28*, 225–242.

Pavlik, B. M & Enberg, A. (2001b) Developing an ecosystem perspective from experimental monitoring programs: II. Ecophysiological responses of a rare geothermal grass to soil water. *Environmental Management 28*, 243–253.

Rajakaruna, N. (2004) The edaphic factor in the origin of plant species. *International Geology Review 46*, 471-478.

Rajakaruna, N., Pope, N., Perez-Orozco, J. & Harris, T. B. (2009) Ornithocoprophilous plants of Mount Desert Rock, a remote bird-nesting island in the Gulf of Maine, USA. *Rhodora 111*, 417-448.

Roberts, P. D. & Pullin, A. S. (2008) The effectiveness of management interventions for the control of *Spartina* species: A systematic review and meta-analysis. *Aquatic Conservation-Marine and Freshwater Ecosystems 18*, 592-618.

Robson, M., Spotts, R., Wade, R. & Erickson, W. (2009) *A case history: Limestone quarry reclamation using fluvial geomorphic design techniques*. American Society of Mining and Reclamation Conference Proceedings. Online: http://www.asmr.us/ Publications/ Conference%20Proceedings/2009/1166-Robson-CO.pdf. Accessed March 31, 2014.

Rufo, L. & de la Fuente, V. (2010) Successional dynamics of the climatophile vegetation of the mining territory of the Rio Tinto basin (Huelva, Spain): Soil characteristics and implications for phytoremediation. *Arid Land Research and Management 24*, 301-327.

Safford, H. D. (2011) Serpentine endemism in the California flora. *Fremontia 38/39*, 32-39.

Safford, H. D., Viers, J. H. & Harrison, S. P. (2005) Serpentine endemism in the California flora: A database of serpentine affinity. *Madroño 52*, 222-257.

Salisbury, F. B. (1954) Some chemical and biological investigations of materials derived from hydrothermally altered rock in Utah. *Soil Science 78*, 277-294.

Salisbury, F. B. (1964) Soil formation and vegetation on hydrothermally altered rock material in Utah. *Ecology 45*, 1-9.

Sawyer, J. (1984) The Lassics. *Fremontia 11*, 15-16.

Schaller, F. (2000) *Reclamation of drastically disturbed lands*. Madison, WI: American Society of Agronomy-Crop Science Society of America-Soil Science Society of America.

Schlesinger, W. H., DeLucia, E. H. & Billings, W. D. (1989) Nutrient-use efficiency of woody plants on contrasting soils in the western Great Basin, Nevada. *Ecology 70*, 105-113.

SER (2004) *SER international primer on ecological restoration*. Online: http://www.ser.org/resources/resources-detail-view/ser-international-primer-on-ecological-restoration. Accessed on March 28, 2014.

SFEISP (2014) *San Francisco Estuary invasive* Spartina *project*. Online: http://www.spartina.org. Accessed March 26, 2014.

Sommers, S. (1984) Edgewood Park. *Fremontia 11*, 19-20.

Stein, B. A., Kutner, L. S. & Adams, J. S. (2000) *Precious heritage: The status of biodiversity in the United States*. Oxford, NY: Oxford University Press.

Stout, R. G. & Al-Niemi, T. S. (2002) Heat-tolerant flowering plants of active geothermal areas in Yellowstone National Park. *Annals of Botany 90*, 259-267.

Stout, R. G., Summers, M. L., Kerstetter, T. & McDermott, T. R. (1997) Heat- and acid-tolerance of a grass commonly found in geothermal areas within Yellowstone National Park. *Plant Science 130*, 1-9.

Tordoff, G. M., Baker, A. J. M. & Willis, A. J. (2000) Current approaches to the revegetation and reclamation of metalliferous mine wastes. *Chemosphere 41*, 219-228.

Tropek, R., Kadlec, T., Karesova, P., Spitzer, L., Kocarek, P., Malenovsky, I., Banar, P., Tuf, I. H., Hejda, M. & Konvicka, M. (2010) Spontaneous succession in limestone quarries as an effective restoration tool for endangered arthropods and plants. *Journal of Applied Ecology 47*, 139-147.

USEPA (2011) *Five year review report for Atlas Asbestos mine superfund site and Coalinga Asbestos mine superfund site, Fresno County, California*. Online: http://yosemite.epa.gov/r9/sfund/r9sfdocw.nsf/3dc283e6c5d6056f88257426007417a2/c34c1b6f902defdb88257922006c744e!OpenDocument. Accessed March 25, 2014.

USEPA (2014a) *United States Environmental Protection Agency. Sulphur Bank Mercury Mine*. Online: http://yosemite.epa.gov/r9/sfund/r9sfdocw.nsf/3dec8ba3252368428825742600743733/08592ed1469e8bd188257007005e9469!OpenDocument. Accessed March 25, 2014.

USEPA (2014b) *United States Environmental Protection Agency. Leviathan Mine*. Online: http://yosemite.epa.gov/r9/sfund/r9sfdocw.nsf/vwsoalphabetic/Leviathan+Mine. Accessed March 25, 2014.

USFWS (1984) *Macdonald's rock-cress recovery plan (*Arabis macdonaldiana *Eastwood).* Portland, OR: United States Fish and Wildlife Service.

USFWS (1985a) *Brady's pincushion cactus (*Pediocactus bradyi*) recovery plan.* Albuquerque, NM: United States Fish and Wildlife Service.

USFWS (1985b) *Recovery plan for salt marsh bird's-beak (*Cordylanthus maritimus *ssp.* maritimus*).* Portland, OR: United States Fish and Wildlife Service.

USFWS (1990a) *Recovery plan for the endangered and threatened species of Ash Meadows, Nevada.* Portland, OR: United States Fish and Wildlife Service.

USFWS (1990b) *Maguire primrose (*Primula maguirei*) recovery plan.* Denver, CO: United States Fish and Wildlife Service.

USFWS (1995a) *Draft Heliotrope milkvetch (*Astragalus montii*) recovery plan.* Denver, CO: United States Fish and Wildlife Service.

USFWS (1995b) *Arizona cliffrose (*Purshia subintegra*) recovery plan.* Phoenix, AZ: United States Fish and Wildlife Service.

USFWS (1995c) *Steamboat buckwheat (*Eriogonum ovalifolium *var.* williamsiae*) recovery plan.* Portland, OR: United States Fish and Wildlife Service.

USFWS (1997) *San Bernardino Mountains carbonate endemic plants recovery plan.* Portland, OR: United States Fish and Wildlife Service.

USFWS (1998) *Recovery plan for serpentine soil species of the San Francisco Bay Area.* Portland, OR: United States Fish and Wildlife Service.

USFWS (2002) *Recovery plan for gabbro soil plants of the Central Sierra Nevada foothills.* Portland, OR: United States Fish and Wildlife Service.

USFWS (2005) *Farallon National Wildlife Refuge. Planning update 2 – Fall 2005. Comprehensive conservation* planning continues. Online: http://www.fws.gov/uploadedFiles/FNWR-update2.pdf. Accessed March 25, 2014.

USFWS (2006a) *Recovery plan for* Camissonia benitensis *(San Benito evening-primrose).* Sacramento, CA: United States Fish and Wildlife Service.

USFWS (2006b) *Recovery plan for* Phlox hirsuta *(Yreka phlox).* Portland, OR: United States Fish and Wildlife Service.

USFWS (2006c) *Sentry milk-vetch* (Astragalus cremnophylax *var.* cremnophylax) *recovery plan.* Albuquerque, NM: United States Fish and Wildlife Service.

USFWS (2014) *Recovery plan for tidal marsh ecosystems of northern and central California.* Sacramento, CA: United States Fish and Wildlife Service.

Utah Rare Plant Guide (2014) *Utah rare plant guide.* Salt Lake City, UT: Utah Native Plant Society. Online: http://www.utahrareplants.org/rpg_species.html#All. Accessed April 13, 2014.

Utah Vascular Plant Atlas (2014) *Digital atlas of the vascular plants of Utah.* Online: http://earth.gis.usu.edu/plants/. Accessed April 14, 2014.

Van Kekerix, L. & Kay, B. L. (1986) *Revegetation of disturbed land in California: An element of mined-land reclamation.* Sacramento, CA: California Department of Conservation.

Vasey, M. (1985) The specific status of *Lasthenia maritima* (Asteraceae), an endemic of seabird-breeding habitats. *Madroño 32*, 131-142.

Weber, D. J. & Hanks, J. (2006) Salt tolerant plants from the Great Basin region of the United States. In: M. A. Khan & D. J. Weber (Eds.). *Ecophysiology of high salinity tolerant plants* (pp. 69-106). Dordrecht: Springer.

Whipple, J. (1984) Mt. Eddy. *Fremontia 11*, 14-15.

Whitcraft, C. R. & Grewell, B. J. (2012) Evaluation of perennial pepperweed (*Lepidium latifolium*) management in a seasonal wetland in the San Francisco Estuary prior to restoration of tidal hydrology. *Restoration Ecology 19*, 32-42.

Williams, R. D. & Schuman, G. E. (1987) *Reclaiming mine soils and overburden in the Western United States: Analytical parameters and procedures*. Ankeny, IA: Soil Conservation Society of America.

Williamson, N. A., Johnson, M. S. & Bradshaw, A. D. (1982) *Mine wastes reclamation*. London: Mining Journal Books.

Wilson, J. L., Ayres, D. R., Steinmaus, S. & Baad M. (2009) Vegetation and flora of a biodiversity hotspot: Pine Hill, El Dorado County, California, USA. *Madroño 56*, 246-278.

Wilson, P. & Norris, D. H. (1989) *Pseudoleskeella* in North America and Europe. *The Bryologist 92*, 387-396.

York, D. (2001) Discovering the endemic plants of Kings River Canyon. *Fremontia 29*, 3-6.

Zedler, J. B. (1996) *Tidal wetland restoration: A scientific perspective and southern California focus*. La Jolla, CA: University of California Press.

Zedler, J. (2000) *Handbook for restoring tidal wetlands*. Boca Raton, FL: CRC Press.

In: Plant Ecology and Evolution in Harsh Environments ISBN: 978-1-63321-955-7
Editors: N. Rajakaruna, R. S. Boyd and T. B. Harris

Chapter 15

PHYTOREMEDIATION AND PHYTOMINING: USING PLANTS TO REMEDIATE CONTAMINATED OR MINERALIZED ENVIRONMENTS

Rufus L. Chaney[1*], Roger D. Reeves[2], Ilya A. Baklanov[3], Tiziana Centofanti[4], C. Leigh Broadhurst[3], Alan J. M. Baker[5], Antony van der Ent[6] and Richard J. Roseberg[7]

[1]USDA-Agricultural Research Service, Crop Systems and Global Change Laboratory, Beltsville, MD, USA
[2]Palmerston North, New Zealand
[3]Department of Civil and Environmental Engineering, University of Maryland, MD, USA
[4]Department of Plant Science; Jordan College of Agricultural Sciences and Technology; California State University-Fresno, CA, USA
[5,6]School of Botany, The University of Melbourne, Victoria, Australia
[6]Centre for Mined Land Rehabilitation, Sustainable Minerals Institute, The University of Queensland, Brisbane, Australia
[7]Crop and Soil Science, Oregon State University, Klamath Falls, OR, USA

ABSTRACT

One type of harsh environment for plants is metal- and metalloid-contaminated or mineralized soils: these exist in most countries due to geological formations or to a history of mining and/or smelting. Depending on soil pH and fertility, metal-rich soils may be barren and eroding into wider areas. Some elements present risk to humans, wildlife, livestock, plants, or soil organisms and require remediation. The engineering approach of removing the contaminated soil is extremely expensive. Thus, alternative methods for *in situ* remediation of element-rich soils have been developed by the agricultural sciences. These methods include phytoextraction (growing plants which accumulate high concentrations of an element in shoots for removal from the field) and phytostabilization (adding soil amendments which convert soil elements into forms with much lower phytoavailability and bioavailability so they no longer pose a risk to the

* Corresponding author: Rufus.Chaney@ars.usda.gov.

environment). Phytomining is a variant of phytoextraction in which the element accumulated in plant shoots has enough value to support farming a hyperaccumulator crop to produce a commercial bio-ore. This chapter reviews these valuable phytotechnologies which have been developed in the last few decades to reduce the costs of alleviating environmental risks of contaminated soils.

INTRODUCTION

Phytoremediation comprises a group of technologies which may be used to reduce risks from elements (metals and metalloids) and biodegradable organic compounds in contaminated or mineralized soils. Phytoremediation includes phytoextraction, the use of plants to remove elements from soils into shoots to decontaminate a soil. If the elements accumulated in plants have no economic value and the biomass, ash or compost of the plant materials must be handled as a hazardous waste and placed into a landfill or recycled, it is simple phytoextraction. Phytomining uses plants to recover soil elements in above-ground biomass which then has value in markets; for example, biomass is ashed and the ash marketed as a high-grade bio-ore. Rather than removing the contaminant, phytostabilization uses plants and soil amendments to convert soil contaminants to forms that are no longer sufficiently bioavailable or phytoavailable that they have adverse effects on plants, animals, or soil organisms. Soil amendments can promote the formation of more strongly adsorbed, precipitated or occluded forms of the contaminants, depending on the chemistry of the specific contaminant and amendments (Chaney et al., 2010; Scheckel et al., 2009). Plant roots can promote formation of less soluble forms of Pb (Cotter-Howells et al., 1999) and plants hold soil in place to prevent erosion, which could cause more extensive contamination. Phytoremediation may also include phytodegradation of soil xenobiotic compounds by plants or rhizosphere microbes (not covered in this review).

Perhaps the first use of soil metal phytostabilization is the work of Bradshaw (1975) and Smith & Bradshaw (1970; 1972) to use minimal soil amendments (fertilizers) to revegetate mine wastes using metal-tolerant ecotypes of native grasses. Gadgil (1969) showed that, by combining application of organic amendments with metal tolerant ecotypes, even better revegetation could be achieved. Baker (1981) reviewed this knowledge and that of Cannon (1960) and Ernst (1974) which communicated to new generations these ideas about metal tolerance and the important distinction between accumulation and exclusion (see Ernst (2006) for a recent review of the evolution of metal-tolerant plants). One difficulty of using metal tolerant grasses in remediation is the need for continuing N fertilization to maintain plant cover. Until recently there were no known legumes with appreciable metal tolerance in acidic Zn contaminated soils. *Anthyllis vulneraria* var. *carpatica* (Pant) Nyman (Fabaceae) has recently been found to tolerate Zn and fix N in Zn/Pb mine wastes in France (Mahieu et al., 2011; Soussou et al., 2013). Even if a metal-tolerant grass could be established with this Zn-tolerant legume, persistence of the plant cover would require additional fertilization (P, K) and would strongly benefit from limestone incorporation if the soil is non-calcareous. N-fixation generates soil acidity over time which increases Zn phytoavailability and could eventually cause phytotoxicity if excess limestone is not included in the phytostabilization practice. The decline in pH and reduction in plant survival and seedling establishment over time is illustrated in forest soils near the Palmerton, PA (USA) Zn smelter (Beyer et al., 2010;

2013). For highly metal contaminated soils it is often necessary to make the soil calcareous to avoid a subsequent drop in pH and loss of vegetative cover, and hence the reversal of the remediation (see Chapter 14).

Inexpensive persistent remediation will usually require the use of soil amendments to alleviate metal toxicity and any nutrient deficiencies to aid plant growth and reduce risks for wildlife and livestock feeding upon the plants. For example, some soil treatment is often needed to convert soil Pb to forms which have lower bioavailability to animals when soil is ingested (see below). For a few elements which are not easily accumulated in plant shoots in phytostabilized soils (*e.g.*, As, F, and Pb), the inadvertent ingestion of soil by wildlife or children is the dominant route for risk (Basta et al., 2005; Chaney, 1983). Chaney (1983) introduced the 'Soil-Plant Barrier' model to summarize the overall patterns of risk from soil elements. Some elements are so insoluble in soil or immobilized in plant roots (*e.g.*, Cr and Pb) that they do not reach food-chain dangerous levels in plant shoots for humans, livestock, or wildlife. Other elements (*e.g.*, As, Cu, F, Mn, and Ni) are sufficiently phytotoxic, and animals sufficiently tolerant, that even plants suffering element phytotoxicity do not cause adverse effects on the most sensitive animal species. Under some soil pH conditions, a small group of elements can be absorbed and translocated by plants from contaminated or mineralized soils to poison livestock, wildlife, or humans (Cd, Mo, and Se).

When an element in the contaminated soil has enough economic value, and plants exist which hyperaccumulate the element, it is possible to establish phytomining to alleviate the environmental risk from such sites and over time to improve the original fertility of the soils to support farming as a byproduct post phytomining. But if the value of the element in biomass is low it cannot pay for the remediation service and companies doing the phytoremedation will need to be paid. In many cases, phytostabilization will be the method of choice due to low costs compared with phytoextraction when the biomass has insufficient economic value. Consideration of the amount of metals in a contaminated soil, and the inability of normal plant species to accumulate high levels of metals without strong phytotoxicity, illustrates why the plants called 'hyperaccumulators' are necessary for functional phytoextraction. Hyperaccumulators accumulate about 100-fold higher metal concentrations than normal plants, and often 1,000-fold higher metal, on mineralized or contaminated soils (Chapter 10; van der Ent et al., 2013).

SOILS WHICH REQUIRE REMEDIATION

Acidic Cu, Ni, Pb, and Zn mine waste and smelter-contaminated soils are often severely phytotoxic and require remediation or revitalization. At many of these locations ecosystems have been destroyed and barren soils are eroded, so rather than 'ecosystem restoration,' we consider 'revitalization' of such sites. In some locations, mine wastes contain pyrite which generates strong soil acidity during its oxidation, causing dissolution of soil Mn and Al minerals to reach phytotoxic levels for crop plants ($pH<5.2$): soil pH may become as low as <3.0 so that combined metal toxicity inhibits all but the most metal-tolerant plant species/ecotypes. Extensive areas of Zn-Pb mine or smelter waste or smelter contaminated soils cause Zn phytotoxicity depending on soil pH and fertility.

Table 1. Estimated removal of Zn and Cd in crop biomass of maize (*Zea mays* L.) at full yield or 50% yield reduction due to Zn phytotoxicity and in biomass of *Noccaea caerulescens*, either the 'Prayon' race, or an improved cultivar derived from southern France populations with higher yield and 10-times higher Cd accumulation

Zn: Presume soil has 2000 ppm Zn=4000 kg Zn (ha•15 cm)$^{-1}$

Crop	Yield t DM ha^{-1}	Zn in Crop			Zn in Ash %
		mg kg^{-1}	kg ha^{-1}	% soil Zn	
Corn (normal)	20	50	1	0.0025	0.5
Corn (Zn toxicity)	10	500	5	0.0125	0.5
Noccaea	2.5	25000	61.2	1.53	40
Noccaea (improved)	5	25000	125	3.12	40

Cd: Presume soil has 20 ppm Cd = 40 kg Cd (ha•15 cm)$^{-1}$

Crop	Yield t DM ha^{-1}	Cd in Crop			Cd in Ash %
		mg kg^{-1}	kg ha^{-1}	% soil Cd	
Corn	20	0.1	0.002	<0.01	0.0002
Corn (Zn toxicity)	10	5	0.05	0.125	0.005
Noccaea ('Prayon')	2.5	200	0.5	1.25	0.4
Noccaea (South France)	2.5	2000	5	12.5	4
Noccaea (Improved)	5	2000	10	25	4

Table 1 shows the concentrations of Zn and Cd in a high biomass forage crop, maize (*Zea mays* L.) compared with the hyperaccumulator *Noccaea* (formerly *Thlaspi*) *caerulescens* (J. & C. Presl) F.K. Meyer (Brassicaceae). Maize forage grown on a normal uncontaminated soil contains about 25 mg Zn kg^{-1} dry matter (DM) and 0.10 mg Cd kg^{-1} DM. If the crop is grown on a contaminated soil with normal ratio of Cd:Zn from smelter or mine waste contamination [about 1 g Cd (200 g Zn)$^{-1}$], it will suffer significant Zn phytotoxicity at about 500 mg Zn kg^{-1} shoot DM (Chaney, 1993), and because of the relationship between Cd and Zn in uptake by maize and most other plant species, only about 5 mg Cd kg^{-1} DM can be reached before Zn phytotoxicity limits Cd phytoextraction. It is evident from Table 1 that a 50% yield-reduced maize crop suffering significant Zn phytotoxicity can only remove a very small amount of Zn and a trivial amount of Cd. No crop plant can do appreciably better in annual removal of Zn or Cd because Zn will reduce yields starting at about 400-500 mg kg^{-1} shoot DM (Chaney, 2010).

The Zn hyperaccumulator *N. caerulescens*, on the other hand, can grow well with up to 25,000 mg Zn kg^{-1} DM; individuals of a southern France race can accumulate 2500 mg Cd kg^{-1} DM from the same soil with normal geogenic Cd:Zn ratio. The original 'Prayon' population from Belgium (studied by many researchers) can accumulate only low levels of Cd compared to southern France plants (Chaney et al., 2000; 2010; Reeves et al., 2001; Schwartz et al., 2006). With the southern France races, annual removal of Cd can be enough to achieve a phytoremediation technology. In addition, lowering soil pH to near 5.5 can significantly increase Cd and Zn accumulation in *N. caerulescens* and hasten phytoextraction of soil Cd (Simmons et al., 2014; Wang et al., 2006; Yanai et al., 2006). Combining improved cultivars of Cd hyperaccumulators with improved agronomy to produce the crop with highest attainable yield of Cd in shoot biomass offers soil Cd remediation at much lower cost than removal and replacement of the surface tillage depth of soil, which is generally considered to cost about US$1 million ha^{-1}. Iwamoto (1999) described engineering remediation of 498 ha of contaminated rice soils in the Jintzu Valley, Toyama, Japan at a cost of about US$2.5 million/ha.

As we have discussed previously, areas of Cd+Zn contaminated rice paddy soils in Asia have caused human Cd disease in at least Japan, China, Thailand, and Vietnam (Chaney et al., 2004; 2007a; 2013). This occurs because rice (*Oryza sativa* L.; Poaceae) is traditionally grown in flooded soils. When fields are drained at flowering, soil Cd can be rapidly converted to phytoavailable forms and soil pH can drop to low levels which favor Cd uptake. Further, polished rice grain is deficient in Ca, Fe, and Zn for human nutrition and these deficiencies cause up to 10-fold higher absorption of Cd by humans. In mammals Cd is mostly absorbed on the DMT1 Fe^{2+} transporter in the duodenum such that deficiency of Fe and Zn strongly increase Cd absorption (see Reeves & Chaney, 2008; Chaney et al., 2013). Recent research has shown that most Cd absorbed by rice roots is transported on NRAMP5 (Ishikawa et al., 2012; Ishimaru et al., 2012; Sasaki et al., 2012), which is a Mn transporter, while in wheat and other crops Cd is accumulated on the Zn transporter such that high Zn strongly inhibits uptake of Cd when the normal geogenic Cd:Zn ratio occurs in soil (*e.g.*, Hart et al., 2005; McKenna et al., 1992). Radiation mutation and selection of the null mutant of NRAMP5 yielded a very low Cd rice genotype that can legally be grown on contaminated paddy soils. If rice is being grown on aerobic soils to reduce accumulation of inorganic As in grain, it is possible that the disabled NRAMP5 genotypes will suffer Mn deficiency. Separately, over-expression of the HMA3 gene increased pumping of Cd into root cell vacuoles and kept rice grain Cd at low concentrations (Ueno et al., 2010); but GMO cultivars are not allowed yet for rice (note that selection of ineffective HMA3 mutants (Murakami et al., 2009; Ueno et al., 2011) allows high Cd transport to rice shoots as discussed below). In addition, subsistence rice farm families consume rice 'home-grown' on their Cd-contaminated soils, leading to extreme Cd exposures. Because contaminated rice soils are responsible for essentially all demonstrated human Cd-disease caused by soil Cd, there is a large need for effective Cd phytoextraction technology.

Another crop which accumulates relatively high levels of Cd which can be accumulated by humans is tobacco (*Nicotiana tabacum* L.; Solanaceae). Tobacco accumulates Cd up to about 25 mg kg^{-1} DM in soils with 1 Cd:100 Zn contamination (such as from smelters, mine wastes, or historic biosolids) before Zn phytotoxicity strongly reduces yield. For example, normal high yielding tobacco crops accumulated 17 mg Cd kg^{-1} DM when grown in a field with mine waste contamination in China (Cai et al., 1990), over 11-25 mg kg^{-1} DM in Pb- and Zn-smelter contaminated soils in Bulgaria (Angelova et al., 2004; Chuldjian & Chaney, unpublished) and up to 70 mg kg^{-1} DM when grown on acidic soils treated with Cd rich biosolids (Mulchi et al., 1987). Tobacco contributes as much or more Cd to the kidneys of smokers than all of their dietary crop foods. Some of the tobacco Cd enters the mainstream smoke and is very effectively absorbed in the lung. It is generally estimated that smoking normal cigarettes with 1 mg Cd kg^{-1} DM at one pack per day from age 20 to 50 doubles the Cd concentration in kidney cortex (Elinder et al., 1976). Thus, both rice and tobacco soils mineralized or contaminated with Cd require remediation (or change in crop grown) to protect human health.

Cd-phytoextraction technology has been sought for rice paddy soils by several research groups. Studies in Japan largely confirmed that most crop plants cannot remove enough Cd to achieve useful phytoextraction (Ishikawa et al., 2006). Ishikawa et al. (2006) clearly show that *Brassica juncea* Czern. (Brassicaceae) has no practical value in phytoextraction because it is not a hyperaccumulator and is not tolerant of accumulated metals. Unfortunately, *N. caerulescens* is not adapted to the tropical climate of these fields. Until the study by Simmons et al. (2014), the use of *N. caerulescens* to remove Cd from rice soils had not been successful. Simmons et al. (2014) found that improving soil drainage by ridge planting, acidification, and application of fungicides allowed survival and effective growth and Cd phytoextraction by southern France genotypes growing in a Thai tropical setting. Alternatively, unusual rice cultivars with an ineffective HMA3 gene for storage of Cd in root vacuoles transport enough Cd to shoots that lowering soil pH and growing a high yield of shoots with up to 100 mg Cd kg^{-1} DM offers a valid phytoextraction technology (Murakami et al., 2009). In contrast to major contaminating metals such as Zn and Pb, Cd concentrations are usually relatively low so that acidifying the soil to near pH 5.5 to obtain rapid annual removal of several kg Cd ha^{-1} and then returning the pH to >6.5 might reduce crop Cd to acceptable levels. With the recognition that perhaps 40,000 ha of land in Japan, and likely more than that in China, require Cd remediation to produce rice which meets the CODEX international limit of 0.4 mg Cd kg^{-1} FW, there is renewed interest in commercial Cd phytoextraction. These estimates of rice land in need of Cd-phytoextraction could grow substantially if rice will need to be produced on aerobic soils to reduce grain levels of As (Zhao et al., 2010).

Other soil Cd contamination cases may also benefit from phytoextraction: these include soils with high Cd:Zn ratio from biosolids, soil contaminated by Cd industries, and some Cd-mineralized marine shale-derived soils with geogenic Cd high enough to cause excessive crop Cd. In these cases other plant species may also be useful. With high rates of metal-rich biosolids, soils may be rich in phytoavailable Cu which can limit *N. caerulescens* growth (McLaughlin & Henderson, 1999). Schwartz et al. (2003) studied Cd phytoextraction from field plots in France where high Cd biosolids had been applied and showed that *N. caerulescens* could significantly decrease Cd accumulation by lettuce (*Lactuca sativa* L.; Asteraceae) post-phytoextraction. Broadhurst et al. (2014) tested a maize "inbred" with unusually strong Cd accumulation which grew well on high metal biosolids-amended soil and appears to offer Cd phytoextraction capability similar to HMA3 mutant rice genotypes. No other plant species has been shown to provide this capability for Cd rich soils with simultaneous high Cu levels and, as noted above, most species claimed to be Cd hyperaccumulators based on spiked soil or nutrient solution tests with Cd salt addition without Zn are not of any use in practical phytoextraction of contaminated soils (van der Ent et al., 2013).

Another approach suggested by some is growth of bioenergy crops such as willow (*Salix* spp.) or maize with comparatively high Cd accumulation and high yield ability compared to most crop plants, but not nearly a Cd hyperaccumulator (*e.g.*, Thewys et al., 2010; Witters et al., 2012). If the bioenergy crop paid for the Cd phytoextraction practice over a long period and the ash or other residue is placed in landfills, it might be a cost-effective alternative for phytoextraction (Thewys et al., 2010). This group conducted "Life Cycle Assessment" to estimate the time required for Cd removal to allow production of vegetable crops.

Several *Sedum* species with true Cd hyperaccumulator ability (*Sedum alfredii* Hance, *Sedum plumbizincicola* X.H. Guo et S.B. Zhou ex L.H. Wu, and *Sedum jinianum* X.H. Guo; Crassulaceae) have been identified in China (Deng et al., 2007; Liu et al., 2011; Wu et al., 2013; Xu et al., 2009; Yang et al., 2004), but none exhibit the extreme Cd accumulation of southern France *N. caerulescens*. These species are taller than *N. caerulescens* and appear to offer higher harvestable annual yields. Several other natural Cd hyperaccumulators which accumulated over 500 mg Cd kg^{-1} DM in the field have been found at a tropical Zn-Cd mine site in Thailand (Phaenark et al., 2009). Study of these species has been limited because they were reproduced only by cuttings. One of the species (*Gynura pseudochina* (L.) DC.; Asteraceae) found was 0.4-1 m tall and accumulated 458 mg Cd kg^{-1} and 6.17 g Zn kg^{-1} DW on a soil with 184 mg Cd kg^{-1} and 16.7 g Zn kg^{-1}, and grew well in the rainy season (this had limited the yield of *N. caerulescens*). Thus the shoot biomass contained a much higher Cd:Zn ratio than the soil, similar to southern France *N. caerulescens*. Khaokaew et al. (2014) confirmed the value of this species for Cd phytoextraction in contaminated rice fields. A tropical fruit tree, carambola (*Averrhoa carambola* L.; Oxalidaceae) or star fruit, which accumulates relatively high levels of Cd has been identified by Li et al. (2010; 2011). Use would require high planting density and rapid planting because the seeds have a very short lifetime after harvest of the fruit. Because of its tropical adaptation, research is continuing to develop this species for practical Cd phytoextraction. Additional searching for natural strong Cd accumulators for tropical soils is needed.

Many scientists have ignored the fundamental definition of hyperaccumulators: the accumulation of an element above some limit for a plant growing in soils where the species occurs naturally (van der Ent et al., 2013); and the usual 100-200 times higher soil Zn than Cd in geological Zn+Cd enrichment. If Zn kills crops with about 500 mg Zn kg^{-1} DM, and a species accumulates Cd and Zn at about the ratio in soil, the plant will reach no higher than 5 mg Cd kg^{-1} DM. Growing plants in Cd-salt spiked soils or nutrient solutions with Cd addition does not test their utility for phytoextraction, and over 10 species have been claimed to be hyperaccumulators of Cd based on this false definition.

Other Phytoextraction Technologies to Protect Human Health

Excessive soil Se has long harmed livestock and wildlife, and human Se toxicity has been observed in China from food crops (Yang et al., 1983). Phytoextraction of soil Se from such soils is also needed to protect the safety of irrigation drainage waters which may harm wildlife (Bañuelos et al., 1997). It is also possible that increased Se in foods could contribute to improved human health, and Se-rich crops might be sold as 'nutraceuticals' (Bañuelos & Dhillon, 2011) or used to replace mined Se salts in livestock feeds (Bañuelos & Mayland, 2000). Interestingly, the ability of the natural Se hyperaccumulators to accumulate high levels of Se in the presence of high levels of soil sulfate is a critical part of the Se hyperaccumulator characteristic (Bell et al., 1992). In the phytoextraction model of Bañuelos & Dhillon (2011), use of several relatively high Se-accumulating crop plants can be a safe practice because sulfate limits Se accumulation to levels which will not be harmful in foods, but gives significant removals from a mineralized or contaminated soil to limit Se risk from irrigation drainage waters.

Soil As can be high from both mineralization and contamination. Most plants accumulate only low levels of As (Zhao et al., 2010), but the fern *Pteris vittata* L. (Pteridaceae) was found to accumulate high levels of As on slightly contaminated soils (Ma et al., 2001). Recent recognition that rice grown in flooded soils commonly accumulated high levels of possibly carcinogenic inorganic As in grain and stover has raised questions about the safety of rice and rice products. Growing rice in aerobic soils strongly reduces grain As but lowers yield, so extensive effort will be required to breed genotypes of rice adapted to aerobic soils which give both high grain yields and quality along with low grain levels of inorganic As and Cd. *Pteris vittata* is a tropical species which can be grown in aerobic rice soils. Several fern species are being tested by research groups to learn if phytoextraction can remove enough As to allow production of rice grain with lower As levels, but no successful field demonstration has yet been reported.

Induced Phytoextraction

Another proposed phytotechnology is 'induced phytoextraction,' in which chelating agents are applied to soils to dissolve soil metals and aid their uptake by plants (*e.g.*, Blaylock et al., 1997). As discussed previously by Chaney et al. (2010), addition of chelating agents to promote plant uptake of soil metals is neither cost effective nor environmentally acceptable. Nowack et al. (2006) provides a thorough review of the environmental risks of using chelating agents to induce phytoextraction: ultimately, more metals are leached than are absorbed by plants. We obtained information to make a new estimate of the cost of applying EDTA for induced phytoextraction. We assume 10 mmol Na_2EDTA kg^{-1} soil and that the EDTA is purchased in truckload (20 t) quantities. The price of technical grade $Na_2EDTA \cdot 2H_2O$ (FW 372 g mol^{-1}) (US\$3.16 kg^{-1} in 2014) was obtained from a major international manufacturer. Assuming 15 cm depth of soil Pb contamination with $2 \cdot 10^6$ kg soil ha^{-1}, one application of Na_2EDTA at 10 mmol kg^{-1} soil costs US\$23,500 ha^{-1}. Induced phytoextraction with EDTA was never a good idea and has not been permitted for over 10 years in the US or the EU.

For gold, application of cyanide or thiocyanate to soils can promote plant uptake, but this cannot be done in open environments, only on leaching pads with plastic liners to collect and treat any leachate (Anderson et al., 2005). A similar result has been reported for Hg contaminated soils where application of thiosulfate may allow significant phytoextraction of soil Hg (Pedron et al., 2013). Although Hg phytovolatilization has been developed (Heaton et al., 1998), emission of soil Hg to the global pool of atmospheric Hg has not become a favored technology.

Phytomining

The issues and concepts involved in phytomining are presented below, with focus on Ni. Ni is the element for which phytomining appears most feasible because of the widespread occurrence and extent of Ni-rich ultramafic soils and mine wastes, the wide variety of Ni-accumulating plants, and the ready market for Ni metal, Ni salts and Ni fertilizers. However, the process is much more complex than simply finding a suitable tract of land and growing hyperaccumulator plants. The central rationale for Ni phytomining is that the Ni concentration of currently mined ore materials is typically 0.8-2.5% Ni, whereas the Ni concentration of certain hyperaccumulator plants is 1-3% Ni in dry leaf tissue (or 8-25% in plant ash). These plants grow on soils with 0.05-0.8% Ni, which would be sub-economic for mining, but the plant ash constitutes an ore material an order of magnitude richer than mined ores. Further, the plant ash is low in Fe and Mn oxides and Mg silicate which make recovery of Ni from lateritic ores complicated and expensive.

A question may be asked about the long-term sustainability of the phytomining process. Over an area with Ni averaging 2500 mg/kg to 30 cm rooting depth, the total Ni present is about 10 t Ni (ha-30 cm)$^{-1}$ (Table 2). A single crop of a hyperaccumulator plant with dry weight 10 t ha^{-1} and 2% Ni yields 200 kg Ni ha^{-1}, which is 2% of this resource. Thus phytomining of the area should be sustainable on a 50-year time scale if the soil/subsoil is turned over periodically and if natural buffering processes replenish the plant-available Ni on the time scale of the phytoextraction process. It is likely that for the same site and pH, Ni concentration in phytomining crop shoots will decline over time as the readily phytoavailable pool is depleted. This condition is site-specific. We infer that *Alyssum* (Brassicaceae) hyperaccumulators also obtain Ni from subsurface soils in serpentine soils because Ni accumulation is higher when these species are grown on serpentine soils than when they are grown on Ni-refinery contaminated soils with similar total Ni concentration where the Ni is limited to the tillage depth (Chaney, unpublished).

The following are the critical steps to consider in establishing a phytomining operation: (1) selection of a suitable land area; (2) selection of suitable plant species; (3) planting technology; (4) harvesting strategies; (5) post-harvest treatment of biomass and marketing; and (6) the economics of the whole process. Some of these factors have been discussed by various authors (Angle et al., 2001; Brooks & Robinson, 1998; Chaney et al., 2000; 2007a; 2010; Li et al., 2003a; b). Unfortunately, only some of the world's ultramafic areas are suitable for arable-style cropping of Ni hyperaccumulator plants, as many exposures have steep and/or rocky topography, and even some of the flatter land may be too stony for mechanical cultivation. Some adjacent colluvial soils with strong Ni enrichment could also be phytomined. Extensive preliminary soil analysis is required to establish the extent of the resource because usual soil survey reports showing soils derived from serpentine parent rocks have little relationship with soil Ni levels or phytoavailability, and adjacent Ni-rich colluvial soils are not recognized as potential Ni resources. The rainfall regime needs to be considered because the profitability of the process is reduced by any need for irrigation. Land ownership factors may add complexity to the arrangements, and long-term commitments of owners as farmers or as lessors of the land need to be established.

Selection of suitable plant species is also not straightforward. The most obvious criteria are the maximum (and typical) Ni concentrations so far found in the plant in its natural environment, and the annual yield of biomass. Experimental work can establish the possibility of enhancement of total uptake through appropriate fertilization. However, the suitability of plant species to the phytomined environment must be taken into account: this includes not only the need to ensure that the species is appropriate to the climatic conditions, but also the need for physical protection of the plants, protection against pests and diseases, and biosecurity issues. In this last respect, we note that many countries (particularly those not sharing a land border with another country) are increasingly strict about the introduction of new species, and even within a country there may be concern about the effects of transferring a species from one region to another. In view of the fact that many ultramafic areas host rare and/or endemic species, the maintenance of existing biodiversity in the face of the possible spread of the phytomining crop species becomes a serious concern.

Table 2. Estimated Ni phytoextraction by maize (*Zea mays* L.) vs. *Alyssum murale* grown as a phytomining crop; assume control soil contains 25 mg Ni kg^{-1} and the Ni-rich soil contains 2500 mg Ni kg^{-1} = 10,000 kg Ni (ha•30 cm)$^{-1}$; assume soil Ni is sufficiently phytoavailable that corn has 50% yield reduction compared to corn grown on similar soil without Ni mineralization. Research has shown that unimproved *Alyssum murale* can easily yield 10 t ha^{-1} with fertilizers, and selected cultivars can exceed 20 g Ni kg^{-1} DM with appropriate soil and crop management on serpentine soils. Most crop plant species suffer ≥25% yield reduction when the shoots contain 100 mg Ni kg^{-1} dry weight. Ni concentration in ash is limited by formation of $NiCO_3$ with only 49% Ni

Crop	Soil	Yield t DM ha^{-1}	Ni in Crop			Ni in Ash %
			mg kg^{-1}	kg ha^{-1}	% soil Ni	
Maize	Control	20	1	0.02	0.01	0.002
Maize (50% Yield)	Ni-rich	10	100	1	0.01	0.2
Wild *Alyssum* in pasture	Ni-rich	3	10,000	30	0.3	10-15
Wild *Alyssum murale* + fertilizer	Ni-rich	10	20,000	200	2	20-25
Alyssum murale cultivar	Ni-rich	20	25,000	500	5	25-30

For several reasons, therefore, there are strong arguments in favor of using species native to the region (or sterile cultivars of non-natives) for any phytomining venture.

From more than 400 known Ni hyperaccumulators about 150 have shown Ni in leaves at concentrations that can exceed 1%; more than 70 of these are from Mediterranean-climate areas, and at least another 70 are from tropical areas. They include herbs (annual and biennial), shrubs (small and large; short- and long-lived perennial), and trees. In some species the high-Ni tissue is a low proportion of total biomass, rendering them less suitable as phytomining candidates.

The following are a few examples of Ni hyperaccumulator species that seem to have the best potential, some of which have already been the subjects of extensive field and pilot-scale trials. From more than 50 hyperaccumulators in the genus *Alyssum*, several species (e.g. *A. murale* Waldst. & Kit., *A. corsicum* Duby, *A. lesbiacum* (Cand.) Rech.f., and *A. pinifolium* (Nyár.) Dudley, among others, native to Turkey, Greece, and the Balkan region) are short-lived perennials that appear appropriate for areas with a Mediterranean-type climate. Another key property of these species is that their height and growth pattern are amendable to mechanical harvest. A few other Mediterranean members of the Brassicaceae may also be suitable, such as *Leptoplax emarginata* (Boiss.) O.E. Schulz (Greece), *Bornmuellera* species (Greece and Turkey) (e.g., Cai et al., 2005), and the largest of the species of Thlaspi such as *T. jaubertii* Hedge (Turkey). In the Asteraceae, useful biomass is also produced by some hyperaccumulator species in *Centaurea* in Turkey (e.g., *C. ptosimopappa* Hayek, *C. ensiformis* P.H. Davis) (Reeves & Adıgüzel, 2008) and by species of *Berkheya* and *Senecio* in South Africa (Morrey et al., 1992). Detailed discussion of experiments relevant to phytomining with *Berkheya coddii* Roessl. and the Italian *Alyssum bertolonii* Desv. has been given by Brooks & Robinson (1998) and Brooks et al. (2001).

Although a list of tropical Ni hyperaccumulators with >1% Ni has been published (Reeves, 2003), they have generally been less intensively studied for their phytomining potential. In tropical regions the largest resources of Ni hyperaccumulators come from the ultramafic floras of Cuba and New Caledonia, although additional species from areas of Indonesia, Philippines, and Malaysia (Sabah) should also be considered. Examples include large shrub or small tree species such as *Rinorea bengalensis* (Wall.) O.K. (Violaceae) and *Dichapetalum gelonioides* (Bedd.) Engl. subsp. *tuberculatum* Leenh. (Dichapetalaceae) from SE Asia, Geissois (Cunoniaceae) species from New Caledonia, and species from the very large genera *Phyllanthus* (Euphorbiaceae) and *Psychotria* (Rubiaceae; Cuba, New Caledonia, S.E. Asia). In many cases we still have insufficient knowledge about important matters for these species such as rates of biomass production and efficient methods of propagation.

Experimental work is needed to optimize planting technology and to compare seed drilling with the planting out of seedlings or the production of rooted cuttings (for shrub species). Planting densities need to be optimized, e.g., by use of Nelder plots (Angle et al., 2001), to achieve the maximum biomass production. Harvest strategies include the harvest of the total aboveground biomass of annual herbs or short-lived small shrubs, renewable harvest of annual growth of perennial shrubs, coppicing of large shrubs or small trees, or the total harvest of large trees.

Decisions also need to be made regarding the post-harvest treatment of biomass. Options include natural or energy-assisted drying, followed by ashing or chemical digestion. Where ashing is chosen, there is the possibility of energy recovery from the biomass combustion. Ash may be transported to be used as feedstock for a conventional smelter or a stand-alone extraction process may be developed, in which the Ni is extracted from unashed or ashed biomass by chemical and/or electrochemical means. Treating Ni-rich ash as a smelter additive appears to be the simplest way to proceed (Chaney et al., 2007a), as it is unlikely that in any one area an operation producing about 200 kg Ni ha^{-1} yr^{-1} could be conducted on such a scale as to make stand-alone ash-processing equipment economically viable.

Ultimately, the economics of the whole process must take into account: (1) the total development and optimization costs through pilot plant stages to full operation; (2) the annual costs of maintaining the process (labor and machinery costs of propagation and harvesting, fertilizer and irrigation if needed, and plant protection measures); and (3) land lease costs, if any. The final analysis is highly dependent on the rather volatile price of Ni on world markets, and on the net value of alternative uses of the land (e.g., for forestry or for grain crops), which can themselves be subject to considerable price fluctuations.

Wheat (*Triticum aestivum* L.; Poaceae) yields in the USA average about 50 bushels per acre or 3.3 t ha^{-1}; prices for the harvested grain crop since 1998 have ranged from about US$2-9 per bushel, depending on the season and the incidence of natural disasters such as flood, storm, and drought. The returns therefore correspond to about US$224-1008 ha^{-1}. For wheat grown on serpentine soils the yield would be suboptimum and the returns would be at the lower end of this range. Recently Ni has sold for about $16-20 kg^{-1}, such that growing a good crop of *Alyssum* (10 t ha^{-1} dry matter), which at 2% Ni contains about 200 kg Ni, would be valued at US$3200-4000 after processing. This indicates the economic viability of Ni phytomining, at least in the situation where processing costs (transport, ashing, refining) are low; this should be the case if the plant ash is treated as a minor but Ni-rich feedstock for a larger-scale conventional smelter operation.

Chaney et al. (1998; 1999; 2007b) obtained patents (expiring in 2015) for practical Ni phytomining using *Alyssum murale*, *A. corsicum*, and other high Ni accumulating species, and reported that *Alyssum* ash was an excellent ore when processed by electric arc furnace

(Chaney et al., 2007a; 2010). Bani et al. (2014) tested methods for phytomining of Ni in Albanian serpentine soils starting with collecting wild plants, then adding fertilizer to wild plants, then applying both fertilizer and herbicides to wild plants, and finally preparing the soil and planting seeds of several Ni hyperaccumulators with phytoextraction yields over 100 kg ha^{-1}. Li et al. (2003b) reported higher Ni concentrations and Ni yields when growing improved cultivars of *Alyssum* generated in our breeding program.

The group of Morel et al. (Barbaroux et al., 2012) also attempted to produce $Ni(NH_4)_2(SO_4)_2$ as a more valuable commercial Ni chemical from the biomass ash rather than only Ni metal. Possible Ni products which could be made from hyperaccumulator biomass ash besides Ni metal will require additional experimentation (Hunt et al., 2014). A minor use for Ni-rich biomass is as an 'organic' Ni fertilizer in remedying plant Ni deficiency, known to occur in pecan trees (Wood et al., 2005; 2006).

Some predicted that Ni availability to hyperaccumulators would follow the patterns known for crop plants, and that the Ni extractable by usual agricultural extractants such as 1.0 M NH_4-acetate would be related to uptake such that extractions with this reagent could predict the utility of Ni phytomining (Robinson et al., 1999a; b). With this extractant, lowering soil pH increases extractable Ni (Figure 1). However, in our studies of agronomic factors in Alyssum Ni phytomining, we learned that lowering pH increased extractable Ni but actually reduced Ni accumulation in *Alyssum* (Kukier et al., 2004; Li et al., 2003a). Further, it has been clearly shown that *Alyssum* hyperaccumulators extract Ni from the same soil labile Ni pool do as other plant species (Massoura et al., 2004).

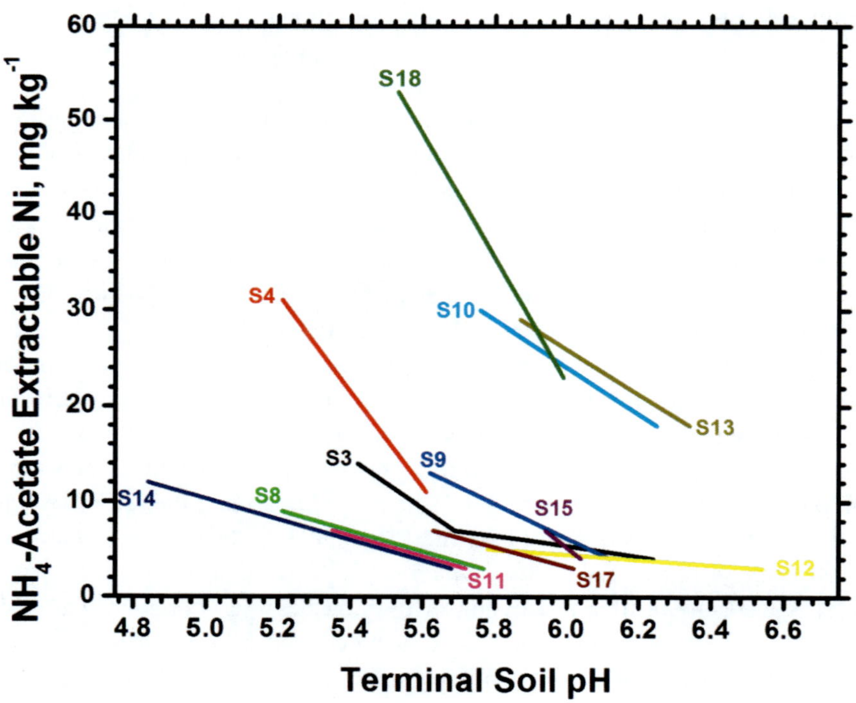

Figure 1. Increase in soil pH decreases 1 M NH_4-acetate extractable Ni in diverse serpentine soils (Chaney et al., unpublished; Li et al., 2003a).

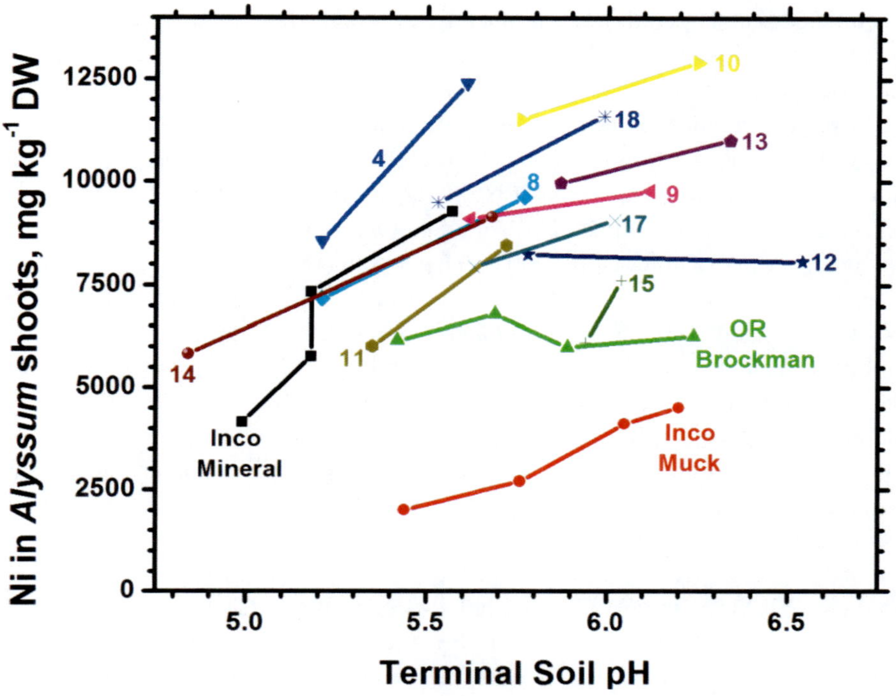

Figure 2. Increasing soil pH increases Ni accumulation by *Alyssum* species grown on serpentine and Ni refinery contaminated soils (Chaney et al., unpublished; Li et al., 2003a).

The pH response on *Alyssum* Ni accumulation is illustrated in Figure 2 which shows that lower pH usually reduces Ni concentration in shoots of *Alyssum* species (a similar pH response was observed for *B. coddii* Ni hyperaccumulation; Chaney et al., unpublished). For soils with very high levels of Fe oxides, raising pH above 6.2 may reduce Ni accumulation because of the stronger binding and occlusion of Ni by Fe oxides at higher pH (Kukier et al., 2004). These data were selected from our study of the effect of soil acidification on Ni hyperaccumulation by *Alyssum* species from 20 Ni-rich smelter contaminated or serpentine soils. Robinson et al. (1999b) reported that for Ni hyperaccumulators, adding chelating agents to the soil actually decreased uptake of Ni in contrast with the use of EDTA to promote Pb uptake noted above.

Several research groups have tested the effect of soil microbes on Ni hyperaccumulation. Abou-Shanab et al. (2003) found that some strains of rhizosphere bacteria could increase yield of *A. murale* shoot Ni even when inoculated into non-sterilized serpentine soil. Cabello-Conejo et al. (2014) reported similar results for *A. pintodasilvae*. Orłowska et al. (2011) reported that various mycorrhiza increased Ni yield in *Berkheya coddii* Dudley grown on a serpentine soil. *Berkheya coddii* was found to have mycorrhiza symbioses in the field by Turnau & Mesjasz-Przybylowicz (2003) which suggests that the relatively low Ni accumulation by this species in studies in soils away from South Africa may have been a result of missing a specific mycorrhizal fungus from *Berkheya*'s native soils. Brassica species do not support mycorrhizae, so the *Alyssum* species reaction with mycorrhizae have not been reported.

It is also conceivable that 'organic' B fertilizers could be produced by phytoextraction on B phytotoxic soils by harvesting B-rich leaves (e.g., Robinson et al., 2007). Anderson et al.

(1999) estimated that Tl could be economically phytomined, and plant species with excellent Tl accumulation and useful yields have been identified (see also LaCoste et al., 1999). No commercial Tl phytoextraction has been reported to date.

Cobalt phytoextraction is theoretically profitable, but no technology has been identified to date. Malik et al. (2000) compared Co accumulation by several species known to accumulate Co. Their work showed that Ni in serpentine soils inhibited Co accumulation by *Alyssum* species, and that optimal Co accumulation occurred at low pH while that of Ni occurred at higher pH (Kukier et al., 2004). Thus one model for Ni and Co phytomining is to maximize Ni removals at higher pH, and then acidify soil to phytomine soil Co. It is also conceivable that land with ^{60}Co contamination could be phytoextracted to remove the radionuclide at far lower cost than removal and hauling of the soil to a radionuclide landfill. The long half-life of ^{60}Co (5.3 yr) does not encourage simply waiting for decay of this radionuclide. Unexpectedly, Tappero et al. (2007) found that although both Co and Ni were accumulated by a root transporter, and pumped into the xylem by another transporter, Co was not transported into epidermal cell vacuoles along with Ni. Cobalt was not accumulated in leaf epidermal cells of *Alyssum* species, but was precipitated outside cells at the end of leaf veins.

As noted in Table 3, the Cu-Co accumulators of central Africa are no longer believed to usefully hyperaccumulate these elements (Faucon et al., 2007), but the plants continue to be excellent bio-indicator or botanical-prospecting plants for Cu-Co deposits (e.g., de Plaen et al., 1982). Other Cu accumulators may reach the newly accepted Cu hyperaccumulator limit of 300 mg kg^{-1} (van der Ent et al., 2013), but they are not useful for phytoextraction of soil Cu because the annual Cu removal would be very small compared to the levels of Cu in contaminated soils needing remediation (Faucon et al., 2009; Kolbas et al., 2014; Peng et al., 2012).

PHYTOSTABILIZATION OF METAL CONTAMINATED SOILS

Zinc-, Cu-, Ni- and Pb-contaminated soils occur in many countries and locations. Because of co-mineralization of Zn-Pb and Zn-Cu ores, Zn-Cd-Pb mixed contamination occurs at literally thousands of locations. If these are alkaline, infertility is the main limitation, although Pb risk through soil ingestion may be important. However, because Zn is usually present at 100-200 times higher concentration than Cd, and Zn in a crop reduces the bioavailability of Cd in a crop, few locations pose Cd risk except through rice and tobacco as discussed above.

Thus phytostabilization of soil Zn and Pb might allow cost-effective remediation of contaminated soils and protect the environment. Some mine wastes or smelter contaminated sites have become extremely acidic from co-occurring pyrite so that high Zn and pH below 4.5 occur together and prevent growth of nearly all plant species (*e.g.,* Beyer et al., 2010; Brown et al., 2003b; Li et al., 2000). It is possible to grow some metal-tolerant ecotypes as discussed above, such as the grasses *Agrostis capillaris* L. (Poaceae) or 'Merlin' red fescue (*Festuca rubra* L.; Poaceae) developed by Smith & Bradshaw (1972), or Ni- and Zn-tolerant *Deschampsia cespitosa* (L.) P. Beauv. (Poaceae; Cox & Hutchinson, 1980; von Frenckell-Insam & Hutchinson, 1993).

Table 3. A selection of plant species which hyperaccumulate elements to over 1% of their shoot dry matter; usually to at least 100-fold levels tolerated by crop species

Element	Plant species	Maximum metal concentration mg kg^{-1} dry wt.	Location collected	Reference
Zn	*Noccaea caerulescens*	39,600	Germany	Reeves & Brooks, 1983a
Cd	*Noccaea caerulescens*	2,910	France	Reeves et al., 2001
Cu[1]	*Aeolanthus biformifolius*	13,700	Zaire	Brooks et al., 1978
Ni	*Phyllanthus serpentinus*	38,100	New Caledonia	Kersten et al., 1979
Co[1]	*Haumaniastrum robertii*	10,200	Zaire	Brooks et al., 1978
Se	*Astragalus racemosus*	14,900	Wyoming, USA	Beath et al., 1937
Mn	*Alyxia rubricaulis*	11,500	New Caledonia	Brooks et al., 1981
As	*Pteris vittata*	22,300	Florida, USA	Ma et al., 2001
Tl	*Biscutella laevigata*	15,200	France	Anderson et al., 1999

[1]Although Cu and Co hyperaccumulation were confirmed in field collected samples, similar concentrations have not been attained in controlled studies and additional research showed that much of the shoot Cu and Co was from soil contamination (Faucon et al., 2007; 2009).

Phytostabilization will usually require incorporation of soil amendments to reduce metal solubility/phytotoxicity (liming), addition of metal sorbents (Fe and Mn oxides), and addition of organic matter with soil microbes (composts or biosolids), along with any other fertilizers needed to reduce soil metal bioavailability or satisfy plant nutrient requirements, and so attain revitalization (Allen et al., 2007; Brown et al., 2005; Chaney et al., 2010; Stuczynski et al., 2007). Soil Pb contamination occurs widely and causes risk to very young children, especially in urban and garden soils, because soil can be carried into homes where very young children might ingest Pb-rich housedusts by hand-to-mouth transfer (Ryan et al., 2004; Scheckel et al., 2013; Zia et al., 2011). The number of children at risk from Pb in mine wastes is trivial compared to those exposed to urban Pb-rich soils from historical automotive exhaust and paint Pb contamination (Scheckel et al., 2013; Zia et al., 2011). Fortunately, addition of phosphate can promote the formation of chloropyromorphite [$Pb_5(PO_4)_3Cl$] which has been shown to have low bioavailability. Ryan et al. (2004) conducted a field test using several P-rich soil amendments to reduce the bioavailability of soil Pb and fed the treated soils to rats, pigs, and humans. The experiment showed that high phosphate application could reduce soil Pb bioavailability to rats, pigs, and humans by nearly 70%. The soil had about 3000 mg Pb kg^{-1}, along with high levels of Zn and corresponding Cd. Making the soil pH neutral prevented any adverse effects of the Zn and Cd, and limited plant accumulation of all three elements (Brown et al., 2004). Further, Scheckel & Ryan (2004) showed that pyromorphite was indeed formed in phosphate treated soils. Besides phosphate-induced Pb inactivation, high Fe biosolids rich in phosphate have also strongly reduced soil Pb bioavailability to animals and Pb bioaccessibility (Brown et al., 2003a; 2004). A comprehensive review of soil Pb risk reduction by formation of chloropyromorphite has recently been published (Scheckel et al., 2013). The most important application of *in situ* Pb inactivation, or soil Pb phytostabilization, is for Pb-rich urban soils. Garden and yard soils are carried into homes on clothing and gardening equipment and become part of the house dust to which young children are exposed. Although interior paint is still the more important source of Pb poisoning of young children, concern about Pb in urban soils has raised the need for extensive phytostabilization of urban soil Pb (Scheckel et al., 2013; Zia et al., 2011).

Phytostabilization of Soil Ni

Smelter emissions and mine wastes in many locations have caused severe Ni phytotoxicity, preventing growth of most plant species. Sulfidic Ni-mine wastes generate strong acidity, which increases Ni^{2+} solubility and phytotoxicity. Large areas with Ni smelter deposition were denuded by the Ni toxicity mixed with SO_2 emissions where Ni deposition occurred in strongly acidic boreal forest areas of Canada (Amiro & Courtin, 1981; Courtin, 1994; Freedman & Hutchinson, 1980; Hutchinson & Whitby, 1977), Russia (Chernenkova & Kuperman, 1999; Helmisaari et al., 1999; Kozlov, 2005; Stjernquist, et al., 1998), and Norway and Finland (Almas et al., 1995). SO_2 caused acute toxicity and killed trees, forest fires often followed, and erosion caused loss of nutrients and organic matter. When SO_2 emissions were controlled to prevent the acute toxicity, soil Ni had accumulated to high enough levels in the very strongly acidic forest soils to prevent regrowth of most species. Highly Ni-tolerant grass ecotypes were selected at these locations (Cox & Hutchinson, 1980) and tree species which coppiced might persist.

Remediation of very strongly acidic, Ni-contaminated soils near smelters has been achieved through phytostabilization. The amount of limestone required depends on soil properties and pH, and dolomitic limestone was more effective than calcitic limestone at Sudbury apparently because both Ca and Mg were leached from the extremely acidic soils after sulfuric acid deposition (Lautenbach, 1987; Winterhalder, 1983). The extensive denuded area surrounding the Sudbury smelters was successfully remediated (phytostabilized) with limestone, fertilizer, and seeding (Gunn et al., 1995; Winterhalder, 1983; 1996). In the project at Sudbury, workers spread soil amendments and seeds on hilly soils to achieve the needed coverage. In general, surface applied limestone can increase pH in deeper soil only over periods of decades due to the slow diffusion of Ca. However, in several case studies surface application of high rates of combinations of organic and alkaline amendments has achieved the needed revitalization of contaminated soils (Brown et al., 2003b; Chaney et al., 2011; Stuczynski et al., 2007; Winterhalder, 1983). It is difficult or impossible to incorporate soil amendments on sloping soils or in forested soils, so leaching of alkalinity (Brown et al., 2003a) into soil profiles is necessary to make phytostabilization successful in such locations. Mixtures of biodegradable organic matter and alkaline materials allowed leaching of alkalinity down the profile. In the case of an asbestos mine waste site in Vermont, USA, the extreme Ca deficiency without Ni phytotoxicity was remediated using surface applied gypsum which yields leachable Ca to correct the extreme Ca deficiency of these serpentinite mine wastes (Chaney et al., 2011).

At another location in Canada, Ni refinery emissions were deposited on neutral to calcareous regional soils near Port Colborne, Ontario. In a small part of this contaminated area, Ni accumulated in acidic muck soils (to 2000 to 4000 mg Ni kg^{-1}) used for vegetable production and caused moderate and then severe Ni phytotoxicity to numerous vegetable crops, but not maize which is much more resistant to soluble soil Ni (Frank et al., 1982; Kukier & Chaney, 2004; McIlveen & Negusanti, 1994). The Ni toxicity of these soils was readily remediated by added limestone; but in the muck soils liming the soil reduced the phytoavailability of soil Mn and induced Mn deficiency, which caused the earlier researchers to believe that liming could not cure the Ni phytotoxicity of these soils. Adding Mn fertilizer with the limestone gave full remediation of these low Mn muck soils (Siebielec et al., 2007).

The volume and area of mine or ore beneficiation tailings at the Sudbury smelters grew over time and remained barren due to infertility and strong soil acidity from sulfide oxidation which increased Ni solubility and phytotoxicity (e.g., Bagatto & Shorthouse, 1999). Fertilization and limestone application achieved revegetation at these sites.

Although laterite and other ultramafic soils can be quite high in Ni, most such sites have limited plant species cover due to very low soil Ca:Mg ratio and low soil P status. These soil conditions cause long-term evolutionary processes which have selected serpentine tolerant species and ecotypes. Because most natural serpentine soils are near neutral pH and contain high levels of Fe oxides, most are not actively Ni phytotoxic (*e.g.*, Zhang et al., 2007). However, if soil pH drops due to soil formation processes or from use of acidifying N-fertilizers, Ni phytotoxicity can also occur in serpentine soils (Anderson et al., 1973; Crooke, 1956; Halstead, 1968; Hunter & Vergnano, 1952). Furthermore, if these soils are fertilized enough, crop plants can be grown. Oat (*Avena sativa* L.; Poaceae) grown on acidic serpentine soils shows Ni phytotoxicity via characteristic banded chlorosis (Hunter & Vergnano, 1952). Liming and addition of any other required nutrients (N, P, K, Mo, and/or B) can prevent Ni phytotoxicity of such Ni-contaminated or -mineralized soils (Kukier & Chaney, 2001; Siebielec et al., 2007) for all plant species (Kukier & Chaney, 2004). The early experiment by Crooke (1956) showed that addition of Na_2CO_3 was as effective as $CaCO_3$ in reversing Ni phytotoxicity in oat growing on acidic colluvial serpentine soil (not deficient in Ca).

Another case of mine waste phytostabilization has been demonstrated for disturbed asbestos mine wastes rich in Ni (>2,000 mg kg^{-1}). These mine wastes are essentially ground serpentinite rock with alkaline pH, intense macronutrient deficiency, and are biologically inert. Chaney et al. (2011) showed that a compost made from livestock manure and yard debris supplied N, P, K, and many other nutrients, plus high rates of gypsum ($CaSO_4 \cdot 2H_2O$) needed to counter the excessive Mg from the $Mg_3Si_2O_5(OH)_4$ mine debris, allowed immediate and persistent revitalization of a site in northern Vermont, USA.

CONCLUSION

Soils with high levels of phytoavailable potentially toxic elements comprise a harsh environment for plants. Evolution has selected several groups of plants on such soils, the excluders which tolerate metals by not absorbing them, and the hyperaccumulators which accumulate high levels of metals. Fuller understanding of metal phytotoxicity in relation to contaminated sites led to development of *in situ* remediation or phytostabilization of soil metals to protect both plants and wildlife. Soil amendments and agronomic considerations allow effective remediation/revitalization of highly contaminated sites to protect the environment. An alternative phytotechnology, phytoextraction, has been developed to remove metals from contaminated soils. Until government enforced soil remediation generates a market for Cd, As, and other phytoextraction methods, these remain in research. However, Ni phytomining can be profitable, growing hyperaccumulator plants which serve as a high grade ore for Ni. Research continues to further develop the technologies and to help us understand how plants achieve such remarkable reactions to soil metals.

REFERENCES

Abou-Shanab, R. A., Angle, J. S., Delorme, T. A., Chaney, R. L., van Berkum, P., Moawad, H., Ghanem, K. & Ghozlan, H. A. (2003) Rhizobacterial effects on nickel extraction from soil and uptake by *Alyssum murale*. *New Phytologist 158*, 219-224.

Allen, H. L., Brown, S. L., Chaney, R. L., Daniels, W. L., Henry, C. L., Neuman, D. R., Rubin, E., Ryan, J. A. & Toffey, W. (2007) *The use of soil amendments for remediation, revitalization and reuse.* Washington, DC: US Environmental Protection Agency, EPA 542-R-07-013.

Almas, A., Singh, B. R. & Sveistrup, T. E. (1995) The impact of the nickel industry in Russia on concentrations of heavy metals in agricultural soils and grass in Sor-Varanger, Norway. *Norwegian Journal Agricultural Science 9*, 61-74.

Amiro, B. D. & Courtin, G. M. (1981) Patterns of vegetation in the vicinity of an industrially disturbed ecosystem, Sudbury, Ontario. *Canadian Journal of Botany 59*, 1623-1639.

Anderson, A. J., Meyer, D. R. & Mayer, F. K. (1973) Heavy metal toxicities: Levels of nickel, cobalt, and chromium in the soil and plants associated with visual symptoms and variation in growth of an oat crop. *Australian Journal of Agricultural Research 24*, 557-571.

Anderson, C. W. N., Brooks, R. R., Chiarucci, A., LaCoste, C. J., Leblanc, M., Robinson, B. H., Simcock, R. & Stewart, R. B. (1999) Phytomining for nickel, thallium and gold. *Journal of Geochemical Exploration 67*, 407-415.

Anderson, C., Moreno, F. & Meech, J. (2005) A field demonstration of gold phytoextraction technology. *Minerals Engineering 18*, 385-392.

Angelova, V., Ivanov, K. & Ivanova, R. (2004) Effect of chemical forms of lead, cadmium, and zinc in polluted soils on their uptake by tobacco. *Journal of Plant Nutrition 27*, 757-773.

Angle, J. S., Chaney, R. L., Baker, A. J. M., Li, Y.-M., Reeves, R., Volk, V., Roseberg, R., Brewer, E., Burke, S. & Nelkin, J. P. (2001) Developing commercial phytoextraction technologies: Practical considerations. *South African Journal of Science 97*, 619-623.

Bagatto, G. & Shorthouse, J. D. (1999) Biotic and abiotic characteristics of ecosystems on acid metalliferous mine tailings near Sudbury, Ontario. *Canadian Journal of Botany 77*, 410-425.

Baker, A. J. M. (1981) Accumulators and excluders -- Strategies in the response of plants to heavy metals. *Journal of Plant Nutrition 3*, 643-654.

Bani, A., Echevarria, G., Sulçe, S. & Morel, J. L. (2014) Improving the agronomy of *Alyssum murale* for extensive phytomining: A five-year field study. *International Journal of Phytoremediation.* In press.

Bañuelos, G. S. & Dhillon, K. S. (2011) Developing a sustainable phytomanagement strategy for excessive selenium in Western United States and India. *International Journal of Phytoremediation 13(S1),* 208-228.

Bañuelos, G. S. & Mayland, H. F. (2000) Absorption and distribution of selenium in animals consuming canola grown for selenium phytoremediation. *Ecotoxicology and Environmental Safety 46*, 322-328.

Bañuelos, G. S., Ajwa, H. A., Mackey, B., Wu, L., Cook, C., Akahoue, S. & Zambruzuski, S. (1997) Evaluation of different plant species used for phytoremediation of high soil selenium. *Journal of Environmental Quality 26*, 637-646.

Barbaroux, R., Plasari, E., Mercier, G., Simonnot, M. O., Morel, J.-L. & Blais, J. F. (2012) New process for nickel ammonium disulfate production from ash of the hyperaccumulating plant *Alyssum murale*. *Science of the Total Environment 423*, 111-119.

Basta, N. T., Ryan, J. A. & Chaney, R. L. (2005) Trace element chemistry in residual-treated soil: Key concepts and metal bioavailability. *Journal of Environmental Quality 34*, 49-63.

Beath, O. A., Eppsom, H. F. & Gilbert, C. S. (1937) Selenium distribution in and seasonal variation of type vegetation occurring on seleniferous soils. *Journal of the American Pharmaceutical Association 26,* 394-405.

Bell, P. F., Parker, D. R. & Page, A. L. (1992) Contrasting selenate-sulfate interactions in selenium-accumulating and nonaccumulating plant species. *Soil Science Society of America Journal 56*, 1818-1824.

Beyer, W. N., Krafft, C., Klassen, S., Green, C. E. & Chaney, R. L. (2010) Relating zinc contamination from smelting to injury to the forest near Palmerton, PA. *Archives of Environmental Contamination and Toxicology 61*, 376-388.

Beyer, W. N., Green, C. E., Beyer, M. & Chaney, R. L. (2013) Phytotoxicity of zinc and manganese to seedlings grown in soil contaminated by zinc smelting. *Environmental Pollution 179*, 167-176.

Blaylock, M. J., Salt, D. E., Dushenkov, S., Zakharova, O., Gussman, C., Kapulnik, Y., Ensley, B. D. & Raskin, I. (1997) Enhanced accumulation of Pb in Indian mustard by soil-applied chelating agents. *Environmental Science and Technology 31*, 860-865.

Bradshaw, A. D. (1975) The evolution of metal tolerance and its significance for vegetation establishment on metal contaminated sites. In: T. C. Hutchinson (Ed.). *Proceedings of the International Conference on Heavy Metals in the Environment 2(II)* (pp. 299-322). Toronto, Canada: CEP Consultants Ltd., Edinburgh.

Broadhurst, C. L., Chaney, R. L., Davis, A. P., Cox, A., Kumar, K., Reeves, R. D. & Green, C. E. (2014) Growth and cadmium phytoextraction by Swiss chard, corn, rice, *Noccaea caerulescens* and *Alyssum murale* in pH adjusted biosolids amended soils. *International Journal of Phytoremediation,* DOI: 10.1080/15226514.2013.828015.

Brooks, R. R. & Robinson, B. H. (1998) The potential use of hyperaccumulators and other plants for phytomining. In: R. R. Brooks (Ed.). *Plants that hyperaccumulate heavy metals* (pp. 327-356). Wallingford & New York: CAB International.

Brooks, R. R., Morrison, R. S., Reeves, R. D. & Malaisse, F. (1978) Copper and cobalt in African species of *Aeolanthus* Mart. (Plectranthinae, Labiatae). *Plant and Soil 50*, 503-507.

Brooks, R. R., Trow, J. M., Veillon, J.-M. & Jaffré, T. (1981) Studies on manganese-accumulating *Alyxia* from New Caledonia. *Taxon 30*, 420-423.

Brooks, R. R., Robinson, B. H., Howes, A. W. & Chiarucci, A. (2001) An evaluation of *Berkheya coddii* Roessler and *Alyssum bertolonii* Desv. for phytoremediation and phytomining of nickel. *South African Journal of Science 97*, 558-560.

Brown, S. L., Chaney, R. L., Hallfrisch, J. G. & Xue, Q. (2003a) Effect of biosolids processing on the bioavailability of lead in urban soils. *Journal of Environmental Quality 32*, 100-108.

Brown, S. L., Henry, C. L., Chaney, R. L., Compton, H. & DeVolder, P. S. (2003b) Using municipal biosolids in combination with other residuals to restore metal contaminated mining areas. *Plant and Soil 249*, 203215.

Brown, S., Chaney, R., Hallfrisch, J., Ryan, J. A. & Berti, W. R. (2004) In situ soil treatments to reduce the phyto- and bioavailability of lead, zinc, and cadmium. *Journal of Environmental Quality 33*, 522-531.

Brown, S., Sprenger, M., Maxemchuk, A. & Compton, H. (2005) Ecosystem function in alluvial tailings after biosolids and lime addition. *Journal of Environmental Quality 34*, 139-148.

Cabello-Conejo, M. I., Becerra-Castro, C., Prieto-Fernández, A., Monterroso, C., Saavedra-Ferro, A., Mench, M. & Kidd, P. S. 2014 Rhizobacterial inoculants can improve nickel phytoextraction by the hyperaccumulator *Alyssum pintodasilvae. Plant and Soil 379*,35-50.

Cai, S., Lin, Y., Zhineng, H., Xianzu, Z., Zhaolu, Y., Huidong, X., Yuanrong, L., Rongdi, J., Wenhau, Z. & Fangyuan, Z. (1990) Cadmium exposure and health effects among residents in an irrigation area with ore dressing wastewater. *Science of the Total Environment 90*, 67-73.

Cannon, H. L. (1960) Botanical prospecting for ore deposits. *Science 132*, 591-598.

Chaney, R. L. (1983) Plant uptake of inorganic waste constituents. In: J. F. Parr, P. B. Marsh & J. M. Kla (Eds.). *Land treatment of hazardous wastes* (pp. 50-76). Park Ridge, NJ: Noyes Data Corp.

Chaney, R. L. (1993) Zinc phytotoxicity. In: A. D. Robson (Ed.). *Zinc in soils and plants* (pp. 135-150). Dordrecht: Kluwer Academic.

Chaney, R. L. (2010) Cadmium and zinc. In: P. Hooda (Ed.). *Trace elements in soils* (pp. 409-439). Oxford: Blackwell.

Chaney, R. L., Angle, J. S., Baker, A. J. M. & Li, Y.-M. (1998) Method for phytomining of nickel, cobalt and other metals from soil. *US Patent No.* 5,711,784.

Chaney, R. L., Angle, J. S., Baker, A. J. M. & Li, Y.-M. (1999) Method for phytomining of nickel, cobalt and other metals from soil. *US Patent No.* 5,944,872.

Chaney, R. L., Li, Y.-M., Angle, J. S., Baker, A. J. M., Reeves, R. D., Brown, S. L., Homer, F. A., Malik, M. & Chin, M. (2000) Improving metal hyperaccumulator wild plants to develop commercial phytoextraction systems: Approaches and progress. In: N. Terry & G. S. Bañuelos (Eds.). *Phytoremediation of contaminated soil and water* (pp. 131-160). Boca Raton, FL: CRC Press.

Chaney, R. L., Reeves, P. G., Ryan, J. A., Simmons, R. W., Welch, R. M. & Angle, J. S. (2004) An improved understanding of soil Cd risk to humans and low cost methods to remediate soil Cd risks. *BioMetals 17*, 549-553.

Chaney, R. L., Angle, J. S., Broadhurst, C. L., Peters, C. A., Tappero, R. V. & Sparks, D. L. (2007a) Improved understanding of hyperaccumulation yields commercial phytoextraction and phytomining technologies. *Journal of Environmental Quality 36,* 1429-1443.

Chaney, R. L., Angle, J. S., Li, Y.-M. & Baker, A. J. M. (2007b) Recovering metals from soil. *US Patent 7,268,273* (Continuation-in-Part of *US Patent No.* 5,711,784).

Chaney, R. L., Broadhurst, C. L. & Centofanti, T. (2010) Phytoremediation of soil trace elements. In: P. Hooda (Ed.). *Trace elements in soils* (pp. 311-352). Oxford: Blackwell.

Chaney, R. L., Newhart, G., Mahoney, M. & Schmeltzer, J. (2011) Ecosystem restoration of the long-barren serpentine mine waste at the Vermont Asbestos Group Mine Site. Abstracts of the 2011 International Phytotechnology Society Conference (Sept. 12-15, 2011. Portland, OR.)

Chaney, R. L., Ryan, J. A. & Reeves, P. G. (2013) Cadmium in soils and its transfer to plants and the human food chain. *Proceedings of the eighth international cadmium conference* (Nov. 10-13, 2011, Kunming, China) (pp. 175-212). Brussels, Belgium: International Cadmium Association.

Chardot, V., Massoura, S. T., Echevarria, G., Reeves, R. D. & Morel, J.-L. (2005) Phytoextraction potential of the nickel hyperaccumulators *Leptoplax emarginata* and *Bornmuellera tymphaea*. *International Journal of Phytoremediation 7*, 323-335.

Chernenkova, T. V. & Kuperman, R. G. (1999) Changes in the spruce forest communities along a heavy metal deposition gradient on Kola Peninsula. *Water Air and Soil Pollution 111*, 187-200.

Cotter-Howells, J. D., Champness, P. E. & Charnock, J. M. (1999) Mineralogy of PbP grains in the roots of *Agrostis capillaris* L. by ATEM and EXAFS. *Mineralogy Magazine 63*, 777-789.

Courtin, G. M. (1994) The last 150 years: A history of environmental degradation in Sudbury. *Science of the Total Environment 148*, 99-102.

Cox, R. M. & Hutchinson, T. C. (1980) Multiple metal tolerance in the grass *Deschampsia cespitosa* (L.) Beauv. from the Sudbury smelting area. *New Phytologist 84*, 631-647.

Crooke, W. M. (1956) Effect of soil reaction on uptake of nickel from a serpentine soil. *Soil Science 81,* 269-276.

de Plaen, G., Malaisse, F. & Brooks, R. R. (1982) The ‘copper flowers’ of Central Africa and their significance for prospecting and archaeology. *Endeavor 6*, 72-77.

Deng, D. M., Shu, W. S., Zhang, J., Zou, H. L., Lin, Z., Ye, Z. H. & Wong, M. H. (2007) Zinc and cadmium accumulation and tolerance in populations of *Sedum alfredii*. *Environmental Pollution 147*, 381-386.

Elinder, C. G., Kjellström, T., Friberg, L., Lind, B. & Linmann, L. (1976) Cadmium in kidney cortex, liver, and pancreas from Swedish autopsies: Estimation of biological half time in kidney cortex, considering calorie intake and smoking habits. *Archives of Environmental Health 31*, 292-302.

Ernst, W. (1974) *Schwermetallvegetation der Erde*. Stuttgart: Fischer.

Ernst, W. H. O. (2006) Evolution of metal tolerance in higher plants. *Forest, Snow and Landscape Research 80*, 251-274.

Faucon, M.-P., Shutcha, M. N. & Meerts, P. (2007) Revisiting copper and cobalt concentrations in supposed hyperaccumulators from SC Africa: Influence of washing and metal concentrations in soil. *Plant and Soil 301*, 29-36.

Faucon, M.-P., Colinet, G., Mahy, G., Ngongo Luhembwe, M., Verbruggen, N. & Meerts, P. (2009) Soil influence on Cu and Co uptake and plant size in the cuprophytes *Crepidorhopalon perennis* and *C. tenuis* (Scrophulariaceae) in SC Africa. *Plant and Soil 317*, 201-212.

Frank, R., Stonefield, K. I., Suda, P. & Potter, J. W. (1982) Impact of nickel contamination on the production of vegetables on an organic soil, Ontario, Canada, 1980-1981. *Science of the Total Environment 26*, 41-65.

Freedman, B. & Hutchinson, T. C. (1980) Long-term effects of smelter pollution at Sudbury, Ontario, on forest community composition. *Canadian Journal of Botany 58*, 2123-2140.

Gadgil, R. L. (1969) Tolerance of heavy metals and the reclamation of industrial waste. *Journal of Applied Ecology 6*, 247-259.

Gunn, J., Keller, W., Negusanti, J., Potvin, R., Beckett, P. & Winterhalder, K. (1995) Ecosystem recovery after emission reductions: Sudbury, Canada. *Water Air and Soil Pollution 85*, 1783-1788.

Halstead, R. L. (1968) Effect of different amendments on yield and composition of oats grown on a soil derived from serpentine material. *Canadian Journal of Soil Science 48*, 301-305.

Hart, J. J., Welch, R. M., Norvell, W. A., Clarke, J. M. & Kochian, L. V. (2005) Zinc effects on cadmium accumulation and partitioning in near-isogenic lines of durum wheat that differ in grain cadmium concentration. *New Phytologist 167*, 391-401.

Heaton, A. C. P., Rugh, C. L., Wang, N. J. & Meagher, R. B. (1998) Phytoremediation of mercury- and methylmercury-polluted soils using genetically engineered plants. *Journal of Soil Contamination 7*, 497-509.

Helmisaari H.-S., Makkonen, K., Olsson, M., Viksna, A. & Mälkönen, E. (1999) Fine-root growth, mortality and heavy metal concentrations in limed and fertilized *Pinus silvestris* (L.) stands in the vicinity of a Cu-Ni smelter in SW Finland. *Plant and Soil 209*, 193-200.

Hunt, A. J., Anderson, C. W. N., Bruce, N., Garcia, A. M., Graedel, T. E., Hodson, M., Meech, J. A., Nassar, N. T., Parker, H. L., Rylott, E. L., Sotiriou, K., Zhang, Q. & Clark, J. H. (2014) Phytoextraction as a tool for green chemistry. *Green Processing and Synthesis 3*, 3-22.

Hunter, J. G. & Vergnano, O. (1952) Nickel toxicity in plants. *Annales of Applied Biology 39*, 279-284.

Hutchinson, T. C. & Whitby, L. M. (1977) The effects of acid rainfall and heavy metal particulates on a boreal forest ecosystem near the Sudbury smelting region of Canada. *Water, Air, and Soil Pollution 7*, 421-438.

Ishimaru, Y., Takahashi, R., Bashir, K., Shimo, H., Senoura, T., Sugimoto, K., Ono, K., Yano, M., Ishikawa, S., Arao, T., Nakanishi, H. & Nishizawa, N. K. (2012) Characterizing the role of rice NRAMP5 in manganese, iron and cadmium transport. *Scientific Reports 2, No. 286*, 1-8.

Ishikawa, S., Ae, N., Murakami, M. & Wagatsuma, T. (2006) Is *Brassica juncea* a suitable plant for phytoremediation of cadmium in soils with moderately low cadmium contamination? – Possibility of using other plant species for Cd-phytoextraction. *Soil Science and Plant Nutrition 52*, 32-42.

Ishikawa, S., Ishimaru, Y., Igura, M., Kuramata, M., Abe, T., Senoura, T., Hase, Y., Arao, T., Nishizawa, N. K. & Nakanishi, H. (2012) Ion-beam irradiation, gene identification, and marker-assisted breeding in the development of low-cadmium rice. *Proceedings of the National Academy of Science (USA) 109*, 19166-19171.

Iwamoto, A. (1999) Restoration of Cd-polluted paddy fields in the Jinzu River Basin – Progress and prospects of the restoration project. In: K. Nogawa, M. Kurachi & M.

Kasuya (Eds.). *Advances in the prevention of environmental cadmium pollution and countermeasures* (pp. 179-183). Kanazawa: Eiko Laboratory.

Kersten, W. J., Brooks, R. R., Reeves, R. D. & Jaffré, T. (1979) Nickel uptake by New Caledonian species of *Phyllanthus*. *Taxon 28*, 529-534.

Khaokaew, S., Nakbanpote, W., Leungprasert, S. & Landrot, G. (2014) Cadmium phytoremediation in a contaminated paddy soil: A field study in Mae Sot District, Thailand. *Abstract AF2800, 18th World Congress of Soil Science*. In press.

Kolbas, A., Marchand, L., Herzig, R., Nehnevajova, E. & Mench, M. (2014) Phenotypic seedling responses of a metal-tolerant mutant line of sunflower growing on a Cu-contaminated soil series: Potential uses for biomonitoring of Cu exposure and phytoremediation. *Plant and Soil 376*, 377-397.

Kozlov, M. V. (2005) Pollution resistance of mountain birch, *Betula pubescens* subsp. *czerepanovii*, near the copper-nickel smelter: Natural selection or phenotypic acclimation? *Chemosphere 59*, 189-197.

Kukier, U. & Chaney, R. L. (2001) Amelioration of Ni phytotoxicity in muck and mineral soils. *Journal of Environmental Quality 30,* 1949-1960.

Kukier, U. & Chaney, R. L. (2004) In situ remediation of Ni-phytotoxicity for different plant species. *Journal of Plant Nutrition 27,* 465-495.

Kukier, U., Peters, C. A., Chaney, R. L., Angle, J. S. & Roseberg, R. J. (2004) The effect of pH on metal accumulation in two *Alyssum* species. *Journal of Environmental Quality 33*, 2090-2102.

LaCoste, C., Robinson, B., Brooks, R., Anderson, C., Chiarucci, A. & Leblanc, M. (1999) The phytoremediation potential of thallium-contaminated soils using *Iberis* and *Biscutella* species. *International Journal of Phytoremediation 1*, 327-338.

Lautenbach, W. E. (1987) The greening of Sudbury. *Journal of Soil and Water Conservation 42*, 228-231.

Li, Y.-M., Chaney, R. L., Siebielec, G. & Kershner, B. A. (2000) Response of four turfgrass cultivars to limestone and biosolids compost amendment of a zinc and cadmium contaminated soil at Palmerton, PA. *Journal of Environmental Quality 29*, 1440-1447.

Li, Y.-M., Chaney, R. L., Brewer, E. P., Angle, J. S. & Nelkin, J. P. (2003a) Phytoextraction of nickel and cobalt by hyperaccumulator *Alyssum* species grown on nickel-contaminated soils. *Environmental Science and Technology 37*, 1463-1468.

Li, Y.-M., Chaney, R. L., Brewer, E., Roseberg, R. J., Angle, J. S., Baker, A. J. M., Reeves, R. D. & Nelkin, J. (2003b) Development of a technology for commercial phytoextraction of nickel: Economic and technical considerations. *Plant and Soil 249*, 107-115.

Li, J. T., Liao, B., Lan, C. Y., Ye, Z. H., Baker, A. J. M. & Shu, W. S. (2010) Cadmium tolerance and accumulation in cultivars of a high-biomass tropical tree (*Averrhoa carambola*) and its potential for phytoextraction. *Journal of Environmental Quality 39*, 1262-1268.

Li, J. T., Liao, B., Zhu, B., Dai, Z. Y., Lan, C. Y. & Shu, W. S. (2011) Characteristics of Cd uptake, translocation and accumulation in a novel Cd-accumulating tree, star fruit (*Averrhoa carambola* L., Oxalidaceae). *Environmental and Experimental Botany 71*, 352-358.

Liu, L., Wu, L., Li, N., Luo, Y., Li, S., Li, Z., Han, C., Jiang, Y. & Christie, P. (2011) Rhizosphere concentrations of zinc and cadmium in a metal contaminated soil after repeated phytoextraction by *Sedum plumbizincicola*. *International Journal of Phytoremediation 13*, 750-764.

Ma, L. Q., Komar, K. M., Tu, C., Zhang, W. H., Cai, T. & Kennelley, E. D. (2001) A fern that hyperaccumulates arsenic. *Nature 409*, 579.

Mahieu, S., Frérot, H., Vidal, C., Galiana, A., Heulin, K., Maure, L., Brunel, B., Lefèbvre, C., Escarré, J. & Cleyet-Marel, J.-C. (2011) *Anthyllis vulneraria/Mesorhizobium metallidurans*, an efficient symbiotic nitrogen fixing association able to grow in mine tailings highly contaminated by Zn, Pb and Cd. *Plant and Soil 342*, 405-417.

Malik, M., Chaney, R. L., Brewer, E. P. & Angle, J. S. (2000) Phytoextraction of soil cobalt using hyperaccumulator plants. *International Journal of Phytoremediation 2*, 319-329.

Massoura, S. T., Echevarria, G., Leclerc-Cessac, E. & Morel, J.-L. (2004) Response of excluder, indicator, and hyperaccumulator plants to nickel availability in soils. *Australian Journal of Soil Research 42*, 933-938.

McIlveen, W. D. & Negusanti, J. J. (1994) Nickel in the terrestrial environment. *Science of the Total Environment 148*, 109-138.

McKenna, I. M., Chaney, R. L. & Williams, F. M. (1992) The effects of cadmium and zinc interactions on the accumulation and tissue distribution of zinc and cadmium in lettuce and spinach. *Environmental Pollution 79*, 113-120.

McLaughlin, M. J. & Henderson, R. (1999) Effect of zinc and copper on cadmium uptake by *Thlaspi caerulescens* and *Cardaminopsis hallerii*. In: W. Wentzel et al. (Eds.). *Extended abstracts from the 5th international conference on the biogeochemistry of trace elements* (July 11-15, 1999, Vienna, Austria) (pp. 886-887). Vienna, Austria: International Society of Trace Element Research.

Morrey, D. R., Balkwill, K., Balkwill, M.-J. & Williamson S. (1992) A review of some studies of the serpentine flora of southern Africa. In: A. J. M. Baker, J. Proctor & R. D. Reeves (Eds.). *The vegetation of ultramafic (serpentine) soils* (pp. 147-157). Andover: Intercept.

Mulchi, C. L., Bell, P. F., Adamu, C. & Chaney, R. L. (1987) Long term availability of metals in sludge amended acid soils. *Journal of Plant Nutrition 10*, 1149-1161.

Murakami, M., Nakagawa, F., Ae, N., Ito, N. & Arao, T. (2009) Phytoextraction by rice capable of accumulating Cd at high levels: Reduction of Cd content of rice grain. *Environmental Science and Technology 43*, 5878-5883.

Nowack, B., Schulin, R. & Robinson, B. H. (2006) Critical assessment of chelant-enhanced metal phytoextraction. *Environmental Science and Technology 40*, 5225-5232.

Orłowska, E., Przybyłowicz, W., Orlowski, D., Turnau, K. & Mesjasz-Przybyłowicz, J. (2011) The effect of mycorrhiza on the growth and elemental composition of Ni-hyperaccumulating plant *Berkheya coddii* Roessler. *Environmental Pollution 159*, 3730-3738.

Pedron, F., Petruzzelli, G., Barbafieri, M. & Tassi, E. (2013) Remediation of a mercury-contaminated industrial soil using bioavailable contaminant stripping. *Pedosphere 23*, 104-110.

Peng, H., Wang-Müller, Q., Witt, T., Malaisse, F. & Küpper, H. (2012) Differences in copper accumulation and copper stress between eight populations of *Haumaniastrum katangense*. *Environmental and Experimental Botany 79*, 58-65.

Phaenark, C., Pokethitiyook, P., Kruatrachue, M. & Ngernsansaruay, C. (2009) Cd and Zn accumulation in plants from the Padaeng zinc mine area. *International Journal of Phytoremediation 11*, 479-495.

Reeves, P. G. & Chaney, R. L. (2008) Bioavailability as an issue in risk assessment and management of food cadmium: A review. *Science of the Total Environment 398*, 13-19.

Reeves, R. D. (2003) Tropical hyperaccumulators of metals and their potential for phytoextraction. *Plant and Soil 249*, 57-65.

Reeves, R. D. & Adıgüzel, N. (2008) The nickel hyperaccumulating plants of Turkey and adjacent areas. *Turkish Journal of Biology 32*, 143-153.

Reeves, R. D. & Brooks, R. R. (1983) Hyperaccumulation of lead and zinc by two metallophytes from mining areas of Central Europe. *Environmental Pollution Series A 31*, 277-285.

Reeves, R. D., Schwartz, C., Morel, J. L. & Edmondson, J. (2001) Distribution and metal-accumulating behavior of *Thlaspi caerulescens* and associated metallophytes in France. *International Journal of Phytoremediation 3*, 145-172.

Robinson, B. H., Brooks, R. R. & Clothier, B. E. (1999a) Soil amendments affecting nickel and cobalt uptake by *Berkheya coddii*: Potential use for phytomining and phytoremediation. *Annals of Botany 84*, 689-694.

Robinson, B. H., Brooks, R. R., Gregg, P. E. H. & Kirkman, J. H. (1999b) The nickel phytoextraction potential of some ultramafic soils as determined by sequential extraction. *Geoderma 87*, 293-304.

Robinson, B. H., Green, S. R., Chancerel, B., Mills, T. M. & Clothier, B. E. (2007) Poplar for the phytomanagement of boron contaminated sites. *Environmental Pollution 150*, 225-233.

Ryan, J. A., Berti, W. R., Brown, S. L., Casteel, S. W., Chaney, R. L., Doolan, M., Grevatt, P., Hallfrisch, J. G., Maddaloni, M. & Mosby, D. (2004) Reducing children's risk from soil lead: Summary of a field experiment. *Environmental Science and Technology 38*, 18A-24A.

Sasaki, A., Yamaji, N., Yokosho, K. & Ma, J. F. (2012) Nramp5 is a major transporter responsible for manganese and cadmium uptake in rice. *The Plant Cell 24*, 2155-2167.

Scheckel, K. G. & Ryan, J. A. (2004). Spectroscopic speciation and quantification of lead in phosphate-amended soils. *Journal of Environmental Quality 33,* 1288-1295.

Scheckel, K. G., Chaney, R. L., Basta, N. T. & Ryan, J. A. (2009) Advances in assessing bioavailability of metal(loid)s in contaminated Soils. *Advances in Agronomy 104*, 1-52.

Scheckel, K. G., Diamond, G. L., Burgess, M. F., Klotzbach, J. M., Maddaloni, M., Miller, B. W., Partridge, B. R. & Serda, S. M. (2013) Amending soils with phosphate as means to mitigate soil lead hazard: A critical review of the state of the science. *Journal of Toxicology and Environmental Health Part B 16*, 337-380.

Schwartz, C., Echevarria, G. & Morel, J.-L. (2003) Phytoextraction of cadmium with *Thlaspi caerulescens*. *Plant and Soil 249,* 27-35.

Schwartz, C., Sirguey, C., Peronny, S., Reeves, R. D., Bourgaud, F. & Morel, J.-L. (2006) Testing of outstanding individuals of *Thlaspi caerulescens* for cadmium phytoextraction. *International Journal of Phytoremediation 8*, 339-357.

Siebielec, G., Chaney, R. L. & Kukier, U. (2007) Liming to remediate Ni contaminated soils with diverse properties and a wide range of Ni concentration. *Plant and Soil 299*, 117-130.

Simmons, R. W., Chaney, R. L., Angle, J. S., Kruatrachue, M., Klinphoklap, S., Reeves, R. D. & Bellamy, P. (2014) Towards practical cadmium phytoextraction with *Thlaspi caerulescens. International Journal of Phytoremediation,* In press. DOI: 10.1080/15226514.2013.876961.

Smith, R. A. H. & Bradshaw, A. D. (1970) Reclamation of toxic metalliferous wastes using tolerant populations of grass. *Nature 227*, 376-377.

Smith, R. A. H. & Bradshaw, A. D. (1972) Stabilization of toxic mine wastes by the use of tolerant plant populations. *Transactions of the Institute of Mining and Metallurgy A81*, 230-238.

Soussou, S., Mahieu, S., Brunel, B., Escarré, J., Lebrun, M., Banni, M., Boussetta, H. & Cleyet-Marel, J.-C. (2013) Zinc accumulation patterns in four *Anthyllis vulneraria* subspecies supplemented with mineral nitrogen or grown in the presence of their symbiotic bacteria. *Plant and Soil 371*, 423-434.

Stjernquist, I., Wihlgard, B., Filiptchouk, A. N. & Strakhov, V. V. (1998) Soil and forest vitality as affected by air pollution on the Kola Peninsula. *Chemosphere 36*, 1119-1124.

Stuczynski, T. I., Siebielec, G., Daniels, W. L., McCarty, G. C. & Chaney, R. L. (2007) Biological aspects of metal waste reclamation with sewage sludge. *Journal of Environmental Quality 36*, 1154-1162.

Tappero, R. V., Peltier, E., Gräfe, M., Heidel, K., Ginder-Vogel, M., Livi, K. J. T., Rivers, M. L., Marcus, M. A., Chaney, R. L. & Sparks, D. L. (2007) Hyperaccumulator *Alyssum murale* relies on a different metal storage mechanism for cobalt than for nickel. *New Phytologist 175*, 641-654.

Thewys, T., Witters, N., van Slycken, S., Ruttens, A., Meers, E., Tack, F. M. G. & Vangronsveld, J. (2010) Economic viability of phytoremediation of a cadmium contaminated agricultural area using energy maize. Part I: Effect on the farmer's income. *International Journal of Phytoremediation 12*, 650-662.

Turnau, K. & Mesjasz-Przybylowicz, J. (2003) Arbuscular mycorrhiza of *Berkheya coddii* and other Ni-hyperaccumulating members of Asteraceae from ultramafic soils in South Africa. *Mycorrhiza 13*, 185-190.

Ueno, D., Yamaji, N., Kono, I., Huang, C. F., Ando, T., Yano, M. & Ma, J. F. (2010) Gene limiting cadmium accumulation in rice. *Proceedings of the National Academy of Sciences (USA) 107*, 16500-16505.

Ueno, D., Koyama, E., Yamaji, N. & Ma, J. F. (2011) Physiological, genetic, and molecular characterization of a high-Cd-accumulating rice cultivar, Jarjan. *Journal of Experimental Botany 62*, 2265-2272.

van der Ent, A., Baker, A. J. M., Reeves, R. D., Pollard, A. J. & Schat, H. (2013) Hyperaccumulators of metal and metalloid trace elements: Facts and fiction. *Plant and Soil 362,* 319-334.

von Frenckell-Insam, B. A. K. & Hutchinson, T. C. (1993) Nickel and zinc tolerance and co-tolerance in populations of *Deschampsia cespitosa* (L.) Beauv. subject to artificial selection. *New Phytologist 125*, 547-553.

Wang, A. S., Angle, J. S., Chaney, R. L., Delorme, T. A. & Reeves, R. D. (2006) Soil pH effects on uptake of Cd and Zn by *Thlaspi caerulescens*. *Plant and Soil 281,* 325-337.

Winterhalder, K. (1983) The use of manual surface seeding, liming and fertilization in the reclamation of acid metal-contaminated land in the Sudbury, Ontario mining and smelting region of Canada. *Environmental Technology Letters 4*, 209-216.

Winterhalder, K. (1996) Environmental degradation and rehabilitation of the landscape around Sudbury, a major mining and smelting area. *Environmental Review 4*, 185-224.

Witters, N., Mendelsohn, R. O., Van Slycken, S., Weyens, N., Schreurs, E., Meers, E., Tack, F., Carleer, F. & Vangronsveld, J. (2012) Phytoremediation, a sustainable remediation technology? Conclusions from a case study. I: Energy production and carbon dioxide abatement. *Biomass and Bioenergy 39*, 454-469.

Wood, B. W., Reilly, C. C. & Nyczepir, A. P. (2005) Mouse-ear of pecan: A nickel deficiency. *HortScience 39*, 1238-1242.

Wood, B. W., Chaney, R. L. & Crawford, M. (2006) Correcting micronutrient deficiency using metal hyper accumulators: *Alyssum* biomass as a natural product for nickel deficiency correction. *HortScience 41*, 1231-1234.

Wu, L. H., Liu, Y. J., Zhou, S. B., Guo, F. G., Bi, D., Guo, X. H., Baker, A. J. M., Smith, J. A. C. & Luo, Y. M. (2013) *Sedum plumbizincicola* X.H. Guo et S.B. Zhou ex L.H. Wu (Crassulaceae): A new species from Zhejiang Province, China. *Plant Systematics and Evolution 299*, 487-498.

Xu, L., Zhou, S., Wu, L., Li, N., Cui, L., Luo, Y. & Christie, P. (2009) Cd and Zn tolerance and accumulation by *Sedum jinianum* in East China. *International Journal of Phytoremediation 11*, 283-295.

Yanai, J., Zhao, F.-J., McGrath, S. P. & Kosaki, T. (2006) Effect of soil characteristics on Cd uptake by the hyperaccumulator *Thlaspi caerulescens*. *Environmental Pollution 139*, 167-175.

Yang, G., Wang, S., Zhou, R. & Sun, S. (1983) Endemic selenium intoxication of humans in China. *American Journal of Clinical Nutrition 37,* 872-881.

Yang, X. E., Long, X. X., Ye, H. B., He, Z. L., Calvert, D. V. & Stoffella, P. J. (2004) Cadmium tolerance and hyperaccumulation in a new Zn-hyperaccumulating plant species (*Sedum alfredii* Hance). *Plant and Soil 25*, 181-189.

Zhang, L., Angle, J. S. & Chaney, R. L. (2007) Do high-nickel leaves shed by the nickel hyperaccumulator *Alyssum murale* inhibit seed germination of competing plants? *New Phytologist 173*, 509-516.

Zhao, F.-J., McGrath, S. P. & Meharg, A. A. (2010) Arsenic as a food chain contaminant: Mechanisms of plant uptake and metabolism and mitigation strategies. *Annual Review of Plant Biology 61*, 535-559.

Zia, M. H., Codling, E. E., Scheckel, K. G. & Chaney, R. L. (2011) In-vitro and In-vivo approaches for the measurement of oral bioavailability of lead (Pb) in soil: A critical review. *Environmental Pollution 159*, 2320-2327.

In: Plant Ecology and Evolution in Harsh Environments ISBN: 978-1-63321-955-7
Editors: N. Rajakaruna, R. S. Boyd and T. B. Harris

Chapter 16

SYNTHESIS AND FUTURE DIRECTIONS: WHAT HAVE HARSH ENVIRONMENTS TAUGHT US ABOUT ECOLOGY, EVOLUTION, CONSERVATION, AND RESTORATION?

Nishanta Rajakaruna[1, 2,*], Robert S. Boyd[3] and Tanner B. Harris[4]

[1]College of the Atlantic, Bar Harbor, ME, USA
[2]Unit for Environmental Sciences and Management, North-West University, Potchefstroom, South Africa
[3]Department of Biological Sciences, Auburn University, Auburn, AL, USA
[4]WRA, Inc., San Rafael, CA, USA

INTRODUCTION

Harsh environments, due to their extreme conditions and unique biota, have piqued human interest over the centuries. Botanists interested in the study of plant diversity are especially drawn to harsh environments because they are frequently characterized by unique plant communities with relatively high proportions of rare and endemic species. Such plant communities, which are often restricted to fragmented islands of habitat, offer exceptional opportunities for exploring biogeographical and ecological theory (Harrison, 2011), including aspects of plant-plant (Davies, 2011; Moore & Elmendorf, 2011) and cross-kingdom (Strauss & Boyd, 2011; Wolf & Thorp, 2011) interactions. Plants found in harsh environments also provide model organisms for the study of adaptation and evolution (O'Dell & Rajakaruna, 2011; Ostevik et al., 2012). Adaptation is a central focus in Darwin's theory of evolution by natural selection, and harsh environments provide prime settings in which to examine the factors and mechanisms driving adaptive evolution (Levin, 2001; Via, 2009). Some of the earliest evidence of rapid evolution in plants has come from studies of plants growing in

* Corresponding author: nrajakaruna@gmail.com.

extreme habitats such as metal-rich mine tailings (Chapters 11, 14; Antonovics et al., 1971; Bradshaw, 1991; O'Dell & Rajakaruna, 2011) and serpentine outcrops (Chapters 6, 11; Kay et al., 2011). Plants found in harsh environments provide unique challenges for conservation (Chapter 14; Thorne et al., 2011) and restoration (Chapters 14, 15; O'Dell & Claassen, 2011) and are especially prone to stressors associated with climate change (Chapters 7, 13; Damschen et al., 2011; 2012). Plants of harsh environments also serve as model organisms for investigating the genetic and physiological bases for tolerance of abiotic stress (Chapters 4-6, 9, 11; Jenks & Hasegawa, 2013; Kantar et al., 2011; Szabados et al., 2011), with important applications in agriculture, restoration, and conservation (Chapters 9, 14, 15). In this chapter, we summarize what harsh environments have taught us about the diversity, ecology, evolution, conservation, and restoration of plants, lichens, and microbes found in harsh environments. Finally, we highlight areas of research needed to expand our understanding of the role that harsh environments may play in generating and maintaining plant and other biotic diversity.

HARSH ENVIRONMENTS AS MODEL SETTINGS FOR STUDIES ON BIODIVERSITY

Alpine summits (Nagy & Grabherr, 2009), polar regions (Thomas et al., 2008), arid deserts (Ward, 2009), remote oceanic islands (Stuessy & Ono, 2007), saline flats (Flowers & Colmer, 2008), acidic bogs (Rydin & Jeglum, 2009), rock outcrops (Anderson et al., 1999), and even wastelands created by human activities such as mining (Walker, 2011), all provide extreme habitats for plants and other biota adapted to harsh abiotic factors. Such habitats are characterized by extremes in temperature, light, water availability, and chemical and physical soil attributes. Adaptation to such environments often leads to population differentiation and subsequent speciation (Kay et al., 2011; O'Dell & Rajakaruna, 2011), thereby generating biodiversity. Harsh environments also often provide a refuge for species which may be at a competitive disadvantage in more benign habitats (Anacker, 2014; Moore & Elmendorf, 2011). Whether through adaptive evolution or exaptation (*i.e.,* ecological filtering), harsh environments often contain a unique assemblage of plants and other biota able to thrive under conditions inhospitable for most other organisms (*i.e.,* extremophiles; Bell, 2012; Pikuta et al., 2007).

Much attention has been focused on microbes as model organisms for the study of the diversity and ecology (Chapters 1, 2; Margesin & Miteva, 2011; Seckbach, 2007; Takai, et al., 2005), physiology and genetics (Chapters 1, 2; Gerday & Glansdorff, 2007), and evolution (Chapters 1, 2; Rampelotto, 2013) of extremophiles. The study of extremophile biology and ecology has shed light not only on other organisms found in extreme environments on Earth, but also has implications for the study of astrobiology (Chapter 1; Cardace & Hoehler, 2011; McCollom & Seewald, 2013). The roles that microorganisms (Chapters 1, 2), as well as terricolous and saxicolous lichens (Chapter 3), play in the ecology of both below- and above-ground habitats via their influence on biogeochemical processes—including weathering, pedogenesis, nutrient cycling, and nutrient acquisition by plants—has also received much attention (Casamayor et al., 2013; Kirchman, 2012; Southworth et al., 2013).

The intimate and inseparable relationship between plants and their substrates results from the need for plants to obtain water and nutrients from the substrate upon which they grow. Thus, it is no surprise that the chemical and physical attributes of the substrate control many aspects of plant diversity, ecology, and evolution. Plants closely associated with harsh substrates have been described as indicators of the minerals and elements found within the substrate, and close observation of such substrate-plant relations has led to biogeochemical prospecting worldwide (Brooks, 1983; Martin & Coughtrey, 1982). Studies of the diversity, physiology, genetics, ecology, and evolution of plants found on extreme substrates have generated much interest in recent years (Jenks & Hasegawa, 2013), particularly the study of plants found on serpentine (Chapters 2, 3, 6, 10, 11, 13-15; Alexander et al., 2007; Brady et al., 2005; Brooks, 1987; Harrison & Rajakaruna, 2011; Roberts & Proctor, 1992), gypsum (Chapters 2, 5, 14; Escudero et al., 2014), dolomite (Chapters 5, 14; Pignatti & Pignatti, 2013), gabbro (Chapter 14; Wilson et al., 2010), metal-rich mine tailings (Chapters 2, 3, 6, 10, 11, 14, 15), and saline soils (Chapters 2, 4, 11, 14). Plants associated with unique geomorphologic features such as mountains (Chapter 7; Clausen, 1951; Körner, 2003; Nagy & Grabherr, 2009) and deserts (Ward, 2009) have also been the subject of much study due to their unique assemblages of plant species possessing adaptations to climatic extremes and other abiotic and biotic stressors. Such plants have also served as models for elucidating mechanisms of convergent evolution, showing how similar functional and phenotypic diversity can be maintained among phylogenetically distinct lineages in response to similar selective pressures (McGhee, 2011; Reich et al., 1997).

Harsh environments have played an important role in generating biodiversity both at the species and community levels, as well as in maintaining biodiversity at varying spatial scales (Yost et al., 2012). Inventories of this biodiversity are still incomplete and continued inventory efforts are needed, particularly given that the current wave of human-mediated extinctions is expected to crest as the sixth major extinction event that has occurred during life's 3+ billion-year history on our planet (Pievani, 2014).

HARSH ENVIRONMENTS AS MODEL SETTINGS FOR STUDIES ON ECOLOGY

American plant ecologist Frank Egler is reputed to have said: "Ecosystems are not only more complex than we think; they are more complex than we can think." Ecological studies are difficult to implement and interpret due to their multivariate nature; multiple factors interact in complex ways making it difficult to determine how specific environmental factors impact organisms. Harsh environments have characteristics that, because they are extreme, amplify their influences on organisms, thereby making it easier to identify their effects. Thus, studies of harsh environments allow us to better understand the importance of specific ecological factors in generating and maintaining taxonomic and functional diversity (Harrison & Rajakaruna, 2011). For example, in California (USA) shrublands dominate in dry areas, but in more benign areas, shrublands are restricted to environments typified by harsh soils (Chapter 8). Competition is a major factor influencing this distribution, limiting shrublands to more stressful habitats under more productive settings. Similarly, serpentine endemism appears to peak under high precipitation (*i.e.*, high productivity) in California (Fernandez-

Going et al., 2013) as well as globally (Anacker, 2011), suggesting that higher competition resulting from greater regional productivity could restrict stress tolerant species to harsh habitats. Recent studies have shown that at the community level, there is greater species, functional, and phylogenetic turnover across serpentine and non-serpentine soil boundaries in California's mesic northwest than in the arid south (Anacker & Harrison, 2012; Fernandez-Going et al., 2013). Recent tests (*e.g.,* Serrano et al., 2014) of this concept (*i.e.*, the Inclusive Niche Hypothesis; Colwell & Fuentes, 1975) have demonstrated the interplay between competition and stress tolerance in determining the distribution of stress-tolerant species. The patchy nature of some harsh environments has allowed the application of island biogeography theory and, more recently, metapopulation and metacommunity theories (Harrison, 2011) and has provided insights into life history traits that may be selected for (or against) in patchy, often harsh, habitats (*e.g.,* Shenk, 2013; Spasojevic et al., 2014).

Adaptation to harsh environments may also involve ecological partners. For example, chapters in this book highlight how specific characteristics of fungi allow them to tolerate harsh conditions, both as the mycobiont in lichens (Chapters 3, 7) and as partners in mycorrhizal interactions (Chapters 2, 5, 7), as well as how their stress tolerance can influence such mutualistic associations. Rodriguez et al. (2008) documented "habitat-adapted symbiosis" in which certain fungal endophytes conferred particular types of stress tolerance (either heat or salt tolerance) on *Leymus mollis* (Poaceae) plants from specific habitats (geothermal or coastal). For example, Rodriguez et al. (2008) found that the fungal endophytes that provided salt tolerance to the coastal grass population conferred salt tolerance on tomato and rice when the endophytes colonized tissues of those hosts. Other symbioses are often not addressed, but likely can be of great importance. As an example, note the possible role of pollinators (*e.g.*, bumblebees) in limiting the northern range extent of legumes (Chapter 7). Antagonists may also play an important role in the distribution of organisms, and stresses conferred from harsh environments may affect the impacts of such antagonists as well. As demonstrated by metal hyperaccumulator plants (Chapter 10), tolerance of a stress (in this case heavy metal stress) can lead to plant adaptations that result in uptake of elevated levels of metals or other generally toxic compounds, which may deter or confer resistance to herbivores or pathogens. Stresses found in harsh environments may magnify the importance of organismal interactions such as mutualism or exploitation, providing opportunities to study the ecological and evolutionary importance of these interactions (Chapter 7; Strauss & Boyd, 2011).

Harsh environments are model settings for closely examining cross-adaptation, which is when a trait that evolved for one function becomes useful as an adaptation for another, and exaptation, which is when non-adaptive traits become adaptive when placed in an alternative context (Barve & Wagner, 2013). Traits that confer tolerance of one type of stress may confer tolerance of other stresses, which may allow species to occupy habitats which would otherwise be inhospitable (Chapters 4, 9, 11). New modeling tools have improved our ability to study exaptation. For example, Barve and Wagner (2013) analyzed bacterial metabolic networks and determined that particular pathways have latent potential as evolutionary innovations (in their study, an evolutionary innovation was defined as a bacterium's potential ability to utilize a particular carbon source for its metabolic reactions). They found that a bacterium able to utilize glucose as a sole carbon source for all its metabolic reactions also may be able to utilize 44 other sole carbon sources. Thus, ability to utilize glucose provided opportunities for a bacterium to utilize other carbon sources for which its metabolism was not adapted. Similar techniques may be applicable to other areas of biology that involve complex networks (*e.g.,* genetic networks or food webs) to reveal potential pathways that are not currently utilized. This approach may reveal the potential for organisms to respond to new environmental stresses: for example, once we have a more complete understanding of how gene networks determine tolerance to particular types of stress (Chapter 9), we may be able to show how the ability to tolerate one stress can pre-adapt an organism to a different stress. Because harsh environments impose strong selection pressures on organisms, they may be good initial subjects for this approach.

One of the major challenges in studying the effects of environmental stress is defining physiological stress and measuring its level in an organism. For example, salinity tolerance is a continuous variable rather than a binary one, and so it is more useful to have a measure that reflects the degree of tolerance to a particular stress rather than reducing tolerance to a binary condition which is either present or absent (Chapter 4). This approach may help to fine-tune our understanding of the effects of stress on plants and other organisms. Connecting specific genes to traits that confer stress tolerance is now possible using model organisms about which we have a profound genetic knowledge (Chapters 9, 11). Ecologists can use these model organisms by comparing the performance of genetic lines with specific genes "turned on" or "off." This single-gene approach is made more complex because stress tolerance often results from multi-gene adaptations (*e.g.*, Agarwal et al., 2013); however, our increasing ability to study genetic networks and to process large datasets will help to deepen our understanding of the genetic basis of adaptation to harsh environments.

An increased understanding of the genetics, physiology, and ecology of harsh environments has many applications in agriculture, conservation, and restoration. The success of both agricultural production and restoration efforts are limited by environmental stresses; alleviating the effects of environmental stress could allow increased success in both of these endeavors. Knowing how stress limits organisms or affects organismal interactions may allow us to manipulate biological systems to favor outcomes we consider to be desirable.

HARSH ENVIRONMENTS AS MODEL SETTINGS FOR STUDIES ON EVOLUTION

Harsh environments provide model settings in which to explore evolutionary questions, as has been highlighted by several chapters in this book. Here we summarize how studies of plants, fungi, and lichens found in harsh environments have contributed to our understanding of broader evolutionary themes which are central to understanding how diversity is generated and maintained. Whether accommodation of harsh environments comes about by ecotypic differentiation or by phenotypic plasticity has long interested ecologists. Turesson (1922) and Clausen et al. (1940) represent two classic works on evolutionary ecology, describing how experimental methods such as common garden and reciprocal transplant studies can be used to examine the roles of phenotypic plasticity and local adaptation in habitat specialization (Wright & Stanton, 2011). These studies, combined with genetic studies, have shown that species found in harsh environments either have genotypes that confer broad tolerance to wide-ranging environmental conditions (Baker, 1965; Parker et al., 2003; Richards et al., 2006) or have genotypes that confer adaptation to specific biotic or abiotic stressors characteristic of the environment in which they occur (Bieger et al., 2014; Leimu & Fischer, 2008; Yost et al., 2012). Although there is much evidence for ecotypic differentiation and species-level endemism among plants found in harsh environments (Chapters 4-6, 9, 11; O'Dell & Rajakaruna, 2011), such specialization is not as common among mycorrhizal fungi (Chapters 2, 7; but see Chapter 5), lichens (Chapters 3, 7; Rajakaruna et al., 2012), or bryophytes (Chapter 12, Briscoe et al., 2009) found in similar environments.

Plants found in extreme environments have often been used to explore the mechanisms that drive the evolution of habitat specialization (Harris & Rajakaruna, 2009). Such studies have revealed that habitat specialists tend to arise via two mechanisms (Chapter 8): neoendemics, arising from nearby, non-specialized relatives via rapid and local speciation (Anacker & Strauss, 2014; Kay et al., 2011; O'Dell & Rajakaruna, 2011), or paleoendemics, resulting from gradual speciation via biotype depletion (Kruckeberg, 1957; Mayer & Baseda, 2010; Mayer et al., 1994). Whether there is directionality in the evolution of habitat specialization (*i.e.,* endemism) has been another avenue of interest that has been addressed by plants growing in extreme environments. The increasing availability of molecular phylogenies has provided us with unparalleled tools to ask questions about the evolutionary dynamics of habitat specialization (Chapters 4, 5; Anacker, 2011). Phylogenetic analyses of serpentine-tolerant plants (*e.g.,* Anacker, 2011; Anacker et al., 2011), halophytes (Chapter 4), and gypsophiles (Chapter 5) have demonstrated that tolerance to serpentine, saline, and gypsum soils, respectively, has been gained numerous times within various groups of angiosperms and has even been lost (although rarely) in some groups. These studies suggest that traits conferring stress tolerance are evolutionarily labile. Differences in the evolutionary dynamics of edaphic tolerance among different lineages may result from the association of stress tolerance traits with other ecophysiological strategies related to environmental stress, with stress tolerance more easily gained in those lineages with certain enabling traits as starting points (Chapters 4, 11); such enabling traits may enhance the capacity of those lineages to evolve other ecophysiological strategies for stress tolerance (see Chapter 4 for a discussion of trait associations for salt tolerance). Research by Anacker et al. (2011) shows that among 23 genera found in California, the direction of serpentine endemism is mostly from non-tolerant and tolerant species to endemic species, with a few reversals from the tolerant or endemic state to the non-tolerant state. Their work suggests that the evolution of edaphic endemism, and possibly other forms of habitat specialization, may represent an evolutionary dead-end. Recent work on *Knautia arvensis* (Caprifoliaceae) suggests that this may not always be the case, as serpentine endemic cytotypes and genotypes are often able to influence surrounding populations by repeated episodes of introgression and polyploidization (Kolář et al., 2012). Similarly, in *Streptanthus* (Brassicaceae), clades with endemism are often quite diverse, with endemism having evolved numerous times (and having been lost at least once). This suggests that endemism is not always an evolutionary dead-end and that endemic species have the potential to radiate further, even within the same habitat (Ivalú Cacho et al., 2014).

Plants found in harsh environments provide ideal models for exploring the parallel (*i.e.,* independent) evolution of traits responsible for adaptation and reproductive isolation (Levin, 2001), as well as parallel speciation (Ostevik et al., 2012; Rajakaruna & Whitton, 2004). Parallel (or repeated) evolution provides strong evidence for the role of ecology as a driver of divergence (Nosil, 2012). Levin (2001) suggests that repeated evolution of traits may not be unusual in taxa that have undergone ecological radiation, particularly in those taxa adapted to environments posing strong divergent selection (*e.g.*, harsh environments). Contrary to these findings, Ostevik et al. (2012) note that there are few studies which show the repeated evolution of traits conferring adaptation and reproductive isolation, citing only 23 potential cases of parallel ecological speciation in plants. However, of the 23 cases they cite, 13 are from plants adapted to harsh edaphic (Brattler et al., 2006a; b; Mayer & Baseda, 2010; Nyberg Berglund et al., 2004; Rajakaruna et al., 2003; Westerbergh & Saura, 1992) or other extreme ecological settings (*e.g.,* dunes, alpine environments, etc.). This suggests that plants found in harsh environments are ideal models for examining the potential for repeated evolution of ecologically driven traits.

The potential evolutionary costs associated with endemism to harsh environments have been another area of recent research. Habitat specialists often appear to be less competitive (Anacker, 2014; Kay et al., 2011) and more susceptible to herbivory when found in "normal" habitats (Kay et al., 2011; Lau et al. 2008; Strauss & Boyd, 2011). To our knowledge, the evolutionary costs of habitat specialization relative to broad tolerance have not been compared using pairs of sister taxa found in such environments. Such information would help to address why some species become endemic to a particular habitat while others maintain genotypes that are able to tolerate wide-ranging habitat conditions.

In this modern era of genomics, we have many tools with which to explore the genetic basis of adaptation to harsh environments (Chapters 4-6, 9, 11; Brattler et al., 2006a; b; Turner et al., 2010; von Wettberg & Wright, 2011), including differences in the nature of adaptation to disparate habitats with similar abiotic stressors (*e.g.*, arid and saline habitats contributing to water stress, saline and serpentine soils contributing to ionic stress, or desert and alpine environments contributing to light and heat stress). The tools at our disposal can reveal the genetic architecture of adaptive traits which confer reproductive isolation (*i.e.,* speciation genes; Chapter 11; Nosil & Schluter, 2011), revealing new insights into the mechanisms by which natural selection can bring about reproductive isolation and speciation. As noted in Chapters 7 and 9, however, our model systems are not representative of all organisms or life history strategies, and we need to continue to add new model systems to allow comparative approaches to explore adaptive genetics.

Two evolutionary principles emerge from the chapters in this volume. One is that traits of organisms can "pre-adapt" them for particular stressors and may promote the evolution of new species or ecotypes in response to local environmental conditions (Chapters 2, 4-6, 9, 11). The other is that investigations of the genetic architecture of adaptation to stressors usually reveal multiple evolutionary origins (Chapters 4, 10). Several chapters (Chapters 4, 5, 9, 10) highlight the usefulness of phylogenetic approaches in explaining the timing and number of adaptive events related to particular stresses. As our phylogenetic understanding continues to be further resolved, these tools will give us improved information about how and when lineages have gained ability to tolerate these stresses. This, in turn, can push our understanding of adaptation deeper into evolutionary time and connect the origins of adaptive traits to other major events (*e.g.*, climatic, tectonic, or astronomic events) that have influenced Earth's biotic and abiotic history.

Whereas much attention has been paid to adaptation and evolution of vascular plants under harsh environmental conditions, it is critical that we continue to expand our knowledge of the diversity and evolution of under-studied groups such as cryptogams (Chapters 3, 12), microbes (Chapters 1, 2; Porter et al., 2011; Springer, 2009), and invertebrates (Chapters 7, 10; Wolf & Thorp, 2011) found in harsh environments, including the roles these organisms may play in the evolution of stress tolerance (Chapter 2) and habitat specialization (Chapters 5, 7) in plants.

HARSH ENVIRONMENTS AS MODEL SETTINGS FOR STUDIES ON CONSERVATION AND RESTORATION

Harsh environments, characterized by extremes in chemical and physical factors, often harbor unique ecotypes, species, and communities (Chapters 2-5, 7, 10, 13, 14). Identifying and cataloging such unique biological elements are first steps toward their conservation (Jacobi et al., 2011; Wulff et al., 2013). Although harsh environments may be difficult for exotic species to colonize, exotics, when established, can create conservation problems in these habitats. Vallano et al. (2012) demonstrated that anthropogenic nitrogen deposition may promote invasion by exotics by increasing the fertility of serpentine soils, and thereby shifting competitive relationships to favor non-native species, particularly non-native annual grasses which have greater nitrogen use efficiency relative to many serpentine-adapted species. Similarly, nitrogen deposition has had drastic effects on native plant diversity in other harsh habitats, especially arid deserts in California (Schneider & Allen, 2012) and nutrient-poor bogs in New England, USA (Gotelli & Ellison, 2002). In addition to invasion by *non-adapted* exotics, sites with harsh soils such as serpentine (Chapter 6), gypsum (Chapter 5), metal-rich mine tailings (Chapter 14), and saline soils (Chapter 4) may be vulnerable to invasion by exotic species which are adapted to similar substrates in other regions of the world. Because they are adapted to extreme substrates within their native range, such species may be better able to colonize extreme habitats elsewhere relative to other non-adapted exotics. As such, adapted exotics may pose a greater threat to extreme habitats. This has caused much concern regarding the use of non-native species adapted to extreme geologies for the purposes of habitat restoration (Chapter 14; Gall & Rajakaruna, 2013; Ma et al., 2013), as well as for phytoremediation and phytomining operations (Chapter 15).

On a broader scale, extreme habitats also pose unique challenges for the management of protected areas (Chapter 14; Gordon et al., 2002; Thorne et al., 2011), including in the development of networks of protected land (*i.e.*, conservation networks) to effectively preserve biodiversity. Conservation networks are critical for protecting species that are restricted to habitat islands such as those formed by serpentine (Chapter 6), gypsum (Chapter 5), and limestone-derived soils (Clements et al., 2006), as well as to protect species found in other spatially isolated or fragmented habitats (Harrison, 2011) such as alpine summits (Chapter 7; Gordon et al., 2002). Although the importance of conservation networks is well known, the development and management of conservation networks requires tremendous coordination among stakeholders and acquisition of lands owned by a range of public and private parties, which may be a limiting factor in their development.

Climate change is a major challenge to the conservation of global biodiversity (*e.g.*, Bellard et al., 2012). The impacts of climate change on biodiversity may be disproportionately detrimental for species with geographically restricted or patchy distributions, such as species restricted to harsh environments (Chapter 13). As the climate changes, habitat for these species may migrate, be significantly altered, or disappear altogether. For example, low islands are vulnerable to sea level rise (Wetzel et al., 2013), and polar regions are particularly susceptible to rising temperatures (Chapter 7; Gormezano & Rockwell, 2013). In alpine settings, upward movement of climate zones may drastically impact species, and in some cases, climate change may cause upper elevation zones to disappear entirely (Chapter 7; Beever et al., 2011). Dispersal ability may be a challenge for species restricted to extreme habitats if climate change causes such habitats to shift to areas where a species cannot disperse, or to areas where it cannot disperse rapidly enough (Chapters 7, 13). Damschen et al. (2012) used species distribution models to examine the risk of extinction for plant species endemic to serpentine outcrops and determined that many factors, including a species' dispersal ability and the geographic distribution of the suitable habitat, can affect this outcome (also see Chapter 7). In addition to affecting habitat distribution, climate change may also affect species interactions, particularly plant-pollinator mutualisms in fragmented habitats (Wolf & Thorp, 2011). Changes in the phenology of plant-pollinator mutualisms may lead to reduced fecundity and a gradual decline in populations. Studies of harsh environments have revealed the potential for rapid evolutionary change in plants on the order of decades to centuries (O'Dell & Rajakaruna, 2011); however, it is unclear how this potential for rapid evolutionary change may affect plant response to climate change. There is some evidence that plant communities on extreme soils change little in response to changes in climate; however, it is not clear how generalized this condition may be (Chapter 13).

Much research has been focused on ecologically appropriate restoration methods for extreme habitats (Chapter 14). Much of our understanding comes from the restoration of saline wetland habitats (*e.g.*, Zedler, 2000), which has resulted from the Clean Water Act and similar wetland protection laws in the United States and elsewhere, as well as from the reclamation of former mines and quarries (*e.g.*, Williams & Schuman, 1987), which has resulted from the Surface Mining Control and Reclamation Act and similar laws in the United States and elsewhere. Restoration efforts have focused on both single-species and ecosystem approaches as well as passive and resource-intensive approaches (Chapter 14). Methods range from low-tech to highly-engineered, including modification of soil physical and chemical properties, restoration of hydrologic regimes, slope engineering, landscape re-contouring, erosion control, and extensive planting (Chapter 14). Long-term monitoring studies and increased reporting of restoration outcomes are expanding our understanding of the restoration of these extreme habitats; however, data remain difficult to access, and disparate monitoring parameters and methods, as well as high variability in site characteristics and land use history, make it challenging to compare results and identify trends among restoration sites. The development of monitoring networks and international standards will make it easier to compare results among restoration sites and will make the data more useful.

Studies on the restoration of extreme habitats have highlighted the need to understand the nature of adaptation or habitat restriction to inform restoration methods (Chapters 11, 13, 14). We are only beginning to understand the role of mycorrhizae in plant establishment and survival in extreme environments, with some evidence that mycorrhizae may confer stress tolerance to their plant partners (Chapters 2, 5, 7). An improved understanding of mycorrhizal interactions in extreme environments is likely to improve the outcome of restoration efforts. Similarly, we have limited knowledge of plant-soil feedback loops and the potential responses of extremophiles to long-term changes in soil chemistry resulting from nitrogen deposition or other anthropogenic pollution (Chapters 13-15), making it difficult to identify appropriate restoration measures. More basic research on habitat function and plant response in extreme environments is needed to develop better restoration approaches, particularly low-cost and environmentally-friendly approaches.

The limited availability of habitat and the patchy distribution of species restricted to extreme environments make them particularly vulnerable to anthropogenic impacts and highlights the importance of conserving such habitats (Chapters 13, 14). Reducing impacts of climate change and increasing the resiliency of plant communities restricted to extreme environments should be a focus of conservation efforts. Conservation networks will be a critical component in strategies to protect these patchily distributed habitats and to buffer them from the effects of climate change. A number of geobotanical preserves and other similar conservation efforts have been established globally (Rajakaruna & Boyd, 2008); however, more work is needed to improve the public's understanding of these botanically rich habitats and the need for their conservation.

The resource needs of our expanding human population continue to increase human impacts on all areas of the planet, including harsh environments that may have experienced little prior impact. Basic inventories of species are still needed in many harsh environments; this is illustrated in several chapters (*e.g.,* Chapters 2, 3, 5, 10) in which important gaps in our knowledge are discussed. Climate change poses a major challenge to life across the globe and increases the importance of conservation efforts that take climate change into account (Chapters 7, 13; Cowie, 2013). Unfortunately, we need improved tools for predicting the effects of climate change on biodiversity; Bellard et al. (2012) summarize the drawbacks of current techniques but conclude that most modeling approaches predict major negative effects on global biodiversity. An improved understanding of the mechanisms by which environmental stresses affect the distribution of species may aid in producing better climate change models and better management approaches for threatened species or communities (Chapter 7, 13-15). Harsh environments are often more easily damaged by human activities and are a unique restoration challenge due to the prevalence of abiotic stresses inherent to the site. As we obtain more complete species and population inventories of harsh environments and gain experience with restoration techniques, we will be better able to restore damaged areas and recover all or part of their former species composition and ecological function.

CONCLUSION

Harsh environments are important biological resources and represent some of the most promising frontiers in the study of ecology and evolution. They are important depositories for a significant portion of life's diversity, they illustrate how organisms respond to environmental challenges, and they provide resources important to both human society and other life on Earth. We can also learn much about basic biological, ecological, and evolutionary principles—including natural selection, adaptation, and coevolution—from studies that focus on plant and other life found in harsh habitats. This knowledge can be employed in numerous biotechnological applications beneficial to human society (*e.g.,* Chapters 11, 15; Agarwal et al., 2013; Peleg et al., 2011). This book builds on the current interest in plants and other organisms found in harsh environments (*e.g.,* Aroca, 2012; Horikoshi & Grant, 1998; Liebezeit et al., 2000; Lubzens et al., 2010; Lüttge et al., 2011) and shows the importance of harsh environments to current and future research in all aspects of plant biology, ecology and evolution, and the conservation and restoration sciences.

REFERENCES

Agarwal, P. K., Shukla, P. S., Gupta, K. & Jha, B. (2013) Bioengineering for salinity tolerance in plants: State of the art. *Molecular Biotechnology 54*, 102-123.

Alexander, E. B., Coleman, R. G., Keeler-Wolf, T. & Harrison, S. (2007) *Serpentine geoecology of Western North America*. New York, NY: Oxford University Press.

Anacker, B. L. (2011) Phylogenetic patterns of endemism and diversity. In: S. P. Harrison & N. Rajakaruna (Eds.). *Serpentine: The evolution and ecology of a model system* (pp. 49-79). Berkeley, CA: University of California Press.

Anacker, B. L. (2014) The nature of serpentine endemism. *American Journal of Botany 101*, 219-224.

Anacker, B. L. & Harrison, S. P. (2012) Historical and ecological controls on phylogenetic diversity in Californian plant communities. *The American Naturalist 180*, 257-269.

Anacker, B. L. & Strauss, S. Y. (2014) The geography and ecology of plant speciation: Range overlap and niche divergence in sister species. *Proceedings of the Royal Society Series B,* DOI: 10.1098/rspb.2013.2980.

Anacker, B. L., Whittall, J., Goldberg, E. & Harrison, S. P. (2011) Origins and consequences of serpentine endemism in the California flora. *Evolution 65*, 365-376.

Anderson, R. C., Fralish, J. S. & Baskin, J. M. (Eds.) (1999) *Savannas, barrens, and rock outcrop plant communities of North America*. Cambridge: Cambridge University Press.

Antonovics, J., Bradshaw, A. D. & Turner, R. G. (1971) Heavy metal tolerance in plants. *Advances in Ecological Research 7*, 1–85.

Aroca, R. (Ed.) (2012) *Plant responses to drought stress: From morphological to molecular features.* New York, NY: Springer.

Baker, H. G. (1965). Characteristics and modes of origin of weeds. In: H. G. Baker & G. L. Stebbins (Eds.). *The genetics of colonizing species* (pp. 147-169). New York, NY: Academic Press.

Barve, A. & Wagner, A. (2013) A latent capacity for evolutionary innovation through exaptation in metabolic systems. *Nature 500,* 203-206.

Beever, E., Chris R., Wilkening, J., Brussard, P. & Mote, P. (2011) Contemporary climate change alters the pace and drivers of extinction. *Global Change Biology 17*, 2054-2070.

Bell, E. (2012) *Life at extremes: Environments, organisms, and strategies for survival.* Oxfordshire: CABI.

Bellard, C., Bertelsmeier, C., Leadley, P., Thuiller, W. & Courchamp, F. (2012) Impacts of climate change on the future of biodiversity. *Ecology Letters 15,* 365-377.

Bieger, A., Rajakaruna, N. & Harrison, S. P. (2014) Little evidence for local adaptation to soils or microclimate in the postfire recruitment of three Californian shrubs. *Plant Ecology and Diversity 7*, 411-420.

Bradshaw, A. D. (1991). The Croonian lecture, 1991: Genostasis and the limits to evolution. *Philosophical Transactions of the Royal Society Series B 333 (1267)*, 289-305.

Brady, K. U., Kruckeberg, A. R. & Bradshaw Jr., H. D. (2005) Evolutionary ecology of plant adaptation to serpentine soils. *Annual Review of Ecology, Evolution, and Systematics 36*, 243–266.

Brattler, M., Baltisberger, M. & Widmer, A. (2006a) QTL analysis of intraspecific differences between two *Silene vulgaris* ecotypes. *Annals of Botany 98*, 411–419.

Brattler, M., Lexer, C. & Widmer, A. (2006b). Genetic architecture of traits associated with serpentine adaptation of *Silene vulgaris*. *Journal of Evolutionary Biology 19*, 1149-1156.

Briscoe, L. R. E., Harris, T. B., Dannenberg, E., Broussard, W., Olday, F. C. & Rajakaruna, N. (2009) Bryophytes of adjacent serpentine and granite outcrops on the Deer Isles, Maine, USA. *Rhodora 111*, 1-20.

Brooks, R. R. (1983) *Biological methods of prospecting for minerals*. New York, NY: John Wiley & Sons.

Brooks, R. R. (1987) *Serpentine and its vegetation: A multidisciplinary approach*. Portland, OR: Dioscorides.

Cardace, D. & Hoehler, T. M. (2011) Microbes in extreme environments: Implications for life on the early Earth and other planets. In: S. P. Harrison & N. Rajakaruna (Eds.). *Serpentine: The evolution and ecology of a model system* (pp. 29-48). Berkeley, CA: University of California Press.

Casamayor, E. O., Triadó-Margarit, X. & Castañeda, C. (2013) Microbial biodiversity in saline shallow lakes of the Monegros Desert, Spain. *FEMS Microbiology Ecology 85*, 503–518.

Clausen, J. (1951) *Stages in the evolution of plant species*. Ithaca, NY: Cornell University Press.

Clausen, J. C., Keck, D. D. & Hiesey, W. H. (1940) *Experimental studies on the nature of species. I. Effects of varied environments on western North American plants.* Washington, DC: Carnegie Institution of Washington Publication 520.

Clements, R., Sodhi, N., Schilthuizen, M. & King, P. (2006) Limestone karsts of southeast Asia: Imperiled arks of biodiversity. *BioScience 56*, 733-742.

Colwell, R. K. & Fuentes, E. R. (1975). Experimental studies of the niche. *Annual Review of Ecology and Systematics 6*, 281-310.

Cowie, J. (2013) *Climate change: Biological and human aspects.* Second edition. Cambridge: Cambridge University Press.

Damschen, E. I., Harrison, S., Anacker, B. L. & Going, B. M. (2011) Climate change and plant communities on special soils. In: S. P. Harrison & N. Rajakaruna (Eds.). *Serpentine: The evolution and ecology of a model system* (pp. 359-381). Berkeley, CA: University of California Press.

Damschen, E. I., Harrison, S., Ackerly, D., Fernandez-Going, B. & Anacker, B. (2012) Endemic plant communities on special soils: Early victims or hardy survivors of climate change? *Journal of Ecology 100*, 1122-1130.

Davies, K. F. (2011). Community invasibility: Spatial heterogeneity, spatial scale, and productivity. In: S. H. Harrison & N. Rajakaruna (Eds.). *Serpentine: The evolution and ecology of a model system* (pp. 237-248). Berkeley, CA: University of California Press.

Escudero, A., Palacio, S., Maestre, F. & Luzuriaga, A. (2014) Plant life on gypsum: A review of its multiple facets. *Biological Reviews,* DOI: 10.1111/brv.12092.

Fernandez-Going, B. M., Harrison, S. P., Anacker, B. L. & Safford, H. (2013) Climate interacts with soil to produce beta diversity in Californian plant communities. *Ecology 94*, 2007-2018.

Flowers, T. J. & Colmer, T. D. (2008) Salinity tolerance in halophytes. *New Phytologist 179*, 945–963.

Gall, J. E. & Rajakaruna, N. (2013) The physiology, functional genomics, and applied ecology of heavy metal-tolerant Brassicaceae. In: M. Lang (Ed.). *Brassicaceae: Characterization, functional genomics and health benefits* (pp. 121-148). New York, NY: Nova Science Publishers.

Gerday, C. & Glansdorff, N. (Eds.) (2007) *Physiology and biochemistry of extremophiles*. Washington, DC: ASM Press.

Gordon, J. E., Dvorák, I. J., Jonasson, C., Josefsson, M., Kociánová, M. & Thompson, D. B. A. (2002) Geo-ecology and management of sensitive montane landscapes. *Geografiska Annaler: Series A, Physical Geography 84*, 193-203.

Gormezano, L. & Rockwell, R. (2013) What to eat now? Shifts in polar bear diet during ice-free season in western Hudson Bay. *Ecology and Evolution 3*, 3509-3523.

Gotelli, N. J. & Ellison, E. M. (2002) Nitrogen deposition and extinction risk in northern pitcher plant, *Sarracenia purpurea*. *Ecology 83*, 2758-2765.

Harris, T. B. & Rajakaruna, N. (2009) *Adiantum viridimontanum, Aspidotis densa, Minuartia marcescens,* and *Symphyotrichum rhiannon*: Additional serpentine endemics from Eastern North America. *Northeastern Naturalist 16*, 111-120.

Harrison, S. P. (2011) Spatial ecology: The effects of habitat patch size, shape, and isolation on ecological processes. In: S. P. Harrison & N. Rajakaruna (Eds.). *Serpentine: The evolution and ecology of a model system* (pp. 297-308). Berkeley, CA: University of California Press.

Harrison, S. P. & Rajakaruna, N. (Eds.) (2011) *Serpentine: The evolution and ecology of a model system*. Berkeley, CA: University of California Press.

Horikoshi, K. & Grant, W. D. (1998) *Extremophiles: Microbial life in extreme environments*. New York, NY: Wiley.

Ivalú Cacho, N., Burrell, A. M., Pepper, A. & Strauss, S. Y. (2014) Systematics and the evolution of serpentine tolerance in the California Jewelflowers *(Streptanthus)* and its allies. *Molecular Phylogenetics and Evolution 72*, 71-81.

Jacobi, C., do Carmo, F. & de Campos, I. (2011) Soaring extinction threats to endemic plants in Brazilian metal-rich regions. *AMBIO 40*, 540-543.

Jenks, M. A. & Hasegawa, P. M. (Eds.) (2013) *Plant abiotic stress*. Second edition. Oxford, UK: John Wiley & Sons.

Kantar, M., Lucas, S. J. & Budak, H. (2011) Drought stress: Molecular genetics and genomics approaches. *Advances in Botanical Research 57*, 445–493.

Kay, K. M., Ward, K. L., Watt, L. R. & Schemske, D. W. (2011) Plant speciation. In: S. P. Harrison & N. Rajakaruna (Eds.). *Serpentine: The evolution and ecology of a model system* (pp. 71-95). Berkeley, CA: University of California Press.

Kirchman, D. L. (2012) *Processes in microbial ecology*. New York, NY: Oxford.

Kolář, F., Fér, T., Štech, M., Trávníček, P., Dušková, E., Schönswetter, P. & Suda, J. (2012) Bringing together evolution on serpentine and polyploidy: Spatiotemporal history of the diploid-tetraploid complex of *Knautia arvensis* (Dipsacaceae). *PLoS ONE 7*, DOI: 10.1371/journal.pone.0039988.

Körner, C. (2003) *Alpine plant life: Functional plant ecology of high mountain ecosystems.* Second edition. Heidelberg: Springer.

Kruckeberg, A. R. (1957) Variation in fertility of hybrids between isolated populations of the serpentine species, *Streptanthus glandulosus* Hook. *Evolution 11*, 185-211.

Lau, J. A., McCall, A. C., Davies, K. F., McKay, J. K. & Wright, J. W. (2008) Herbivores and edaphic factors constrain the realized niche of a native plant. *Ecology 89*, 754–762.

Leimu, R. & Fischer, M. (2008) A meta-analyisis of local adaptation in plants. *PLoS ONE 3*, DOI: 10.1371/journal.pone.0004010.

Levin, D. A. (2001) The recurrent origin of plant races and species. *Systematic Botany 26*, 197-204.

Liebezeit, G., Dittmann, S. & Kröncke, I. (Eds.) (2000) *Life at interfaces and under extreme conditions. Developments in hydrobiology Vol. 151*. New York, NY: Springer.

Lubzens, E., Cerdà, J. & Clark, M. S. (Eds.) (2010) *Dormancy and resistance in harsh environments.* New York, NY: Springer.

Lüttge, U., Beck, E. & Bartels, D. (Eds.) (2011) *Plant desiccation tolerance.* New York, NY: Springer.

Ma, Z., Gan, X., Choi, C.-Y. & Li, B. (2013) Effects of invasive cordgrass on presence of Marsh Grassbird in an area where it is not native. *Conservation Biology 28*, 150-158.

Margesin, R. & Miteva, V. (2011) Diversity and ecology of psychrophilic microorganisms. *Research in Microbiology 162*, 346-361.

Martin, M. H. & Coughtrey, P. J. (1982) *Biological monitoring of heavy metal pollution: Land and air*. Dordrecht, The Netherlands: Springer

McCollom, T. M. & Seewald, J. S. (2013) Serpentinites, hydrogen, and life. *Elements 9.2*, 129-134.

McGhee, G. R. (2011) Convergent evolution: Limited forms most beautiful (Vienna Series in Theoretical Biology). Cambridge, MA: The MIT Press.

Mayer, M. S. & Beseda, L. (2010) Reconciling taxonomy and phylogeny in the Streptanthus glandulosus complex (Brassicaceae). Annals of the Missouri Botanical Garden 97, 106-116.

Mayer, M. S., Soltis, P. S. & Soltis, D. E. (1994) The evolution of the Streptanthus glandulosus complex (Cruciferae): Genetic divergence and gene flow in serpentine endemics. American Journal of Botany 81, 1288-1299.

Moore, K. A. & Elmendorf, S. C. (2011) Plant competition and facilitation in systems with strong environmental gradients. In: S. P. Harrison & N. Rajakaruna (Eds.). *Serpentine: The evolution and ecology of a model system* (pp. 223-236). Berkeley, CA: University of California Press.

Nagy, L. & Grabherr, G. (2009) The biology of alpine habitats. Oxford: Oxford University Press.

Nosil, P. (2012) Ecological speciation. Oxford, UK: Oxford University Press.

Nosil, P. & Schluter, D. (2011) The genes underlying the process of speciation. Trends in Ecology and Evolution 26, 160-167.

Nyberg Berglund, A. B., Dahlgren, S. & Westerbergh, A. (2004) Evidence for parallel evolution and site-specific selection of serpentine tolerance in Cerastium alpinum during the colonization of Scandinavia. New Phytologist 161, 199–209.

O'Dell, R. E. & Claassen, V. P. (2011) Restoration and revegetation of harsh soils. In: S. P. Harrison & N. Rajakaruna (Eds.). Serpentine: The evolution and ecology of a model system (pp. 383-413). Berkeley, CA: University of California Press.

O'Dell, R. E. & Rajakaruna, N. (2011) Intraspecific variation, adaptation, and evolution. In: S. P. Harrison & N. Rajakaruna (Eds.). *Serpentine: The evolution and ecology of a model system* (pp. 97-137). Berkeley, CA: University of California Press.

Ostevik, K. L., Moyers, B. T., Owens, G. L. & Rieseberg, L. H. (2012) Parallel ecological speciation in plants? International Journal of Ecology 2012, DOI: 10.1155/2012/939862.

Parker, I. M., Rodriguez, J. & Loik, M. E. (2003) An evolutionary approach to understanding the biology of invasions: Local adaptation and general-purpose genotypes in the weed Verbascum thapsus. Conservation Biology 17, 59–72.

Peleg, Z., Apse, M. P. & Blumwald, E. (2011) Engineering salinity and water-stress tolerance in crop plants: Getting closer to the field. Advances in Botanical Research 57, 405–443.

Pievani, T. (2014) The sixth mass extinction: Anthropocene and the human impact on biodiversity. Rendiconti Lincei 25, 85-93.

Pignatti, E. & Pignatti, S. (2013) *Plant life of the dolomites: Vegetation structure and ecology*. New York, NY: Springer.

Pikuta, E. V., Hoover, R. B. & Tang, J. (2007) Microbial extremophiles at the limits of life. *Critical Reviews in Microbiology 33*, 183-209.

Porter, S. S., Stanton, M. L. & Rice, K. J. (2011) Mutualism and adaptive divergence: Co-invasion of a heterogeneous grassland by an exotic legume–rhizobium symbiosis. *PLoS ONE 6*, e27935 DOI: 10.1371/journal.pone.0027935.

Rajakaruna, N. & Boyd, R. S. (2008) The edaphic factor. In: S. E. Jorgensen & B. Fath (Eds.). *The encyclopedia of ecology*. Volume 2 (pp. 1201–1207). Oxford: Elsevier.

Rajakaruna, N. & Whitton, J. (2004) Trends in the evolution of edaphic specialists with an example of parallel evolution in the *Lasthenia californica* complex. In: Q. C. B. Cronk, J. Whitton, R. H. Ree & I. E. P. Taylor (Eds.). *Plant adaptation: Molecular biology and ecology* (pp. 103-110). Ottawa: NRC Research Press.

Rajakaruna, N., Baldwin, B. G., Chan, R., Desrochers, A. M., Bohm, B. A. & Whitton, J. (2003) Edaphic races and phylogenetic taxa in the *Lasthenia californica* complex (Asteraceae: Heliantheae): An hypothesis of parallel evolution. *Molecular Ecology 12*, 1675-1679.

Rajakaruna, N., Knudsen, K., Fryday, A., O'Dell, R. E., Pope, N., Olday, F. C. & Woolhouse, S. (2012) Investigation of the importance of rock chemistry for saxicolous lichen communities of the New Idria serpentinite mass, San Benito County, California, USA. *The Lichenologist 44*, 695-714.

Rampelotto, P. H. (Ed.) (2013) Special issue "Extremophiles and extreme environments." *Life 3*, 1-517.

Reich, P. B., Walters, M. B. & Ellsworth, D. S. (1997) From tropics to tundra: Global convergence in plant functioning. *Proceedings of the National Academy of Sciences (USA) 94*, 13730 –13734.

Richards, C. L., Bossdorf, O., Muth, N. Z., Gurevitch, J. & Pigliucci, M. (2006) Jack of all trades, master of some? On the role of phenotypic plasticity in plant invasions. *Ecology Letters 9*, 981–993.

Roberts, B. A. & Proctor, J. (1992) *The ecology of areas with serpentinized rocks: A world view*. Dordrecht: Kluwer.

Rodriguez, R. J., Henson, J., van Volkenburgh, E., Hoy, M., Wright, L., Beckwith, F., Kim, Y.-O. & Redman, R. S. (2008) Stress tolerance in plants via habitat-adapted symbiosis. *The ISME Journal 2*, 404-416.

Rydin, H. & Jeglum, J. K. (2009) *The biology of peatlands.* New York, NY: Oxford University Press.

Schneider, H. E. & Allen, E. B. (2012) Effects of elevated nitrogen and exotic plant invasion on soil seed bank composition in Joshua Tree National Park. *Plant Ecology 213*, 1277-1287.

Seckbach, J. (Ed.) (2007) *Algae and cyanobacteria in extreme environments*. Cellular origin, life in extreme habitats and astrobiology series (Book 11). Dordrecht: Springer.

Serrano, H. C., Antunes, C., Pinto, M. J., Máguas, C., Martins-Loução, M. A. & Branquinho, C. (2014) The ecological performance of metallophyte plants thriving in geochemical islands is explained by the Inclusive Niche Hypothesis. *Journal of Plant Ecology,* DOI: 10.1093/jpe/rtu0007.

Shenk, J. (2013) Evolution of limited seed dispersal ability on gypsum islands. *American Journal of Botany 100,* 1811-1822.

Southworth, D., Tackaberry, L. E. & Massicotte, H. B. (2013) Mycorrhizal ecology on serpentine soils. *Plant Ecology and Diversity 7*, 445-455.

Spasojevic, M. J., Damschen, E. I. & Harrison, S. P. (2014) Patterns of seed dispersal syndromes on serpentine soils: Examining the roles of habitat patchiness, soil infertility and correlated functional traits. *Plant Ecology and Diversity 7*, 401-410.

Springer, Y. P. (2009) Edaphic quality and plant-pathogen interactions: Effects of soil calcium on fungal infection of a serpentine flax. *Ecology 90*, 1852–62.

Strauss, S. Y. & Boyd, R. S. (2011) Herbivory and other cross-kingdom interactions on harsh soils. In: S. P. Harrison & N. Rajakaruna (Eds.). *Serpentine: The evolution and ecology of a model system* (pp. 181-199). Berkeley, CA: University of California Press.

Stuessy, T. F. & Ono, M. (Eds.) (2007) Evolution and speciation of island plants. Cambridge: Cambridge University Press.

Szabados, L., Kovács, H., Zilberstein, A. & Bouchereau, A. (2011) Plants in extreme environments: Importance of protective compounds in stress tolerance. *Advances in Botanical Research 57*, 105–150.

Takai, K., Moyer, C. L., Miyazaki, M., Nogi, Y., Hirayama, H., Nealson, K. H. & Horikoshi, K. (2005) *Marinobacter alkaliphilus* sp. nov., a novel alkaliphilic bacterium isolated from subseafloor alkaline serpentine mud from Ocean Drilling Program Site 1200 at South Chamorro Seamount, Mariana Forearc. *Extremophiles 9.1*, 17–27.

Thomas, D. N., Fogg, G. E., Convey, P. Fritsen, C. H., Gili, J.-M., Gradinger, R., Laybourn-Parry, J., Reid, K. & Walton, D. W. H. (2008) *The biology of polar regions.* Second Edition. New York, NY: Oxford University Press.

Thorne, J. H., Huber, P. R. & Harrison, S. P. (2011). Systematic conservation planning: Protecting rarity, representation, and connectivity in regional landscapes. In: S. P. Harrison & N. Rajakaruna (Eds.). *Serpentine: The evolution and ecology of a model system* (pp. 309-328). Berkeley, CA: University of California Press.

Turesson, G. (1922) The genotypical response of the plant species to the habitat. III. The species and the variety as ecological units. *Hereditas 3*, 100-113.

Turner, T. L., Bourne, E. C., von Wettberg, E. J., Hu, T. T. & Nuzhdin, S. V. (2010) Population resequencing reveals local adaptation of *Arabidopsis lyrata* to serpentine soils. *Nature Genetics 42*, 260–263.

Vallano, D. M., Selmants, P. C. & Zavaleta, E. S. (2012) Simulated nitrogen deposition enhances performance of an exotic grass relative to native serpentine grassland competitors. *Plant Ecology 213*, 1015-1026.

Via, S. (2009) Natural selection in action during speciation. *Proceedings of the National Academy of Sciences (USA) 106 Supplement 1*, 9939-9946.

von Wettberg, E. J. & Wright, J. (2011) Genomic approaches to understanding adaptation. In: S. P. Harrison & N. Rajakaruna (Eds.). *Serpentine: The evolution and ecology of a model system* (pp. 139-153). Berkeley, CA: University of California Press.

Walker, L. R. (2011) *The biology of disturbed habitats*. New York, NY: Oxford University Press.

Ward, D. (2009) *The biology of deserts*. New York, NY: Oxford University Press.

Westerbergh, A. & Saura, A. (1992) The effect of serpentine on the population structure of *Silene dioica* (Caryophyllaceae). *Evolution 46*, 1537–1548.

Wetzel, F., Beissmann, H., Penn, D. & Jetz, W. (2013) Vulnerability of terrestrial island vertebrates to projected sea-level rise. *Global Change Biology 19*, 2058-2070.

Williams, R. D. & Schuman, G. E. (1987) *Reclaiming mine soils and overburden in the Western United States: Analytical parameters and procedures*. Ankeny, IA: Soil Conservation Society of America.

Wilson, J. L., Ayres, D. R., Steinmaus, S. & Baad, M. (2010) Vegetation and flora of a biodiversity hotspot: Pine Hill, El Dorado County, California, USA. *Madroño 56*, 246-278.

Wolf, A. T. & Thorp, R. (2011) Plant-pollinator interactions in naturally fragmented habitats. In: S. P. Harrison & N. Rajakaruna (Eds.). *Serpentine: The evolution and ecology of a model system* (pp. 275-296). Berkeley, CA: University of California Press.

Wright, J. W. & Stanton, M. L. (2011) Local adaptation in heterogeneous landscapes: reciprocal transplant experiments and beyond. In: S. P. Harrison & N. Rajakaruna (Eds.). *Serpentine: The evolution and ecology of a model system* (pp. 155-179). Berkeley, CA: University of California Press.

Wulff, A., Hollingsworth, P., Ahrends, A., Jaffré, T., Veillon, J.-M., L´Huillier, L. & Fogliani, B. (2013) Conservation priorities in a biodiversity hotspot: Analysis of narrow endemic plant species in New Caledonia. *PLoS One 8*, DOI: 10.1371/journal.pone.0073371.

Yost, J. M., Barry, T., Kay, K. M. & Rajakaruna, N. (2012) Edaphic adaptation maintains the coexistence of two cryptic species on serpentine soil. *American Journal of Botany 99*, 890-897.

Zedler, J. (2000) *Handbook for restoring tidal wetlands*. Boca Raton, FL: CRC Press.

Editors' Contact Information

Dr. Nishanta Rajakaruna,
College of the Atlantic,
105 Eden Street
Bar Harbor, ME 04609, US
Tel: 207-801-5731
Email: nrajakaruna@gmail.com

Dr. Robert S. Boyd,
101 Rouse Life Sciences Bldg
Biological Sciences,
Auburn University,
Auburn, AL 36849, US
Tel: 334-844-1626
Email: boydrob@auburn.edu

Tanner B. Harris,
WRA, Inc.,
2169-G East Francisco Blvd,
San Rafael, CA 94901, US
Tel: 415-454-8868 x121
Email: harris@wra-ca.com

Species Index

A

B

C

D

E

F

N

O

P

Q

R

S

T

U

V

W

X

Z

SUBJECT INDEX

A

B

C

G

H

I

J

L

M

N

O

P

Q

R

S

T

U

V

W

X

Z